DECOMMISSIONING HEALTH PHYSICS

A Handbook for MARSSIM Users

SECOND EDITION

DECOMMISSIONING HEALTH PHYSICS

A Handbook for MARSSIM Users

SECOND EDITION

Eric W. Abelquist

CRC Press
Taylor & Francis Group
Boca Raton London New York

CRC Press is an imprint of the
Taylor & Francis Group, an **informa** business

CRC Press
Taylor & Francis Group
6000 Broken Sound Parkway NW, Suite 300
Boca Raton, FL 33487-2742

First issued in paperback 2019

© 2014 by Taylor and Francis Group, LLC
CRC Press is an imprint of Taylor & Francis Group, an Informa business

No claim to original U.S. Government works

ISBN-13: 978-1-4665-1053-1 (hbk)
ISBN-13: 978-0-367-86713-3 (pbk)

Visit the Taylor & Francis Web site at
http://www.taylorandfrancis.com

and the CRC Press Web site at
http://www.crcpress.com

Contents

Preface, xix

Acknowledgments, xxi

Author, xxiii

CHAPTER 1 ■ Introduction: Current Issues in
Decommissioning and MARSSIM Overview   1

1.1   DECOMMISSIONING OVERVIEW   2

1.2   MARSSIM OVERVIEW   4

1.3   CLEARANCE OF MATERIALS OVERVIEW   8

QUESTIONS   10

SUGGESTED READING   10

CHAPTER 2 ■ Decommissioning Project Overview and
Regulatory Agency Interface   11

2.1   DECOMMISSIONING OPTIONS   11

2.2   DECOMMISSIONING PROJECT PHASES   16

2.3   RADIOLOGICAL SURVEYS PERFORMED DURING
DECOMMISSIONING   18

2.4   REGULATORY AGENCY INTERFACE WHEN
DESIGNING MARSSIM SURVEYS   21

  2.4.1   NRC Standard Review Plans   22

  2.4.2   Verification Process   25

    2.4.2.1   Review of Decommissioning
Documentation   27

    2.4.2.2   Confirmatory Analysis of Laboratory Samples   29

2.4.2.3 IV Survey Activities 30

2.4.3 Verification Program Experiences and Lessons
Learned 31

QUESTIONS 38

SUGGESTED READING 39

CHAPTER 3 ▪ Characterization Surveys and the DQO Process 41

3.1 INTRODUCTION 41

3.2 DQOs PROCESS AND APPLICATION
TO CHARACTERIZATION 42

3.2.1 EPA's DQOs Process 43

3.2.2 Example Application of DQO Process: Use Default
or Site-Specific DCGL? 45

3.2.3 Example Application of DQO Process: Waste
Disposal Characterization 47

3.3 DECOMMISSIONING OBJECTIVES OF
CHARACTERIZATION: DATA NEEDED
TO MAKE DECISIONS 52

3.4 CHARACTERIZATION SURVEY DESIGN
AND CONSIDERATIONS 54

3.5 CHARACTERIZATION SURVEY ACTIVITIES 57

3.5.1 Structure Surveys 58

3.5.2 Land Area Surveys 59

3.5.3 Other Measurements/Sampling Locations 61

3.5.4 Data Quality Assessment for Characterization 62

3.6 CHARACTERIZATION SURVEY RESULTS
TO SUPPORT MARSSIM FSS DESIGN 63

QUESTIONS AND PROBLEMS 65

CHAPTER 4 ▪ Guidelines and Dose-Based Release Criteria 67

4.1 HISTORIC RELEASE CRITERIA AND GUIDANCE
DOCUMENTS 69

4.2 DOSE-BASED RELEASE CRITERIA AND NRC'S
DECOMMISSIONING RULEMAKING 75

4.3 CLEARANCE OF MATERIALS AND ANSI N13.12 83

QUESTIONS AND PROBLEMS   83

SUGGESTED READING   84

CHAPTER 5 ■ Exposure Pathway Modeling: DCGLs
                and Hazard Assessments   85

5.1   SCREENING VERSUS SITE-SPECIFIC: WHEN IS IT
      TIME TO GO BEYOND THE SCREENING DCGLs?   87

5.2   EXPOSURE PATHWAY MODELING: SCENARIOS,
      PATHWAYS, AND PARAMETERS   89

      5.2.1   NRC's Policy and Guidance Directive PG-8-08   90

      5.2.2   NUREG/CR-5512 and NUREG-1549   92

      5.2.3   Pathway Modeling Parameters   95

5.3   MODELING CODES   97

      5.3.1   RESRAD and RESRAD-BUILD Models   98

      5.3.2   DandD Model   100

5.4   DETERMINATION OF DCGLs AND AREA FACTORS   103

      5.4.1   Dose Modeling to Obtain DCGLs   103

              5.4.1.1   DCGLs Using the DandD Model   104

              5.4.1.2   DCGLs Using the RESRAD-BUILD Model   110

              5.4.1.3   DCGLs Using the RESRAD Model   111

      5.4.2   Modeling to Obtain Area Factors   114

5.5   HAZARD ASSESSMENTS: AN EARLY APPLICATION
      OF DOSE-BASED RELEASE CRITERIA?   118

      5.5.1   Hazard Assessment for Contaminated Roof Panels
              at Uranium Site   119

              5.5.1.1   Dose to Building Employee   121

              5.5.1.2   Dose to Demolition Worker   122

      5.5.2   Hazard Assessment for Contaminated
              Underground Pipes at Sealed Source Facility   128

              5.5.2.1   Inhalation Dose   130

              5.5.2.2   Ingestion Dose   131

              5.5.2.3   External Radiation Dose   132

QUESTIONS AND PROBLEMS   133

CHAPTER 6 ■ Preliminary Survey Design Concerns and Application of DCGLs — 135

6.1   DIRECT APPLICATION OF DCGLs — 136

6.2   USE OF DCGLs FOR SITES WITH MULTIPLE RADIONUCLIDES — 138

 6.2.1   Use of Surrogate Measurements — 141

  6.2.1.1   *Surrogates for Soil Concentrations* — 141

  6.2.1.2   *Surrogates for Surface Activity* — 149

  6.2.1.3   *Surrogates for Exposure Rate* — 151

 6.2.2   Gross Activity DCGLs for Surface Activity — 152

 6.2.3   Use of the Unity Rule — 156

QUESTIONS AND PROBLEMS — 158

CHAPTER 7 ■ Background Determination and Background Reference Areas — 161

7.1   BACKGROUND REFERENCE AREAS AND RELATED CONSIDERATIONS — 162

7.2   SURFACE MATERIAL BACKGROUNDS — 165

7.3   SCENARIO B SURVEY DESIGN: INDISTINGUISHABLE FROM THE BACKGROUND — 167

7.4   SCENARIO A VERSUS SCENARIO B FSS DESIGNS — 174

QUESTIONS AND PROBLEMS — 179

CHAPTER 8 ■ Survey Instrumentation Selection and Calibration — 181

8.1   CALIBRATION FOR SURFACE ACTIVITY MEASUREMENT INSTRUMENTS — 182

8.2   OVERVIEW OF SURVEY INSTRUMENTATION AND PHILOSOPHY OF INSTRUMENT SELECTION — 184

8.3   SURVEY INSTRUMENTATION FOR SURFACE ACTIVITY MEASUREMENTS AND SCANNING — 187

 8.3.1   Overview of Field Survey Instruments — 187

 8.3.2   Conventional Survey Instrument Types — 189

  8.3.2.1   *ZnS Detector: Alpha Measurements* — 189

		8.3.2.2	GM Detector: Beta Measurements	190
		8.3.2.3	Plastic Detector: Beta and Gamma Measurements	191
		8.3.2.4	Dual Phosphor Detectors: Alpha and Beta Measurements	192
		8.3.2.5	Gas Proportional Detector: Alpha or Beta Measurements	193
		8.3.2.6	NaI Scintillation Detectors: Gamma Radiation	195
	8.3.3	Advanced Survey Instrument Types		196
		8.3.3.1	In Situ Gamma Spectrometers: Gamma Radiation Measurements	197
		8.3.3.2	Automated Scanning and Measurement Systems	203
	8.3.4	Environmental Effects on Survey Instrument Operation		205
8.4	DETERMINATION OF INSTRUMENT EFFICIENCY FOR SURFACE ACTIVITY MEASUREMENTS			208
8.5	SURVEY INSTRUMENTATION FOR EXPOSURE RATE MEASUREMENTS			215
	8.5.1	Pressurized Ionization Chambers		216
	8.5.2	Micro-R and Micro-Rem Meters		216
8.6	LABORATORY INSTRUMENTATION			218
	QUESTIONS AND PROBLEMS			220
CHAPTER 9 ■ Detection Sensitivity: Static and Scan MDCs				223
9.1	CRITICAL LEVEL AND DETECTION LIMIT			224
9.2	STATIC MDC			236
9.3	SCAN MDC			239
	9.3.1	Signal Detection Theory for Scanning		239
	9.3.2	Decision Processes of the Surveyor (Human Factors)		241
	9.3.3	Scan MDCs for Structure Surfaces		244
		9.3.3.1	Scan MDCs on Structure Surfaces for Alpha Radiation	245

        9.3.3.2 Scan MDCs on Structure Surfaces for Beta Radiation   250

    9.3.4   Scan MDCs for Land Areas   255

    9.3.5   Scan MDCs for Multiple Contaminants for Structure Surfaces and Land Areas   266

    9.3.6   Empirically Determined Scan MDCs   274

  QUESTIONS AND PROBLEMS   277

CHAPTER 10 ■ Survey Procedures and Measurement Data Interpretation   281

  10.1 SURFACE-ACTIVITY MEASUREMENTS   282

    10.1.1   Surface Efficiency (ε_s) 283

    10.1.2   Building Material-Specific Backgrounds   290

  10.2 SCANNING BUILDING SURFACES AND LAND AREAS   294

  10.3 GAMMA SPECTROMETRY ANALYSES FOR SOIL   298

    10.3.1   Calibration Standards and Geometries for Soil Analyses   298

    10.3.2   Interpreting Gamma Spectrometry Data   299

  QUESTIONS AND PROBLEMS   303

CHAPTER 11 ■ Radiological Hot-Spot Survey Considerations   305

  11.1 INTRODUCTION: CURRENT HOT-SPOT APPROACH   305

  11.2 HOT-SPOT LIMITS IN SOIL   306

    11.2.1   External Radiation Pathway   309

        11.2.1.1 RESRAD Calculation of Area Factor   309

        11.2.1.2 MicroShield Calculation of Area Factor   310

        11.2.1.3 Receptor Located Some Distance from the Hot Spot   311

        11.2.1.4 A Realistic Hot-Spot Dose Assessment   312

        11.2.1.5 External Radiation Pathway Results   315

    11.2.2   Inhalation Exposure to Resuspended Soil Pathway   319

        11.2.2.1 RESRAD Area Factor Approach for Inhalation Exposure Pathway   319

11.2.2.2 Calculation of Inhalation Pathway Dose Based on First Principles   323

11.2.2.3 Proposal to More Realistically Assess Hot-Spot Dose   324

11.2.2.4 Inhalation Pathway Results   326

11.2.2.5 Inhalation Pathway Conclusions   329

11.2.3 Ingestion-Based Environmental Pathways   330

11.2.3.1 Direction Ingestion of Soil   330

11.2.3.2 Ingestion of Drinking Water   331

11.2.3.3 Ingestion of Plant Products Grown in Contaminated Soil   332

11.2.3.4 Ingestion-Based Pathway Conclusions   333

11.2.4 Conclusion: Hot-Spot Limits in Soil   334

11.3 HOT-SPOT LIMITS ON BUILDING SURFACES   335

11.3.1 External Radiation Pathway   335

11.3.1.1 RESRAD-BUILD Area Factor Approach for External Radiation Pathway   336

11.3.1.2 MicroShield Area Factor Calculation   337

11.3.1.3 Receptor Location 1 m Distance from the Hot Spot   337

11.3.1.4 External Radiation Pathway Conclusions   339

11.3.2 Inhalation Pathway   340

11.3.3 Ingestion Pathway   341

11.3.4 Conclusion: Hot-Spot Limits on Building Surfaces   343

11.4 BAYESIAN STATISTICAL APPROACH TO ASSESS HOT SPOTS   343

11.4.1 Bayesian Statistical Approach   345

11.4.2 Bayesian Hot-Spot Assessment Using Robust *t* Distribution   347

11.4.3 Demonstrating Compliance with the Hot-Spot Limits   351

11.4.4 Conclusions for Bayesian Statistical Approach   352

QUESTIONS AND PROBLEMS   353

CHAPTER 12 ■ Statistics and Hypothesis Testing 355

12.1 BASIC POPULATION STATISTICS AND CONFIDENCE
INTERVAL TESTING 356

12.1.1 Basic Statistics 356

12.1.2 Confidence Interval Testing 359

12.2 DATA DISTRIBUTIONS 360

12.2.1 Binomial Distribution 360

12.2.2 Poisson Distribution 363

12.2.3 Normal Distribution 364

12.2.4 Student's *t* Distribution 366

12.3 HYPOTHESIS TESTING 367

12.3.1 Hypothesis Testing Fundamentals and Examples 368

12.3.2 Chi-Square Test: A Hypothesis Test for Evaluating
Instrument Performance 373

12.4 BAYESIAN STATISTICS 376

QUESTIONS AND PROBLEMS 381

SUGGESTED READING 382

CHAPTER 13 ■ MARSSIM Final Survey Design and Strategies 383

13.1 FSS PROTOCOLS PRIOR TO MARSSIM 384

13.1.1 NUREG/CR-2082 Guidance 384

13.1.2 NUREG/CR-5849 Guidance 387

13.2 OVERVIEW OF MARSSIM SURVEY DESIGN 392

13.2.1 Sign Test Example: Co-60 in Soil 393

13.2.1.1 *Derived Concentration Guideline Levels* 393

13.2.1.2 *Sign Test: Determining Numbers of Data
Points* 394

13.2.1.3 *Determining Data Points for Areas
of Elevated Activity* 398

13.2.2 WRS Test Example: Uranium and Thorium in Soil 402

13.2.2.1 *Derived Concentration Guideline Levels* 403

13.2.2.2 *WRS Test: Determining Numbers of Data
Points* 403

13.2.2.3 *Determining Data Points for Areas of Elevated Activity* 407

13.3 SURFACE-ACTIVITY MEASUREMENTS: WILCOXON RANK SUM TEST OR SIGN TEST? 409

13.3.1 Surface-Activity Measurements 410

13.3.2 WRS Test for Surface-Activity Assessment 411

13.3.3 Sign Test for Surface-Activity Assessment 413

13.3.4 Simulation Study Conceptual Design 419

13.4 COMPARISON OF MARSSIM AND NUREG/CR-5849 FSSS FOR NUCLEAR POWER PLANT DECOMMISSIONING PROJECTS 420

13.5 ANNOTATED MARSSIM EXAMPLES 425

13.5.1 Example 1: Class 1 Interior Survey Unit 425

13.5.1.1 *Survey Instrumentation* 426

13.5.1.2 *WRS Test Sample Size Determination* 427

13.5.2 Example 2: Class 2 Interior Survey Unit 429

13.5.2.1 *Gross DCGLs* 429

13.5.2.2 *Survey Instrumentation* 430

13.5.2.3 *Sign Test Sample Size Determination* 431

13.5.3 Example 3: Class 1 Exterior Survey Unit 433

13.5.3.1 *Modified DCGL* 434

13.5.3.2 *Sign Test Sample Size Determination* 434

13.5.4 Example 4: Class 1 Interior Survey Unit with Multiple Contaminants 436

13.5.4.1 *Gross Activity DCGLs and Area Factors* 436

13.5.4.2 *Instrumentation, Static MDC, and Scan MDC* 438

13.5.4.3 *Sample Size Determination* 440

13.6 MARSSIM FSS DESIGN STRATEGIES: UNDERSTANDING THE POWER CURVE 443

13.7 RANKED SET SAMPLING 452

QUESTIONS AND PROBLEMS 455

CHAPTER 14 ■ MARSSIM Data Reduction   459

14.1 DATA QUALITY ASSESSMENT FOR THE SIGN TEST
      FOR Co-60 IN SOIL   460

      14.1.1 Review of the DQOs   460

      14.1.2 Preliminary Data Review   462

      14.1.3 Selection of Statistical Test   464

      14.1.4 Verification of Statistical Test Assumptions   464

      14.1.5 Perform Statistical Test and Draw Conclusions
              from the Data   465

14.2 DATA QUALITY ASSESSMENT FOR THE WRS TEST
      FOR URANIUM AND THORIUM IN SOIL   469

14.3 WHAT IF THE SURVEY UNIT FAILS?   473

      14.3.1 Why Survey Units Fail   473

      14.3.2 Double Sampling   474

QUESTIONS AND PROBLEMS   477

CHAPTER 15 ■ Clearance of Materials   479

15.1 CLEARANCE: A CONTROVERSIAL HISTORY   480

15.2 MARSAME AND DQOs FOR THE RELEASE
      OF MATERIALS   482

15.3 RELEASE CRITERIA, PROCESS KNOWLEDGE, AND
      OTHER SURVEY DESIGN CONSIDERATIONS   484

      15.3.1 Solid Material Description and Survey Units   487

      15.3.2 Process Knowledge   491

      15.3.3 Inaccessible Areas   492

      15.3.4 Nature of Contamination   494

      15.3.5 Material Classification   494

              15.3.5.1 Class 1 Solid Materials   496

              15.3.5.2 Class 2 Solid Materials   496

              15.3.5.3 Class 3 Solid Materials   497

      15.3.6 Application of Release Guidelines   497

15.4 DETECTION LIMITS FOR MATERIAL RELEASE SURVEYS   499

      15.4.1 Static MDCs   501

15.4.2 Scanning-Based MDCs   503

      15.4.2.1 *Hand-Held Detector Scan MDCs*   504

      15.4.2.2 *Conveyor Survey Monitor Scan MDCs*   505

15.5 CLEARANCE SURVEY APPROACHES   507

15.5.1 Background Radiation Levels for Clearance Measurements   508

15.5.2 Clearance Survey Activities: Measurement and Sampling Methods   509

15.5.3 MARSSIM-Type Clearance Survey Design   510

15.5.4 Scanning-Only Clearance Survey Design   514

      15.5.4.1 *Scan-Only Using Conventional Survey Instrumentation*   515

      15.5.4.2 *Conveyor Survey Monitors*   516

15.5.5 *In Toto* Clearance Survey Design   518

      15.5.5.1 In Situ *Gamma Spectrometry*   519

      15.5.5.2 *Volume Counters*   521

QUESTIONS AND PROBLEMS   522

SUGGESTED READING   523

CHAPTER 16 ■ Decommissioning Survey Applications at Various Facility Types   525

16.1 URANIUM SITES   526

16.1.1 Nature of the Contaminant   527

16.1.2 Field Measurements   528

16.1.3 Laboratory Measurements   530

16.2 THORIUM AND RADIUM SITES   532

16.2.1 Nature of Contaminants   532

16.2.2 Field Measurements   533

16.2.3 Laboratory Measurements   535

16.3 POWER REACTOR   537

16.3.1 Nature of Contaminants   537

16.3.2 Field Measurements   538

16.3.3 Laboratory Measurements   539

16.4 UNIVERSITY/RESEARCH FACILITIES 539

   16.4.1 Nature of Contaminants 540

   16.4.2 Field Measurements 541

   16.4.3 Laboratory Measurements 542

QUESTIONS AND PROBLEMS 543

CHAPTER 17 ■ FSS Reports and Measurement Uncertainty 545

17.1 FSS REPORT CONTENT 546

17.2 REPORTING SURVEY RESULTS: MEASUREMENT OF
UNCERTAINTIES AND ERROR PROPAGATION 550

   17.2.1 Instrument Efficiency 556

   17.2.2 Surface Efficiency 557

QUESTIONS AND PROBLEMS 558

CHAPTER 18 ■ Practical Applications of Statistics
to Support Decommissioning Activities 559

18.1 TESTS FOR DATA NORMALITY 560

   18.1.1 Shapiro–Wilk (W Test) 560

   18.1.2 D'Agostino Test 562

18.2 APPLICATIONS OF STATISTICS IN
DECOMMISSIONING: COMPARISON OF DATA SETS 564

   18.2.1 t Test with Unequal Variances: Evaluating
Automated Soil Sorter Performance 566

   18.2.2 Pairwise t Test: Evaluating "Wet" versus Processed
Gamma Spectroscopy Results 569

   18.2.3 Confirmatory Analyses Using Nonparametric
Statistics 572

18.3 CASE STUDY: COMPARING Cs-137
CONCENTRATION IN CLASS 3 AREA WITH
BACKGROUND REFERENCE AREA USING BOTH t
TEST AND WRS TEST 575

   18.3.1 Two-Sample t Test 576

      18.3.1.1 Survey Design 577

18.3.1.2 *Survey Implementation and Data Reduction* 582

18.3.1.3 *Survey Design with Different Null Hypothesis* 585

18.3.2 Mann–Whitney Test: Comparing Cs-137 Concentrations in a Class 3 Area with a Background Reference Area 590

18.3.2.1 *Survey Implementation and Data Reduction* 592

QUESTIONS AND PROBLEMS 594

CHAPTER 19 ■ International Decommissioning Perspectives 595

REFERENCES, 599

SOLUTIONS TO SELECTED QUESTIONS AND PROBLEMS, 607

APPENDIX A: RADIONUCLIDE AND NATURAL DECAY SERIES CHARACTERISTICS, 629

APPENDIX B: MARSSIM WRS AND SIGN TEST SAMPLE SIZES (FROM THE MARSSIM TABLES 5.3 AND 5.5), 645

APPENDIX C: EXAMPLE DECOMMISSIONING INSPECTION PLAN FOR FINAL STATUS SURVEY PROGRAM, 649

INDEX, 657

Preface

This book is intended to serve as a valuable resource for decommissioning professionals, particularly those charged with planning and implementing radiological surveys in support of decommissioning. The book equips the reader with proven strategies for demonstrating to regulators and the stakeholder community that contaminated sites can be released for other beneficial uses. This goal is achieved through detailed derivations and discussion of technical bases and illustrated through real-world examples.

The book articulates clearly the technical issues that arise during decommissioning projects—including the application of statistics for survey design and data reduction, selection of survey instrumentation and detection sensitivity, final status survey procedures, and dose modeling to translate release criteria to measurable quantities—and presents solutions for navigating the complexity inherent in designing and implementing MARSSIM and MARSAME surveys. Case studies and worked examples are used extensively to clarify the technical concepts presented. Examples and problems are provided in many of the chapters, and detailed solutions are furnished in the appendix. Finally, the specific survey issues related to uranium, thorium, power reactor, and university/research facilities are discussed in detail and include effective strategies for streamlining final status surveys.

This second edition of *Decommissioning Health Physics* is an extensive revision of the first edition. Significant changes in this new edition include the following:

- Chapter 3 on characterization was extensively revised to reflect the recent guidance in ANSI N13.59 on the use of DQOs for planning surveys.

- A new chapter on hot-spot assessment (Chapter 11) discusses the dosimetric significance of hot spots when designing surveys and proposes a novel approach for establishing hot-spot limits.

- A new chapter on the clearance of materials (Chapter 15) highlights aspects of the MARSAME manual and includes information contained in draft NUREG-1761 to provide an integrated view of releasing materials.

- The revised statistics chapter (Chapter 12) includes an introduction to Bayesian statistics and double sampling and ranked set sampling statistical approaches.

- Decommissioning regulations and guidance documents have been updated throughout the book.

- The survey instrumentation used to support decontamination and decommissioning (D&D) surveys have been updated, including the expanded coverage of *in situ* gamma spectrometer.

- Numerous case studies and examples have been added throughout the book.

It is my sincere hope that this book will benefit MARSSIM (and MARSAME) users by being a resource that comprehensively describes the technical aspects of designing and executing radiological surveys in support of decommissioning.

Acknowledgments

My knowledge and expertise in helping decommissioning professionals successfully perform radiological surveys comes from the experiences I have had while working with many outstanding professionals in the ORAU Survey Program. Since 1980, the ORAU Survey Program has been the leading provider of characterization and independent verification surveys of environmental cleanup, building public trust and confidence in the cleanup of contaminated sites. ORAU's survey team leverages its continually evolving environmental and decommissioning survey procedures to solve challenging decommissioning survey problems—through the application of innovative field survey instrumentation, laboratory analyses of media samples, and development and delivery of MARSSIM/MARSAME training courses. To this end, numerous ORAU Survey Program procedures and protocols have been referenced in preparing this book.

Since the publication of the first edition, I have received a number of constructive comments and suggestions for improvements. In particular, I would like to thank Jack McCarthy for his valuable review of the first edition. I am also grateful to others (especially Bill Lipton and Bryan Werner) who took the time to point out technical errors in the text.

Many of my friends and colleagues at ORAU contributed to the preparation of the first edition of this book. I am sincerely grateful for the experiences and expertise they shared with me to prepare this text. I would like to offer my very special thanks to Dr. Larry Miller (University of Tennessee), who was my major professor and a tremendous help in ensuring that I completed my PhD dissertation. Thanks to my friend and colleague, Dr. Alexander Williams, who has been a venerable champion and supporter of independent verification of cleanup. Finally, I am grateful for the too-numerous-to-mention technical reviews and debates with my friend and fellow "survey aficionado," Tim Vitkus. I appreciate your tireless efforts for more than two decades to continuously improve and keep the ORAU Survey Program at the vanguard.

It has been a sincere pleasure working with the publishing team at Taylor & Francis Group on this project. Special thanks to Robert Sims, Marsha Pronin, Rachel Holt and the editorial and production staffs of T&F, with particular praise to Syed Mohamad Shajahan for his excellent copyediting.

Thanks to my best friend and greatest advocate, my beautiful wife Sandy. Her unwavering support has encouraged me through many "career projects." And for understanding my many hours at the computer, I would like to thank my children, Alyssa, Elizabeth, and Gunnar.

Above all, I give thanks to God for His many blessings—"I can do all things through Him who strengthens me" (Philippians 4:13).

Author

SUMMARY

Dr. Eric W. Abelquist, executive vice president of Oak Ridge Associated Universities (ORAU) and deputy director of the Oak Ridge Institute for Science Education, is responsible for working in collaboration with the president/CEO and other vice presidents to oversee organizational best practices, program and business unit leadership, and community relations. He works directly with the president/CEO to formulate organizational strategic objectives and manage key strategic initiatives. Abelquist also advises the president/CEO on scientific and engineering issues that advance scientific research and education opportunities.

Abelquist began his more than 20-year career at ORAU as a project leader for Independent Environmental Assessment and Verification (IEAV), overseeing a team of health physics technicians and conducting characterization and independent verification surveys at the Department of Energy's (DOE) and Nuclear Regulatory Commission's (NRC) sites. He later worked for many years as the associate director of IEAV where, most notably, he contributed to the development and implementation of the Multiagency Radiation Survey and Site Investigation Manual (MARSSIM). For several years Abelquist served as the vice president and director of IEAV, where he oversaw ORISE's radiochemistry laboratory, conducted training courses in radiation sciences, and managed radiological surveys, environmental assessments, and independent verification of cleanup projects involving DOE's and NRC's D&D programs. Abelquist continues to provide technical assistance in various aspects of decommissioning surveys. He also published a textbook, *Decommissioning Health Physics: A Handbook for MARSSIM Users* (Institute of Physics Publishing, Bristol, UK) in 2001.

EDUCATION

- PhD, Nuclear Engineering, University of Tennessee, Knoxville, Tennessee, 2008

- MS, Radiological Sciences and Protection, University of Massachusetts, Lowell, Massachusetts, 1991

- BS, Radiological Health Physics, University of Massachusetts, Lowell, Massachusetts, 1989

CERTIFICATIONS AND SECURITY CLEARANCE

- Certified Health Physicist

- U.S. Department of Energy Security "Q" Clearance

RELATED EXPERIENCE

ORAU

Executive Vice President, ORAU and Deputy Director, ORISE
2009–Present

Responsible for day-to-day operations of ORAU's programmatic business lines, working with the CEO and other vice presidents to oversee organizational best practices, program and business unit leadership, and community relations.

Vice President and Director, IEAV
2002–2009

Responsible for strategic vision, planning and direction of the ORISE program that includes 60 employees with an annual budget of $8 million. Successfully directed program activities that included independent verification of cleanup at DOE and NRC sites, radiation sciences training, and health physics projects.

ORAU, IEAV

Associate Director
1996–2002

Successfully led technical direction of our program while working effectively with the program director on overall planning and direction. Led a team to develop new and updated guidance and standards for characterization and final status surveys.

Participated in the development and implementation of MARSSIM—the Multi–Agency

Radiation Survey and Site Investigation Manual, NUREG–1575, and training—the first manual and training to consolidate all the procedures and regulations of four federal agencies with major D&D efforts (DOE, NRC, the Environmental Protection Agency, and the Department of Defense). All four agencies have endorsed this training and manual as the standard for all of their radiation survey activities.

ORAU, IEAV
Project Leader
1993–1996

Responsible for managing a team of health physics technicians. Tasked with performing characterization and independent verification surveys at DOE and NRC sites.

Nuclear Energy Services
Project Manager
1992–1993

Managed D&D projects encompassing all aspects of operational health physics. Served as radiation safety officer for NRC materials license. Prepared decommissioning work plans, final survey plans, and technical reports.

DOE Pinellas Plant
Advanced Health
Physicist
1991–1992

Responsible for the health physics training program, stack sampling tritium effluent, the internal and external dosimetry program, and decommissioning tritium process equipment.

Brookhaven National Lab
DOE Fellow
January 1991–June 1991

Projects included a whole-body counting mission in the Marshall Islands, a failed-fuel detection program for the medical research reactor, environmental pathway analysis for dose assessment from air and water effluent and stack sampling accelerator effluent.

Harvard University
Health Physics Intern
1989–1991

Performed master's thesis research: Stack Sampling Effluent from a Radioactive Waste Incinerator; conducted radiation surveys and operated nuclear instrumentation.

PROFESSIONAL ASSOCIATIONS

- Oak Ridge Chamber, Executive Committee

- United Way of Anderson County, Board of Directors (2011–present)

- Leadership Oak Ridge (Class of 2009)

- American Academy of Health Physics

- Health Physics Society (HPS)

- HPS Ask the Expert Web Editor for Decommissioning Topic

- Taught a course as adjunct professor at University of Tennessee, "Radiological Characterization of Facilities Undergoing Decontamination and Decommissioning"

SELECTED PUBLICATIONS

Abelquist E.W. and Cragle, D.L. *Beryllium Contamination Assessment of Oak Ridge National Laboratory (ORNL) Building 9201-2*, September 2006.

Chapman, J.A., Boerner, A.J., and Abelquist, E.W. *Spatially-Dependent Measurements of Surface and Near-Surface Radioactive Material Using In Situ Gamma Spectrometry (ISGRS) for Final Status Surveys*, ORISE report, November 15, 2006.

Characteristics of the Radiological Environment in Preparation for and in Support of D&D, Chapter 6 in Decommissioning and Restoration of Nuclear Facilities, Health Physics Society 1999 Summer School.

Clearance Survey Costs—Technical Bases for Developing Survey Costs (draft), October 2002, revised February 2004 (not published).

Dickson, H., Abelquist, E., and Karam, A., *Radiation Safety without Borders Guidance Manual*, Draft HPS Publication, 2002.

Guest Editorial, Operational Radiation Safety, June 2003.

HPS/ANSI N13.49, *Performance and Documentation of Radiological Surveys*, August 6, 2001.

Nonparametric Statistics, article in 2003 PTP Brochure.

NUREG-1507, *Minimum Detectable Concentrations with Typical Radiation Survey Instruments for Various Contaminants and Field Conditions*, June 1998.

NUREG-1761 (draft), *Radiological Surveys for Controlling Release of Solid Materials*, July 2002.

Scan MDCs for Multiple Radionuclides in Class 1 Areas, Operational Radiation Safety, June 2003.

Scrap Metal Characterization, Technical Solutions Report ORO-1, U.S. DOE Technical Solutions Team, June 2003.

AWARDS/ACCOMPLISHMENTS

- President, East Tennessee Chapter, HPS, 2000–2001

- 2003 HPS Elda E. Anderson Award

- Board of Directors, National HPS, 2003–2006

- Founder and First President of HPS Decommissioning Section

- DOE Operational Health Physics Fellow, 1989–1991

- HPS/ANSI N13.49, Chair—Performance and Documentation of Ionizing Radiation Surveys

- HPS/ANSI N13.59, Chair—Characterization in Support of Decommissioning Using the Data Quality Objectives Process, 2008

CHAPTER 1

Introduction

Current Issues in Decommissioning and MARSSIM Overview

I T HAS BEEN 13 years since *Decommissioning Health Physics* was first published in 2001. Decommissioning projects continue to be performed throughout the United States and all over the world. Many lessons have been learned. The Department of Energy's (DOE) Environmental Management (EM) program has claimed numerous successes as sites are cleaned up—for example, Rocky Flats, Oak Ridge, Fernald, and Miamisburg (Mound). The Nuclear Regulatory Commission (NRC) continues to make progress in removing sites from its Site-Decommissioning Management Plan (NRC 2012a). The United Kingdom Nuclear Decommissioning Authority (NDA) is progressing with the D&D (decontamination and decommissioning) of facilities at Sellafield, ensuring the surveillance and maintenance of the Magnox reactors, and remediating the Dounreay site. These examples are a small sample of the worldwide D&D footprint.

The major benefit of decommissioning is that "the levels of radioactive material at the site are reduced to levels that permit termination of the license and use of the site for other activities, rather than leaving the radioactive contamination on the site so that it could adversely affect public health and safety and the environment in the future" (NRC 1998d, p. 2). Obviously, decommissioning is considered when the facility has become obsolete and is no longer considered operationally viable—that is, when it has become a "surplus facility." This is occurring at many sites around the world as nuclear weapon plants reduce their footprints and as nuclear power plants shut down.

Some nuclear power reactors opt for D&D when faced with upgrades or refurbishments that are too costly to maintain the facility operational. Indeed, the deregulation of the electric utilities was a significant driver for a number of nuclear power plants to decommission units, resulting in several utilities deciding to decommission their reactor years before the expiration of their operating license. Utilities continue to grapple with the difficult decision of determining whether nuclear power can compete financially with other forms of energy—particularly as advances in fracking keep natural gas incredibly inexpensive.

1.1 DECOMMISSIONING OVERVIEW

Decommissioning has become more technically challenging because of the move to dose-based release criteria and the commensurate use of MARSSIM (*Multi-Agency Radiation Survey and Site Investigation Manual*; NRC 2000a). But in reality, it has always been a rigorous exercise to demonstrate to the public and regulators that contaminated sites have been sufficiently cleaned to achieve agreed-upon release criteria. Technical issues that often arise during decommissioning projects include the application of statistics for survey design and data reduction, selection of survey instrumentation and detection sensitivity, final status of survey procedures, and pathway modeling to translate the release criteria into measurable quantities.

Decommissioning is a complex activity that involves characterizing the contaminated areas, remediating those areas that exceed the acceptable contamination guidelines, and performing radiological surveys to demonstrate that the site has been successfully cleaned up. In the United States, this activity is regulated by the U.S. NRC, the U.S. Environmental Protection Agency (EPA), as well as the individual states.

An important aspect of decommissioning is determining "how clean is clean enough." As mentioned above, the NRC and the EPA are the two principal federal agencies responsible for the cleanup and decommissioning of radioactively contaminated sites. The NRC's release criteria for unrestricted release are promulgated in Subpart E of 10 CFR 20.1402; they include a dose limit to an average member of the critical group of 25 mrem/y, and that the residual radioactivity has been reduced to levels that are as low as reasonably achievable (ALARA). The EPA's release criteria are risk based rather than dose based. Specifically, the EPA uses an acceptable lifetime excess cancer risk of 10^{-6}–10^{-4} to assess whether a site should be released or not. Typically, individual states use the same release criteria as the NRC, although in some states, more restrictive release criteria have been adopted—for example, Connecticut has a release criterion of 19 mrem/y, New Jersey uses 15 mrem/y, and Massachusetts has adopted 10 mrem/y. The DOE has a basic dose limit of 100 mrem/y for members of the public from all sources, and for a single source such as a decommissioning site has stated that NRC's 25 mrem y^{-1} is reasonable (DOE 2002).

A common feature of the regulatory release criteria mentioned above is that they are not measurable quantities, at least not directly. This is the role of dose modeling—to translate the dose- or risk-based release criteria into measurable concentrations of radioactivity in the soil and on building surfaces. Dose modeling considers how future receptors might be exposed to residual radioactivity that remains following the decommissioning of a site or a building. The specific exposure scenarios such as the residential farmer or building occupant scenarios are postulated, and environmental pathways that are commensurate with each scenario are used to calculate or translate the release criterion into a measurable quantity. These measurable quantities are called derived concentration guideline levels (DCGLs). So, demonstrating compliance with DCGLs is the same thing as demonstrating compliance with the release criteria.

Various software tools exist to facilitate dose modeling in support of decommissioning. The most widely used modeling codes in the decommissioning industry are likely to be RESRAD for soil areas and RESRAD-BUILD for building surfaces, both written and maintained by Argonne National Laboratory. These software tools allow the quick calculation of DCGLs by modeling the transport of radionuclides through the environment to the future receptor via various pathways such as direct external radiation, ingestion of drinking water, plant and animal products, and inhalation of contaminated dust. The modeling parameters associated with each of the pathways are needed to perform these calculations. These parameters can be classified as physical (e.g., resuspension factor), metabolic (e.g., breathing rate) or behavioral (e.g., time spent on gardening), and they can be default or site-specific values. Therefore, pathway modeling considers the various scenarios and exposure pathways to convert the dose or the risk into measurable concentrations.

RESRAD and RESRAD-BUILD are used to calculate DCGLs that equate to the appropriate release criteria for the site. This is performed by modeling the unit concentration (e.g., 1 pCi/g for soil) for a particular radionuclide, and then calculating the receptor dose based on the defined scenario(s), exposure pathways, models, and parameter distributions. The dose that results for unit concentration is then scaled to the dose-based release criterion (e.g., 25 mrem/y) to directly calculate the radionuclide concentration (DCGL) that corresponds to the release criterion. It is important to note that this radionuclide concentration is typically taken to be more or less uniformly distributed over the survey unit (i.e., in the order of 1000–10,000 m²).

The single-most important contemporary issue affecting decommissioning surveys has been the move toward dose-based release criteria. For more than 25 years, the release criteria for surface activity levels were obtained from the NRC's Regulatory Guide 1.86. These guideline levels were loosely based on dose, but were focused more on survey instrument capabilities at that time (Slobodien 1999). So, this change resulted in a departure from the relatively simple guidelines that were largely based on field measurement capabilities, to criteria that may require extensive pathway modeling to determine guidelines. Chapter 4 provides a more complete picture of the historic release criteria used and the transition to the current dose-based criteria. This paradigm shift to dose-based decommissioning criteria places the emphasis on pathway modeling to translate the dose criterion (e.g., NRC regulations provide for a 25 mrem/y unrestricted release criterion) into measurable quantities of surface activity and radionuclide concentrations in the soil, called DCGLs.

It is interesting to summarize the major technical advances in decommissioning surveys that were either derived or applied by the MARSSIM, or were contemporary with the development of the MARSSIM. These advances include (1) the use of dose- or risk-based release criteria and DCGLs, (2) the application of the data quality objectives (DQO) process, (3) the scan sensitivity and the methodology to design the final status surveys (FSSs) to deal with the potential hot spots, (4) the application of nonparametric statistics in the hypothesis-testing framework, and (5) the use of international guidance documents (i.e., ISO-7503) to address surface activity measurements and their technical defensibility. It is precisely this added complexity in designing and implementing MARSSIM FSSs that provided the impetus for this book.

1.2 MARSSIM OVERVIEW

Radiological surveys in support of decommissioning are planned at the same time as DCGLs are being developed for the site. MARSSIM, which stands for the Multiagency Radiation Survey and Site Investigation Manual, is the industry standard for decommissioning surveys. It has been the buzzword in the D&D arena since the document was published in December 1997. The MARSSIM is a consensus document prepared by the EPA, DOE, NRC, and Department of Defense (DOD) to provide consistent methods for conducting radiation surveys and investigations at potentially contaminated sites. The details particular to a specific agency were largely omitted from the MARSSIM; instead, agency-specific guidance is provided in documents that reference the MARSSIM such as NRC's NUREG-1757 series and DOE's O 458.1.

The MARSSIM is geared toward the FSS—although it does discuss the other survey types—and uses the DQO process to help plan the FSS. The DQO process should be used whenever there are decisions to be made by the survey designer that impact the number of samples needed to address the particular survey question. In the case of FSSs, the question is generally "Does the residual radioactivity in the survey unit comply with radiological release criteria?" Contrast this to the past survey design practices that were based on a prescriptive sample size—that is, NUREG/CR-5849 guidance recommended that four soil samples in each 100 m^2 that affected the survey unit must be collected and analyzed. There is not much to be gained using the DQO process when the end result of a planning process has already been specified; in other words, how can one take advantage of the flexibility offered by the DQO process when the sample size is prescriptive? Chapter 3 details the flexibility that the MARSSIM provides through the DQO process. In one sense, the DQO process is the underlying survey design framework adopted by the MARSSIM.

The MARSSIM's popularity is due to the broad agency support it has received for providing a uniform methodology for conducting radiological surveys to support decommissioning. The MARSSIM provides guidance on the planning, implementation, and evaluation of decommissioning radiological surveys—historical site assessment (HSA), scoping, characterization, and FSSs. It is geared toward the FSS—that demonstrates that dose-based or risk-based release criteria for decommissioning sites have been satisfied. A brief description of the types of MARSSIM survey follows.

The HSA is not a survey per se. It can be described as an effort to collect as much background information on the site as possible. Examples of HSA information include site inspection reports, routine operational survey reports, documentation of off-normal occurrences and effluent releases, and interviews with the former employees. The objectives of the HSA are to identify the potential sources of contamination, differentiate areas of different contamination potential, and provide input to scoping and characterization survey designs. The scoping and characterization surveys are built upon the HSA data by collecting both random and judgmental samples from all potential areas of concern. The objectives of these preliminary surveys are to determine the nature and extent of contamination to allow effective planning for remediation and waste-disposal activities, as well as to provide site data for the dose-modeling input for site-specific DCGLs, and input to the FSS design.

The MARSSIM provides many details on the FSS design. The initial steps are to identify the contaminants and to classify all site areas according to the contamination potential—with the underlying premise being that the greater the contamination potential, the greater the survey coverage (i.e., greater scan and sampling density). The areas that have no reasonable potential for residual contamination are classified as the non-impacted areas. These areas have no radiological impact from site operations and are typically identified early in decommissioning. The areas with a reasonable potential for residual contamination are classified as the impacted areas. The impacted areas are further subdivided into one of the three classifications (NRC 2000a):

- *Class 1 areas*: Areas that have or had prior to remediation, a potential for radioactive contamination (based on site-operating history) or known contamination (based on the previous radiation surveys) above the DCGL. Simply stated, class 1 areas are likely to have hot spots.

- *Class 2 areas*: Areas that have or had prior to remediation, a potential for radioactive contamination or known contamination, but are not expected to exceed the DCGL.

- *Class 3 areas*: Any impacted areas that are not expected to contain any residual radioactivity or are expected to contain levels of residual radioactivity at a small fraction of the DCGL, based on site- operating history and the previous radiation surveys.

Once classified as class 1, class 2, and class 3 areas, each area is further divided into survey units based on the guidance offered in the MARSSIM. A survey unit is a physical area consisting of the structure or land area of specified size and shape for which a separate decision will be made as to whether or not that area exceeds the release criterion. Survey units range in size from 2000 to 10,000 m² or more for land areas and 100–1000 m² or more for building surfaces.

The FSS consists of two general activities—radiological scanning to identify any elevated radiation levels in the survey unit, and random systematic sampling over the survey unit (soil samples for land areas and surface activity measurements for building surfaces). Two statistical tests are used to plan and evaluate the FSS sampling data—Wilcoxon rank sum (WRS) test when the contaminants are present in the natural background, and the sign test when the contaminants are not present in the background. The number of data points needed to satisfy these nonparametric tests is based on the $DCGL_W$ value (or modified DCGL for multiple contaminants), the expected standard deviation (σ) of the contaminant in the background and in the survey unit, DQO inputs that include the acceptable probability of making type I and type II decision errors and the lower bound of the gray region (LBGR). More specifically, the statistical sample size is based on the relative shift, given by Δ/σ, where $\Delta = DCGL_W - LBGR$, and the selected decision errors. As can be seen in MARSSIM (Table 5.3) for the WRS test, for relative shifts less than 1, the sample size becomes quite large, whereas the relative shifts greater than 3 yield much lower sample sizes.

An important part of the MARSSIM flexibility is exercised in the balance between the type II error (failing a survey unit that does satisfy the release criteria) and the selected sample size. A decision concerning this balance is best achieved through the use of a power curve that plots the probability of the survey unit passing as a function of the contaminant concentration. Depending on the survey planner's expectation of the residual contamination concentration, the power curve is an excellent planning tool to help balance the sample size with the risk of failing a "clean" survey unit. This survey design strategy is further elaborated in Chapter 13.

A second evaluation is performed on judgmental samples that were collected at likely areas of contamination or were based on the scanning results. These judgmental samples are commonly referred to as "hot spots" (radionuclide concentrations that exceed the DCGL) identified in the survey unit. This design aspect helps to address the public's concern that hot spots may be missed during the FSS—for example, the common rhetorical question goes something like "How do you know that a hot spot has not been left behind that my child might stumble upon in the future playground constructed at this site?" Therefore, sampling on a specified grid size, in conjunction with surface scanning, are used to obtain an adequate assurance level that small locations of elevated radioactivity will still satisfy DCGLs—that are dose based. Consequently, the sample size for class 1 survey units is at least the statistical sample size, but may be increased based on the necessary scan sensitivity to detect hot spots. Chapter 11 describes an alternative approach for modeling the receptor dose due to hot spots—using first principles to assess how environmental pathways and parameters are impacted by hot-spot source terms.

Once the number of samples (either surface activity measurements or soil samples depending on the site area) is determined based on the DQOs, an integrated survey plan is prepared. Simply stated, the integrated survey plan combines the sample size determined for each survey unit with the selection of survey instruments and scan coverage for each survey unit. Of course, there are many other considerations in the survey plan—but in a nutshell, the integrated survey plan covers survey unit identification, sample size, selected analytical techniques and survey instrumentation, and scan coverage. One final aspect for a FSS plan is the use of investigation levels. Investigation levels are established to identify the additional actions necessary to examine the measurement and sample results that may indicate that the survey unit has been improperly classified.

Upon completion of the FSS, the WRS or sign test is used to test the data against the null hypothesis. The purpose is to determine if the mean (or median) of the contaminant distribution satisfies the release criteria. The elevated measurement comparison is then performed to demonstrate that hot-spot concentrations do not exceed the $DCGL_{EMC}$ for small areas. Both tests must be satisfied for the survey unit to pass.

1.3 CLEARANCE OF MATERIALS OVERVIEW

One emerging concern related to decommissioning is the recycle and the reuse of materials from decommissioning activities. Enormous quantities of metals and concrete are expected to be generated from the decommissioning of the commercial nuclear reactors and weapons complex sites across the United States. To address this concern, the multiagency radiation survey and assessment of materials and equipment manual (MARSAME) was published in 2009 as a MARSSIM supplement (NRC 2009). Chapter 15 covers the release of materials, based on MARSAME, which expands the scope of MARSSIM to include methods and processes to support the disposition of materials and equipment. As with MARSSIM, the MARSAME does not provide the clearance criteria.

NUREG-1640, Radiological Assessments for Clearance of Equipment and Materials from Nuclear Facilities (NRC 1999c), provides the technical basis for the NRC to base regulatory standards for the clearance of equipment and materials with residual radioactivity from regulatory control. The methods described in NUREG-1640 address both surficially contaminated equipment and volumetrically contaminated scrap materials under equipment reuse and recycle scenarios. The clearance of materials under an equipment reuse scenario will likely include items that have more value due to their function as opposed to their material composition, such as power tools and other specialty equipment.

NUREG-1640 reports dose factors (in mrem/y per pCi/cm^2) used for deriving the clearance levels for numerous radionuclides for steel, copper, aluminum, and concrete materials. The exposure pathways included in the dose evaluation are external, inhalation, and secondary ingestion. Interestingly, the recycled material was evaluated using material flow models and dose assessment models based on probabilistic methods. Probabilistic methods ensure that the input parameters are modeled as distributions instead of point values that result in distributions for the output values (i.e., radionuclide concentrations and dose factors). Again, Chapter 15 addresses the various release criteria and survey methodology for the clearance of materials.

To conclude, decommissioning continues to be a significant discipline and a livelihood for many health physicists. The Health Physics Society's Decommissioning Section was formed in 1998 and continues to provide a valuable forum for the practicing D&D professionals. According to Section Bylaws, the primary objectives are to disseminate the information related to decommissioning activities, including regulatory rulemakings on decommissioning, decontamination techniques and advances in survey instrumentation and procedures, while providing a forum for the discussion of decommissioning issues. The Decommissioning Section web page is available at http://www.hpschapters.org/sections/decommissioning/index.php. The D&D Science Consortium (DDSC) was established in 2002, a few years after the Decommissioning Section was established in 1998; it provides regular updates on D&D news in the United States and around the world. Larry Boing, ANL has voluntarily provided D&D news updates on a weekly basis since the inception of the DDSC (2002) (http://www.orau.gov/ddsc/).

QUESTIONS

1. What is the underlying survey design framework adopted by the MARSSIM?

2. List some of the major advances in decommissioning surveys brought about with the MARSSIM process.

3. What are the current issues in decommissioning in the clearance of materials?

4. Name three important aspects of NUREG-1640's technical basis for the clearance of materials.

5. What inputs from the DQO process to design the FSSs affect the sample size?

6. What were some impacts with the transition of release criteria from Regulatory Guide 1.86 to dose-based criteria?

7. What is the null hypothesis for each survey unit using the MARSSIM process?

8. State the general steps in the MARSSIM FSS design.

SUGGESTED READING

MARSSIM
MARSAME
NUREG-1757 (three volumes)
NUREG-1640
Hypothesis testing chapter in your favorite statistics text

CHAPTER 2

Decommissioning Project Overview and Regulatory Agency Interface

O NCE IT HAS BEEN decided to finally shutdown an operating facility, whether it be a small R&D laboratory building or a power reactor station, the planning for decommissioning begins in earnest. Deciding not to decommission a facility once it has been shutdown is usually not an option—and definitely the case for nuclear reactors. So one of the first considerations is to select the most appropriate decommissioning alternative: DECON (DECONtamination), SAFSTOR (SAFe STORage), or ENTOMB (ENTOMBment). While these designations are typically used for decommissioning nuclear reactors; nearly all material licensees select the DECON option. In the next two sections these decommissioning options are defined and their attributes presented. The remainder of this chapter focuses specifically on the DECON option, and the FSSs and independent verification (IV) process used in conjunction with DECON.

2.1 DECOMMISSIONING OPTIONS

DECON is the most popular option because it settles the site decommissioning concerns once and for all. It includes complete facility decommissioning and license termination (or similar shutdown status) immediately following final shutdown. Specifically, the facility is decontaminated and/or dismantled to levels that permit release; contaminated equipment is either decontaminated or removed as radioactive waste. In the case of shutdown nuclear power reactors, the reactor is defueled—that is, the spent fuel is placed in the spent fuel pool. This is the predominant source of risk that exists at shutdown reactors during decommissioning. Many times the fuel is transferred from the spent fuel pool to a more stable configuration—the on-site independent spent fuel storage installation or ISFSI (independent spent fuel storage installation). Spent fuel is expected to be stored in dry casks in on-site ISFSIs until a long-term solution is implemented for high-level radioactive waste. Currently, the United States is evaluating options for a long-term solution for managing and disposing of the nation's spent nuclear fuel and high-level radioactive waste. The proposed high-level radioactive waste disposal site at Yucca Mountain (Nevada) continues to be shrouded in uncertainty.

Licensees usually opt for the spent fuel storage in an ISFSI because it allows them to complete the decommissioning process on the reactor facility and to ultimately terminate the reactor license; a separate license is issued for the ISFSI. One of the early considerations for power reactor D&D projects is where to site the ISFSI and whether to perform a FSS on this land area designated for the ISFSI. This was the case for both the Fort St. Vrain and Maine Yankee D&D projects. A complicating aspect of the FSS at an early stage in the D&D project is that elevated background radiation levels from various radioactive source terms make it very difficult to effectively scan the land area.

DECON is often selected because it provides greater certainty about the availability and disposal costs of radioactive waste facilities. Another frequently cited attribute of DECON is that it quickly removes a facility from future liability issues. A notable disadvantage, as compared to the other decommissioning options, is the subsequent higher worker and public doses since there is less time for radioactive decay to remove some of the source term. One more topic to discuss before we leave the DECON option is the concept of building "rubblization."

As defined in the February 14, 2000 NRC Commission Paper, rubblization involves (1) removing all equipment from buildings, (2) some decontamination of building surfaces, (3) demolishing the above-grade part of the structure into concrete rubble, (4) leaving the below-grade structure in place, (5) placing the rubble into the below-grade structure, and (6) covering, re-grading, and landscaping the site surface (NRC 2000d). Exposure pathway modeling is an important technical detail that must be addressed in conjunction with the decision to pursue rubblization. The Commission paper stated that possible exposure scenarios could include (1) a concrete-leaching scenario that impacts groundwater, (2) resident farmer scenario, (3) excavation scenario, and (4) intruder scenario.

This procedure has been somewhat controversial because it calls for radioactive waste generated from decommissioning activities to be buried on-site, as opposed to the standard practice of shipping to a licensed disposal facility. Some stakeholders have expressed concern that rubblization constitutes nothing more than a low-level waste disposal facility, and that if approved, will result in a proliferation of low-level waste sites. And as such, these stakeholders contend that rubblization sites should be licensed under the same regulations as a LLW site. Others rightly argue that rubblization, by approving the disposition of elevated contamination at a D&D site, is a departure from past decommissioning practices. However, given the dose-based nature of the NRC's decommissioning rulemaking and solely from a technical standpoint, as long as the scenarios and exposure pathways meet the release criteria, rubblization may be a viable option.

The anticipated bottom line is that these "rubblization scenarios" allow for higher acceptable surface activity guidelines than would be allowed for the conventional building occupancy scenario. For instance, Attachment 1 of the Commission Paper cites an acceptable guideline of approximately 100,000 dpm/100 cm^2 for Co-60 and Cs-137 based on the rubblization conceptual model, while the surface contamination values for the building occupancy scenario are 7100 and 28,000 dpm/100 cm^2 for Co-60 and Cs-137, respectively. Chapter 5 provides a more complete description of pathway modeling and exposure scenarios.

SAFSTOR involves maintaining the facility in a safe condition over a number of years, through performance of surveillance and maintenance (S&M) activities, followed by D&D. SAFSTOR may be considered deferred dismantlement, while DECON would be prompt dismantlement. To prepare a nuclear power plant for SAFSTOR, the spent fuel is removed from the reactor vessel (as it is in DECON) and radioactive liquids are drained from systems and components—leaving the facility in a stable condition.

The Department of Energy's "Implementation Guide for Surveillance and Maintenance during Facility Transition and Disposition" (DOE 1999) provides a reasonable overview of what is meant by S&M activities, stating that surveillance includes "activities to be performed to determine the operability of critical equipment, monitor radiological conditions, check safety-related items, provide for facility security controls, and assess facility structural integrity." The "maintenance" part includes those activities necessary to sustain the property in a condition suitable for its intended purpose, which is primarily to contain the residual radioactivity within the facility, and to minimize its further spread, while awaiting full decontamination. Routine maintenance activities are required to ensure the proper functioning of radiological monitoring equipment, facility systems such as heating and air conditioning units, and overhead cranes, to name a few. Obviously, these S&M activities can accrue significant costs in a short period of time.

The principal advantage of SAFSTOR is the substantial reduction in radioactivity and radiation levels, and thus worker doses, as a result of radioactive decay. A major disadvantage of SAFSTOR is that most (if not all) of the personnel familiar with the facility will not be able to assist in the actual dismantlement and decontamination of the plant. Furthermore, the site undergoing S&M activities will not be able to be used for more productive uses during SAFSTOR. Finally, the clincher for not undertaking the SAFSTOR option is that following the storage period it is quite possible that a radioactive waste disposal site will not be available. Frequently, multi-unit power reactor sites will select SAFSTOR for a shutdown reactor, while other units continue to operate (NRC 1998d). However, a competing consideration is whether there will be room in the spent fuel pool for the continued generation of spent fuel from the operational units.

The third decommissioning option, ENTOMB, is the method by which the contaminated portions of the facility are encased in a structurally long-lived material, which usually means concrete. The concept behind ENTOMB is to isolate the contamination from the environment, mainly by keeping water out of the containment. The most probable source of exposure would be from the inadvertent leakage of contamination from the structure. Another possible scenario is the inadvertent intruder, but this seems somewhat far-fetched—can you imagine someone inadvertently trespassing into the Chernobyl sarcophagus? This option may not be feasible if the contamination levels are not expected to be reduced to levels that allow release over some reasonable period of time.

On the basis of studies performed by the Pacific Northwest National Laboratory (PNNL), the radiation dose rate to an intruder after 60 years was 1800 rem/y, after 135 years it was 100 mrem/y, and after 160 years it was 25 mrem/y. This is consistent with a study performed by Nuclear Electric of the United Kingdom, who have proposed safe storage period of 135 years for their Magnox reactors and advanced gas-cooled reactors.

The proposed technical approach in the United Kingdom is to "entomb by grouting the internal void spaces, sealing up and capping over the reactor structure which is then covered with a mound of sea-dredged sand that is capped with re-vegetated top soil" (NRC 1999d, p. 9).

The only wastes likely to be generated with ENTOMB are those associated with post-shutdown deactivation activities, such as spent fuel removal. S&M activities are performed until the radioactivity decays to levels satisfying the release criteria, so this option will result in reduced worker and public doses. There has been renewed interest in ENTOMB because of the very real possibility that some D&D projects may not have low-level waste disposal options—long considered to be the trigger for serious consideration of this option. Realization of ever escalating disposal costs, which are expected to only increase as the major share of total decommissioning costs, reactor licensees have engaged their regulators to open dialog on ENTOMB. In response, the NRC is considering a rulemaking that examines the ENTOMB option in more detail.

Under current NRC regulations, decommissioning actions must be completed within 60 years of ceasing operations. This effectively places a limit on ENTOMB, as well as reducing the likelihood of it being seriously considered (i.e., dose rate still too high after 60 years of ENTOMB). The NRC is evaluating the option to move the 60-year limit to 100 years, or possibly even to 300 years. Attachment 2 of the NRC's Commission Paper on the ENTOMB option provides a very informative discussion of this decommissioning option titled "Viability of the Entombment Option" as an Alternative for Decommissioning Commercial Nuclear Power Reactors (NRC 1999d). The expectation is that at the conclusion of the entombment period, sufficient radioactive decay will have reduced residual radioactivity to levels that permit unrestricted release. What is not discussed is whether a FSS of some sort is necessary to ensure that the entombment period was successful in reducing radioactivity levels.

While this is feasible from a technical basis, a major hurdle with ENTOMB is to convince the regulator that even if institutional controls fail, the intruder barriers would remain effective to maintain public doses within release criteria (i.e., 25 mrem/y).

Another obstacle in the implementation of this option is that some of the contaminants that would be entombed are very long-lived activation products, including Nb-94, Ni-59, Ni-63, and C-14 that largely reside in the activated metals. These Greater Than Class C (GTCC) wastes, as currently regulated, are not generally acceptable for near-surface disposal and must be disposed of in a geologic repository as defined in NRC 10 CFR Part 60 (NRC 1999d). The entombment option for these GTCC wastes would require an amendment to the Low-Level Radioactive waste Policy Amendment Act of 1985.

For the remainder of this chapter—and this book for that matter—the focus will be on the DECON option.

2.2 DECOMMISSIONING PROJECT PHASES

Decommissioning projects are performed to remove sufficient radioactive and hazardous contamination from the facility so that decommissioning criteria are satisfied. The ultimate goal of the decommissioning process is the assurance that future uses of the decommissioned facility and/or land areas will not result in undue risk to human health and the environment from unacceptable levels of radiation and/or radioactive materials.

A typical decommissioning project may be divided into four phases: assessment, development, operations, and closeout (DOE 1994). It should also be mentioned that up to five decommissioning survey types may be performed throughout a typical decommissioning project: scoping, characterization, remedial action support, final status, and confirmatory. Each of these surveys will be further defined in a subsequent section, but it is important to recognize the specific goal of each survey type based on the decommissioning phase in which it is performed.

The assessment phase consists of performing scoping and characterization surveys to provide an assessment of the radiological and hazardous material conditions at the site. Survey information needed for this preliminary assessment includes the general radiation levels and gross levels of residual contamination on building surfaces and in environmental media. Scoping survey tasks include investigatory surface scanning, limited surface activity measurements and sample collection. Radiological characterization surveys are often performed to determine the nature and extent of radiological contamination at the site. Characterization survey activities often involve the detailed assessment of various types of building and environmental media, including building surfaces, surface and subsurface soil, surface water, and groundwater. Characterization surveys of building surfaces and structures include surface scanning, surface activity measurements, exposure rate measurements, and sample collection. Characterization surveys for surface and subsurface soils involve procedures for determining the horizontal and vertical extent of radionuclide concentrations in soil. Chapter 3 introduces the DQO process that is the foundation of characterization survey planning.

The development phase involves the engineering and planning for the project, and results in the production of the decommissioning plan or license termination plan. The decommissioning plan makes use of the characterization data to make estimates of personnel exposure, radioactive waste volume, and decommissioning costs. A schedule for decommissioning is also prepared at this time, and should make allowances for contingencies. The release criteria are also finalized during the development phase. In summary, an outline of typical decommissioning project plan includes the following elements (DOE 1994): Introduction; Facility, History, Characterization, and Status; Alternative Selection (SAFSTOR, ENTOMB, DECON); Decommissioning Activities (e.g., decontamination techniques); Program Management; Worker and Environmental Protection—Health Physics Program (operational surveys) and ALARA Practices; Waste Management; and Final Survey Plan (e.g., based on NUREG/CR-5849, MARSSIM).

It should be emphasized that this last technical area is the focus of this text—to provide engineers and health physicists the tools necessary for planning and designing FSSs. FSSs need to be carefully planned to ensure that compliance with release criteria can be demonstrated with the desired confidence level. MARSSIM provides detailed guidance for preparing and conducting FSSs.

The operations phase includes activities related to the physical decommissioning of the facility, including decontamination, dismantlement of plant systems and equipment, demolition, soil excavation, and waste disposal. The operations phase very well may be the most challenging phase—designing and implementing the tasks associated with decontamination and dismantlement of the facility. These remediation activities must be well-planned, as they may involve specialized techniques such as hydrolazing, underwater plasma arc cutting, and a variety of chemical, physical, electrical and ultrasonic processes to decontaminate equipment and surfaces. For example, the decommissioning of a nuclear power plant involves removing the spent fuel from the reactor vessel, dismantling systems and components associated with the reactor vessel and primary loop and cleaning up contaminated materials. Oftentimes, contaminated materials are size-reduced to fit in waste containers rather than being decontaminated.

Operational health physics support, site health and safety, and environmental protection play paramount roles during this stage. Support activities include air sampling during decontamination activities with a potential for generating airborne contamination or monitoring of radiation dose levels for work in high radiation areas. Remedial action support surveys are performed to assess the success of the decontamination and excavation activities. Also, the potential for re-contaminating previously remediated areas must be constantly assessed and guarded against.

The closeout phase is the final phase of a decommissioning project and begins at the completion of the physical decommissioning, when it is anticipated that the site meets the stated decommissioning release criteria. Closeout consists of FSSs, IV activities, and completion of the documentation that certifies that the facility has met the decommissioning criteria. IV, performed by the regulatory agency, includes reviews of all the characterization and FSS documentation and independent survey measurements. In essence, IV is a quality assurance (QA) function that increases uniformity and consistency among decommissioning projects, enhances public credibility, and increases the probability of complete remediation and documentation (further detailed in Section 2.4).

2.3 RADIOLOGICAL SURVEYS PERFORMED DURING DECOMMISSIONING

While not really a survey type, the HSA is an integral first step in the planning stages for a decommissioning project. MARSSIM Chapter 3 describes in sufficient detail the conduct of an HSA. This involves collecting various types of historical information for the purpose of identifying potential sources of contamination and differentiating areas of different contamination potential. The kinds of information to collect and review can include licenses and permits, environmental monitoring reports, audit and inspection reports, and operational survey records, especially those that indicate spills or other nonroutine effluent releases. It is also advisable to interview current and past employees, as much of a site's history never gets documented. The major "deliverable" of an HSA is having site diagrams showing classification of areas: Impacted (e.g., class 1, 2, or 3) and nonimpacted. The HSA also provides input to scoping and characterization survey designs. It should be anticipated that there may be data deficiencies as a result of the HSA, and filling these data gaps is precisely the purpose of scoping and characterization surveys. Chapter 3 provides a more complete discussion on HSA activities.

The scoping and characterization surveys are sometimes referred to as preliminary surveys as they provide the information needed to remediate the site and to design the FSS. Remember from the previous section that these surveys comprise the assessment phase and are performed to provide an assessment of the radiological and hazardous material conditions at the site. The scoping survey expands upon HSA data by collecting judgmental (i.e., biased) samples in potential areas of concern. In general, the objectives of the scoping survey are to provide input to characterization survey design, support classification of site areas (following initial HSA classification), and to identify site contaminants and their variability.

It is important to note that the scoping survey is performed only after the HSA results indicate that there is a reasonable potential for contamination. Sometimes the HSA results indicate an extremely low probability of contamination, yet the decision may be to perform a scoping survey— just in case the HSA missed something of significance and/or to provide additional assurance that contamination is not present. Thus, it is highly possible that the scoping survey may result in no contamination being identified. So once the scoping survey has been planned and implemented, there are two possible outcomes based on the scoping survey results.

The first possible outcome is that the judgmental sampling performed during the scoping survey identifies contamination. The survey then provides general extent of contamination in the area and this information is used to plan for characterization. The second outcome is when contamination is not identified during the scoping survey. In this case, a characterization survey is not performed, but rather a class 3 area FSS. That is, prior to releasing the site or area it is necessary to supplement the judgmental scoping survey with an unbiased, statistically based FSS. The relatively minimal survey coverage of a class 3 FSS is appropriate in this case.

As mentioned previously, the DQO process drives the characterization survey design. DQOs ensure that the focus is on collecting only those data needed to address the characterization decision. The primary objective (decision) for characterization surveys is to determine the nature and extent of contamination. Characterization survey design is built upon the survey results from both the HSA and scoping survey. The characterization survey is different than other decommissioning surveys in that there can be a multitude of survey objectives. Other characterization objectives may include evaluating decontamination techniques (e.g., use of strippable latex coatings versus chemical decontamination of surfaces or scabbling), determining information for site-specific parameters used in pathway modeling (e.g., resuspension factors, soil porosity), and evaluating remediation alternatives (e.g., unrestricted versus restricted release criteria). The MARSSIM guidance for characterization objectives concentrates on obtaining information necessary for FSS design, such as estimating the projected radiological status at the time of the FSS. This may include identifying the radionuclides present and their concentration ranges and variances, and reevaluating the initial survey unit classifications. The contaminant standard deviation is necessary for calculating the FSS sample size.

For many areas, the survey design should consider balance between characterization and remediation. That is, it may not make good sense to characterize the contaminant distribution to the nearest 0.1 m of land area if gross remediation techniques like digging with a backhoe are to be employed. Also, in areas where no remediation is anticipated, characterization data may be used as final status data, so it is prudent to plan accordingly.

The remedial action support survey provides a real-time assessment of the effectiveness of decontamination efforts. This survey relies on a simple radiological parameter—for example, surface scans, direct measurements of surface activity, soil analyses in the field—to determine when an area/survey unit is ready for the FSS. This survey may also be needed to update the projected radiological status at the time of FSS that was established with the characterization survey (i.e., once an area is remediated, the characterization data are likely to be rendered obsolete).

The MARSSIM provides the most detailed guidance for FSSs. The straightforward objective for the FSS is to demonstrate that residual radioactivity in each survey unit satisfies release criteria. As aptly put by the NRC, "the FSS is not conducted for the purpose of locating residual radioactivity; the historical site assessment and the characterization survey perform that function" (NRC 2006a, p. A-1). Guidance is rooted in past experience with misused FSS efforts.

An overview of FSS activities includes classifying the site areas, dividing these areas into survey units, and collecting sufficient survey data to demonstrate compliance with release criteria. As discussed in the previous chapter, two nonparametric statistical tests are used to plan and evaluate FSS data—WRS and sign tests—using a hypothesis testing approach. The FSS design relies on DQOs to set acceptable decision errors for the hypothesis test. Additionally, the sample size for each survey unit is determined based on $DCGL_W$, contaminant standard deviation, scan MDC, and area factors (based on dose modeling). Planning for the FSS should include early discussions with regulator concerning logistics for the regulator's confirmatory survey.

The confirmatory (or verification) survey validates FSS and is performed by regulatory agency or an independent third party. This survey provides data to substantiate results of FSS and may include a review of survey procedures and results, and performance of independent field measurements (typically the confirmatory survey is less than 10% of the FSS effort). It is critical to the legitimacy of the confirmatory survey that the party performing this survey be completely independent of the party performing the FSS, so as not to constitute a conflict of interest.

To review, the radiological surveys performed in support of decommissioning include scoping, characterization, remedial action support, final status, and confirmatory. Scoping and characterization surveys identify the nature and extent of contamination and support the assessment phase. The remedial action support survey demonstrates that the remediation in the operations phase was complete, and the FSS and confirmatory support the closeout phase of a decommissioning project.

2.4 REGULATORY AGENCY INTERFACE WHEN DESIGNING MARSSIM SURVEYS

It goes without saying that D&D contractors are going to do their best in cleaning up the site and demonstrating that release criteria have been met. But just the same, the regulator has the responsibility (to the public) to ensure that the project has been performed properly, and will therefore pay close attention to all phases of the decommissioning. Regulators provide guidance to assist D&D contractors, such as regulatory guides and standard review plans (SRPs), and also perform verification to ensure that the D&D contractor has adequately decommissioned the site.

Both regulators and D&D contractors should ensure that stakeholders—residents, local business owners, local government officials, and others are involved during all phases of the decommissioning project. For example, the public is very interested in the cleanup criteria (particularly if institutional and engineering controls are involved) and whether the site cleanup will involve an on-site waste disposal facility. Stakeholders should be involved in upfront project planning and communication with the public should be frequent during the lifecycle of the project. They should feel confident that they are partners in the decommissioning project's decision making. It is no surprise that projects can be delayed or even derailed when stakeholder involvement is stifled, or when they are brought into the process after many of the D&D decisions impacting them have been made. Finally, stakeholders want assurance that the cleanup criteria established for the project have been met, and that future use of the property can proceed without concern for residual contamination. IV of cleanup provides that important public assurance.

Perhaps, the most common form of regulatory interface during the D&D project is the confirmatory or verification process performed by the regulator. This verification survey may be performed by the regulators themselves, or they may hire an independent contractor to provide technical assistance, usually with the resources to provide survey activities. It should be emphasized that the D&D contractor also stands to gain from the regulatory oversight process. The regulator and D&D contractor, with their separate perspectives, provide a sort of checks and balances on each other. It is not effective for the D&D contractor to spend more resources than necessary to attain release criteria, and of course, it is very important for the D&D contractor to achieve the release criteria. The regulator helps to ensure that this balance is met through the verification process.

One final thought. When decisions arise during planning for the FSS, it is generally a good idea to err on the side of conservatism. These opportunities may arise during technical discussions on area classification (perhaps class 1 versus class 2 survey units), calculation of instrument MDCs, or in responding to an investigation level that has been exceeded. This recommendation may seem more costly at face value, but in reality, time saved in regulatory discussions (debates) is generally well worth the conservative tack taken. Now this does not mean to promise the regulator that as an added step you will use a toothbrush to decontaminate every square inch of the turbine building, but it does mean to carefully consider the inclination to take a hard stance on some D&D issues, which may not add up in your favor.

2.4.1 NRC Standard Review Plans

The NRC's Consolidated Decommissioning Guidance (NUREG-1757 series) and the Standard Review Plan for Evaluating Nuclear Power Reactor License Termination Plans (NUREG-1700, Rev. 1) provide a wealth of decommissioning guidance. While the SRP is chiefly intended to be used by the NRC staff in conducting their reviews to provide consistency and standardization in NRC's oversight responsibilities, it of course represents an excellent guide or checklist for licensees and D&D contractors—like knowing what is going to be on the exam. NUREG-1757 largely covers the same territory as NUREG-1700, but goes into expansive detail on dose modeling, compliance with the ALARA condition in the license termination rule, and FSSs.

The SRP is formatted in a systematic fashion with sections that include (1) area of review, (2) acceptance criteria, (3) evaluation of findings, (4) implementation, and (5) references. The acceptance criteria sections will probably be the most helpful for D&D contractors, as they provide the technical bases for determining acceptability for program elements. Both NUREG-1757 and 1700 suggest that NRC staff review characterization results, remaining remediation activities, and the FSS plan. Specifically, the NUREG-1700 SRP provides guidance on the following LTP sections: general information; site characterization; identification of remaining site dismantlement activities; remediation plans; final radiation survey plan; compliance with the radiological criteria for license termination; update of the site-specific decommissioning costs; and supplement to the environmental report.

For the most part, the acceptance criteria for each of the above categories will not be discussed; rather, the detailed information of most of these sections is the subject of the technical information in this book. One area that will be discussed is the compliance with radiological criteria for license termination.

The SRP indicates that the licensee must clearly indicate the purpose or goal of the license termination—whether the licensee is seeking unrestricted release or restricted release (or conditional release as it is sometimes called). The details of the various release criteria are largely covered in Chapter 4 in this book. Here it is intended to provide a discussion of the regulatory advice offered by the NRC, as contained in the SRPs. NUREG-1700 states that the license termination plan should "describe in detail the methods and assumptions used to demonstrate compliance" with release criteria. This should include a description of the determination of DCGLs and modeling used to establish the DCGLs, and the FSS used to demonstrate compliance with the DCGLs. As you would imagine, the guidance contains some fairly standard material—mostly the subject of the chapters in this book.

An interesting development in decommissioning is that site-specific modeling information may be used to derive DCGLs. This area is covered in Chapter 5. Regarding this topical area, the NRC requests that the licensee should justify the site-specific parameter values used in the modeling. This is a relatively new technical area in the D&D field for many decommissioning practitioners. Many sites that cannot live with the default screening models will likely be performing experimentation to determine parameter values like k_d, soil porosity, and resuspension factors for alpha contamination.

One other technical area that is specifically addressed in the SRP is restricted release. Of course, all things being equal, a licensee will much sooner desire an unrestricted release versus a restricted release of a D&D site. The unrestricted release is a more definitive release action, it comes with no restrictions on future uses of the site and more importantly, it offers the licensee greater freedom from future liability from a formerly contaminated site. Restricted release, as can be implied from its name, comes with a number of conditions on the release of the site. The SRP provides a number of stipulations to consider when contemplating restricted release:

- Sites requesting restricted release will be included in the NRC's Site Decommissioning Management Plan (SDMP).

- Licensee must demonstrate that reductions of residual radioactivity to unrestricted levels are not feasible because it would (1) result in net public or environmental harm, or (2) levels are already ALARA.

- Dose to average member of the critical group is less than 25 mrem/y with restrictions in place.

- Licensee has made provisions for legally enforceable institutional controls.

- If institutional controls fail, the annual dose would not exceed 100 mrem, or under specified conditions, would not exceed 500 mrem.

- Licensee describes the site end use for areas proposed for restricted release.

The standard review plan also outlines the importance of identifying stakeholders in the restricted release decommissioning process. NUREG 1757 (vol. 1, p. 14-3) states that the licensee must provide for "participation by a broad cross-section of community interests in obtaining the advice." Stakeholders may include state and local agencies, other Federal agencies, environmental organizations, and adjacent land owners to the decommissioning site. The burden is on the licensee to demonstrate that it has "provided for a comprehensive, collective discussion on the issues by the participants represented" (NRC 2006b, p. 14-3).

2.4.2 Verification Process

The verification process is an independent evaluation of the final radiological conditions at the decommissioned site, as well as the processes used to achieve these final conditions, to assure that the site release criteria have been appropriately applied and met. Verification is an important aspect of the decommissioning process, largely because it provides a second look at the D&D process and offers the public added assurances that the project has been properly performed. The verification process satisfies the following objectives: (1) to provide independent, third-party overview of the decommissioning project, (2) to evaluate the adequacy of the documentation, (3) to substantiate the credibility of the radiological survey procedures, and (4) to validate the accuracy of field measurements and laboratory analytical techniques.

It has been a long-standing policy of regulatory agencies to perform independent, third-party verifications of the adequacy and effectiveness of decommissioning projects. DOE recently issued DOE O 458.1 (2011) that states that "if the real property is to be transferred to the public, or managed by another agency/entity other than DOE or a new facility constructed, an independent verification plan will be prepared and independent verification surveys and sample analysis will be conducted to verify compliance ...."

Over time the verification process has evolved. Recently, much attention has been given to streamlining activities to reduce the cost of verification for larger decommissioning projects. One such streamlining activity—termed in-process decommissioning inspections—has emphasized the early involvement of verification in the decommissioning process. The in-process decommissioning inspection may be described as the result of the DQO process being applied to the former manner of conducting verification surveys at the conclusion of FSSs.

This is in contrast to the historical verification process where the verification survey was performed upon completion of the D&D contractor's FSS. The verification coverage was typically 1–10% of the FSS measurements. Two basic problems arise with this back-end verification approach: (1) if the D&D contractor has a good track record for performing FSSs,

1–10% may be too many verification measurements (and not consistent with the DQO process), and (2) if substantial problems are identified during the verification, it is rather late in the decommissioning process to resolve issues smoothly. These problems associated with back-end verification practices led to the streamlining initiative of in-process decommissioning inspections. Essentially, this involves several inspections of the decommissioning process to evaluate specific elements of the D&D contractor's activities, focusing on the FSS. The activities inspected are quite similar to those evaluated with the previous verification paradigm, with the noticeable exception that potential problem areas are identified much earlier in the decommissioning process. Appendix C provides an example of the FSS program elements addressed during an in-process decommissioning inspection.

Thus, it is important that the IV contractor (IVC) be integrated into the early planning stages of the decommissioning project. This will ensure that the verification process proceeds smoothly and allows for timely identification of potential problems. For example, document reviews of the decommissioning plan and FSS plan should be performed prior to the commencement of decommissioning operations. The logic is straightforward: if inappropriate survey procedures are identified prior to their implementation, they may be revised at minimal cost and expenditure of time. However, if inappropriate survey procedures are identified following their implementation, then it may be necessary to re-survey, with significant increases in cost and time.

Verification activities may be implemented at two different levels depending on the complexity of the decommissioning project. The simpler type of verification is called a type A (or "limited") verification. Activities comprising a type A verification typically include document reviews, data validation, and sometimes confirmatory sample analyses (e.g., analysis of the D&D contractor's archived samples). A type B (or "full") verification includes IV survey activities, such as scanning and soil sampling, in addition to those activities included in the type A verification. The complexity may be driven by the objective of the remedial action (e.g., restricted or unrestricted release), the nature of the radiological contaminants and their form, and results of previous verification activities. For example, some contamination situations are quite straightforward to remediate and verification may not be deemed necessary—for example, laboratory rooms contaminated with C-14. Also, if a positive track record has been established via past verification, it is much more likely to have reduced scope verification, and the converse is true if the track record indicates poor performance. The level of verification required is determined by the regulatory agency with consideration of the IVC's recommendations.

The following sections cover these verification activities in greater detail. Verification activities may be performed by an IVC, and in some cases they are performed by the regulators themselves, usually depending on the overall scope of the decommissioning project. Of all the verification activities discussed in the next few sections, by far the most important activity is the review and critical technical evaluation of the D&D contractor's decommissioning documentation.

2.4.2.1 Review of Decommissioning Documentation

The regulatory agency typically initiates the verification process by forwarding documents related to the decommissioning project to the IVC for review and preparation of a comment letter. All documents associated with decommissioning activities include decommissioning plans, characterization reports, decommissioning reports, and probably most important, FSS plans and reports. Sometimes it is appropriate to review D&D contractor survey and laboratory procedure manuals. The IVC uses these documents to develop a better understanding of the decommissioning project. Comments on the documentation are provided to the regulatory agency—and usually to the D&D contractor through the regulatory agency—for resolution prior to document finalization. As mentioned in the preceding section, the NRC's Standard Review Plan may be used to provide consistency of D&D contractor document reviews.

The FSS documentation embodies most of the important items that are covered in a document review. Therefore, the discussion of document reviews will be based primarily on the FSS documentation. The document review is usually facilitated by dividing the review into four categories: (1) general decommissioning information, (2) statistical survey design, (3) survey instrumentation and procedures, and (4) survey findings and results.

General decommissioning information reviewed includes the background information (e.g., decommissioning objective), site/facility description, operating history, identification of major contaminants, decommissioning activities performed, and waste disposal practices. The site/facility description is reviewed to determine that the general locations of residual activity to be decontaminated have been identified, including locations of any spills or accidents involving radioactive materials. The operational history is reviewed to identify the processes performed and to ensure that all potential contamination pathways have been addressed. For example, suppose that the operational history indicates that liquid radioactive effluent was discharged to a septic field, a thorough survey of the septic field would be indicated.

Review of the decommissioning activities performed is necessary to provide the IVC with an update of the current site condition, following any remediation. Waste disposal practices are reviewed to determine their impact on the contamination status of the facility. Identification of the major contaminants is reviewed to ensure consistency with the stated operational history and subsequent justification for DCGL determination. The release criteria and DCGL determination are reviewed to determine that they have been approved by the responsible regulatory agency and that they have been properly applied (e.g., conditions for implementing DCGLs, such as use of area factors).

The FSS design review includes selection of background reference areas and background level determinations, area classifications, survey unit identification, and survey measurements—for example, surface scans, surface activity measurements, and media sampling and analysis. Since release criteria are stated in terms of background levels (release criteria values generally do not include background levels), it is important to identify how the background levels were established for the site. The area classifications and survey unit identification are reviewed relative to the contamination potential in each surveyed area. The statistical basis for sample size is compared to the DQOs by evaluating the statistical power of the hypothesis test. Survey measurements are reviewed to ensure that they are appropriate for the expected radioactive contaminants, considering the relative mix of radionuclides as well. Specifically, the review assures in part that all potential radiations have been measured, that scanning techniques are capable of detecting locations of elevated activity and that inaccessible areas have been addressed.

The philosophy behind the selection of instrumentation and procedures for use is reviewed. Specifically, the review assures that the instruments selected are appropriate for the desired survey measurement. Instrument calibration procedures are reviewed to ensure that the calibration source is representative of the radioactive contaminant(s) (i.e., similar source geometry and radiation energy) and that instrument response checks are performed at the specified frequency and fall within the established acceptance criteria. The instrument and surface efficiencies as defined by

International Standard ISO-7503 used for surface activity measurements are evaluated. The calculational expression for instrument sensitivity, typically stated as the minimum detectable concentration (MDC), is reviewed and the instrument sensitivity relative to the $DCGL_W$ is determined. Scan MDCs are also assessed relative to the $DCGL_{EMC}$.

The survey results review includes data interpretation, statistical evaluation and comparison of results with release criteria. Data interpretation is reviewed to ensure that calculational methods used to convert survey data to units consistent with DCGL units are appropriate and accurate, paying close attention to efficiencies and conversion factors. The data quality assessment is reviewed to determine that the appropriate statistical test was applied to the survey results and that the assumptions of the test were validated. The comparison of results in the survey unit with DCGLs is reviewed to ensure that results of survey measurements and sampling were in compliance with the release criteria. Graphical representations of data are reviewed to evaluate trends and to assess the underlying distributions of the data.

Another important subject area reviewed is the D&D contractor's use and implementation of investigation levels. Investigation levels are stated in the FSS plan and suggest specific actions when certain radiological measurement triggers are exceeded. FSS reports are reviewed for the use of investigation levels to assess whether the D&D contractor performed appropriate follow-up actions, and whether the investigation procedures performed are in compliance with the FSS plan. Miscellaneous items reviewed include (1) indication of measurement and sampling locations, (2) indication of remediated areas, (3) data provided for all areas addressed during the FSS, and (4) a statement that release criteria have been met.

Data validation is performed to ensure that the data collected during the final survey are of known and acceptable quality. The data are typically validated by comparing with the DQOs established for the FSS (see Chapter 3 for a discussion on DQOs), and evaluating whether there was sufficient statistical power to satisfy the release criteria. Data validation involves the review of both laboratory and field survey data packages to determine the accuracy and to what extent the reported data conform to the QA plan.

2.4.2.2 Confirmatory Analysis of Laboratory Samples

Confirmatory analysis, a type A verification activity, is the comparison of results of the same type of laboratory analyses, performed on either a split sample or the same sample analyzed by both the IVC and the D&D

contractor. The goal is to identify any potential discrepancies in analytical results between the D&D contractor's laboratory and the IVC's laboratory. It is preferable to perform the analyses on the same sample to avoid the inherent problems with split samples not being representative of one another—that is, it is very common for split sampling to result in heterogeneous samples.

Ideally, the D&D contractor will have a number of archived samples available for selection by the IVC; at least 5–10 samples are collected to facilitate the confirmatory analysis. Procedures for comparison of results are determined prior to laboratory analysis and typically should agree within the expected statistical deviations of the analysis methods used. It is important to ensure that a similar calibration geometry exists for both laboratories (desire equivalent amount of the media sample to be analyzed by both labs) and the IVC lab analysis should be aware of the potential reduction in radionuclide concentration via radioactive decay and/or volatility of the radionuclide (e.g., Tc-99).

2.4.2.3 IV Survey Activities

The guidance offered in this section may be considered typical for many verification surveys; the true scope of verification is determined and agreed upon by the regulator and verification contractor. IV survey activities may include surface scans, measurements of surface activity, exposure rate measurements, and radionuclide concentrations in media (e.g., soil, water, and sediments). The IVC establishes background levels for the site by collecting survey measurements and samples in background reference areas that are unaffected by site operations. Typically, the IVC will use the existing reference grid system established by the D&D contractor.

One of the first activities performed by the IVC is to evaluate the D&D contractor in performing surface activity measurements. In all, 10–20 locations representing a range of surface activity levels—perhaps values from background to several times the $DCGL_W$—are identified for side-by-side measurements. Both the D&D contractor and the IVC make separate measurements of the surface activity level at each location. These side-by-side measurements are then compared to determine the source of any systematic biases that may be present. Side-by-side measurements may also be performed for other measurements, such as exposure rates or radionuclide concentrations in soil.

The IVC conducts surface scans in areas where the D&D contractor has completed their FSSs. These scans help to assess the performance of the

D&D contractor in identifying elevated areas of concern. It is important to know how many the D&D contractor identified so that the number found during verification is in proper context. This requires the D&D contractor to clearly record the number of hot spots that were identified in each survey unit. Consider the following hypothetical example. The IVC finds one or two hot spots in a survey unit where the D&D contractor has documented that five hot spots were found and appropriately dispositioned. This scenario provides more assurance to the regulator that the FSS was successful when compared to a second scenario that has the IVC finding two or three hot spots, and the D&D contractor finding none.

Measurements for surface activity are performed at locations identified by scanning about 1–10% of the survey units, depending on the scope of the verification. For each survey unit selected, the number of random measurements should be sufficient to determine the mean surface activity or soil concentration with the prescribed level of confidence.

The verification survey data are statistically evaluated and compared principally to the FSS results and release criteria (DCGLs). While individual measurements may be compared to the $DCGL_{EMC}$ and the mean of a number of measurements in the survey unit to the $DCGL_W$, it is critical to remember that the objectives of the verification survey are focused on confirming the D&D contractor's FSS. Locations that exceed the DCGLs are addressed (1) by notifying the D&D contractor and regulatory agency of the verification results, and (2) allowing the regulator to decide whether supplemental guidelines (hazard assessment) are warranted or whether the D&D contractor must perform additional decontamination and survey activities. The important message is that the IVC only provides the confirmatory data; it is the decision of the regulator to determine an appropriate course of action based on these results.

The IVC prepares and submits the draft verification survey report to the regulatory agency for review and comment. The IVC makes recommendations as to the accuracy of the D&D contractor's FSS measurements and it appears that release criteria have been satisfied. Comments received from the regulatory agency are incorporated into the report and the final verification report is issued. As one might expect, the verification process has identified a number of recurring problems over the years.

2.4.3 Verification Program Experiences and Lessons Learned

The verification process provides a mechanism for identifying potential problems associated with a decommissioning project. Experience with the verification process has resulted in the identification of problems in all phases of the decommissioning process. General deficiencies observed have included poorly defined survey scopes (i.e., not all contaminated site areas were identified), contaminants not being identified, inaccessible areas not being addressed, the use of inappropriate survey instrumentation and misapplication of release criteria.

DOE requested that ORAU prepare a "lessons learned" document to summarize IV activities performed at several DOE sites (Bailey 2008). The DOE established a good practice policy stating "Independent Verification should be an integral part of site restoration and cleanup projects" (DOE 2006). The lessons learned included the following:

- IV should be integrated into the planning stages of D&D.

- IV of onsite remediation and FSS activities should be coordinated and if possible implemented in parallel with the contractor to minimize schedule impacts.

- IV should not be a substitute for routine contractor QA activities.

- IV recommendations often improve the contractor's FSS procedures and results, and increase the probability of complete remediation and documentation.

- IV can validate the D&D contractor's methods and FSS results; public credibility concerning DOE operations and cleanup objectives are greatly enhanced.

A lesson learned captured from DOE's cleanup of the Rocky Flats (RF) site pertained to the need to follow through with IV. ORAU's IV surveys in the 903 Lip Area at RF had identified 13 hot spots that contained contamination above the cleanup criteria or action level of 50 pCi of plutonium (Pu-239/240) per gram of soil. Contamination in these 13 hot spots ranged from 65 to 425 pCi g of soil. The D&D contractor remediated the soil areas containing these hot spots. However, the remaining IV scope was terminated. The GAO found that DOE's implementation of verification was not completed, and therefore DOE missed an opportunity to independently verify the sufficiency of several aspects of this cleanup project. GAO reported that "DOE decided to eliminate parts of the planned independent review of the accuracy of contractor-conducted scans for remaining radiological contamination because DOE officials decided that the likely results would not justify the completion of an independent review" (GAO 2006, p. 39).

Another lesson learned or best practice relates to the best way to implement the verification process itself—the streamlining of the verification process that produced in-process decommissioning inspections. Appendix C illustrates possible lines of inquiry to use when performing an in-process decommissioning inspection. This streamlining format evolved during the verification of nuclear power reactor decommissioning projects. It was apparent to many regulators that waiting until the licensee's FSS was completed prior to initiating verification (especially the lines of inquiry that address FSS design and procedures) could be disastrous. As an example, a technical issue that needs attention early in the FSS design process is how to address the hard-to-detect nuclides (e.g., Fe-55, Ni-63, H-3) that exist in the radionuclide mix at a power reactor. If there was a technical flaw in the licensee's hard-to-detect nuclide (HTDN) design—either in the modification of the DCGLs to account for HTDN or in the measurement and assessment of HTDN—and it goes unnoticed all throughout the FSS, only to be discovered during verification. Well, hopefully the point is clear, it would not go over well with any of the parties involved in the decommissioning process.

A common deficiency involves the misapplication of surface activity DCGLs for radionuclides of decay series (e.g., thorium and uranium) that emit both alpha and beta radiation. Specifically, the problem is relying on the use of alpha surface contamination measurements to demonstrate compliance with DCGL values. Experience has shown that beta measurements typically provide a more accurate assessment of thorium and uranium contamination on most building surfaces, due to problems inherent in measuring alpha contamination on rough, porous, and/or dusty surfaces due to the unknown degree of alpha radiation attenuation. Such surface conditions cause significantly less attenuation of beta particles than alpha particles. Beta measurements therefore, may provide a more accurate determination of surface activity than can be achieved by alpha measurements. The relationship of beta radiation to the thorium or uranium activity must be considered when determining the surface activity for comparison with the DCGLs. This technical area is covered in Chapter 6, where another calibration technique is introduced that involves evaluating the complete nature of alpha and beta radiations from all radionuclides in the decay series, and determining the detector's response to all of these radiations.

Another FSS program deficiency involves the classification of areas according to contamination potential. One particular example was the clearance of piping from a facility being decommissioned. D&D project personnel had mistakenly grouped contaminated process piping with noncontaminated piping. This occurred because the D&D project staff were not aware of the differences in radiological condition of the various piping sections—that is, some sections of the piping were used solely for nitrogen or dry air supply and had an extremely small chance of being contaminated, while piping that was similar in appearance was used for the process gas system and was intimately in contact with the radioactive materials. Because this difference in operational function of the piping was not recognized by some D&D project staff, potentially contaminated piping was grouped with the essentially "radiologically-clean" piping. Furthermore, because the lot of material was thought to consist entirely of the "radiologically-clean" piping, the survey design was statistically based with less than 50% of the material actually surveyed. The improper grouping of process piping in this "radiologically clean" piping survey unit was identified during verification surveys of a small percentage of the piping. According to the occurrence report, a contributing cause was the lack of a more thorough D&D worker understanding of the facility's past operational history, particularly an understanding and appreciation of the equipment that they were dismantling. This led to the inappropriate determination of a material survey unit containing diverse materials; materials that were likely to be "radiologically-clean" commingled with materials that had a much higher probability of being contaminated.

A number of deficiencies related to the implementation of MARSSIM have been identified during the verification process. One of the more obvious shortcomings was identified during the review of FSS plans. Some D&D contractors who were familiar with the FSS guidance in NRC's NUREG/CR-5849 prepared plans that were supposedly consistent with MARSSIM—it appeared that they simply replaced "NUREG/CR-5849" with "MARSSIM." This is one technical deficiency that is rather easily identified.

Another MARSSIM-related deficiency relates to the FSS design for multiple radionuclides. As discussed in Chapter 6, the design for multiple radionuclides requires a modification of DCGLs to account for the radionuclides present, such as the use of surrogates, gross activity DCGLs or the unity rule. A design deficiency detected during the verification process was a situation where the D&D contractor allowed each of the three radionuclides to be present in the soil at their respective DCGLs. This survey design potentially resulted in survey units that were three times the release criteria (because each individual DCGL is the radionuclide concentration that equates to the release criterion).

In summary, the verification process provides for a complete review of all decommissioning documents, IV survey activities, and thorough and complete decommissioning documentation to certify that the site satisfies release criteria. The verification process provides the timely identification of discrepancies (if initiated early in the decommissioning process), increases uniformity and consistency among decommissioning projects, enhances public credibility, and increases the probability of complete remediation and documentation.

The following case study is a detailed narrative of an all too common misuse of verification—using it as a replacement for the FSS data. This case study takes the form of a report (or perhaps "open letter") to the regulator following a number of failed discussions concerning the proper use of verification.

CASE STUDY   A COMMON MISUSE OF THE VERIFICATION PROCESS

The objectives of the verification process are to provide independent document reviews and radiological survey data, for use by the regulator in evaluating the adequacy and accuracy of the D&D contractor's procedures and FSS results. While this statement will likely not cause any disagreement among all parties involved, a closer inspection into the intent of this statement may reveal that some D&D projects may not be adequately verified. This narrative discusses some of the potential problems with the implementation of cleanup projects and recommends solutions for overcoming these problems.

Once a site has been identified as needing remediation, it is placed on a schedule and planning begins for characterization, decontamination, FSSs, and IV activities. Ultimately, the site will be certified as having met release criteria established for the site. Because the certification is used to demonstrate to the public that the cleanup has been effective and complete, the verification process serves as an important QA function to ensure that remediation and FSSs have been conducted appropriately.

It is of utmost importance to understand the role of the IVC in these cleanups and the certification process—that is, to critically evaluate the remedial actions performed and to add further credibility to the FSS and survey report. Specifically, the objective of IV is to validate the accuracy and completeness of the D&D contractor's field survey measurements and the appropriateness of the survey procedures followed, resulting in an independent assessment of site radiological conditions. The primary emphasis of the verification is reviewing what the D&D contractor did, and not on the independent data collected by the IVC. In fact, IVC activities may be either type A or type B, with type A activities consisting solely of document reviews (no field activities whatsoever). The primary purposes of both types of verifications are to confirm the adequacy of the procedures and methods used by the D&D contractors and to verify the results of the decommissioning activities.

The following discussion further illustrates the proper role of the IVC. The D&D contractor submits the FSS report to the regulator that provides data for the final radiological condition of the site and demonstrates that the established release criteria have been met. It should be recognized that this survey report alone must certify the true conditions of the site and demonstrate that release criteria have been met. It is this report that covers the full breadth and depth of the survey performed and provides proof that the site should be docketed as fully remediated. Thus, the final status report is the principal document that the regulator should rely on when making decisions concerning the site's release from further consideration and certifying to the public that the dose limits have been attained.

Because of the importance of these decisions, the regulator has often requested that a QA/QC function be incorporated into the cleanup and certification process—IV of the cleanup and survey results contained in the FSS report. It is important to understand that the verification does not supplement the survey report, but rather, adds further credibility to the conclusions reached by the D&D contractor in its FSS report. Thus, whenever the IVC is able to confirm the results and conclusions of the D&D contractor, the regulator is provided increased confidence in making the decision to release the site.

An analogy is offered to illustrate the proper roles of the IVC and the D&D contractor. A company has developed a new food product and is petitioning the U.S. Food and Drug Administration (FDA) to allow the food product to be sold to the public. The basis of its claims that the product is safe is a 200-page report that details exhaustive testing and presents results that demonstrate compliance with all of the FDA's product testing guidelines. The company has convinced itself that there is no reason why the product should not be consumed by the public. The FDA thoroughly reviews the company's petition and product report, and as a matter of policy, conducts limited testing (verification) of the product. The FDA obtains results consistent with those reported by the company and makes a decision to allow sales of the product, largely based on the results presented by the company. Just as the FDA may perform a greater verification coverage in certifying that new food products are safe for sale, the IVC typically does the same for certifying that sites are clean and the point is that in both cases the bulk of the data for basing a decision comes from the company or D&D contractor; the verification of the primary reports serve to provide added evidence that adequate public safety has been achieved.

While the critical functions of the IVC include both document review and independent survey activities, by far, the most important verification activity performed is the document review. Document review helps to ensure that the D&D contractor used proper survey procedures, appropriate instrumentation, correct interpretation of survey results and comparisons to DCGLs. Many times, if the document review identifies no significant deficiencies, the regulator may decide that on-site independent survey activities are not warranted.

Furthermore, even when the IVC performs field activities, only a small percentage of the D&D contractor-surveyed area can be surveyed by the IVC. It would be extremely inefficient and expensive to have the IVC completely duplicate the FSS, besides no QA/QC function should call for 100% verification of initial results (in reality, if 100% verification was deemed necessary, serious consideration should be given to finding a new D&D contractor). Thus, since the IVC can only perform survey activities over a relatively small portion of the site, the IVC results do not nearly have the same importance or weight as those of the D&D contractor in expressing the final radiological condition of the site. Instead, the IVC results are compared to those reported by the D&D contractor and are used to support the conclusions reached by the D&D contractor in its FSS report. However, it cannot be overstated that the most important function of the IVC is the up-front review of the D&D contractor's documentation, with the FSS plan and report review being the most critical document reviewed. The fact is if the IVC can technically review the FSS report and support its methodologies and conclusions, it is very unlikely that the public will find deficiencies with the decommissioning documentation.

> The most important function of the IVC is the up-front review of the D&D contractor's documentation, with the FSS plan and report review being the most critical documents reviewed.

A serious problem facing some D&D projects today is the reversed levels of importance given to the FSS report and the verification survey report. Sometimes verification activities are performed prior to the completion of a FSS report. Without the FSS as a basis of comparison for IVC measurements, the IVC survey results are incorrectly compared to the DCGLs, and the entire focus is shifted to one of the IVC performing a second survey that also demonstrates compliance with the release criteria. This is not the intended use of the IVC—it is inefficient and costly, and most importantly, does not add much to the overall certification of the site. The D&D contractor should, at a minimum, prepare an interim survey report and provide it to the IVC for review prior to independent site activities. It is essential that the D&D contractor arrives at its own decision that the site is clean, and further, be ready to defend his decision. The regulator should demand this of the D&D contractor, as the FSS report is the primary document on which it is making a decision to release a site. The IVC's critical review and comments serves to strengthen the FSS report.

A related problem can crop up if the D&D contractor has not convinced himself that the release criteria have been met. In fact, the problem has been acute on some D&D projects where the D&D contractor focused primarily on remediation, paying very little attention to the FSS. In extreme cases, the D&D contractor prefers to yield its responsibility to the IVC in making the decision that the release criteria have been met. In these instances, too much weight has been given to the results of IVC to release a site—it has to be based on D&D contractor's results. A variation of this problem occurs when the D&D contractor obtains FSS results (i.e., raw data), and expects immediate verification. The important step of evaluating their results and preparing a report is bypassed. [Now in some cases immediate verification of raw survey results is warranted, such as when the site is in an unstable condition (open trenches), and expedited actions are required to secure the unstable condition].

So what can be done to ensure that the proper mechanisms are in place for certifying a site? It is suggested that the regulator reemphasizes the importance of the FSS through performance-based measures. That is, providing incentives for the D&D contractor to perform satisfactory FSSs and properly documenting the conditions of the site. Similarly, when the D&D contractor is not performing satisfactory FSSs and reports, the message should be clear from the regulator that it is not acceptable. For example, potential deficiencies identified by the IVC in FSS procedures should be discussed between the involved parties (regulator, D&D contractor, and IVC), and possible solutions clearly communicated. The bottom line is that the FSS is the primary basis document for demonstrating compliance with release criteria. The verification process serves to strengthen the credibility of the FSS report and its conclusions.

QUESTIONS

1. What are the four phases of a decommissioning project according to DOE's Decommissioning Handbook?

2. Briefly state the purpose of each of the five radiological surveys performed in support of decommissioning.

3. When might ENTOMB be an attractive decommissioning alternative relative to DECON?

4. What are the major differences between the FSS and the confirmatory survey? Can the same organization perform both surveys?

5. While characterization survey data are frequently used to establish the standard deviation of the contaminant(s) for FSS planning, what data should be used to determine the standard deviation if the

particular area has been remediated? [Hint: The standard deviation used for the FSS planning should represent the radiological status of the site area at the time that the FSS is performed.]

6. State four conditions necessary for an NRC licensee to terminate their license using restricted release.

7. State the six principal elements of the rubblization decommissioning option as outlined by the NRC. What are the possible exposure scenarios with rubblization?

8. Name three common D&D program deficiencies identified via verification.

9. What was the primary driver for regulator's decision to streamline the verification process via in-process decommissioning inspections, a change from the back-end verifications?

SUGGESTED READING

DOE Decommissioning Handbook
NUREG-1628
NUREG-1700
NUREG-1757

CHAPTER 3

Characterization Surveys and the DQO Process

T HE DECOMMISSIONING INDUSTRY CONTINUES to refine and expand guidance for conducting D&D-related activities. Characterization is certainly no exception. Recognizing the need for characterization guidance the ANSI/HPS N13 Standards Committee sponsored a national standard (ANSI N13.59) titled, "Characterization in Support of Decommissioning" using the Data Quality Objectives Process (2008). This recent addition to the bookshelf of many D&D professionals provides a technical approach for designing characterization surveys using the DQO process.

3.1 INTRODUCTION

Guidance for planning and implementing characterization surveys has been limited over the years. The NRC's Branch Technical Position on Site Characterization in 1994 (NRC 1994d) offered early guidance to aid decommissioning professionals with the physical characterization of contaminated sites. This document was useful for those D&D sites performing characterization to develop site-specific information for modeling code inputs. Again, while it provided relevant information, it was somewhat short on practical guidance to direct characterization planning efforts.

Characterization planning activities should consider public opinion, government policy, financial restrictions, and the availability of waste disposal capacity (IAEA 1997), as these ultimately affect the selection of the most appropriate decommissioning scenario/alternative. DECON is a very popular option because it entails complete facility decommissioning immediately following final shutdown. SAFSTOR involves maintaining the facility in a safe condition over a number of years, through performance of S&M activities, followed by D&D. For immediate dismantlement and decontamination (i.e., DECON alternative), extensive characterization is usually required. However, under the SAFSTOR alternative where deferral of decommissioning is intended, the initial characterization may be less comprehensive.

A recent development in the decommissioning guidance arena is NRC's NUREG-1757 vol. 2, rev. 1 (2006) Consolidated Decommissioning Guidance Characterization, Survey, and Determination of Radiological Criteria. NUREG-1757 mentions that characterization surveys may be performed to satisfy a number of specific objectives, and provides several examples. However, detailed characterization guidance is absent, as indicated by the NUREG itself—"the scope of this volume precludes detailed discussions of characterization survey design for each of these objectives, and therefore, the user should consult other references for specific characterization survey objectives ...." (NRC 2006a). Section 2.3 of NUREG-1757 discusses the use of characterization data for FSS data provided that the data are of sufficient quality. This is a reference to the DQOs process.

Guidance in MARSSIM also refers to several decommissioning objectives that characterization surveys can address, essentially deferring to DQOs in the design of characterization survey efforts. Moreover, the DQO process transcends MARSSIM—it is the process for solving environmental or decommissioning sampling problems. This chapter discusses characterization in the context of DQOs and decommissioning in general.

3.2 DQOs PROCESS AND APPLICATION TO CHARACTERIZATION

Characterization in support of decommissioning often consists of obtaining the necessary information to develop the decommissioning plan, or license termination plan, in the case of nuclear reactor decommissioning. The decommissioning plan uses the information acquired from the assessment phase of the decommissioning project—for example, the detailed assessment of the radiological and hazardous material conditions at the site—to plan for remedial activities. As discussed in the following section, there are many possible reasons for performing characterization activities.

The DQO process should be followed to ensure that the characterization plan will provide the necessary data to answer the particular study question in the most cost-effective manner.

3.2.1 EPA's DQOs Process

The U.S. Environmental Protection Agency (EPA) developed "Guidance on Systematic Planning Using the Data Quality Objectives Process" (EPA QA/G-4 2006) as the recommended planning process for collecting environmental data. The DQO process is focused on *making decisions* (e.g., choosing between two alternatives). For example, the DQO process makes it possible to objectively decide when to end the characterization phase, that is, once sufficient data have been collected. Importantly, the DQO process ensures that the investigator understands the study objectives, defines the appropriate type of data, and specifies acceptable levels of potential decision errors. EPA's QA/G-4 is well worth a careful read as it is the basis for establishing the quality and quantity of data needed to support characterization decisions.

In summary, DQOs ensure that an adequate amount of data, with sufficient quality, is collected. The process is both flexible and iterative, and ultimately results in an efficient and effective characterization survey design. An explanation of what's involved in each of the seven steps of the DQO process is highlighted in bulleted fashion below (EPA 2006).

Step 1: State the Problem

- Give a concise description of the problem
- Identify the leader and members of the planning team
- Develop a conceptual model of the environmental hazard to be investigated
- Determine available resources—budget, personnel, and schedule

Step 2: Identify the Goal of the Study

- Identify principal study question(s)
- Consider alternative outcomes or actions that can occur upon answering the question(s)
- For decision problems, develop decision statement(s), organize multiple decisions

- For estimation problems, state what needs to be estimated and key assumptions

Step 3: Identify Information Inputs

- Identify types and sources of information needed to resolve decisions or produce estimates

- Identify the basis of information that will guide or support choices to be made in later steps of the DQO process

- Select appropriate sampling and analysis methods for generating the information

Step 4: Define the Boundaries of the Study

- Define the target population of interest and its relevant spatial boundaries

- Define what constitutes a sampling unit

- Specify temporal boundaries and other practical constraints associated with sample/data collection

- Specify the smallest unit on which decisions or estimates will be made

Step 5: Develop the Analytic Approach

- Specify appropriate population parameters for making decisions or estimates

- For decision problems, choose a workable Action Level and generate an "If … then … else" decision rule which involves it

- For estimation problems, specify the estimator and the estimation procedure

ANSI/N13.59 (2008) adds that step 5 should provide a statement as to exactly how the data that will be collected will be used to decide between alternative actions. As part of the decision rule, the statistical parameter of interest is specified, along with the value(s) at which action will be taken (e.g., the release criteria, waste acceptance criteria, action level).

Step 6: Specify Performance or Acceptance Criteria

- For decision problems, specify the decision rule as a statistical hypothesis test, examine consequences of making incorrect decisions from the test, and place acceptable limits on the likelihood of making decision errors

- For estimation problems, specify acceptable limits on estimation uncertainty

Step 7: Develop the Detailed Plan for Obtaining Data

- Compile all information and outputs generated in steps 1 through 6

- Use this information to identify alternative sampling and analysis designs that are appropriate for your intended use

- Select and document a design that will yield data that will best achieve your performance or acceptance criteria

The DQOs process is not only the basis of the MARSSIM and MARSAME, but is also useful for solving all kinds of D&D technical problems, whether designing a sampling regime for a particular D&D survey or selecting a particular survey instrument. Additionally, there are several decisions to be made during the FSS design that can be facilitated by the DQO process. One such DQO decision is whether to use default DCGLs when planning the FSS, or perform modeling to determine site-specific DCGLs.

3.2.2 Example Application of DQO Process: Use Default or Site-Specific DCGL?

Consider an example that highlights the formalized steps of the DQO process. This example helps one to make a decision on whether to use generic modeling versus site-specific modeling for establishing DCGLs for soil contamination. This decision balances the impact of having a relatively small DCGL value—and the resulting increased FSS effort—versus a larger DCGL value (and reduced FSS effort). The desired larger DCGL value is achievable through site-specific modeling, which requires greater resources to generate the necessary modeling inputs. The decision is whether the reduced FSS effort is worth the additional resources to generate a larger DCGL.

For example, let us assume that the default DCGL$_W$ for a radionuclide is 6 pCi/g and results in an expensive FSS. It may be worthwhile to develop site-specific parameters for modeling to increase the DCGL$_W$ for that radionuclide to 15 pCi/g. The larger DCGL value results in fewer FSS samples being necessary and a more economical FSS. The DQO process provides a formal approach for making this decision.

Step 1: State the Problem

Need to perform an FSS within a soil area to demonstrate compliance with release criteria.

Step 2: Identify the Decision

The decision is whether to use default parameters to generate soil concentration DCGLs versus site-specific parameters to generate DCGLs. In other words, are the total costs using the default modeling parameters less than the total costs using site-specific modeling inputs in the development of the DCGL$_W$ for each soil contaminant present?

Step 3: Identify Inputs to the Decision

- A number of inputs can be identified for making the decision.

- Screening DCGL value for specific soil contaminants from NRC Federal Register notice, or simply running modeling codes using default parameters.

- Critical pathways and parameters—necessary to identify what site-specific parameters are to be evaluated. For example, if soil porosity and k_d are important to the groundwater pathway, then the determination of these parameters may be specified as an explicit objective for the characterization survey. Costs for determining these parameters are assessed at this step, and likely will include sample collection and analytical costs. Note: The *RESRAD Data Collection Handbook* (ANL 1993a) provides information on the type of samples that are necessary for evaluating specific parameters.

- FSS costs using the screening values for DCGLs (these will typically be more restrictive than site-specific modeling, therefore, the FSS costs are greater) versus FSS survey costs using the less restrictive site-specific DCGLs.

Step 4: Define the Study Boundaries

Select the areas to sample or study site-specific parameters (again, consider the *RESRAD Data Collection Handbook*). Ensure that the areas evaluated for these site-specific parameters are representative of all the possible survey areas on the site.

Note: The dose modeling information—for example, scenarios, exposure pathways, and parameters—should be well-described.

Step 5: Develop Decision Rule

If total costs (FSS plus generic modeling) using default DCGLs are less than the total costs (FSS plus characterization data to support site-specific modeling) using site-specific modeling, then use default modeling; otherwise, use site-specific modeling. This example points out that the DQO process can be used for cost–benefit analyses.

Step 6: Limits on Decision Errors

Errors involved in estimating survey costs and modeling costs, including site-specific inputs, are assessed. It can be quite complicated to estimate survey costs, as it would necessarily entail the preparation of a detailed survey plan (design) for each case in step 5.

Step 7: Optimize Design

If the cost difference determined in step 5 is significant, then the decision is clear cut. If cost difference is minimal, it becomes more important to review assumptions made throughout the DQO process. This includes looking at all of the inputs developed during step 3.

3.2.3 Example Application of DQO Process: Waste Disposal Characterization

This example illustrates the use of the DQO process to design characterization of a building for waste disposal considerations. Specifically, Building J-23 contains several hot cells and is planned for demolition. Building demolition requires precautions to ensure that radioactive releases are minimized and that the structural integrity of the hot cells is maintained. The radioactive source term needs to be adequately characterized, and then removed or stabilized to achieve the goal of minimal environmental release and appropriate waste disposal certification.

It is necessary to have an understanding of the physical, chemical, and radiological hazards associated with Building J-23—hazards that are assessed via characterization data. Radiation levels, nature of contaminants, building condition, available records, data adequacy, among others, drive the effort level necessary for the characterization decision. Once the building is in a physically safe condition, building debris, equipment, and miscellaneous components are characterized to determine waste disposition (Note: clearance of materials is addressed in Chapter 15).

The interior and exterior of Building J-23, and remaining systems and hot cells contained within are contaminated at levels above the waste acceptance criteria (WAC) limits. Further, contamination levels associated with Building J-23 do not allow for safe and controlled open-air demolition. Remaining liquids, loose contamination, and friable asbestos that exist in the building's process systems must be stabilized or removed prior to demolition.

Characterization planning for Building J-23 can be performed using the DQO process. The following steps illustrate how the characterization survey design (and data needs) is driven by identifying the necessary decisions.

Step 1: State the Problem

Data are needed to prepare the demolition plan, evaluate personal protection equipment (PPE) requirements, segregate waste, consistent with chemical and radiological criteria, and ensure that the appropriate waste acceptance criteria for the disposal site are satisfied.

Step 2: Identify the Decisions

The decisions in this example focus on the necessary radiological data to identify areas and components that require removal or encapsulation prior to demolition, and the radiological/chemical data to develop estimated source term that might be released during demolition and disposal site waste concentrations.

Examples of specific decisions include:

- Do any structural surfaces and/or system components require remediation to meet the disposal site WAC limits?

- Can debris be appropriately and adequately segregated for waste disposal? For instance, consider the ability to segregate radiological material from hazardous and asbestos-containing materials.

- What is the volume of structural debris and system components for disposal at the disposal site(s)?

- Can contamination remaining on structural surfaces and within system components be stabilized (using foaming, fixatives, etc.) to allow for safe demolition, or is additional remediation required?

- Can structural surfaces, system components, and hot cells be characterized and/or remediated post-demolition?

- How many samples are required for each decision?

- Will a multi-phase characterization plan need to be developed?

Step 3: Identify Input to Decisions

- HSA and scoping survey data from available survey yields results that provide information on the contaminants of concern, both concentration and distribution

- WAC limits for radiological, chemical, and asbestos-containing material

- Accessibility of structural surfaces and system components for characterization and/or remediation; post-demolition; nature of inaccessible areas

- Structural integrity of hot cells and whether or not remediation can be performed pre- or post-demolition

- Calculated action level (for airborne concentrations) for personnel and environmental protection for open-air demolition based on modeling the potential releases of radioactive material during demolition

Step 4: Define the Study Boundaries

Survey unit or area grouping based on process knowledge—areas with similar nature and extent of contamination; areas identified within Building J-23 according to operational history, structural condition/integrity, that is, ventilation ductwork and liquid effluent piping systems

Step 5: Develop a Decision Rule

If structural surfaces and/or system components exhibit widespread contamination levels greater than the WAC, then plan for remediation and/or stabilization of contamination; otherwise, prepare the materials for disposal without D&D.

If total estimated source term less is than 5 Ci, then plan to perform open-air demolition using misting/spraying to control releases; otherwise, consider stabilization techniques to immobilize contaminants and/or perform demolition within a confined structure or enclosure equipped with HEPA filtration.

Step 6: Specify Decision Errors

Decision errors are based on the WAC limits—the DQO team determined that it is acceptable to be within ±20% of the WAC limit and have a 50% chance of being outside the upper or lower bounds on the estimate. The team also debated decision errors on the total source term for open-air demolition, ultimately agreeing that the number of samples needed for making the WAC decision is likely sufficient to calculate the total source term.

Step 7: Optimize the Survey Design

Multiple DQO sessions were scheduled to work through process and to develop successful characterization plan. Characterization surveys for Building J-23 were planned to be performed in two phases:

- Phase 1 to support demolition decision (open-air, or not)

- Phase 2 to support disposal decisions (WAC compliance)

Characterization survey data should be sufficient for making decisions concerning whether contamination levels (i.e., from radionuclides, beryllium, asbestos, lead, and PCBs) are adequate for (1) open-air demolition, and (2) disposal (WAC). To that end, characterization design specified the number of samples, location and type of samples, type of measurement and collection equipment, and scanning requirements.

The DQOs team decided to base the characterization sample size on that needed to make the WAC decision. The phase 1 effort would require data collection to support the open-air demolition decision, while the phase 2 effort would support the WAC decision. The sample size n for both decisions (based on the WAC decision) was determined using an equation to estimate the mean for a specified confidence level and error (d), for a normally distributed sample (Walpole and Myers 1985):

$$n = \frac{z^2 s^2}{d^2} \tag{3.1}$$

where z is the standard normal deviate that corresponds to a given probability (50% in our example yields z of 0.67) and s is the sample standard deviation (1.6 pCi) for U-238 from preliminary scoping data (from step 3 HSA and scoping survey data, U-238 is known to be the radionuclide driving the sample size determination). This equation provides the sample size needed to estimate the mean for a specified confidence level and error ($d = 0.2$); the sample size was 29 for each survey unit and area grouping. The calculation is shown as follows:

$$n = \frac{z^2 s^2}{d^2} = \frac{(0.67)^2 (1.6)^2}{(0.2)^2} = 29$$

Therefore, the characterization survey design included both radiological and chemical samples and other miscellaneous measurements, including area radiation measurements, air sampling, scanning and surface activity measurements (fixed and removable), and miscellaneous media samples and laboratory analyses. Chemical sampling was performed for asbestos, PCBs, lead, and beryllium.

Ultimately, the characterization survey design addressed the Phase 1 demolition decisions:

- Building accessibility issues assessed

- Dose rate measurements to map radiation levels in building

- Systematic scanning to identify residual radioactivity deposits of concern (for alpha, beta, and gamma radiation)

- Smears and miscellaneous sampling to identify transuranic and other contaminants

- Visual observations to remove sources of PCBs, lead, and asbestos

The characterization survey plan also resulted in Phase 2 disposal decisions:

- Segregate building debris into waste lots in controlled lay-down area

- Scanning and surface activity measurements to determine radioactivity levels and concentrations on building debris

- Smears and miscellaneous sampling as needed

- Sampling for chemicals to ensure WAC compliance for beryllium, PCBs, lead, and asbestos

The DQO approach for Building J-23 characterization resulted in identifying the necessary data to make decisions concerning demolition and waste debris disposal. The characterization design provided a comprehensive sampling and analysis plan based on DQOs that were tailored to answer the specific D&D and disposal decisions. In addition to the phased approach for this project, some characterization activities were necessarily sequenced based on timing of decontamination and/or source removal in nearby locations. This allowed the survey team to address accessibility issues, including building areas that were significantly impacted by radiation "shine." Overall, the DQO process for characterization offered an effective and efficient plan that saved time and money.

3.3 DECOMMISSIONING OBJECTIVES OF CHARACTERIZATION: DATA NEEDED TO MAKE DECISIONS

Perhaps the most common objective of characterization is to provide data on the nature and extent of contamination at the facility/site. Depending on data needs, characterization may provide data on: (1) the identification and distribution of potential contamination—including radiological, nonradiological hazardous (asbestos, PCBs, RCRA materials) in buildings, structures, and other site facilities; (2) the concentration and distribution of these contaminants in surface and subsurface soils; (3) the distribution and concentration of contaminants in surface water, groundwater, and sediments; and (4) the distribution and concentration of contaminants in other impacted media such as vegetation, sludge, or paint.

The accuracy needed for characterization depends on the decisions to be made using the characterization data. This is where the DQO process really shines. The design of the site characterization survey is based on the specific DQOs for the information to be collected, and is planned using the HSA and scoping survey results. As the examples illustrate in the previous section, the DQO process focuses the characterization design on the data needed for making decisions. This approach allows the characterization sampling plan to be developed in a straightforward fashion—by stating the decisions to be made, the decision rules that govern the actions to be taken, and considering the decision errors associated with the data. The resulting DQO-based sampling plan ensures that only those data needed, with commensurate data quality, are collected.

Examples of characterization decisions—and the necessary data for making the decision—include: (1) determining the nature and extent of radiological contamination to decide whether the contamination exceeds release criteria, (2) assessing projected waste volumes generated from remediation activities, (3) evaluating remediation project alternatives (e.g., unrestricted use, restricted use, on-site disposal), (4) providing parameter inputs of sufficient accuracy for pathway analysis and dose or risk assessment models (e.g., RESRAD) for determining site-specific derived concentration guideline levels (DCGLs), (5) evaluating cleanup technologies, (6) identifying the appropriate radiological controls during the performance of D&D activities, and (7) providing input to FSS design (further details in Section 3.6). Characterization is also essential for the classification of wastes to demonstrate that transportation and disposal criteria have been met (e.g., waste acceptance criteria). Multiple characterization objectives/decisions can be covered by compound DQOs developed during the planning stage (see example in previous section). Data collection needs should be integrated so that collecting data for one objective may also satisfy another.

As noted above, characterization surveys may be performed to determine site-specific information for the development of DCGLs. For example, vegetation samples may be collected as part of a characterization survey to make a decision related to the uncertainty in the soil-to-plant transfer factor at the site. The relationship between radionuclide concentrations in plants and those in soil, aptly called the soil-to-plant transfer factor, is used in many models to develop site-specific DCGLs. Other parameters to be assessed during characterization include the spatial distributions of contamination, various soil properties, chemical character of the radioactive contamination, among others. These site-specific parameters (and their uncertainty) are input into the pathway modeling code to generate DCGLs applicable to the specific facility/site.

Another potential application for characterization data is the establishment surrogate ratios between radionuclides for FSS design. For sites with hard-to-detect radionuclides, it may be advantageous to measure one radionuclide to estimate the concentration of the hard-to-detect radionuclide. This is a common strategy for FSS design. During the characterization survey a number of samples are collected, and both the measured radionuclide and the one to be inferred (during the FSS) are analyzed. HSA data may provide initial indications concerning the spatial variability of contaminant concentrations; these data may indicate whether the contamination is likely to be relatively uniform or somewhat heterogeneous. The ratio of measured-to-inferred radionuclide is calculated for each sample. A fair amount of variability in this ratio is not uncommon. The DQO process can be used to determine the number and location of samples to establish the surrogate ratio. For example, step 6 of the DQO process might specify that the sample size should be sufficient to maintain the variability in the surrogate ratio to less than 20%. In this case the spatial variability of both the measured and inferred radionuclide is an important factor in determining sample size. Section 6.2.1 provides additional information on the development and use of surrogates.

ANSI N13.59 states that additional uses of the characterization data may be identified after the characterization survey has been performed—that is, the characterization data may become necessary to adequately define the source term for a risk assessment, or may need to be used to perform dose assessments for workers who were previously employed at the site. According to N13.59, the source term should be described consistent with assumptions of the land use and exposure scenarios, and consider spatial parameters such as the areas over which concentrations may be averaged, sizes of elevated concentrations that may be significant, and the depths or thicknesses which should be used for sampling protocols. Again, the data needs in each situation should be driven by the characterization decisions to be made.

3.4 CHARACTERIZATION SURVEY DESIGN AND CONSIDERATIONS

The design of the site characterization survey proceeds from the specific DQOs for the data to be collected to make the decision(s). Various characterization decisions were addressed in the previous section—and these decisions help answer questions such as how detailed should the characterization be, will the characterization be performed in phases, or can the characterization survey data in some areas be used as FSS data? An effective way to reduce characterization costs is to gather as much information as possible from the HSA. And note that the scoping survey, consisting of judgmental radiological measurements and sampling, builds on the information obtained from the HSA, and also serves to validate the HSA data.

The HSA is conducted at the beginning of a decommissioning project. The HSA provides a starting point by assembling existing site information concerning the radiological (and nonradiological) status of the site. It provides information on the expected contamination potential of facility/site areas, the nature of the contaminants, and the types of media potentially contaminated. The HSA draws on many sources of information, including the site operational history, environmental reports, process knowledge, routine health physics surveys, and interviews with current and former site workers. The NRC's NUREG-1700 summarizes the HSA as identifying "all locations, inside and outside the facility, where radiological spills, disposals, operational activities, or other radiological accidents/incidents that occur or could have resulted in contamination of structures, equipment, laydown areas, or soils (subfloor and outside area)" (NRC 2003b). Both MARSSIM and ANSI N13.59 provide additional considerations (and examples) for performing these assessments.

The operational history includes records of site conditions prior to operational activities, operational activities of the facility, effluents and on-site disposal, and significant incidents—including spills or other unusual occurrences—involving the spread of contamination around the site and on areas previously released from radiological controls. Process knowledge includes a description of the processes and materials used in the facility or at the site; it allows one to establish the potential for contamination as well as narrowing the list of potential contaminants (DOE 1994). A good example of process knowledge is the information that facility surfaces were routinely painted to fix alpha contamination in place to minimize personnel intakes.

The HSA review should include other available resources, such as site personnel, former workers, and local residents. Historic aerial photographs and site location maps may be particularly useful in identifying potential areas of contamination and subsequent migration pathways. Radioactive material licenses and results of operational surveys also provide valuable information. For instance, the types and quantities of materials that were handled and the locations and disposition of radioactive materials may be determined from a review of the radioactive materials license. Contamination release and migration routes, such as natural drainage pathways as well as areas that are potentially affected and are likely to contain residual contamination should be identified. Further, the types and quantities of materials likely to remain onsite, considering radioactive decay, should be determined.

The overall characterization approach should identify the potential contaminants of concern—both radiological and nonradiological (hazardous) materials. It is important to recognize that the contaminants of concern for many decommissioning projects extend beyond the familiar radiological constituents, and often include hazardous materials as well (PCBs, asbestos, mercury, and other heavy metals). The characterization survey decisions may dictate the consideration of Resource Conservation and Recovery Act (RCRA) and Comprehensive Environmental Response, Compensation, and Liability Act (CERCLA) requirements, as well as the implication that hazardous materials have on waste disposal issues.

A graded approach to characterization may be considered when the objective of the characterization effort is to determine the nature and extent of contamination. In this case, the characterization survey should clearly identify those portions of the site (e.g., soil, structures, and water) that have been affected by site activities and are potentially contaminated. The survey should also identify the portions of the site that have not been affected by these activities, and which may be classified as nonimpacted. In general, the graded approach to characterization might include minimal characterization of materials and locations that are clearly contaminated and that which require decontamination or removal as radioactive waste, as well as for those materials and locations not likely to be contaminated at all. More extensive characterization efforts should be performed for materials and locations that have a potential for contamination, but where the contamination levels are unknown relative to release criteria. The graded approach provides a mechanism to optimize the limited time and resources available for characterization (Brosey and Holmes 1996).

A special case of the graded approach is when the characterization is performed in phases. A phased approach is warranted when further characterization must wait until decontamination or source removal occurs. For example, the presence of remaining radiological source terms in one area of the site may impact the radiation levels in a number of adjoining areas. It may be impossible, or at least impractical, to perform characterization activities in areas that are influenced by this radiation "shine." Rather, the characterization is performed over a longer period, progressing as decommissioning activities remove radioactive components and materials. This situation is common at decommissioning nuclear power reactor sites, where complete characterization in areas adjacent to the reactor building may be scheduled only after removal of the reactor vessel and other highly radioactive source terms.

Finally, characterization design and its implementation should be flexible—characterization efforts and scope may be increased or decreased based on survey findings. That is, once implementation of the characterization plan has been initiated, deviations from the plan may be warranted if contamination levels encountered are significantly different than anticipated (e.g., from HSA and scoping surveys). In this situation, DQO process should be revisited in light of the new information to assess whether the characterization sampling frequency should be modified. The point is that the decisions being made during characterization may be based on iterative data collection efforts. Simply stated, the survey should be sufficiently flexible to accommodate unexpected information/data generated from the characterization—so that the necessary decisions can be made.

3.5 CHARACTERIZATION SURVEY ACTIVITIES

Characterization survey activities often involve the detailed assessment of various types of building and environmental media, including building surfaces, surface and subsurface soil, sediment, surface water, and groundwater. These survey activities include surface scanning, surface activity measurements, exposure rate measurements, and environmental sample collection. Again, the reader is referred to Chapter 10 and ANSI N13.59 (Section 6.0) for a more complete description of characterization survey activities.

The HSA data should be used to identify the potentially contaminated media onsite. Identifying the media that may contain contamination is useful for preliminary survey unit classification and for planning subsequent survey activities. The NRC states in NUREG-1700 that site characterization should be sufficiently detailed to determine the "extent and range of contamination of structures, systems (including sewer systems, and waste management systems), floor drains, ventilation ducts, piping and embedded piping, rubble, contamination on and beneath paved parking lots ...." The list goes on, but you get the idea, all media should be addressed in the characterization survey.

Selection of survey instrumentation and analytical techniques are typically based on knowledge of the appropriate DCGLs, because remediation decisions are made based on the level of the residual contamination as compared to the DCGL. The IAEA points out that characterization should not rely on the use of a single assessment method, but requires the joint use of calculations (e.g., depth of neutron activation), *in situ* measurements and sampling and analyses (IAEA 1997).

Many times, the HSA information available is simply incomplete or even inaccurate—especially considering that plant and equipment drawings are not always of sufficient quality. In these instances, "walk downs" of the facility help to identify facility conditions that may require special decommissioning considerations and provide valuable information on the current state of piping systems, ductwork and other equipment, including piping size, length, type, and physical location (Brosey and Holmes 1996). The walk down may also include the scoping survey, which builds on the HSA and involves biased measurements and samples in suspect areas (Vitkus 1998).

3.5.1 Structure Surveys

Surveys of building surfaces and structures include surface scanning, surface activity measurements, exposure rate measurements, and sample collection (e.g., smears, subfloor soil, water, paint, and building materials). Both *in situ* survey instrumentation and analytical laboratory equipment and procedures are selected based on their detection capabilities for the expected contaminants and their quantities. Instrument selection is also based on the DQO process—sometimes the instrument is only required to identify gross levels of contamination during a scan, and at other times the instrument must be able to detect contamination at levels below the release criteria. Background surface activity and radiation levels for the area should be determined from appropriate background reference areas, and should include assessments of surface activity on building surfaces, exposure rates, and radionuclide concentrations in various media. Because different surface materials have various amounts of naturally occurring radioactivity, each surface type expected to be encountered during the characterization survey should be assessed.

Scans should be conducted in areas likely to contain residual activity, based on the results of the HSA and scoping survey. Figure 3.1 shows the preparation of a floor monitor used for building surface scans. Both systematic and judgmental surface activity measurements are performed. Judgment direct measurements are performed at locations of elevated direct radiation (hot spots), as identified by surface scans, or at locations likely to contain elevated levels of residual radioactivity—such as in sewers, air ducts, storage tanks, septic systems, and on roofs of buildings.

Exposure rate measurements and media sampling are performed as necessary. Miscellaneous samples may include various media, such as concrete, paint, sludge, and residue in drains, and are collected from locations

FIGURE 3.1   Floor monitor prep for building surface scans. (Courtesy of ORAU.)

likely to accumulate contamination (cracks in the floor, horizontal surfaces, drains, etc.). For example, subfloor soil samples may provide information on the horizontal and vertical extent of contamination. Similarly, concrete core samples are necessary to evaluate the depth of activated concrete in the biological shield. This information allows for the estimation of the amount of activated concrete to be removed from the shield, and therefore predicts the volume of concrete waste materials generated (Brosey and Holmes 1996).

3.5.2 Land Area Surveys

Characterization surveys for surface and subsurface soils and media involve employing techniques to determine the lateral and vertical extent of contamination and radionuclide concentrations in the soil. This may be performed using either sampling and laboratory analyses, or *in situ* gamma spectrometry analyses, depending on the detection capabilities of each methodology for the expected contaminants and concentrations. Note that *in situ* gamma spectrometry analyses or any direct surface measurement cannot easily be used to determine vertical distributions of radionuclides. An attractive use of *in situ* gamma spectrometry during characterization is to document the absence of specific gamma-emitting radioactivity. That is, count sufficiently long to show that contaminant concentration is less than some specified value (fraction of DCGL, or consistent with background levels). A combination of direct measurements and samples is recommended to meet the objectives of the survey.

Radionuclide concentrations in background soil samples should be determined for an appropriate number of soil samples that are representative of the soil in terms of soil type and depth. It is important that the background samples be collected in nonimpacted areas. Sample locations should be documented using reference system coordinates, if appropriate, or fixed site features. Standard reference system grid spacing for open land areas is 10 m (NRC 1992a). This spacing is somewhat arbitrary and is chosen to facilitate in determining survey unit locations and evaluating hot-spot areas.

Surface scans for gamma radiation should be conducted in areas likely to contain residual activity. Beta scans may be appropriate if the contamination is near the surface and represents the prominent radiation emitted from the contamination, such as depleted uranium. The sensitivity of the scanning technique should be appropriate to meet the DQOs.

Both surface and subsurface soil and media samples may be necessary. Subsurface soil samples should be collected where surface contamination is present and where subsurface contamination is known or suspected (based on HSA data)—buried tanks, underground process and sewer lines, and so on. Boreholes should be constructed to provide samples representing subsurface deposits. Borehole logging may be used to measure radiation levels as a function of depth. N13.59 states that judgmental sampling is a more cost-effective subsurface sampling approach compared to random or systematic sampling. For example, boreholes might be drilled at locations suspected to have been used for radioactive waste burials. Furthermore, HSA results may indicate the locations of underground tanks and piping, especially those suspected of leaking—offering useful information concerning locations for subsurface soil assessments.

An interesting development in subsurface assessment is the NRC's NUREG/CR-7021 on geospatial modeling and decision framework (2012b). This framework, titled "A subsurface decision model for supporting environmental compliance," proposes a methodology to extend MARSSIM's surface soil scope into the subsurface. The approach combines sampling, modeling, and decision analysis to develop a contamination concern map (CCM) that is updated as new information is input in the framework. Thus, the conceptual site model for the subsurface represents an evolving picture of the extent, location, and degree of contamination relative to release criteria—moving from a somewhat qualitative beginning (i.e., limited HSA data) to a quantitative, highly detailed final CCM. Stay tuned to this innovative development for subsurface assessments.

3.5.3 Other Measurements/Sampling Locations

Surface water and sediment sampling may be necessary depending on the potential for these media to be contaminated—for example, sampling should be performed in areas of runoff from waste storage and processing operations, at plant outfall locations, both upstream and downstream of the outfall, and any other areas likely to contain residual activity. The contamination potential depends on several factors, including the proximity of streams, ponds, and other surface water bodies to the site, size of the drainage area, total annual rainfall and other meteorological conditions, and spatial and temporal variability in surface water flow rate and volume (NRC 2000a).

Groundwater sampling may be necessary depending on the local geology and potential for subsurface contamination. Characterization of groundwater contamination should determine the extent and distribution of contaminants, and rates and the direction of groundwater migration. This may be performed by designing a suitable monitoring well network. The actual number and location of monitoring wells depends on the size of the contaminated area, the type and extent of the contaminants, the hydrogeologic system, and the objectives of the monitoring program. The NRC's Draft Branch Technical Position on Site Characterization for Decommissioning (NRC 1994d) provides excellent guidance for establishing a groundwater sampling program. Appendix F of NUREG-1757 vol. 2, Ground and Surface Water Characterization reaffirms the Branch Technical Position (NRC 2006a), and lists a number of objectives that the characterization of groundwater contamination should address:

- Extent and concentration distribution of contaminants

- Source (known or postulated) of radioactive contaminants to ground water

- Rate(s) and direction(s) of contaminated groundwater migration

- Location of groundwater plume and concentration profiles (i.e., maximum concentration in the vertical and lateral extent)

NUREG-1757 identifies a number of high potential indicators for groundwater contamination at decommissioning sites, such as (1) unlined lagoons, pits, canals, or surface-drainage ways that received radioactively contaminated liquid effluent, (2) septic systems, dry wells, or injection wells that received radioactively contaminated liquid effluent, (3) storage tanks, waste tanks, and/or piping (above or below ground) that held or transported radioactively contaminated fluids and are known to have leaked, (4) liquid or wet radioactive waste buried onsite, and (5) containerized-liquid waste, stored exterior to buildings, that has leaked.

3.5.4 Data Quality Assessment for Characterization

EPA's Guidance on Data Quality Assessment (EPA 2000), Data Quality Assessment (DQA) is "the scientific and statistical evaluation of data to determine if the data are of the right type, quality, and quantity to support their intended use." For the type of characterization decisions described in Section 3.3, this means analyzing the data that were collected from building surfaces, land areas and other media, and making the decisions listed in the decision rules.

The DQA process involves five steps that begin with reviewing the planning documentation and end with an answer to the question posed at the beginning of the characterization. N13.59 provides an informative description of example activities that should occur during each of the five DQA steps.

- Review DQOs and sampling design from DQO process

- Conduct preliminary data review

- Select statistical test

- Verify assumptions

- Draw conclusions from the data

The characterization data collected during the DQO process are used to answer the pertinent characterization decisions. In each case, the characterization data are used to make decisions within the overall context of the decommissioning project. Recall the decision in our first example: whether to use default parameters to generate soil concentration DCGLs versus site-specific parameters to generate DCGLs. The DQA process yields a decision as to whether a site-specific DCGL should be used; this was the result of collecting data, executing statistical tests, interpreting the outcome of the statistical test, and making a decision. The second example contained a series of decisions, one being whether structural surfaces and/or system components require remediation to meet the disposal site WAC limits. Again, the DQA process is used to assess the data, consider the errors, and make a statistically supported decision.

3.6 CHARACTERIZATION SURVEY RESULTS TO SUPPORT MARSSIM FSS DESIGN

The primary objective or focus of most characterization surveys is to generate sufficient data to plan for remedial actions. A secondary, but no less important objective is to use characterization data to support the design of FSSs. The MARSSIM FSS depends on various facility/site information collected during the HSA and refined based on results of the scoping and characterization surveys.

The characterization survey can provide information on variations in the contaminant distribution in the survey area. The contaminant variation in each survey unit contributes to determining the number of data points based on the statistical tests used during the FSS. Additionally, characterization data may be used to justify reclassification for some survey units (e.g., from class 1 to class 2). In general, the specific objectives for providing characterization information to the FSS design include: (1) estimating the projected radiological status at the time of the FSS, which will help in the selection of Type I and Type II error rates, (2) evaluating potential background reference areas, (3) reevaluating the initial classification of survey units, and (4) selecting instrumentation based on the necessary MDCs. Many of these objectives are satisfied by determining the specific nature and extent of contamination of structures and environmental media. Also note that remediation performed in an area subsequent to that area being characterized would likely alter the data used for FSS planning (e.g., the contaminant variability is probably less following remediation).

The MARSSIM survey design recognizes that all areas of the facility/site do not have the same potential for residual contamination and, accordingly, do not require the same level of survey coverage to achieve the established release criteria. The first level of classification is to divide the site into impacted and nonimpacted areas. Nonimpacted areas have no reasonable potential for residual contamination and do not require any level of survey coverage to satisfy release criteria. These areas have no radiological impact from site operations and are usually used as background reference areas.

Impacted areas—areas that have some potential for containing contaminated material—are further subdivided into one of three classifications (NRC 2000a):

- Class 1: Areas that have, or had, a potential for radioactive contamination or known contamination. Examples of class 1 areas include: (1) site areas previously subjected to remedial actions, (2) locations where leaks or spills are known to have occurred, (3) former burial or disposal sites, and (4) waste storage sites. Areas that are contaminated in excess of the $DCGL_W$ prior to remediation should be classified as class 1 areas.

- Class 2: These areas have, or had, a potential for radioactive contamination or known contamination, but are not expected to exceed the $DCGL_W$. To justify changing an area's classification from class 1 to class 2, the existing data (from the HSA, scoping surveys, or characterization surveys) should provide a high degree of confidence that no individual measurement would exceed the $DCGL_W$. Examples of areas that might be classified as class 2 for the FSS include: (1) locations where radioactive materials were present in an unsealed form (e.g., process facilities), (2) potentially contaminated transportation routes, (3) areas downwind from stack release points, and (4) areas on the perimeter of former contamination control areas.

- Class 3: Any impacted areas that are not expected to contain any residual radioactivity, or are expected to contain levels of residual radioactivity at a very small fraction of the $DCGL_W$, based on site operating history and previous radiological surveys. Examples of areas that might be classified as class 3 include buffer zones around class 1 or class 2 areas, and areas with very low potential for residual contamination but insufficient information to justify a nonimpacted classification.

One question that usually arises is "what exactly is the difference between a nonimpacted and class 3 impacted area?" In some ways, it comes down to a subjective determination of whether there is no reasonable potential for contamination (nonimpacted) versus a very small potential for contamination (class 3). In practice this estimation of contamination potential can be exceedingly difficult to determine. It is usually based on the results of the HSA, and is typically a subjective determination to a large degree. In fact, areas that may be legitimately classified as nonimpacted, supported by a strong technical argument, may end up as class 3 areas due to public perception or for political reasons. However, this must be put into proper context. While no survey activities are necessary in nonimpacted areas, the level of survey effort in class 3 areas is relatively minor, and therefore, it may be prudent to accept disputed areas as class 3 even though a legitimate argument can be made for a nonimpacted classification. In fact, it may be reasonable to adopt an approach of choosing the more restrictive classification whenever there is a reasonable debate between two classifications—that is, select class 2 if the argument is between the area being classified as class 2 or class 3.

The information collected during the characterization survey should be used to reevaluate the initial classification of areas performed during the HSA. For example, if positive contamination is identified in a class 3 area, an investigation and reevaluation of that area should be performed to determine if the class 3 area classification is appropriate. Typically, the investigation will result in part or all of the area being reclassified as class 1 or class 2. For a class 2 area, if survey results identify the presence of residual contamination exceeding the $DCGL_W$ or suggest that there may be a reasonable potential that contamination is present in excess of the $DCGL_W$, an investigation should be initiated to determine if all or part of the area should be reclassified to class 1. Thus, accuracy of the classification of site areas is critically dependent on the quality of the characterization data.

In some cases where no remediation is anticipated, for example, site areas expected to have contamination levels well below the release criteria, the results of the characterization survey may indicate compliance with DCGLs established by the regulatory agency. When planning for the potential use of characterization survey data as part of the FSS, the characterization data must be of sufficient quality and quantity for that use. Specifically, the measuring and sampling techniques and survey instrumentation should be able to attain the selected DQOs. The MARSSIM recommends that the MDC of field and laboratory instruments used during the FSS be 10–50% of the appropriate $DCGL_W$. Do not confuse this MARSSIM recommendation with the scan MDC requirement of being able to detect the $DCGL_{EMC}$ in class 1 survey units.

QUESTIONS AND PROBLEMS

1. Explain the important role characterization plays in the decommissioning process. What are the potential pitfalls of inadequate characterization on a decommissioning project?

2. Under what specific conditions is it possible to use data obtained from the characterization survey for the FSS data?

3. State three uses of *in situ* gamma spectrometry during site characterization.

4. State five characterization decisions that can be addressed by characterization surveys.

5. Use the seven-step DQO process to frame the experimentation required and the decisions needed to determine whether scabbling or chemical decontamination is the best way to remediate surficially contaminated concrete.

6. Consider how the data objectives vary depending on the characterization decision needed for the following examples. First, the decision is to determine the depth of Th-232 contamination in the Burn Pit area that measures approximately 300 m^2. Second, the decision is to determine specifically the soil porosity and k_d needed to model the migration of Th-232 into the vadose zone. (Hint: Addressing the first decision involves analysis of Th-232 concentrations at various depths in the Burn Pit, while the second decision likely requires specific soil analyses to characterize the soil.)

CHAPTER 4

Guidelines and Dose-Based Release Criteria

DECOMMISSIONING IN THE UNITED STATES and abroad has increased steadily since the early 1990s. Cleaning up the Cold War legacy sites in both the United States and the United Kingdom have served to maintain the brisk pace of decommissioning. The deregulation of electricity markets and availability of inexpensive natural gas may hasten the shutdown of nuclear reactors. The fallout from the Fukushima Daiichi reactor incident not only created a multi-decade cleanup program in Japan, but has widespread ramifications across the world as some countries consider downsizing or eliminating their nuclear reactor portfolio (e.g., Germany). So while the number of decommissioning projects has increased over the past couple of decades, it is important to remember that decommissioning projects have been performed since the inception of nuclear programs for military, power, and research applications.

From the mid-1990s through the first years of the new millennium, regulatory agencies led by the NRC, reshaped the decommissioning landscape in terms of release criteria and regulatory guidance. These efforts included the NRC's license termination rule in 1997 and were supported by several NUREGs and regulatory guides, DOE's and EPA's decommissioning rulemaking efforts, ANSI N13.12, and of course, MARSSIM. The pace of issuing new regulatory guidance has slowed a bit over the last 10 years, but not by much. Significant publications over this time have included the NRC's three-volume Consolidated Decommissioning Guidance (NUREG-1757), the MARSAME and DOE's Order 458.1 Radiation Protection of the Public and the Environment. Finally, the NCRP's decommissioning report (146) offers a comprehensive overview on the EPA's and NRC's approaches to risk management in remediation of contaminated sites (NCRP 2004).

Decommissioning professionals are challenged to stay abreast of the latest decommissioning guidance documents, and their current state of revision. In addition to those mentioned above, a sampling of decommissioning-related guidance includes the Standard Review Plan for Evaluating Nuclear Power Reactor License Termination Plans (NUREG-1700, Rev. 1), Decommissioning of Nuclear Power Reactors (Regulatory Guide 1.184), Final Generic Environmental Impact Statement on Decommissioning of Nuclear Facilities (NUREG-0586), Standard Format and Content of License Termination Plans for Nuclear Power Reactors (Regulatory Guide 1.179), Decommissioning Power Reactor Inspection Program (IMC 2561), Residual Radioactive Contamination from Decommissioning Parameter Analysis NUREG/CR-5512 (4 vols.), and Decommissioning Oversight and Inspection Program for Fuel Cycle Facilities and Materials Licensees (IMC 2602). Fortunately, many of these references are accessible on the NRC web site (http://www.nrc.gov/reading-rm.html).

While the decommissioning industry has made monumental technological advances and has undergone numerous changes in release criteria along the way, it is worthwhile to have an appreciation of the regulatory guidance and guidelines that served as precursors to the current dose-based and risk-based decommissioning criteria. In fact, many of these historical guidelines are still being used by the State and Federal regulators today. For example, D&D contractors are still using these guidelines at DOE sites and at NRC D&D projects that had their decommissioning plan approved prior to the issuance of the license termination rule in 1997. This chapter opens with an overview of the historical decommissioning guidance employed at many decommissioning projects.

Release criteria are numerical guidelines for surface activity, volumetric activity, and direct radiation levels that are considered acceptable by the regulatory agency, for a given set of conditions and applications. Release criteria are usually provided by the regulatory agency, and in all cases, must be approved by the appropriate regulatory agency. These guidelines are stated in terms of residual radioactivity or radiation levels that are in excess of the background levels. Consequently, an important aspect of decommissioning surveys over the years has been to adequately assess the background levels at a facility or site. The term "guideline" has been used for many years in the decommissioning arena, and only recently has been replaced. MARSSIM refers to these numerical guidelines as DCGLs.

Two broad categories of release criteria can be considered—unrestricted release and restricted release. Basically, unrestricted use means that there are no restrictions on the future uses of the site. This definition implies that the facility can be used by the general public without restrictions or controls to limit exposure to radiation or radioactive materials. The site may be used for other business, residence, or even dismantled. Compliance with the historical guidelines used prior to the dose-based rulemaking was understood to signify unrestricted use. Restricted use means that conditions are placed on the future uses of the site. Common restrictions include legally enforceable institutional controls such as deed restrictions describing what the property can and cannot be used for. The NRC's NUREG-1757, vol. 2 describes conditions for restricted release in further detail. NUREG-1757 categorizes decommissioning projects according to complexity; for example, Decommissioning Group 6 is defined as "licensed material was used in a way that resulted in residual radiological contamination of building surfaces, and/or soils, and possibly ground water. The licensee demonstrates that the site meets restricted use levels derived from site-specific dose modeling" (NRC 2006a, p. 1-3).

4.1 HISTORIC RELEASE CRITERIA AND GUIDANCE DOCUMENTS

Probably the most recognizable historic release criteria are those found in Regulatory Guide 1.86. This and other early decommissioning guidance provided an expectation that no contamination should remain following decommissioning activities if it could easily be removed with a "bucket and brush." That is, reasonable effort should be expended to remove residual radioactivity, such as washing, wiping, and vacuuming surfaces. This may be construed as the first application of the ALARA principle to decommissioning. Guidance in Regulatory Guide 1.86 cautions that coverings, such as paint, cannot be used to demonstrate compliance with surface activity guidelines. Furthermore, the guidance states that contamination in pipes should be accessed at all possible entry points, especially the low points. If contamination could be expected to be present in a pipe, and the pipe interior surfaces were generally inaccessible, it was assumed that the contamination levels exceeded the guidelines.

Historic release criteria were largely generic in nature, but there were examples of site-specific guidelines issued by regulators. Generic release criteria can be defined as guidance provided by regulatory agencies that did not account for site-specific characteristics. Examples of generic release criteria are the Atomic Energy Commission's Regulatory Guide 1.86, "Termination of Operating Licenses for Nuclear Reactors" (AEC 1974) and DOE Order 5400.5, "Radiation Protection of the Public and the Environment" (DOE 1990). (Note that the DOE Order 5400.5 was superseded in 2011 by DOE O 458.1 under the same title.) The AEC Regulatory Guide 1.86 is commonly referred to as NRC Regulatory Guide 1.86 (which will also be the convention in this text), though the NRC did not exist prior to 1975. In contrast, site-specific criteria are usually derived by the licensee or stakeholder using various scenarios and site characteristics (e.g., depth of contamination, size of contaminated area, depth to groundwater depth, etc.). Site-specific release criteria are usually based on a risk- or dose-based criterion, such as 25 mrem/y, and depends on modeling (e.g., RESRAD or D&D) to translate the dose criterion to a measurable guideline.

The historic regulatory guidance documents were not dose based. Rather, the guidelines provided in the Regulatory Guide 1.86 were generally based on considerations related to the detection capabilities of commercially available survey instruments at that time (early 1970s). NRC guidance included Regulatory Guide 1.86 for reactor licensees and "Guidelines for decontamination of facilities and equipment prior to release for unrestricted use or termination of license for byproduct, source, or special nuclear material" for nonreactor licenses (NRC 1987). In some instances, especially when compared to the DCGLs based on dose for alpha emitters, one may reminisce about these historic surface activity guidelines, and the simplicity offered by having only four groupings of radionuclides. Table 4.1 provides the Regulatory Guide 1.86 surface activity guidelines and conditions for implementation. Removable surface activity guidelines are not shown in Table 4.1, but are 20% of the average surface activity guidelines for each grouping. An excellent discussion on NRC's decommissioning guidance can be found in Chapter 2 of the Health Physics Society's 1999 Summer School Text, "Decommissioning and Restoration of Nuclear Facilities" (Slobodien 1999).

The application of the surface activity guidelines shown in Table 4.1 does require a degree of explanation. First, it is important to understand that surface activity levels are allowed to be averaged over 1 m^2, but no surface activity levels can exceed the maximum surface activity specified for a 100 cm^2 area. Additionally, some of the radionuclide groupings, particularly for decay series, can be complicated. For example, uranium

TABLE 4.1   Regulatory Guide 1.86 Surface Contamination Criteria

Radionuclide	Average Surface Activity in 1 m² (dpm/100 cm²)	Max Surface Activity in 100 cm² (dpm/100 cm²)
U-nat, U-235, U-238, and associated decay products	5000	15,000 α
Transuranics, Ra-226, Ra-228, Th-230, Th-228, Pa-231, Ac-227, I-125, I-129	100	300
Th-nat, Th-232, Sr-90, Ra-223, Ra-224, U-232, I-126, I-131, I-133	1000	3000
Beta–gamma emitters (nuclides with decay modes other than alpha emission or spontaneous fission) except Sr-90 and others noted above	5000	15,000

surface activity guidelines are listed in terms of alpha activity. This has historically been a challenge because of the difficulty of measuring the easily attenuated alpha radiation on most building surfaces. Chapter 6 provides an alternative approach that has been employed at many D&D projects, for example, using beta measurements to demonstrate compliance.

The uranium surface activity guidelines are applicable to both uranium ore (when entire uranium decay series is present) and processed uranium, while the Ra-226 and Th-230 surface activity guidelines are applicable to the mill tailings. The difficulty arises when both Ra-226 and uranium ore are potentially present. In this circumstance it is difficult to establish whether the appropriate guideline would be 5000 dpm/100 cm² for uranium ore, or 100 dpm/100 cm² for the more restrictive Ra-226 and Th-230 guideline. The standard solution has been to evaluate the U-238 to Ra-226 ratio, and from that result determine if the two radionuclides are in equilibrium. If the Ra-226 level clearly exceeds U-238, then the more conservative Ra-226 guidelines would apply. Finally, the NRC has ruled that the guidelines for thorium apply independently to both alpha and beta measurements of surface contamination. That is, the surface activity guideline value is the same for both alpha and beta radiation, and both should be measured and compared to the guideline for thorium to demonstrate compliance.

One of the shortcomings with the Regulatory Guide 1.86 guidance is that the groupings are quite broad. For example, both Co-60 and the largely innocuous H-3 have the same surface activity guideline (5000 dpm/100 cm²), yet when they are present at equal activities, the

Co-60 easily delivers thousands of times more dose. To counter this discrepancy at some D&D sites, higher site-specific surface activity guidelines for H-3 and Fe-55 were approved by NRC to reflect the significantly lower risk posed by H-3 and Fe-55. Specifically, the average surface activity guideline for these radionuclides was increased to 200,000 dpm/100 cm² at both the Shoreham and Fort St. Vrain power reactor decommissioning projects.

Shifting gears to volumetric contamination guidelines, the NRC's Branch Technical Position (BTP), "Disposal or onsite storage of thorium or uranium wastes from past operations" (NRC 1981b) provides the guidelines for unrestricted release of uranium and thorium in soil. These guidelines are referred to as disposal Option 1:

Natural thorium (sum of Th-232 and Th-228)	10 pCi/g
Uranium ore (sum of U-234, U-235, and U-238, when entire decay series is present)	10 pCi/g
Depleted uranium (sum of U-234, U-235, and U-238)	35 pCi/g
Enriched uranium (sum of U-234, U-235, and U-238)	30 pCi/g

The guidelines for disposal Option 1 are "set sufficiently low that no member of the public is expected to receive a radiation dose commitment from disposed materials in excess of 1 millirad per year to the lung or 3 millirads per year to the bone from inhalation and ingestion, under any foreseeable use of the material or property" (NRC 1981b, p. 46 FR 52062).

This BTP also discussed three other disposal options, but Option 2 was the only other option used by NRC. Option 2 covered the burial of contaminated soils in an on-site disposal cell. The guidelines for disposal Option 2 were much higher—for example, 50 pCi/g for natural thorium and 100 pCi/g each for depleted and enriched uranium. This option was allowed provided that it could be demonstrated that the buried materials would be stabilized in place and not be transported from the site. Specifically, Option 2 was somewhat similar to a restricted or conditional release; the contaminated materials were required to be buried under prescribed conditions for constructing the on-site disposal cell, but did not require land-use restrictions following license termination.

The NRC issued Policy and Guidance Directive FC 83-23 on November 4, 1983, titled "Guidelines for Decontamination of Facilities and Equipment Prior to Release for Unrestricted Use or Termination of Licenses for Byproduct, Source, or Special Nuclear Material" (NRC 1983). In essence,

FC 83-23 provided a summary of the various guidelines that NRC staff had been using up to that point in time. This Directive included: (1) surface activity guidelines from Regulatory Guide 1.86, with the addition of surface limits for average and maximum radiation levels of 0.2 and 1.0 mrad/h at 1 cm for beta and gamma emitters; (2) indoor exposure rate guideline of 5 μR/h above background and outdoor exposure rate guideline of 10 μR/h above background, both at 1 m above the surface; and (3) soil concentration limits for uranium and thorium from the 1981 BTP, supplemented by soil concentration limits for Pu-239 and Am-241 at 25 and 30 pCi/g, respectively. FC 83-23 also stated that soil concentration limits for other radionuclides will be determined on a case-by-case basis.

The NRC has also issued guidance documents, including letters and memoranda that addressed general and site-specific guidelines and interpretation (e.g., how to apply thorium surface activity guidelines and release criteria for specific decommissioning projects). For example, the NRC sent a letter to Stanford University establishing guidelines for radiation levels resulting from contamination and activation at a research reactor being decommissioned. These guidelines for exposure rates were set at 5 μrem/h at 1 m, with the intent being to limit exposure to individuals to no more than 10 mrem/y.

The other Federal agencies involved in issuing release criteria are the Department of Energy and the Environmental Protection Agency. First, the DOE issued DOE Order 5400.5 in 1990, that provides generic release criteria for building surfaces, equipment, and so on, that were adapted from NRC's Regulatory Guide 1.86. Initially, DOE Order 5400.5 did not have surface contamination limits for transuranics, but in a clarification memo in 1995, DOE adopted the transuranic criteria (DOE 1995).

DOE Order 5400.5 also provides release criteria for soil contaminated with Ra-226, Ra-228, Th-230, and Th-232. The guidelines and conditions for each of these contaminants are as follows: 5 pCi/g, averaged over the first 15 cm of soil below the surface and 15 pCi/g, averaged over 15-cm thick layers of soil more than 15 cm below the surface. Note: If both Th-230 and Ra-226 are present, then the sum-of-the-ratios should be applied to effectively decrease each radionuclide's allowable limit. These guidelines represent allowable residual concentrations above the background averaged across any 15-cm thick layer to any depth and over any contiguous 100 m² surface area. Further, if the average concentration in any surface or below-surface area, less than or equal to 25 m², exceeds the authorized limit of guideline by a factor of $(100/A)^{1/2}$, where A is the area or the elevated region in square meters, limits for "hot spots" are also applicable. This concept is now referred to as an area factor in MARSSIM.

The Order also makes provision for the use of supplemental limits for D&D circumstances that indicate that the generic guidelines are not appropriate. These supplemental limits must still satisfy the basic dose limits (i.e., 100 mrem/y). Justification for using supplemental limits should be well-documented and appropriate approvals obtained. Supplemental limits are often implemented when (1) remedial actions pose a clear and present risk of injury to D&D workers or members of the public, (2) continued remediation negatively impacts structural integrity, and (3) remediation efforts entail excessive costs that are not commensurate with the resulting dose reduction. Indeed, they have been primarily used in situations where the application of generic guidelines was unnecessarily restrictive and costly. The hazard assessment examples in Chapter 5 illustrate typical dose calculations that may be used to justify the use of supplemental limits.

DOE Order 458.1, which replaced DOE 5400.5 in 2011, approves the use of specific dose constraints for the release of property with the potential to contain residual radioactivity. For real property, the dose constraint is 25 mrem/y above the background. At the same time, O 458.1 allows the use of pre-approved authorized limits such as the Ra-226 and Ra-228 soil concentrations and the surface activity guidelines in DOE O 5400.5. So DOE O 458.1 provides a bridge between the historic guidelines and the dose-based release criteria.

One more historic guidance document will be discussed in this section, but recognize that there are a number of other Federal and State documents that relate to decommissioning release criteria that are not mentioned here. This guidance document relates to groundwater contamination. Groundwater can be an important environmental medium when assessing the possible exposure pathways at a D&D site. For example, contaminants in groundwater can be spread to plants and animals when groundwater is used as an irrigation source, or groundwater can deliver radiation dose directly to humans when a well is installed in a contaminated aquifer. EPA's 40 CFR Part 141, "National Primary Drinking Water Standards for Radionuclides," provides some guidance on the acceptable levels of radioactivity in drinking water. It is important to recognize that the EPA drinking water standards are applicable to public drinking water systems, rather than groundwater concentrations, and are enforced at the drinking water tap. The standards provide for maximum contaminant levels of 5 pCi/L for combined Ra-226 and Ra-228, 15 pCi/L for gross alpha activity, and a limit for beta-gamma emitters based on 4 mrem/y. We will now turn our discussion to the NRC's dose-based release criteria.

4.2 DOSE-BASED RELEASE CRITERIA AND NRC'S DECOMMISSIONING RULEMAKING

The fundamental objective of an FSS is to demonstrate that the established release criteria have been met. Therefore, one of the single most important aspects of FSS planning is to have a clear understanding of the decommissioning release criteria that apply to a particular D&D project. For years, D&D professionals used the well-known historic guidelines mentioned in the previous section for planning and implementing FSSs. Now, these same D&D professionals are quickly becoming aware of new decommissioning release criteria for building surfaces and land areas. How did this all come about?

Let us begin with the NRC, who for years terminated licenses using a mix of guidelines, none of which were primarily dose based. Considering that nearly all of the radiation protection standards are dose- or risk-based, the NRC decided that their new decommissioning rule would be dose based. (Note: NRC did consider a number of other decommissioning options such as a return to background and use of best available technology.) Around the same time, the EPA was also working on a cleanup rule. The EPA based their proposed cleanup criteria on their acceptable lifetime risk range of $10^{-6}-10^{-4}$; the result was an EPA-proposed 15 mrem/y criterion. This proposed rule never went final, and was withdrawn in December 1996. In 1994, the NRC proposed decommissioning criteria of 15 mrem/y for all exposure pathways, and 4 mrem/y for groundwater, in an apparent attempt to satisfy EPA. Not to be outdone, the DOE was also working on a proposed decommissioning rule in 10 CFR 834, "Environmental Radiation Protection Program." DOE's 10 CFR Part 834, issued as a draft for comment, proposed a 30 mrem/y limit for single site, which was generally assumed to satisfy the 100 mrem/y basic dose limit for all sources to the public. This effort was superseded by the substantive revision to 5400.5 that culminated with the issuance of DOE O 458.1, which has a real property dose constraint of 25 mrem/y, consistent with the NRC's LTR.

The NRC promulgated decommissioning criteria in Subpart E, "Radiological Criteria for License Termination" 10 CFR Part 20 in July 21, 1997 (NRC 1997a). Under Subpart E, a licensee may terminate a license for unrestricted use, if the residual radioactivity that is distinguishable from background radiation results in a total effective dose equivalent to an average member of a critical group that does not exceed 25 mrem/y, and the residual radioactivity has been reduced to levels that are as low as reasonably achievable. The implementation date for this rule was August 20, 1998, with a 1-year grandfather period. The NRC initially issued Draft Regulatory Guide DG-4006 (NRC 1998e) that provided regulatory positions on dose modeling, FSSs, ALARA, and restricted use scenarios. Next, the NRC developed a SRP for decommissioning that among other things, addressed areas of excessive conservatism, particularly in the D&D screening code and adopted a probabilistic approach to calculate the total effective dose equivalent (TEDE) to the average member of the critical group. This SRP was published as NUREG-1727 (NRC 2000b). Finally, the NUREG-1727 guidance was rolled-up into the 3-volume NUREG-1757 series.

Therefore, the dose-based release criteria in NRC's decommissioning rulemaking have superseded the guidelines in Regulatory Guide 1.86. It is interesting to see how the guidelines in Regulatory Guide 1.86 actually relate to dose. The NRC performed dose calculations using RESRAD and documented these in its *NMSS Handbook for Decommissioning Fuel Cycle and Material Licensees* (NRC 1997b). The following results were obtained for each of the groupings at the average surface activity guidelines in 1.86:

U-nat, U-235, and U-238	13 mrem/y
Ra-226, Ra-228, transuranics	0.2 mrem/y
Th-nat, Th-232, Sr-90	28 mrem/y
Beta-gamma emitters	20 mrem/y

Therefore, it is remarkable that three of the groupings are generally consistent with the NRC's 25 mrem/y dose criterion. The estimated dose of 0.2 mrem/y for the Ra-226, Ra-228, transuranic grouping indicates that the Regulatory Guide 1.86 guideline of 100 dpm/100 cm^2 for these radionuclides should be increased by roughly a factor of 100 to yield the same dose. Of course, the modeling assumptions used by NRC are a major factor in these results.

NUREG-1757, vol. 1, rev. 2 (Appendix B) provides acceptable license termination screening values of common radionuclides for building surface contamination (reproduced in Table 4.2) (NRC 2006b). These screening DCGLs correspond to an unrestricted release dose criterion of 25 mrem/y, and were derived using the D&D, Version 2 and its default input parameters.

TABLE 4.2 Acceptable License Termination Screening
Values for Building Surface Contamination

Radionuclide	Acceptable Screening Level (dpm/100 cm^2)
H-3	1.2E+08
C-14	3.7E+06
Na-22	9500
S-35	1.3E+07
Cl-36	5.0E+05
Mn-54	32,000
Fe-55	4.5E+06
Co-60	7100
Ni-63	1.8E+06
Sr-90	8700
Tc-99	1.3E+06
I-129	35,000
Cs-137	28,000
Ir-192	74,000

The D&D screening model is described in Chapter 5. The DCGL values in Table 4.2 correspond to surface concentrations of radionuclides contamination that would be deemed in compliance with the license termination rule. Note that Table 4.2 does not include screening values for alpha emitters.

One of the significant observations when comparing to the Regulatory Guide 1.86 guidelines is the lack of radionuclide groupings. It appears that the NRC will not continue the long-standing practice of grouping radionuclides that deliver comparable doses. Therefore, the MARSSIM user must be knowledgeable of the possible ways that multiple radionuclides can be handled in the FSS design (Chapter 6).

NRC issued a second Federal Register Notice dated December 7, 1999, in which the NRC noted several areas where D&D, Version 1, was overly conservative. The explanation provided for this conservatism was that Version 1 used a common default parameter set for all radionuclides, rather than being tailored for each radionuclide. NRC corrected the excessive conservatism in Version 2.0 of the D&D code by using default parameter values based on the specific radionuclides being modeled; this methodology is described in NUREG/CR-5512 vol. 2 (NRC 2001). NRC staff has calculated DCGLs for surface soil concentrations using version 2.0 of D&D. These values correspond to an annual dose of 25 mrem and are presented in Table 4.3 (reproduced from Table B.2, NUREG-1757, vol. 1).

TABLE 4.3   NRC Screening Values (DCGLs) for
Soil Contamination (pCi/g)

H-3	1.1 E+02
C-14	1.2 E+01
Na-22	4.3 E+00
S-35	2.7 E+02
Cl-36	3.6 E−01
Ca-45	5.7 E+01
Sc-46	1.5 E+01
Mn-54	1.5 E+01
Fe-55	1.0 E+04
Co-57	1.5 E+02
Co-60	3.8 E+00
Ni-59	5.5 E+03
Ni-63	2.1 E+03
Sr-90	1.7 E+00
Nb-94	5.8 E+00
Tc-99	1.9 E+01
I-129	5.0 E−01
Cs-134	5.7 E+00
Cs-137	1.1 E+01
Eu-152	8.7 E+00
Eu-154	8.0 E+00
Ir-192	4.1 E+01
Pb-210	9.0 E−01
Ra-226	7.0 E−01
Ra-226+C	6.0 E−01
Ac-227	5.0 E−01
Ac-227+C	5.0 E−01
Th-228	4.7 E+00
Th-228+C	4.7 E+00
Th-230	1.8 E+00
Th-230+C	6.0 E−01
Th-232	1.1 E+00
Th-232+C	1.1 E+00
Pa-231	3.0 E−01
Pa-231+C	3.0 E−01
U-234	1.3 E+01
U-235	8.0 E+00
U-235+C	2.9 E−01
U-238	1.4 E+01
U-238+C	5.0 E−01

TABLE 4.3   (continued) NRC Screening Values
(DCGLs) for Soil Contamination (pCi/g)

Pu-238	2.5 E+00
Pu-239	2.3 E+00
Pu-241	7.2 E+01
Am-241	2.1 E+00
Cm-242	1.6 E+02
Cm-243	3.2 E+00

One condition on using these DCGL values is that the radionuclide contamination be limited to surface soil (e.g., top 15–30 cm).

Note that the "+C" designation in Table 4.3 indicates a value for a radionuclide with its decay progeny present in equilibrium. The values are concentrations of the parent radionuclide, but account for contributions from the complete chain of progeny in equilibrium with the parent radionuclide.

The application the DCGLs provided in Tables 4.2 and 4.3 for natural decay series radionuclides, such as thorium and uranium, deserves mention. For example, most sites are not contaminated with just one isotope of uranium, but rather with total uranium that is comprised of U-238, U-235, and U-234 at particular isotopic ratios. That is, a D&D site may have processed natural uranium, depleted or enriched uranium as its contamination. Therefore, it is necessary to understand how the total uranium DCGL can be "built" from the individual DCGLs for U-238, U-235, and U-234. While this is a technical area covered in Chapter 6, it is beneficial to introduce it here as well.

Let us assume that the site contaminant is enriched uranium, at an average enrichment of 1.2%. Alpha spectrometry analyses were performed to determine the uranium activity fractions for 1.2% enriched uranium, which resulted in 0.3725, 0.029, and 0.598, respectively, for U-238, U-235, and U-234. In addition to these three isotopes of uranium, their immediate progeny are also present in this mixture. The immediate progeny of U-238 include Th-234 and Pa-234 m, both assumed to be in secular equilibrium with U-238 due to their short half-lives. The immediate progeny of U-235 is Th-231, while U-234 has no immediate progeny that are present due to the long half-life of Th-230. It is reasonable to assume that in Table 4.3, DCGLs for U-238 and U-235 include the dose contributions from Th-234 and Pa-234 m and Th-231, respectively. (Note: Whenever radionuclide DCGLs are referenced or derived using a modeling code, it is critical to determine whether or not their progeny were included in the parent DCGL, and if so, exactly which ones.)

The uranium DCGLs from Table 4.3 are 14 pCi/g for U-238, 8 pCi/g for U-235, and 13 pCi/g for U-234. The DCGLs for U-238 and U-235 with the "+C" designation are not used in this example because the entire decay series is not present.

The DCGL for 1.2% enriched uranium can be determined now, based on the individual DCGLs for each uranium isotope and the isotopic ratios shown above. The gross activity DCGL equation presented in Chapter 6 can be used for this calculation:

$$1.2\% \text{ enriched uranium DCGL} = \frac{1}{((0.3725/14) + (0.029/8) + (0.598/13))}$$
$$= 13.1 \text{ pCi/g}$$

This DCGL applies to the sum of U-238, U-235, and U-234, when the uranium isotopic ratios are as calculated for 1.2% enriched uranium. The DCGL for U-238 by itself for 1.2% enriched uranium can easily be calculated by multiplying its isotopic fraction (0.3725) by the 1.2% enriched uranium DCGL, which equals 4.9 pCi/g for U-238. To summarize, the FSS could be designed using either a DCGL of 13.1 pCi/g for total uranium, or 4.9 pCi/g for U-238, in both cases assuming that the contaminant is 1.2% enriched uranium.

The NRC provided additional information in a Federal Register Notice on June 13, 2000 concerning the use of screening values (default DCGLs) to demonstrate compliance with release criteria. In this FRN, the NRC referenced vol. 3 of NUREG/CR-5512, "Residual Radioactive Contamination from Decommissioning, Parameter Analysis, Draft Report for Comment," (NRC 1999a). This supplemental information provided licensees with the conditions that apply to the previously published DCGLs listed in Tables 4.2 and 4.3. The conditions for demonstrating compliance with *surface soil DCGLs* include, in part:

- Residual radioactivity is contained in the top layer of the surface soil (i.e., a thickness of ~15 cm).

- Unsaturated zone and the groundwater are initially free of radiological contamination.

- Vertical saturated hydraulic conductivity at the specific site is greater than the infiltration rate.

The conditions for demonstrating compliance with *building surface DCGLs* include, in part:

- Residual radioactivity is contained in the top layer of the building surface (i.e., there is no volumetric contamination).

- Fraction of removable surface contamination does not exceed 10%.

For this last point, the NRC explains that when the fraction of removable contamination is undetermined or greater than 10%, licensees may assume that 100% of the surface contamination is removable, and therefore the screening values should be decreased by a factor of 10.

The NRC also states in this FRN that the latest version of the D&D code may be used, without modification of the default values, to derive DCGLs for radionuclides not listed in Tables 4.2 and 4.3. The NRC has approved the use of NUREG/CR-5512, vol. 3 to determine acceptable DCGLs. Specifically, Table 5.19 (using a $P_{crit} = 0.90$) may be used for building surface activity DCGLs and Table 6.91 (using a $P_{crit} = 0.10$) may be used for surface soil DCGLs. It is worthwhile to note that the DCGLs provided in Table 4.2 are the same values as those found in Table 5.19 of NUREG/CR-5512, vol. 3 for a P_{crit} of 0.90. Some of the additional radionuclide DCGLs found in Table 5.19 that were not included in the November 18, 1998 FRN, includes alpha emitters. Table 4.4 provides a listing of default surface activity DCGLs from Table 5.19 of NUREG/CR-5512, vol. 3.

It is assumed that the "+C" designation in Table 4.4 indicates a value for a radionuclide with its decay progeny present in equilibrium—that is, the values are concentrations of the parent radionuclide, but account for contributions from the complete chain of progeny in equilibrium with the parent radionuclide.

Table 6.91 in NUREG/CR-5512, vol. 3 provides a number of soil concentration DCGLs to supplement the values in Table 4.3. The DCGLs in Table 6.91 for a P_{crit} of 0.10 are essentially the same as the soil concentration DCGLs reported in the December 7, 1999 FRN (Table 4.3). The only difference is due to how the radionuclides with a "+C" designation are reported. For example, the DCGL for Ra-226+C in the December 7, 1999 FRN (shown in Table 4.3) is given as 0.6 pCi/g, while the DCGL for Ra-226+C in Table 6.91 is 5.45 pCi/g. At first glance, one is very likely to conclude that they represent different DCGLs. But alas, the NRC appears to be guilty of inconsistently using the designation "+C." That is, in Table 4.3 the designation

TABLE 4.4   NRC Default DCGLs for Building Surface Contamination

Cs-134	1.27E+04 dpm/100 cm²
Eu-152	1.27E+04 dpm/100 cm²
Eu-154	1.15E+04 dpm/100 cm²
Ra-226	1120 dpm/100 cm²
Ra-226+C	315 dpm/100 cm²
Th-230	36.9 dpm/100 cm²
Th-232	7.31 dpm/100 cm²
Th-232+C	6.03 dpm/100 cm²
U-234	90.6 dpm/100 cm²
U-235	97.6 dpm/100 cm²
U-238	101 dpm/100 cm²
U-238+C	19.5 dpm/100 cm²
Pu-238	30.6 dpm/100 cm²
Pu-239	27.9 dpm/100 cm²
Am-241	27.0 dpm/100 cm²

Source: U.S. Nuclear Regulatory Commission. *Residual Radioactive Contamination from Decommissioning: Parameter Analysis.* NUREG/CR-5512, vol. 3; Washington, DC; 1999a.

"+C" refers to the concentration of the parent radionuclide given that the complete chain of progeny are present and in equilibrium with the parent. In Table 6.91, it can be inferred that the "+C" designation refers to the sum of the parent and its progeny that are in equilibrium. The proof of this is that Ra-226 and its progeny consist of nine radionuclides, and the DCGL of 5.45 pCi/g divided by nine equals the 0.6 pCi/g reported in Table 4.3. Furthermore, the DCGLs for Th-230+C (10 radionuclides in decay series), Th-232+C (10 radionuclides in decay series), and U-238+C (14 radionuclides in series) are 5.78 pCi/g, 11 pCi/g, and 7.13 pCi/g, respectively. When each of these values is divided by the number of radionuclides in the series the DCGL values in Table 4.3 are obtained. This is a perfect example of how important it is to understand what the DCGL actually represents.

To summarize, decommissioning release criteria have been evolving over the past few decades, and it is important to have a clear understanding of the past and present release criteria. For many D&D projects, the release criteria are now dose based, as opposed to the former guidelines found in guidance documents such as Regulatory Guide 1.86. This presents an added level of complexity because the release criterion in units of mrem/y must be translated to measurable quantities called DCGLs. The determination of DCGLs from release criteria in mrem/y requires dose modeling, which is the subject of Chapter 5.

4.3 CLEARANCE OF MATERIALS AND ANSI N13.12

For many years, the same Regulatory Guide 1.86 surface activity guidelines that were used to release buildings, have also been used to release materials, equipments, and items (or what DOE refers to as nonreal material). On June 30, 1998, the NRC Commission directed the NRC staff to develop a dose-based regulation for clearance of equipment and materials having residual radioactivity. However, this rulemaking effort was subsequently terminated in favor of releasing materials on a case-by-case basis using existing surface contamination values (Regulatory Guide 1.86). The reader is encouraged to refer to Chapter 15 for a more detailed discussion on the clearance of materials and related criteria.

The Health Physics Society, under the auspices of ANSI, began the technical evaluation of clearance in 1964, when the Society was not even 10 years old. Release guidelines considered in the early drafts of the clearance standard were based primarily on detection limits achievable by contemporary instrumentation, with the potential doses that may result considering a secondary concern. Interestingly enough, one of these early N13.12 drafts was consistent with the surface contamination limits that were published by the U.S. Atomic Energy Commission as Regulatory Guide 1.86, Termination of Operating Licenses for Nuclear Reactors in 1974.

This standard on the clearance of materials—prepared under the auspices of the Health Physics Society's Standards Committee—was published in August 1999. ANSI N13.12 (1999), "Surface and Volume Radioactivity Standards for Unconditional Release," provides risk-based release criteria and survey methodologies for the unrestricted release of items or materials that may contain residual levels of radioactivity. It contains a primary dose criterion (1 mrem/y) and derived screening levels for groups of similar radionuclides, for both surface and volume contamination.

It is interesting to note that DOE O 458.1 also specifies a clearance criterion of 1 mrem/y for personal property.

QUESTIONS AND PROBLEMS

1. Discuss how the former guidelines compare to the recently issued dose-based criteria. Which categories of radionuclides took a "hit" with the move to dose-based criteria?

2. What is meant by the "bucket and brush" decontamination philosophy?

3. Assume that the site contaminant is processed natural uranium, with uranium activity fractions of 0.485, 0.022, and 0.493, respectively, for U-238, U-235, and U-234. Using the DCGLs in Table 4.3 for U-238, U-235, and U-234, calculate the total uranium DCGL and U-238 DCGL for processed natural uranium.

4. Explain how the $DCGL_W$ for Th-232+C can be equal to 1.1 and 11 pCi/g at the same time, and both be correct.

5. What is the primary dose criterion used in ANSI N13.12 for the clearance of materials?

SUGGESTED READING

NUREG-1757 3-vol. series
NMSS Handbook for Decommissioning Fuel Cycle and Material Licensees
ANSI N13.12
HPS 1999 Summer School Text—"Decommissioning and Restoration of Nuclear Facilities"

CHAPTER 5

Exposure Pathway Modeling

DCGLs and Hazard Assessments

M ODELING TO UNDERSTAND THE fate and transport of contaminants in the environment is an important aspect of decommissioning. Much progress has been made over the past decade in developing computational tools to model contaminant transport through environmental media such as subsurface soil and groundwater. Certainly, advances in geochemistry and hydrology have led to a better understanding of the physical environment. Consider the Advanced Simulation Capability for Environmental Management (ASCEM) project established by the DOE's EM program. ASCEM provides predictive capabilities for the fate and transport of contaminants using an integrated set of tools for advanced visualization, data manipulation, and uncertainty quantification for decision making. Advanced modeling of proposed cleanup activities, such as the Hanford tank farms, will enable performance and risk assessments for cleanup and closure activities across the DOE complex—modeling innovations that result in better cleanup decisions being made by regulators and various D&D stakeholders.

This chapter will acquaint the reader with some of the basics in how dose modeling intersects with decommissioning planning. While this chapter will by no means cover all aspects of dose modeling, it will provide a good start—understanding the importance of modeling exposure pathways and scenarios, determining whether to expend resources on site-specific parameters, and calculating DCGLs and area factors. Precisely because this is such a complex part of decommissioning, it is recommended that the decommissioning team include someone designated as the "modeling expert." This role becomes increasingly important as the scope and complexity of the decommissioning project increase. For starters, the modeling expert should be well versed in both the DandD and RESRAD modeling codes, and their supporting documentation, such as the four-volume NUREG/CR-5512 series, NUREG-1549, and the RESRAD manuals. And more recently, Chapter 5 of NUREG-1757 Volume 2 and Appendix I, "Technical Basis for Site-Specific Dose Modeling Evaluations," present technical details for performing NRC staff evaluations of the licensee's dose modeling (very useful to know what the regulator is looking for). Appendix I, in particular, covers the technical approaches, procedures, criteria, and guidance for compliance demonstration with the dose criteria in 10 CFR Part 20, Subpart E.

Dose modeling is still a nascent activity in the decommissioning planning process. The impetus for dose modeling began in the late 1990s with the NRC's decommissioning rulemaking that provides a dose basis for unrestricted release. As discussed in the previous chapter, regulatory agencies such as the NRC or DOE establish radiation dose standards (e.g., NRC's 25 mrem/y unrestricted release criterion) to assure that residual radioactivity will not result in individuals being exposed to unacceptable levels of radiation and/or radioactive materials. Dose calculations in this context are to potentially exposed members of the public. The NRC applies the potential dose to the average member of the critical group—which is formally defined as "the group of individuals reasonably expected to receive the greatest exposure to residual radioactivity for any applicable set of circumstances" (NRC 1997). Therefore, the average member of the critical group is an individual who is assumed to represent the most likely exposure situation based on prudently conservative exposure assumptions and parameter values within the model assumptions (NRC 1998c).

Another term that may be encountered is the "reasonably maximally exposed" individual, used by EPA in their proposed radiation site cleanup regulation. In that proposed, but subsequently withdrawn, rulemaking, EPA intended to protect (to a level of 15 mrem/y) the reasonably maximally exposed (RME) individual in the population located on or near a previously contaminated site that has been released for public use after undergoing remediation. The RME is defined as the individual receiving the radiation exposure experienced by the 95th percentile and above of the population at a released site (i.e., the upper 5% exposure level for individuals at the site) (draft EPA 40 CFR Part 196).

Regardless of how the potentially exposed public is defined, these radiation dose standards are not measurable quantities. To demonstrate compliance with release criteria, we need to calculate residual radioactivity levels that can be measured and that correspond to the release criteria by the analysis of various pathways and scenarios. These derived levels are called DCGLs, and are presented in terms of surface or volume activity concentrations. DCGLs refer to average levels of radiation or radioactivity above appropriate background levels. DCGLs are applicable to building surfaces and soil areas, which are expressed in units of activity per surface area (typically in dpm/100 cm^2) and in units of activity per unit of mass (typically in pCi/g), respectively.

5.1 SCREENING VERSUS SITE-SPECIFIC: WHEN IS IT TIME TO GO BEYOND THE SCREENING DCGLs?

Simply stated, if screening values are used in the modeling codes, then it is not incumbent upon the licensee to obtain detailed information about the site to support site-specific pathway modeling. However, the NRC does state in NUREG-1700 that the license termination plan contain "justification for the use of the default scenarios and parameters for the DandD code consisting of a statement that no other conditions are reasonably expected to exist at the site except for those incorporated in the default scenarios and modeling assumptions, that would cause a significant increase in the calculated dose." The downside is that DCGLs based on conservative screening values can be quite restrictive (i.e., DCGL values are low)—significantly impacting the final status survey sample size and corresponding survey costs. (Note: It may be worthwhile to review the DQO example that illustrates this site-specific modeling decision in Section 3.2.2.)

According to NUREG-1549, licensees using a screening model should expect to comply with more restrictive DCGLs, but would do so based on their decision not to expend resources for a more realistic dose assessment. For licensees with more complex situations (or for those who choose to employ more realistic analyses), the dose assessment methodology ensures that as more site-specific data are incorporated, the uncertainty is reduced and the resulting DCGL is larger (less restrictive). Therefore, if one opts to provide site-specific parameters for modeling inputs, then it is necessary to characterize the site to determine the appropriate values. While this can be a resource-intensive process, the end result will usually be larger DCGLs and a better understanding and appreciation of the physical nature of the site/facility. This can pay dividends during the MARSSIM survey design, in particular for the strategies involved in using prospective power curves (Chapter 13).

A quick comparison of the DCGL obtained using screening values to expected site contamination levels and/or background is the first step. If the screening DCGL is sufficiently higher than the contamination level expected at your site, use the screening value—why make things difficult? If uranium, thorium, or radium are your "contaminants du jour," then it is very likely that the screening DCGLs are near background—and makes it very difficult and expensive to demonstrate compliance without determining site-specific DCGLs. This is one general category of sites where it is expected that the site-specific modeling will be performed. The good old days of total thorium guidelines at 10 pCi/g, or uranium guidelines around 30 pCi/g are over, times have changed and we now have leaner and meaner DCGLs for these sites, not to mention the MARSSIM approach.

Let us consider the kind of parameters that may be changed based on site-specific parameters if we used the RESRAD model. Assume that we have to develop DCGLs for a parcel of land. The regulator has informed us that the resident farmer scenario will apply. The nature of the contaminant is mixed fission products. The first parameter that could be changed is the size of the contaminated area. The default in RESRAD is 10,000 m^2. Another common change is the depth of contamination—the default is 2 m in RESRAD, but let us say the characterization data suggests that the average depth is about 20 cm. The length of the survey unit (contaminated soil zone) parallel to the aquifer flow may need to be changed due to the size and shape of the actual site. K_d values and other saturation zone characteristics may not be appropriate for the type of soil present. These are simply a few of the types of parameters that may need to be evaluated when considering the use of site-specific DCGLs.

In fact, in some situations, the screening DCGL may be on the order of background variability. Not only should one contemplate the need for site-specific modeling, but the scenario B approach should be reviewed. In any event, the decision to use default DCGLs or site-specific DCGLs is one best solved by the DQO process. The next section provides a more in-depth view of scenarios, pathways, and parameters that are considered in dose modeling.

5.2 EXPOSURE PATHWAY MODELING: SCENARIOS, PATHWAYS, AND PARAMETERS

While the equations used for pathway modeling oftentimes appear quite involved, the concept that they convey is rather straightforward. That is, pathway modeling equations simply provide a mechanism to calculate the expected activity in various environmental media that result from the transport from an initial source term (e.g., soil concentration), as a function of time. For example, given an initial surface activity on building surfaces, how are the potential doses delivered? To determine the dose, the possible exposure pathways must be evaluated—direct radiation, inhalation, and ingestion—as well as the physical parameters used to calculate the transportation of radioactivity for each pathway. In this surface activity example, one of the primary exposure pathways is inhalation, with the resuspension factor being the principal parameter that accounts for the inhalation dose.

In some instances, the D&D contractor might justify the elimination of certain exposure pathways from the modeling effort. For example, let us suppose that the credible scenarios include reindustrialization, recreational, and nearby residential land use. During the process of evaluating pathways for each media type, the D&D contractor determines that the future use of drinking water from a shallow well on-site is not credible. That is, the D&D contractor proposes not to consider the potential dose from residual radioactivity in shallow groundwater on-site because there is not a credible scenario for a member of the critical group to drink this water. Therefore, the D&D contractor supports this decision by documenting that (1) the municipal drinking water supply is of much higher quality than the shallow groundwater on-site, and (2) higher-quality water supplies from area springs and deeper aquifers make it exceedingly unlikely that the difficult-to-access, poor-quality shallow groundwater would ever be tapped. Now of course this action must be approved by the responsible regulatory agency.

In the next few sections, we will explore the various scenarios, exposure pathways, and parameters that are central to assessing the radiation dose from residual radioactivity in different media. There are a number of standard scenarios that have been used over the past decade, and continue to be used as the basis for the popular pathway modeling codes. Each scenario consists of at least one or two prominent exposure pathways, depending of course upon the specific radionuclides that are responsible for delivering the bulk of the dose.

5.2.1 NRC's Policy and Guidance Directive PG-8-08

The NRC prepared the Policy and Guidance Directive PG-8-08, "Scenarios for Assessing Potential Doses Associated with Residual Radioactivity," in May 1994. Its purpose was to promote consistency in the increasing number of dose assessments being performed by the NRC staff, specifically for soil contamination at licensed facilities. These dose assessments were being prepared by licensees in support of their requests for license termination. The NRC wanted to ensure that the same exposure scenarios were being considered for each review.

In a way, PG-8-08 can be thought of as an interim step in the NRC's goal to an integrated dose modeling package that would eventually include functional computer codes (i.e., DandD) to estimate doses to critical populations. For example, the NRC had recently published NUREG/CR-5512, Volume 1, which details the four primary exposure scenarios (NRC 1992b):

- The building renovation scenario (surface contamination)

- The building occupancy scenario (surface contamination)

- The drinking water scenario (groundwater contamination)

- The residential scenario (volume contamination)

But the NRC recognized that computer codes were not yet available to estimate doses from surface contamination (RESRAD-BUILD was under development), and further, viewed the NUREG/CR-5512 scenarios as primarily screening tools that may indicate the need for more detailed analyses. For these stated reasons, and perhaps the fact that NRC staff routinely used the RESRAD code and that each of the scenarios below could be assessed using RESRAD, the NRC defines three exposure scenarios in PG-8-08:

- Scenario A Worker

- Scenario B Resident

- Scenario C Resident farmer

Before we define each of these scenarios, it is necessary to consider the contaminant source term in PG-8-08. Basically, the assumed source term is a homogeneously distributed contamination within an uncovered soil layer of cylindrical shape. The contaminated soil is underlain by an uncontaminated unsaturated zone of some specified thickness, and further below by a saturated zone. Thus, dose can be delivered directly by the external radiation pathway, or indirectly from the migration of radionuclides via resuspension, plant uptake, direct ingestion, or transport to the groundwater. Given the nature of this soil source term, the PG-8-08 scenarios are defined next.

Scenario A can be briefly defined as representing a worker performing light industrial activities in a building that is situated on a larger contaminated soil zone. The worker spends no more than 2000 h on-site and could be potentially exposed to external radiation and inhalation of radioactive material from contaminated soil.

Scenario B represents a homeowner who spends most of the time on-site, but works at an off-site location. This may be thought of as the "Resident Farmer—Light" scenario—in the sense that the resident potentially receives some dose from the contaminated soil, but not nearly as much as one who also works and farms at the same location. The resident may be potentially exposed to external radiation, and inhalation and ingestion of airborne radioactive material.

Scenario C represents the RME individual, someone who both lives and works on-site. The resident farmer lives, works, drinks well water, grows crops, and raises livestock on-site. The resident farmer in a sense maximizes the dose that could be delivered by the contaminated source term, being potentially exposed to the external radiation, and by inhalation and ingestion of airborne radioactive material, by ingesting groundwater produced from beneath the site, and by eating produce grown on the site. This scenario is very similar to the residential scenario defined in NUREG/CR-5512.

The NRC states that these three scenarios adequately bind the range of potential doses to future workers or residents at a formerly licensed site. Scenario A represents the most probable scenario at industrial sites, while scenario B covers the situation where an industrial site is converted back to residential uses. Scenario C provides the most conservatism, as the potential dose assumes that the individuals not only live on the site but also produce a large portion of their food on-site, as well as obtain drinking groundwater from the site.

PG-8-08 provides a number of modeling and parameter assumptions that continue to be widely used. First, the NRC states that they will estimate potential doses associated with these scenarios for up to 1000 y. The NRC explains that estimating doses beyond 1000 y is not worthwhile due to the large uncertainties associated with future conditions. Another interesting aspect of this guidance is the relationship of the size of the contaminated area with the fraction of the diet used in scenario C. The maximum fraction of the on-site diet (assumed to be 50%) is based on a contaminated garden size of 1000 m². PG-8-08 suggests that the fraction of the contaminated diet should be decreased linearly as the size of the contaminated area is decreased from 1000 m². This consideration has important implications when the outdoor survey unit size is being modeled to determine DCGLs.

The parameters referenced in PG-8-08 are largely the default parameters from either the RESRAD code or NUREG/CR-5512. While the default parameters are recommended in the absence of site-specific parameter values, two parameters specifically identified as being candidates for site-specific information are the hydraulic conductivity of the saturated zone and the thickness of the unsaturated zone. The hydraulic conductivity of soil can be defined as the soil's ability to transmit water when submitted to a hydraulic gradient (ANL 1993a). The default values in PG-8-08 for hydraulic conductivity include 10 m/y for the unsaturated zone and 100 m/y for the saturated zone.

5.2.2 NUREG/CR-5512 and NUREG-1549

The foreword to NUREG/CR-5512 states that the intent of the exposure scenarios is to account for the vast majority of the potential future uses of lands and structures, while discounting a small fraction of highly unlikely future-use scenarios. In this manner, the most probable dose via these "standard" scenarios is likely overestimated to a degree, but not as much if these highly unlikely scenarios were considered. To further simplify, the scenarios in NUREG/CR-5512 err on the side of conservatism for the most likely scenarios, while at the same time do not take account of the absolute worst-case scenarios that occur highly infrequently.

How realistic are the exposure scenarios described in NUREG/CR-5512? It is important to understand up-front that the scenarios employed in dose modeling do not purport to calculate actual doses to real people. We used the term "worst case" to describe somewhat farfetched scenarios. That is, we would assume higher than probable use factors—for example, the time that an individual can be exposed to the source term. We would select physical parameters that would maximize the dose, within the reasonableness of expected parameter ranges—and contentedly state (hopefully) something similar to the common report phrase: "even with these very conservative assumptions, the calculated dose is less than 0.01 mrem/y."

We might safely conclude that in many cases, the scenarios used are not altogether realistic.

The NRC in NUREG-1549 (NRC 1998c) provides useful guidance on the pathway analysis process:

- Compile a list of exposure pathways applicable to a particular type of contaminated site. This list should be constructed using the guidance in NUREG/CR-5512.

- Categorize the general media types that are contaminated (e.g., soil, surface contamination on building surfaces, groundwater).

- Screen out pathways for each media type that does not apply to the site.

- Identify the physical processes pertinent to the pathways for the site.

- Separate the list of exposure pathways into unique pairs of exposure media—source to groundwater, surface contamination to airborne contamination, and so on. Determine the physical processes that are relevant for each exposure media pair and combine the processes with the pathway links.

- Reassemble exposure pathways for each source type, using the exposure media pairs as building blocks, therefore associating all the physical processes identified with the individual pairs with the pathway links.

No wonder we use computer models—there is insufficient time to do this all by hand! But seriously, this also points out the situation that if you are not aware of what the scenario and pathways are modeling, you may be getting entirely unrealistic results. Modeling parameters are discussed in the next section. Historically, conservative assumptions have been part and parcel of the dose assessment process, due in no small measure to the uncertainty regarding appropriate parameter values.

Let us take a closer look at the exposure pathway equations used to calculate the radiation dose to the average member of the critical group from building surface activity. Perhaps by performing a hand calculation of the radiation dose that results from a certain residual radioactivity level, we can better appreciate the operations of a modeling code.

Suppose that the surface is contaminated with Am-241 at an average level of 27 dpm/100 cm². This value is chosen because we know that

27 dpm/100 cm² of Am-241 equates to 25 mrem/y from NUREG/CR-5512, Volume 3 (Table 4.4 of this book). We will attempt to calculate the radiation dose from this residual radioactivity by hand, and compare the result to the DandD, Version 1 modeling code. The DandD, Version 1 should result in a $DCGL_W$ of 27 dpm/100 cm² for Am-241 as well, since the values in NUREG/CR-5512, Volume 3 were obtained from the DandD, Version 1 model in the first place.

The dose from Am-241 on building surfaces is expected to be predominantly from the inhalation pathway given the nature of radiation emissions from this radionuclide (alpha and low-energy gamma emissions). Therefore, the inhalation dose will be calculated first from the surface activity level, using the following pathway equation:

$$\text{Dose}_{\text{inhalation}} = (A_s)(\text{RF})(\text{breathing rate})(\text{time})(\text{DCF})$$

where A_s is the surface activity level, RF is the resuspension factor, and DCF is the dose conversion factor for inhalation. We will use the same RF and breathing rate as that used in the DandD model, 1.42E–5 m⁻¹ and 1.4 m³/h, respectively. Table 2.1 of Federal Guidance Report No. 11 (EPA 1988) provides the inhalation DCF for Am-241 as 0.444 mrem/pCi. The DandD model assumes that the individual is exposed to this surface activity for 97.46 effective 24-h days (from 9-h work days, 5 days/week, for 52 weeks). Thus, the inhalation dose from the Am-241 on building surfaces is calculated as

$$\text{Dose}_{\text{inh,Am}} = \left(27 \frac{\text{dpm}}{100\,\text{cm}^2}\right)\left(1.42\,\text{E}{-}5\,\text{m}^{-1}\right)\left(1.4 \frac{\text{m}^3}{\text{h}}\right)\left(97.5 \frac{\text{day}}{\text{y}}\right)$$
$$\times \left(\frac{0.444\,\text{mrem}}{\text{pCi}}\right)\left(\frac{10{,}000\,\text{cm}^2}{1\,\text{m}^2}\right)\left(24 \frac{\text{h}}{\text{day}}\right)\left(\frac{1\,\text{pCi}}{2.22\,\text{dpm}}\right)$$
$$= 24.86 \frac{\text{mrem}}{\text{y}}$$

While the radiation dose from the inhalation pathway is extremely close to the target of 25 mrem/y, we need to determine the ingestion dose pathway contribution. The ingestion pathway dose can be calculated by

$$\text{Dose}_{\text{ingestion}} = (A_s)(\text{ingestion rate})(\text{time})(\text{DCF})$$

where the ingestion rate is defined as the effective transfer rate for the ingestion of removable surface contamination transferred from surfaces, to hands, then to mouth in units of m²/h (NUREG/CR-5512, Volume 1). The DandD model assumes an ingestion rate of 1.11E–5 m²/h. The DCF for ingestion can be obtained from Table 2.2 of Federal Guidance Report No. 11 (EPA 1988). The ingestion DCF for Am-241 is 3.64E–3 mrem/pCi. The ingestion dose from the Am-241 surface contamination is calculated as

$$
\begin{aligned}
Dose_{inh,Am} &= \left(27\frac{dpm}{100\,cm^2}\right)\left(1.11E-5\frac{m^2}{h}\right)\left(97.5\frac{day}{y}\right)\left(\frac{3.64E-3\,mrem}{pCi}\right) \\
&\times \left(\frac{10,000\,cm^2}{1m^2}\right)\left(24\frac{h}{day}\right)\left(\frac{1\,pCi}{2.22\,dpm}\right) = 0.11\frac{mrem}{y}
\end{aligned}
$$

Last, and indeed least, the external radiation dose from Am-241 is expected to be much smaller than that for either the inhalation or ingestion pathways due to the dose significance of the alpha emission in both of these pathways, versus the minimal dose significance from the 60 keV gamma emission. Federal Guidance Report No. 12, Table III.3 (EPA 1993) provides dose coefficients for exposure to contaminated ground surfaces. The DCF for Am-241 is 3.21E4 mrem/y per µCi/cm². The external dose for Am-241 is the product of the surface activity level, exposure time and DCF, and is given by

$$
\begin{aligned}
Dose_{inh,Am} &= \left(27\frac{dpm}{100\,cm^2}\right)\left(\frac{3.21E4\,mrem/y}{\mu Ci/cm^2}\right) \\
&\times \left(\frac{1\,y}{365\,day}\right)\left(97.5\frac{day}{y}\right)\left(\frac{1\,\mu Ci}{2.22E6\,dpm}\right) = 1.04E-3\frac{mrem}{y}
\end{aligned}
$$

Therefore, the total dose from these three pathways is 24.97 mrem/y. As expected, the inhalation pathway comprises 99.56% of the total radiation dose. The hand calculation certainly confirmed the result that the DCGL$_W$ based on 25 mrem/y is 27 dpm/100 cm². Hopefully, this straightforward exercise provides added confidence in the nature and use of modeling codes.

5.2.3 Pathway Modeling Parameters

There are two key documents that the designated "modeling expert" should become familiar with regarding modeling parameters: (1) Data Collection Handbook to Support Modeling the Impacts of Radioactive Material in Soil, ANL/EAIS-8 (ANL 1993a) and (2) Residual Radioactive

Contamination from Decommissioning, Parameter Analysis, NUREG/CR-5512, Volume 3 (NRC 1999a). The Data Collection Handbook is a "how to" book and contains measurement methodologies for determining parameter values for the RESRAD code. The handbook provides parameter definitions, typical ranges and variations, and measurement methodologies for more than 50 modeling parameters. Examples of parameters include soil density, hydraulic conductivity and gradient, inhalation rate, thickness of the contaminated zone, and the fraction of time spent indoors on-site. It would be wise to have this handbook nearby when planning and designing characterization activities that include the determination of site-specific parameters.

NUREG/CR-5512, Volume 3 recognizes three general types of modeling parameters: behavioral, metabolic, and physical parameters. Behavioral parameters can be defined as those parameters that depend on the characteristics of the critical group. For example, behavioral parameters include the time that individuals spend in various locations in on-site buildings and land areas, area of land used for gardening, and consumption rates for fruit, grains, seafood, milk, and water. The only metabolic parameter considered in Volume 3 is the breathing rate, which is usually a function of being either indoors (light activity) or outdoors (moderate activity or gardening). Physical parameters describe the physical characteristics of the site and can be determined by site-specific data collection or by citing relevant data in the literature, such as the annual rainfall amounts at the D&D site. Common examples of physical parameters include the resuspension factor in a building, thickness of the soil contamination layer, crop yields, moisture content of soil, and soil density.

The reader should recognize that there is certainly a good deal of uncertainty associated with each of these parameter values. Historically, the average parameter value was used in modeling codes. However, over the past decade or so there has been a concerted effort by the NRC and DOE to add probabilistic sampling routines to their modeling codes. For example, the NRC contracted with ANL to provide probabilistic parameter selection for RESRAD, Version 6. Specifically, parameter values are sampled from a probability distribution that describes the variability in the parameter value. Again, the "modeling expert" should study NUREG/CR-5512, Volume 3, as it goes into great detail on this subject of probabilistic parameter distributions. A valuable strategy is to determine which parameters for a specified scenario are important—that is, sensitive to small changes in parameter values—as this is a critical input to the DQO process of evaluating which parameters might be worth the effort of determining site-specific values.

Let us focus on one modeling parameter to better understand the importance of parameter value selection and its impact on dose. Our example will be the distribution coefficient, K_d, which is frequently cited as an example of a parameter value that varies significantly from one site to another. Because of its variability, it may be worthwhile to perform site-specific assessments of the distribution coefficient. As defined in the RESRAD manual, the distribution coefficient (in units of mL/g) is the ratio of the solute concentration in soil to that in solution under equilibrium conditions (ANL 1993b). The physical processes involved include leachability and solubility, and account for the mechanism in which the radioactive material is desorbed from the particulate matter it was initially adhered to and then enters the soil solution. The K_d is important in the context of predicting the amount of radioactive material that moves through the soil and enters the groundwater. Therefore, the importance of this parameter is directly tied to the groundwater pathway, and its importance in the overall delivery of dose. Simply stated, the easier it is for the radioactive material to move through the soil and enter the groundwater, the greater the opportunity for the groundwater pathway to be a dominant dose contributor. High values of K_d mean that the radionuclide is relatively immobile in the soil, while low values of K_d indicate that the radionuclide readily moves through soil into the groundwater. Both the physical and chemical soil characteristics as well as the chemical species of the radionuclide can affect the K_d value.

5.3 MODELING CODES

RESRAD, and DandD to a lesser degree, are currently the most popular choices for dose modeling. These codes will be used exhaustively to determine DCGLs for decommissioning projects. Over the years there has been a discussion concerning a multiagency effort to produce a modeling code, or at least to recommend a code—similar to the Federal agencies behind MARSSIM, MARLAP, and MARSAME. It should also be recognized that the results from running these modeling codes are version-specific. Given that these codes are frequently revised, it is necessary to check the version of the modeling code, and state it clearly when reporting the results of the modeling runs.

In this section, we will attempt to provide an overview of each of these codes, specifically pointing out some of the major differences between the two codes. Perhaps the best document to consult concerning the differences between RESRAD and DandD is an NRC document, NUREG/ CR-5512, Volume 4 (NRC 1999b). In fact, this NUREG states that the fundamental difference between the two codes is that RESRAD is a general-purpose environmental dose assessment model, while DandD is specifically designed to model the four scenarios described in NUREG/ CR-5512, Volume 1. Undoubtedly, the most significant similarity between these two codes is that they are both free!

Before we get concerned with the details of these modeling codes, it is important to remember that their output is only as good as the input data. Certainly, the determination of DCGLs requires defensible data on the nature of the radioactive source term. The characterization survey is frequently the source of this information. Characterization data must provide an adequate representation of the radionuclide mixture and its distribution at the site. Therefore, it is critical that the DQOs for the characterization survey account for the specific nature of the contaminants present. For example, what is the expected depth of the radionuclide contamination? What is the average enrichment for the uranium contamination? One tool commonly used to develop the source term, and radionuclide mixture, is the waste disposal manifests based on requirements of 10 CFR Part 61. That is, for shipping purposes, it is necessary for the operational facilities to know and report the waste profile—for example, nature and quantity of each radionuclide present. The same characterization data used for 10 CFR Part 61 might be able to be used to develop the source term for modeling purposes. Chapter 6 provides an example of how the waste characterization data might be used in this context.

5.3.1 RESRAD and RESRAD-BUILD Models

RESRAD is actually a family of codes, with the RESRAD code for land areas being the centerpiece. It is highly recommended that those genuinely interested in these modeling codes (particularly for the "modeling experts") participate in a course offered by the model developers at Argonne National Laboratory. We will now introduce some of the aspects of the RESRAD codes. Later sections in this chapter will go into further detail as we use these codes to develop DCGLs and area factors.

First, it may be helpful to have a little background perspective on the use of RESRAD at NRC-licensed sites. The NRC notes in NUREG-1757, Volume 2 that as compared to the DandD screening model, RESRAD and RESRAD-BUILD are codes used for site-specific analyses. They caution that because these codes have predefined conceptual models, it is important for the licensee to "demonstrate that key site features and conditions are consistent with the modeling assumptions within the codes or, where they are not consistent, the analysis may not result in an underestimation of potential doses" (NRC 2006a). Furthermore, specific site features and conditions that may be incompatible with this generic representation are listed in Appendix I, Table I.6 (NRC 2006a). The NRC advises that appropriate justification be provided for using the computer code when any of these site features or conditions are present—for example, sites with highly heterogeneous radioactivity, wastes other than soils (e.g., slags and equipment), and sites that have chemicals or a chemical environment that could facilitate radionuclide transport.

The primary scenario in RESRAD-BUILD is that of the office worker. This is considered to be a long-term scenario, which involves direct radiation, inhalation, and ingestion exposure pathways. This modeling code is certainly more complex than the corresponding scenario in the DandD code, but one cannot help but wonder if the complexity offered is really useful. With RESRAD-BUILD, the building can be divided into three rooms, along with controls on ventilation between the rooms, and with the outside air. Of course, this complexity helps with the movement of loose contamination that can become airborne and therefore move throughout the rooms of the building. But is this realistic? The radiological conditions at many building D&D projects at the time of final status survey should include very little loose or airborne contamination.

The RESRAD-BUILD code also provides flexibility through the use of a coordinate system to establish the location of the contaminant source and the receptor individual. Again, is this a useful feature? How often can one be expected to know with any accuracy the future receptor locations? Perhaps, one can run the code several times at different locations to evaluate the sensitivity of various receptor locations, assuming that the source term location is relatively fixed. This might be useful in setting up the room for future office space, particularly the locations of desks and other work station locations, and high traffic areas. In contrast, the DandD code offers a straightforward source term that is modeled as a uniform plane source. In this situation, the receptor location does not matter, and the receptor dose is the same at every location.

Finally, in RESRAD-BUILD, not only can the user provide the location and number of discrete sources, but can also define certain source characteristics that impact the receptor dose. These include the removal fraction, time for source removal, release fraction of material to the indoor air, and the direct ingestion rate. Another plus for RESRAD-BUILD is that the size of the contaminated area can be varied, which allows the calculation of area factors—something that is either impossible or very difficult for DandD.

The RESRAD code has been used by many D&D professionals for more than a decade. The principal application of RESRAD is to calculate the dose rate to a receptor from a specified source term, considering a number of exposure pathways. The pathways include external gamma, inhalation, agricultural (plant, meat, and milk ingestion), soil ingestion, aquatic foods, drinking water, and radon. Each of these pathways can be turned off, provided that sufficient justification exists for not considering a specific exposure pathway.

A somewhat minor difference that is worth to be noted is that the RESRAD codes provide the modeling results in units of mrem/y that represents an instantaneous dose at a specific time. DandD on the other hand calculates a 1-y integrated dose—probably to be consistent with the NRC's decommissioning rulemaking that provides for a 25 mrem/y annual dose for unrestricted release.

5.3.2 DandD Model

The NRC released its screening computer code DandD, Version 1.0 on August 20, 1998, and stated that the DandD code, when used with default parameters, is an acceptable method for licensees to calculate screening values and to demonstrate compliance with the unrestricted release criteria. It should not be overlooked that the NRC intended the DandD model to be used for screening calculations only. That is, the NRC fully anticipated that pathway analysis/dose assessment codes other than DandD would more than likely be necessary for some D&D sites.

Appendix I of NUREG-1757, Volume 2 provides a comprehensive overview of the DandD model. This model was specifically developed for screening analyses to provide a conservative representation of processes and conditions expected for many sites. Appendix I also addresses the excessive conservatism in DandD, Version 1.0, explaining that it was created as a deterministic screening code, with a single set of default parameters to demonstrate compliance with the dose limit. NRC notes examples of excessive conservatism, such as in the resuspension factor; NUREG-1720 was developed as a technical basis document for revisiting the RF (NRC 2002b).

DandD, Version 2.0 was produced to provide for a probabilistic treatment of dose assessments. Simply stated, while deterministic analysis uses single-parameter values for every variable in the code, a probabilistic approach assigns parameter distributions to certain variables, and the code randomly selects the values for each variable from the parameter distribution each time it calculates the dose (NRC 2006a). NUREG/CR-5512, Volume 2 titled "Residual Radioactive Contamination from Decommissioning" is the User's Manual DandD, Version 2.1 (2001). Currently, the only acceptable generic screening code is DandD, Version 2.

The DandD model has four possible scenarios that can be run. These include building occupancy and building renovation for surface contamination on building interiors, and residential occupancy and drinking water scenarios for land areas (Figure 5.1). The residential scenario for Cs-137 can be used to illustrate some important features of the DandD code (Version 1.0). Let us assume that a unit concentration of 1 pCi/g was entered into the code and that all the default parameter values were used. The DandD model estimates a maximum dose of 28.3 mrem/y that occurs at 6 y post D&D. This results in a $DCGL_W$ of 0.88 pCi/g based on 25 mrem/y (the next section will show this in more detail), which seems pretty conservative. Upon closer inspection, the aquatic pathway is responsible for 84% of the dose, while the external

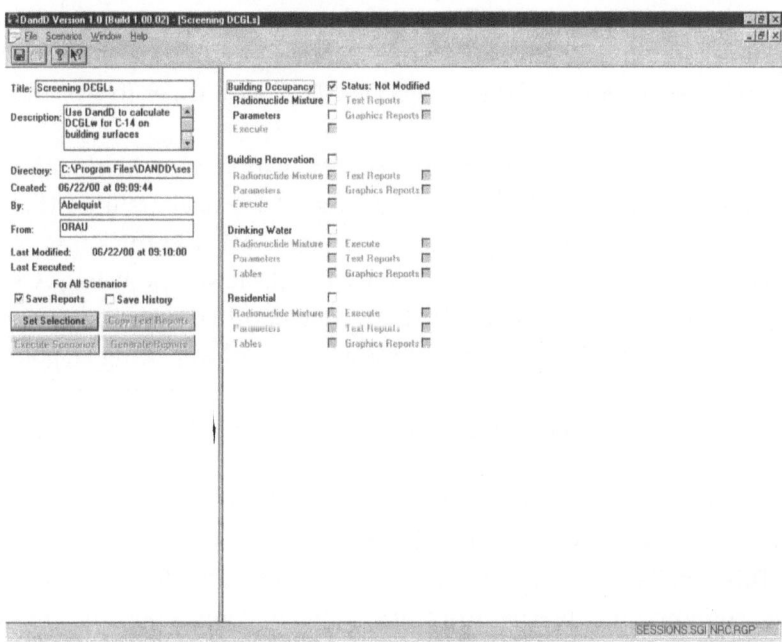

FIGURE 5.1   Main data entry screen for the DandD, Version 1.0 model.

radiation pathway is less that 0.5%. This seems somewhat unexpected because Cs-137 is usually expected to deliver most of its dose through the external radiation pathway.

The dose breakdown shows that 99.51% is from Cs-137 and 0.49% is from its daughter Ba-137 m. Remember that Ba-137 m decays via the popular 662 keV gamma emission, and that Cs-137 is a beta emitter. The aquatic pathway includes the transfer of radioactivity from the soil to groundwater to fish. The dose delivered by eating fish would be dominated by the beta emitted from Cs-137, which delivers the bulk of the internal dose, rather than the Ba-137 m gamma.

Again, upon closer inspection, we see that the reason the aquatic pathway is the driver for Cs-137 is likely due to the size of the unsaturated zone. DandD assumes that it is 1.23 m thick. That is, it does not take much time at all for the soil contamination to reach the groundwater, only having to travel 1.23 m. The code provides an unsaturated zone thickness range between 0.3 and 320 m. It is expected that thicker unsaturated zones would tend to reduce the importance of the aquatic pathway, while stressing the external radiation pathway. To evaluate this conjecture, let us run the DandD code with an unsaturated zone of 20 m. The peak dose now occurs during the first year and is only 2.21 mrem/y, which translates into a $DCGL_W$ of 11.3 pCi/g, much less conservative than the default. An understanding of the way DandD models groundwater flow is helpful to the explanation of the higher DCGL allowed (refer to Appendix I, NUREG-1757, Volume 2).

Apparently, DandD averages the contamination over the thickness of the unsaturated zone. That is, the groundwater model assumes that the average contamination exists instantaneously over the thickness of the unsaturated zone. Thus, the thicker the zone, the lower the average Cs-137 concentration that is considered to be in contact with the groundwater. The thicker unsaturated zone results in the aquatic pathway being responsible for only 6.93% of the dose, while the external radiation pathway soars to 66% of the dose. This is also consistent with the dose breakdown between Cs-137 and Ba-137 m, which now has Cs-137 at 33.9% and Ba-137 m at 66.1%—which makes sense considering that the external radiation pathway is based on the gamma from Ba-137 m.

This detailed example of Cs-137 points out both the importance of understanding the internal workings of the modeling code and the considerable importance of site-specific parameters. We will now turn to the use of modeling codes to calculate DCGLs and area factors.

5.4 DETERMINATION OF DCGLs AND AREA FACTORS

In this section, we will provide a number of illustrative examples for calculating DCGLs and area factors. These examples cover the use of the DandD, RESRAD, and RESRAD-BUILD modeling codes. Through these examples we will see which scenario drives the determination of the DCGL in particular cases, and when the radiation dose reaches a maximum.

5.4.1 Dose Modeling to Obtain DCGLs

The $DCGL_W$, based on pathway modeling, is the uniform residual radioactivity concentration level within a survey unit that corresponds to the release criterion. The $DCGL_{EMC}$ is the residual radioactivity concentration present in smaller areas of elevated activity (i.e., hot spots) that also corresponds to the same release criterion. The survey unit sizes selected should be generally consistent with the size of contaminated areas used in the modeling to obtain the $DCGL_W$. It can be shown that the $DCGL_W$ varies with contaminated area sizes from 2000 to 10,000 m^2, and how Co-60 is insensitive to area changes but that radionuclides that deliver dose primarily by total inventory (e.g., uranium, C-14, etc.) are affected. A reasonable compromise for these radionuclides might be to base the $DCGL_W$ on the largest anticipated survey unit size. This may be a bit conservative, but likely more desirable than having to account for a number of different DCGLs, each based on a different contaminated area size.

Dose assessments to the potentially exposed population using one of the computer models discussed previously usually begins by calculating the dose due to unit activity on building surfaces (1 dpm/100 cm^2) or in soil (1 pCi/g). The $DCGL_W$ based on a particular dose criterion, say 25 mrem/y is determined by direct ratio. For example, assume that the dose from 1 pCi/g of Cs-137 using RESRAD, with default parameters, was 1.76 mrem/y. Then the DCGL based on 25 mrem/y is simply 25 mrem/y divided by 1.76 mrem/y per pCi/g, or 14 pCi/g.

Lastly, there is a specific $DCGL_{EMC}$ for each particular hot spot area— for example, if the hot spot area for a particular radionuclide is 10 m^2, the $DCGL_{EMC}$ may be 32 pCi/g, and if the hot spot for the same radionuclide was now confined to only 3 m^2, the $DCGL_{EMC}$ may be 85 pCi/g (note that the smaller the size of the hot spot area, the higher the radionuclide concentration may be that equates to the release criterion). This increase in the allowable concentration in the smaller area is called the area factor. Again, dose modeling is used to determine the magnitude of these area factors as a function of the contaminated area size—Chapter 11 goes into significant detail on hot spot assessment and the calculation of DCGL$_{EMC}$. We will now begin our tour of the various modeling codes that can be used to determine DCGLs.

5.4.1.1 DCGLs Using the DandD Model

First of all, allow me to build your confidence by acknowledging that the DandD model is very simple to use. We will take a detailed look at how the DandD, Version 1.0 model can be used to determine DCGLs, carefully describing each step in the modeling process. Specifically, we will determine the DCGL$_W$ for C-14. The first step is to create a new session from the File button on the opening DandD screen. Click "OK" and on the screen shown in Figure 5.2, ensure that only Building Occupancy is active.

Next click the Radionuclide Mixture box under Building Occupancy—click on C-14.

Now click on the Execute box, which is found under Parameters and Radionuclide Mixture. The text message indicates that the modeling run has been completed. Click "OK" and then "Done." Now click on "Text Reports." We will look at the NRC Report, so click on "View Report." The NRC Report for this run is ready for inspection (Figure 5.3).

The maximum annual dose from 1 dpm/100 cm^2 is reported as 6.83E−6 mrem. Therefore, the DCGL$_W$ can be calculated via a direct ratio to determine the surface activity level that equates to 25 mrem/y

$$DCGL_W = \frac{25\,mrem/y}{(6.83E{-}6\,mrem/y)/(1\,dpm/100\,cm^2)} = 3.66E6\frac{dpm}{100\,cm^2}$$

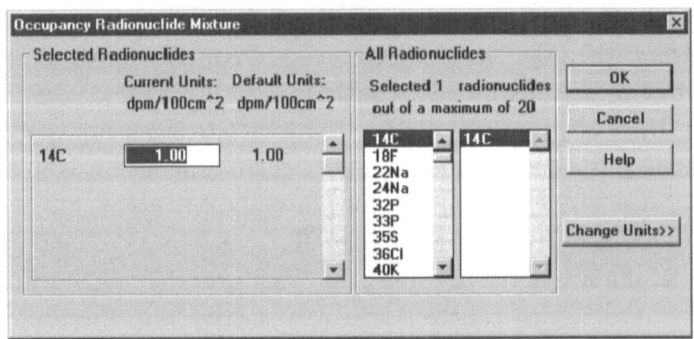

FIGURE 5.2 Entering the radionuclide mixture for the building occupancy scenario.

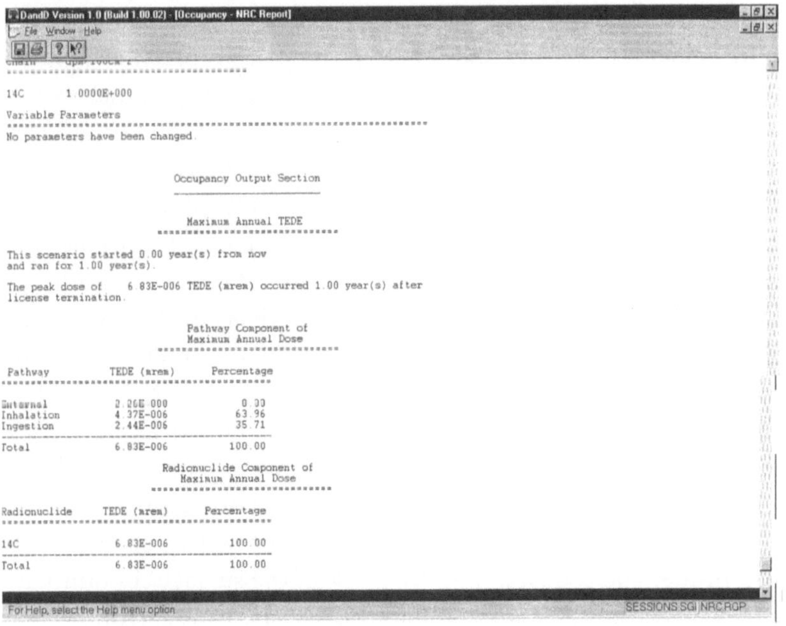

FIGURE 5.3   NRC report from the DandD, Version 1.0 modeling run.

This result is consistent with the value of NRC published in the November 18, 1998 Federal Register Notice of 3.7E6 dpm/100 cm^2.

To consider another example, we can calculate the DCGL$_W$ for Sr-90 (or some other radionuclide) on building surfaces, discussing some of the modeling assumptions used in the process. As before, the process starts by inputting unit surface activity (1 dpm/100 cm^2) for the radionuclide of concern. In this case, one has to carefully understand how DandD handles decay chains. Specifically, Sr-90 decays to an energetic beta-emitting decay product, Y-90, which has a 64-h half-life. Surprisingly, DandD only assumes that progeny with half-lives less than 9 h are in equilibrium with its parent. So, in this case, 1 dpm/100 cm^2 is entered for both Sr-90 and Y-90. (Note: RESRAD considers any progeny with a half-life shorter than six months to be in equilibrium with the parent radionuclide. This makes much more sense for most D&D situations.)

For this example let us use the default parameters used by the code. Clicking on the DandD Application Help, the values of building occupancy values can be viewed. A major parameter for the building occupancy is the time spent in the building. The DandD model assumes a building occupancy of 9 h/day for 52 weeks, which equates to 97.5 effective 24-h days. The default breathing rate is 1.40 m^3/h, which is pretty standard.

One of the most significant building occupancy parameters is the resuspension factor—which relates the surface contamination level to the airborne contamination level. This factor is a highly sensitive parameter that impacts the inhalation dose calculation. It is the fraction of the surface activity that becomes airborne and ultimately respirable, and is simply the ratio of the airborne concentration of contamination to the surface concentration of contamination. The RF is affected by a number of physical factors that include type of disturbance, intensity of disturbance, time since deposition, nature of the surface, particle size distribution, climatic conditions, type of deposition, chemical properties of the contaminant, surface chemistry, and building geometry and physical characteristics (NRC 2002b). The resuspension factor in DandD is 1.42E–5 m^{-1}, but this parameter can easily vary by two or more orders of magnitude. The code assumes that 10% of the surface contamination is removable, and this is worked into the resuspension factor.

The DandD code breaks the total dose delivered into individual pathways that deliver dose. For surface activity on building surfaces these include direct radiation, inhalation, and inadvertent ingestion. The code, at least for the building occupancy scenario, provides the highest dose that occurs in the first year. It seems that the other scenarios in DandD provide the highest dose, averaged over a given year that occurs over a 1000 y time frame. (Note: This is consistent with the NRC's decommissioning rulemaking time frame and is in contrast to RESRAD's determination of instantaneous doses in mrem/y at a specific time.)

Output from Sr-90 run—based on unit activity input
Peak dose in 1 y: 2.89E–3 mrem/y (based on DandD, Version 1.0), which is summarized by the pathway

External	0.27%
Inhalation	93.60%
Ingestion	6.13%

The dose ratio between the two radionuclides is given as

Sr-90	98.71%
Y-90	1.29%

Finally, the bottom line can be determined—that is, what is the $DCGL_W$? This is simply calculated by dividing the dose rate (in mrem/y) for the unit activity into 25 mrem/y:

$$DCGL_W = \frac{25\,mrem/y}{(2.89\,E-3\,mrem/y)/(1\,dpm/100\,cm^2)} = 8700\,\frac{dpm}{100\,cm^2}$$

As you can see, the calculation of DCGL screening values is rather straightforward.

One issue with the DandD, Version 1.0 that has received much criticism is the extreme conservatism employed in the screening values for alpha emissions. The derived surface activity levels for most alpha-emitting radionuclides calculated using DandD, Version 1.0 were unrealistically low, and nearly impossible to measure with reasonable counting times. For example, the screening concentrations equivalent to 25 mrem/y for Th-232, U-238, and Am-241 resulted in DCGLs of 7.3, 101, and 27 dpm/100 cm², respectively (refer to Table 4.4).

Let us take a look at the results for Th-232 screening DCGLs using DandD, Version 1.0.

Output from Th-232 (in equilibrium with progeny) run—based on unit activity input

Peak dose in 1 y: 4.18 mrem/y (based on DandD, Version 1.0), which is summarized by the pathway

External	0.08%
Inhalation	99.78%
Ingestion	0.14%

The code interprets the selection of "Th-232+C" as the entire Th-232 series. Ninety-nine percent of the total dose is delivered by two radionuclides in the series—Th-232 (82%) and Th-228 (17%). This is consistent with the fraction of dose delivered by the inhalation pathway.

Finally, the $DCGL_W$ can be calculated by dividing the dose rate (in mrem/y) for the unit activity into 25 mrem/y:

$$DCGL_W = \frac{25\,mrem/y}{(4.18\,mrem/y)/(1\,dpm/100\,cm^2)} = 6\,\frac{dpm}{100\,cm^2}$$

Yes, 6 dpm/100 cm²; this is not a typo! Obviously, this is not readily measurable using conventionally available instrumentation. In all fairness though, this is the DCGL$_W$ for Th-232 when it is in equilibrium with its progeny, which includes six alpha emitters. Therefore, the DCGL$_W$ in terms of alpha emissions is actually 6 times 6 dpm/100 cm², or 36 dpm/100 cm². Similarly, one could keep the DCGL$_W$ at 6 dpm/100 cm², and recognize that the instrument efficiency would be calculated based on six alphas. In other words, if the total alpha efficiency for a single alpha emitter is 10%, then the total alpha efficiency in this situation is 60%. This issue is further examined in Chapters 6 and 16.

The principal reason for this ridiculously low DCGL$_W$ value is the default parameter for the resuspension factor. That is, the level of dose received from a unit surface activity for alpha emitters is fundamentally related to the inhalation pathway, which of course depends on how much activity gets into the air (resuspension). Therefore, resuspension factors are very important parameters in the calculation of building surface DCGLs, and particularly so for alpha emitters because they impart the bulk of their dose via the inhalation pathway.

The DandD code was run with Pu-239 as the contaminant and the resuspension factor was varied to illustrate its effect on DCGLs for alpha emitters on building surfaces. The default resuspension factor in DandD, Version 1.0 is 1.42E–5 m⁻¹. The Pu-239 DCGL$_W$ based on 25 mrem/y using the default resuspension factor is 28 dpm/100 cm². The code also indicated that the inhalation pathway was responsible for delivering 99.5% of the dose.

The resuspension factor was then increased by 10% to 1.56E–5 m⁻¹, and the DandD code yielded a Pu-239 DCGL$_W$ of 10% lower (25 dpm/100 cm²). Thus, a 10% increase in the resuspension factor resulted in a 10% decrease in the DCGL. To further evaluate the relationship of the resuspension factor to the Pu-239 DCGL$_W$, the code was run with the resuspension factor both increased and decreased by a factor of 10. The results were as follows:

Resuspension Factor	Pu-239 DCGL$_W$
1.42E–4 m⁻¹	2.8 dpm/100 cm²
1.42E–5 m⁻¹ (default)	28 dpm/100 cm²
1.42E–6 m⁻¹	270 dpm/100 cm²

The evidence is convincing; there is a strong direct correlation between the resuspension factor and the Pu-239 DCGL$_W$. This should be incentive enough to seriously consider the effort involved in determining a site-specific resuspension factor. Again, the underlying reason for the strong correlation is because of the paramount role of the inhalation pathway for delivering an alpha emitter's dose. Lest there be any doubt concerning this point, the resuspension factor can also be varied for a known "external radiation pathway radionuclide," Cs-137.

The DandD code was run for Cs-137 using the default resuspension factor; the $DCGL_W$ based on 25 mrem/y was 28,000 dpm/100 cm² (only coincidental that it is exactly 1000 times greater than the Pu-239 $DCGL_W$). The DandD code provided the following pathway components of dose:

External radiation	86.1%
Inhalation	7.4%
Ingestion	6.5%

The resuspension factor was then increased by a factor of 10, to 1.42E–4 m⁻¹. The DandD code yielded a lower Cs-137 $DCGL_W$, not close to the factor of 10 reduction as seen for the Pu-239 DCGL. Rather, the Cs-137 $DCGL_W$ was reduced to 16,800 dpm/100 cm² (dropped by a factor of 1.7). Interestingly, with this increase in the resuspension factor, the rising importance of the inhalation pathway is evident (see the new pathway components of dose):

External radiation	51.7%
Inhalation	44.4%
Ingestion	3.9%

To address these challenges, NRC reevaluated the resuspension factor and identified research data that illustrated the variability of the resuspension factor. The NRC staff analyzed the available literature data on the RF and considered more realistic assumptions based on two facilities undergoing decommissioning. On the basis of this analysis and reevaluation, the NRC published NUREG-1720 that recommended using an RF value of 10^{-6} m⁻¹ in the screening analysis of the inhalation dose calculation for the building occupancy scenario (NRC 2002b). However, the NRC also cautions that the publication of NUREG-1720 should not be construed as approval of, or agreement with, the information contained in the report.

In summary, the resuspension factor is a critical parameter in dose modeling because of its paramount role in the determination of inhalation dose. Individual radionuclides are impacted to varying degrees by the resuspension factor, but none greater than alpha emitters.

5.4.1.2 DCGLs Using the RESRAD-BUILD Model

RESRAD-BUILD can also be used to generate surface activity DCGLs. Version 2.37 was used to derive DCGLs for Cs-137 on building surfaces. The default room size in this model is 36 m^2, which may be considered somewhat small for many survey unit sizes. For the generation of DCGLs, this value was increased to 100 m^2. The source contamination area is also assumed to be 100 m^2. As mentioned in the previous section, RESRAD-BUILD allows the user to specify both the source and dose receptor locations. It is conservative to establish the receptor at the center of the source; to facilitate this, the source coordinates were established at 1, 1 m, and 0 (these are x, y, and z coordinates), while the receptor was positioned at 1, 1, and 1 m—effectively 1 m above the center of the source. RESRAD-BUILD permits volume, area, line, and point source geometries; the source was modeled as an area in this calculation. The input activity was 1 pCi/m^2, which equals 2.22E–2 dpm/100 cm^2. The model was then run and the dose rate evaluated at an initial time 0 y and at 1 y. Remember that this is one of the fundamental differences between DandD and RESRAD— DandD provides doses averaged or integrated over 1 y, while RESRAD reports instantaneous dose rates.

Output screen from RESRAD-BUILD that summarizes source information

Source: 1

Location: Room: 1 x: 1.00 y: 1.00 z: 0.00 (m)

Geometry: Type: Area Area:1.00E+02 (m^2) Direction: z

Pathway:

Direct Ingestion Rate: 0.000E+00 (h^{-1})

Fraction released to air: 1.000E–01

Removable fraction: 5.000E–01

Time to Remove: 3.650E+02 (day)

Contamination: Nuclide Concentration (pCi/m^2)

Cs-137 1.000E+00

The dose rate for the given input of Cs-137 at time $t = 0$ was 1.25E–5 mrem/y. The dose rate at time $t = 1$ y was less than this dose rate; therefore, the DCGL$_W$ will be conservatively calculated at the initial evaluation time. This is shown in the following equation:

$$DCGL_W = \frac{25\,mrem/y}{(1.25E-5\,mrem/y)/(2.22E-2\,dpm/100\,cm^2)}$$
$$= 44,000\,\frac{dpm}{100\,cm^2}$$

For comparison to DandD, RESRAD-BUILD was used to generate the surface activity DCGL$_W$ for Pu-239. Again, the default room size was changed to 100 m², as was the source contamination area. Source and dose receptor locations were as described above. The input activity was 1 pCi/m (2.22E–2 dpm/100 cm²), and the model was run. The resulting dose rate was 4.11E–3 mrem/y at $t = 0$, which translates to a Pu-239 DCGL$_W$ based on 25 mrem/y of 135 dpm/100 cm². The RESRAD-BUILD DCGL$_W$ for Pu-239 is nearly a factor of 5 larger than that from DandD (28 dpm/100 cm²), certainly a result of more conservative resuspension factor used in DandD. Another point of comparison, the DandD model had the inhalation pathway accounting for 99.5% of the dose, while RESRAD-BUILD had the inhalation pathway at 97.8%.

An interesting modeling note: The dose rate as $t = 1$ y using RESRAD-BUILD for Pu-239 is very close to zero. This occurs because the inhalation pathway becomes zero as a result of the "lifetime" parameter being set at 365 days. That is, the entirety of the removable fraction is removed from time $t = 0$ to the time specified as "lifetime," which in the default case is 365 days. Therefore, at $t = 1$ y, there is no more material to be removed, and the only viable pathway is external radiation—which of course is vanishingly small for most alpha-emitting radionuclides.

5.4.1.3 DCGLs Using the RESRAD Model

The RESRAD code can be used to generate soil concentration DCGLs by inputting unit activity concentration (e.g., 1 pCi/g), and running the code to determine the resultant dose rate. This dose factor in mrem/y per pCi/g can then be divided into the release criterion to yield the DCGL.

To illustrate the use of this code, RESRAD, Version 5.95 was used to derive DCGLs for Th-232 and processed natural uranium (U-238, U-234, and U-235 at natural isotopic ratios, along with their immediate progeny). Again, an important point that can easily be overlooked is the consideration of how the code handles the short-lived progeny. Appendix A provides a description of the Th-232 series, which consists of Th-232 and 10 progeny (two of the progeny have branching fractions). To determine the $DCGL_W$ for Th-232, one must be very clear to state the expected equilibrium of the decay series. For example, if it is known that the entire Th-232 series is in equilibrium, then the RESRAD model source term must reflect that fact. Conversely, if the series is not in secular equilibrium, then the specific degree of disequilibrium must be entered into the RESRAD model.

Let us consider the case of secular equilibrium in the series. In this scenario, Th-232, Ra-228, and Th-228 would all be entered at 1 pCi/g. This is because RESRAD assumes that the parent radionuclide is in secular equilibrium with its progeny whenever the progeny half-lives are less than 6 months. Therefore, the Th-232 entry accounts for just itself, the Ra-228 entry accounts for itself and its progeny Ac-228, and the Th-228 entry accounts for itself as well as the remainder of the decay series. For the case of disequilibrium, Th-232 might be entered at 1 pCi/g, and Ra-228 and Th-228 at somewhat less concentrations that describe their disequilibrium condition.

To input the processed natural uranium into RESRAD, it is necessary to know the uranium isotopic ratios. This can be obtained by performing alpha spectrometry on a representative sample, or using generally accepted isotopic ratios for natural uranium. For example, natural uranium isotopic ratios are approximately 0.485 for U-238, 0.493 for U-234, and 0.022 for U-235. This fractional source term is entered directly into RESRAD; the short-lived progeny of U-238 and U-235 are covered by their respective parents.

The next item to consider is the default contaminated zone parameters, which include a 10,000 m^2 contaminated zone area, a 2 m thickness of the contaminated zone, and a 100 m length of the contaminated zone that is parallel to the aquifer flow. These defaults are often changed to better reflect site conditions. For example, let us suppose that the particular survey unit under consideration is about 3000 m^2, and characterization data suggest that only surface contamination is present (within the top 15 cm). In this case, the soil concentration DCGLs should reflect the known site conditions—by modifying the contaminated zone area to 3000 m^2, and the thickness of the contaminated zone to 15 cm. The length of the contaminated zone that is parallel to the aquifer flow should also be modified to be consistent with the contaminated zone area; this is done by simply taking the square root of the contaminated zone area (equals 55 m in this example).

Now, there may be a number of pathway and parameter changes that may need to be made to reflect site conditions. Perhaps some pathways can be turned off (e.g., no possibility for groundwater to be tapped into), or maybe a cover layer of clean soil should be modeled. While making use of site-specific data is generally advantageous in the determination of DCGLs, the discussion of additional site-specific parameter changes is beyond the scope of this book. Just be aware that this is an area ripe with flexibility that has a direct bearing on the magnitude of the final status survey.

Once the source term has been entered, and any site-specific modifications have been made, the RESRAD code can be run. The first contaminant evaluated was the Th-232 decay series in secular equilibrium. The source term entered was 1 pCi/g for each of Th-232, Ra-228, and Th-228. The contaminated zone parameters were modified as discussed in the above paragraph. The resulting dose rate was 8.929 mrem/y, which was the maximum dose rate that occurred at time t equal to zero. Note: The DCGLs are generally based on the maximum dose that occurs over some period of time, for example, 1000 y. The model was at a number of calculation times up to 1000 y. The $DCGL_W$ for Th-232, given that it is in secular equilibrium with all of its progeny, for a release criterion of 25 mrem/y is calculated as

$$DCGL_W = \frac{25\,mrem/y}{(8.929\,mrem/y)/(1\,pCi/g)} = 2.80\,\frac{pCi}{g}$$

Similarly, the processed natural uranium $DCGL_W$ can be determined by running RESRAD. This was accomplished by entering the following source term: 0.485 pCi/g for U-238, 0.493 pCi/g for U-234, and 0.022 pCi/g for U-235. The same contaminated zone parameters were used. The resulting dose rate was 6.8E−2 mrem/y, which also represented the maximum dose rate and that occurred at time t equal to zero. The $DCGL_W$ for processed natural uranium, defined by the specific isotopic ratios modeled, for a release criterion of 25 mrem/y is calculated as

$$DCGL_W = \frac{25\,mrem/y}{(6.8E-2\,mrem/y)/(1\,pCi/g)} = 368\,\frac{pCi}{g}$$

If the sum of the U-238, U-234, and U-235 concentrations equals 368 pCi/g, then the resulting dose is at the release criterion of 25 mrem/y. Now it may be desirable to use a readily available surrogate for processed natural uranium—U-238. Using U-238 as a surrogate for processed natural uranium is beneficial because it can be readily measured using gamma spectrometry, while U-234 is not measurable using gamma spectrometry. In this case, the U-238 $DCGL_W$ is determined by multiplying the processed natural uranium $DCGL_W$ by the U-238 isotopic fraction:

$$\text{U-238 DCGL}_W = 0.485 \times 368 \frac{\text{pCi}}{\text{g}} = 178 \frac{\text{pCi}}{\text{g}}$$

This U-238 DCGL would then be used in the final status survey design and data reduction.

5.4.2 Modeling to Obtain Area Factors

The NRC acknowledges in NUREG-1757, Volume 1 that the DandD model is not appropriate for modeling small limited areas of contamination; therefore, other codes or approaches are necessary to develop $DCGL_{EMC}$ values. These "site-specific" analyses would likely be performed using the RESRAD family of modeling codes. Chapter 11 on radiological hot spot considerations provides an extensive discussion on the use of modeling codes to develop $DCGL_{EMC}$ values.

What about the issue of area factors? First of all, it is important to note that DandD, Version. 1.0 does not allow for the adjustment of modeling area to calculate area factors. RESRAD and RESRAD-BUILD can be used to calculate area factors. In fact, some have used RESRAD to calculate area factors that would be close to DandD area factors, if they could be determined using DandD. This was accomplished by revising the RESRAD defaults to simulate DandD outputs. The Maine Yankee license termination plan (LTP) conservatively used only the direct radiation pathway in RESRAD to generate area factors. They argued that direct radiation is a known physical process that is well handled in both codes, and that it can be represented in RESRAD in a manner that is applicable to DandD.

To obtain area factors, the RESRAD code can be used to calculate the dose for a given input activity and the default contaminated area size (i.e., 10,000 m²). Then, the code is run for successively smaller contaminated area sizes and the resultant dose rates recorded. The dose rate for the smaller contamination area will always be at least as big as that for the default contaminant size. The area factor for a specific contaminant area is simply the dose rate for the smaller contaminant area by the initial dose rate for the default contaminant area. The calculation of area factors can be performed for the desired number of contaminant areas.

In addition to the contaminant area size, the only other parameter that is changed during the determination of area factors is the length of the contaminant area parallel to the aquifer. It may also be argued that the fraction of food originating from these smaller contaminant zones should also be changed. Or perhaps, like the Maine Yankee logic, the area factors should be based only on the direct radiation exposure pathway (an idea expanded upon in Chapter 11).

The following examples illustrate the calculation of area factors using RESRAD-BUILD to generate area factors for building surfaces. Essentially, the area factors are determined by calculating the $DCGL_W$ based on a source area of 100 m² (see $DCGL_W$ determination above), and then running the code for a number of smaller contamination areas, keeping all other parameters constant. This was done for Cs-137 and the results are illustrated in Table 5.1.

Now an interesting situation arises for area factors when multiple radio-nuclides are present on a building surface. As we will discuss in detail in Chapter 6, the primary strategy for dealing with multiple radionuclides on building surfaces is to estimate the relative fractions of each individual radionuclide, and to use this information along with the individual radio-nuclide DCGLs to calculate a gross activity $DCGL_W$. Further, Chapter 9 details how scan MDCs may be calculated for multiple radionuclides. The question becomes: What area factor table should be used in conjunction with the gross activity $DCGL_W$? The best answer may be to calculate

TABLE 5.1   Area Factors for Cs-137 Based on RESRAD-BUILD Model

Source Area (m²)	Dose Rate (mrem/y)	Area Factor	$DCGL_W$
100	1.25E−5	1	44,000 dpm/100 cm²
36	8.26E−6	1.51	
25	7.05E−6	1.77	
16	5.72E−6	2.19	
9	4.22E−6	2.96	
4	2.53E−6	4.94	
1	8.45E−7	14.8	

area factors specific to the contaminant mixture present on the building surface. Consider an example to illustrate this concept.

A concrete for the surface on the refuel floor of a reactor facility is contaminated with Co-60, Cs-137, and Sr-90. A number of smear samples have been analyzed from this area and the following fractional amounts have been noted:

Co-60	30%
Cs-137	50%
Sr-90	20%

The project engineer has been tasked with deriving an area factor table for this specific mixture.

The first step in this evaluation may be to calculate the individual radionuclide DCGLs for these contaminants, and at the same time, generate individual area factor tables like that done for Cs-137 above. These are shown in Tables 5.2 and 5.3.

The next step is to enter the radionuclides at their respective fractions into the RESRAD-BUILD code. Thus, the contaminants were entered as follows: 0.3 pCi/m² for Co-60, 0.5 pCi/m² for Cs-137, and 0.2 pCi/m² for Sr-90—a total activity of 1 pCi/m² was input. The resulting dose rate was 2.4E–5 mrem/y, which yields a gross activity $DCGL_W$ of 23,100 dpm/100 cm², for this specific radionuclide mixture. Finally, the area factor table specific to this mixture is calculated and shown in Table 5.4.

This area factor table should be used in conjunction with the gross activity $DCGL_W$ for MARSSIM survey planning and data reduction purposes. One can also see the similarities of the individual area factor tables of Co-60 and Cs-137 with this composite area factor table. In fact, comparing each individual area factor table with this one, both Co-60 and Cs-137

TABLE 5.2 Area Factors for Co-60 Based on RESRAD-BUILD Model

Source Area (m²)	Dose Rate (mrem/y)	Area Factor	$DCGL_W$
100	4.85E–5	1	11,400 dpm/100 cm²
36	3.35E–5	1.45	
25	2.88E–5	1.68	
16	2.35E–5	2.06	
9	1.75E–5	2.77	
4	1.06E–5	4.58	
1	3.54E–6	13.7	

TABLE 5.3   Area Factors for Sr-90 Based on RESRAD-BUILD Model

Source Area (m²)	Dose Rate (mrem/y)	Area Factor	DCGL$_W$
100	1.60E−5	1	34,400 dpm/100 cm²
36	5.89E−6	2.72	
25	4.14E−6	3.86	
16	2.69E−6	5.95	
9	1.55E−6	10.3	
4	7.17E−7	22.3	
1	1.88E−7	88.9	

are conservative (i.e., smaller than the composite), while Sr-90 provides larger area factors (less conservative). Given the fact that the composite area factors are only as accurate as the radionuclide mixture entered, one alternative might be to use the area factor table for the most restrictive radionuclide—Co-60 in this example. By doing this, one can be assured that the area factors used in survey design and data reduction will not yield nonconservative results if the assumed mixture varies in a nonconservative manner.

Outdoor soil area factors are determined using the RESRAD code. As stated previously, RESRAD is used to calculate the dose rate for a given input activity (1 pCi/g) and the default contaminated area size (i.e., 3000 m²). The code is then run for successively smaller contaminated area sizes and the resultant dose rates recorded. The area factor for a specific contaminant area is simply the dose rate for the smaller contaminant area by the initial dose rate for the default contaminant area. Tables 5.5 and 5.6 illustrate area factor tables for Th-232 and processed natural uranium.

TABLE 5.4   Area Factors for Radionuclide Mixture of Co-60 (30%), Cs-137 (50%), and Sr-90 (20%) Based on RESRAD-BUILD Model

Source Area (m²)	Dose Rate (mrem/y)	Area Factor	DCGL$_W$
100	2.4E−5	1	23,100 dpm/100 cm²
36	1.5E−5	1.60	
25	1.3E5	1.85	
16	1.0E5	2.14	
9	7.7E−6	3.12	
4	4.6E−6	5.22	
1	1.5E−6	16.0	

TABLE 5.5   Area Factors for Th-232 in Secular Equilibrium with Its Progeny

Source Area (m²)	Dose Rate (mrem/y)	Area Factor	DCGL$_w$
3000	8.929	1	2.8 pCi/g
300	7.498	1.19	
100	6.577	1.36	
30	5.002	1.78	
10	3.393	2.63	
3	1.626	5.49	
1	0.7212	12.4	

TABLE 5.6   Area Factors for U-238—Used a Surrogate for Processed Natural Uranium

Source Area (m²)	Dose Rate (mrem/y)	Area Factor	DCGL$_w$
3000	6.800E−2	1	178 pCi/g
300	4.985E−2	1.36	
100	4.233E−2	1.61	
30	3.286E−2	2.07	
10	2.352E−2	2.89	
3	1.269E−2	5.36	
1	7.041E−3	9.66	

These area factors are calculated using the same RESRAD inputs used to calculate their DCGLs in the previous section.

The area factor tables shown in Tables 5.5 and 5.6 are for specific RESRAD modeling inputs and source terms—Th-232 in secular equilibrium and U-238 as a surrogate for processed natural uranium.

5.5 HAZARD ASSESSMENTS: AN EARLY APPLICATION OF DOSE-BASED RELEASE CRITERIA?

Sometimes, during the course of a decommissioning project, it becomes obvious that the agreed-upon release criteria cannot be obtained without serious complications. These complications usually relate to worker health and safety. For example, continued scabbling on a concrete support pillar may seriously undermine the structural integrity of the building or remediation that requires workers to be precariously perched in an overhead boom to decontaminate roof trusses which may not seem reasonable compared to the minimal dose reduction achieved. In fact, it sometimes comes down to the monetary expense of achieving release criteria—that is, continuing to clean up to the release criteria is so expensive that the commensurate reduction in personal dose makes it impractical to do so. In these situations, it is necessary for the D&D contractor to justify the use of alternative release criteria through the use of a hazard assessment, which is a detailed site-specific scenario analysis to estimate the dose and then compare it to the costs and risks involved with the proposed action. It is very much like a cost versus benefit analysis.

Regardless of the specifics, the first step of a hazard assessment begins with an assessment of the source term. Once the source term has been characterized, the possible exposure scenarios are evaluated. Many times these scenarios are very similar to those considered for generic dose modeling, such as building occupancy or building renovation. But, unlike the generic scenarios in dose modeling, there may be a number of very specific scenarios (e.g., dose from a contaminated pipe) as illustrated in the next two examples. Finally, exposure pathways and parameters (e.g., breathing rate, dust loading factors) are evaluated and the resultant dose calculated. In conclusion, similarly to generic modeling to obtain DCGLs, hazard assessments begin with an estimated source term that is converted to dose via scenarios and exposure pathways.

While there are a myriad of scenarios where hazard assessments may be needed, two examples are provided in the following sections. These examples show the importance of different pathways in delivering dose. These examples only show the dose calculation portion of the hazard assessment; in reality, the D&D contractor would perform an assessment of economic costs and health and safety risks to compare to the projected doses.

5.5.1 Hazard Assessment for Contaminated Roof Panels at Uranium Site

During the characterization survey of Building 5, an old processing building, processed uranium contamination was identified in numerous overhead roof locations. The characterization was subsequently expanded to include surface activity measurements both on the exterior surface of the roof and on the underside of the roof, while perched in a cherry-picker. Upon completion of the characterization, the D&D project manager concluded that while there was contamination present that exceeded the release criteria (i.e., 5000 dpm/100 cm^2), the removal of the corrugated roof panels would certainly pose a substantial safety risk to the D&D workers. The project manager decides to task the health physicist with performing a hazard assessment of the contaminated roof panels—the desired outcome being to demonstrate that removing the roof panels does not result in significant dose reduction to justify the financial costs or safety risks to workers.

The health physicist plans to review the detailed characterization survey results and then formulate specific scenarios and exposure pathways to calculate the projected dose depending upon whether the roof panels are removed or left in place. The characterization report provides an adequate description of both the size and construction of the Building 5 roof, as well as the nature and results of characterization survey activities.

The roof consists of approximately 160 roof panels on each side of the roof. However, from the collected data, it appears that only the east side of the roof is contaminated. This is consistent with the HSA results, which indicate that the uranium processing primarily occurred in the east bay of the building. Each roofing panel measures approximately 1 m × 1.5 m, and when used in the construction of the roof, each panel overlaps approximately 10 cm of the adjacent roof panel on all four sides.

As indicated previously, the contamination is from processed uranium operations. Therefore, the radionuclides include U-238 and its immediate two beta-emitting progeny (Th-234 and Pa-234 m), U-235 (with its progeny Th-231), and U-234. Twenty direct measurements of surface activity using a Geiger-Mueller (GM) pancake detector were performed over the exterior of the eastern roof. The surface activity levels ranged from 800 to 14,000 dpm/100 cm², with an average level of 2600 dpm/100 cm². Even more surface activity measurements were performed on the interior portion of the roof panels. Ten surface activity measurements were performed on a total of 30 of the 160 roof panels in the east bay, for a total of 300 surface activity measurements. The surface activity levels ranged from 2800 to 60,000 dpm/100 cm², with an average level of 16,000 dpm/100 cm². This indicated that the surface activity levels were substantially higher on the interior side of the roof (which of course makes sense). The development of the source term for this hazard assessment is not yet complete.

One of these panels was removed to better evaluate the nature of the contamination identified between the overlapping panels. It was determined at that time that not only was dust trapped between the overlapping panels, but also between the roof panels and purlins (part of the roof support structure). A total of 280 g of a powder-like dust was collected from the removal of this single panel. The material was placed in a calibrated geometry and analyzed via gamma spectrometry. The result was 62 pCi/g of U-238; the concentrations of the other uranium isotopes can be determined by their relationship to U-238. The total amount of this contaminated dust can be estimated by multiplying the dust collected under one roof panel by the total number of roof panels: (280 g) × (160 panels) = 44,800 g of dust. This calculation is conservative in at least one regard: due to the fact that each of the panels does not have four adjacent sides without double counting some of the dust material. This simplification is considered to be a reasonable approach though, since it is generally accepted that assumptions should tend to be in a conservative direction. Given this characterization of the source term, the health physicist is now ready to consider the potential doses.

5.5.1.1 Dose to Building Employee

The health physicist decides to consider the potential dose to the future building employee. This scenario corresponds to the situation where the roofing panels are left in place. What are the likely pathways for delivering dose to a building occupant? Inhalation? Not likely, since the roof is about 10 m above floor level, it is unlikely for the loose contamination to be disturbed by the employee. Ingestion? Again, not likely because the contamination will be held in place by the roof panels as long as they are not removed. Direct radiation? This exposure pathway should be considered further. The source term calculated above could be used to estimate the direct radiation exposure to the future building worker, but this may not be the best approach. Would exposure rate measurements at a number of likely future office locations not provide a more straightforward approach for assessing the direct radiation pathway? Absolutely.

> For hazard assessments, and dose modeling in general, it is always better to measure than to calculate, whether it be the dose quantity itself, or some parameter value that leads to dose.

The health physicist realizes that additional measurements are warranted to assist in the calculation of dose to the future building occupant. Ten exposure rate measurements using a pressurized ionization chamber are performed 1 m above the floor at a number of locations in the east bay. The exposure rates range from 9.7 to 11.6 µR/h, with an average of 10.8 µR/h. Additionally, exposure rate measurements are performed in an adjacent building on site that has similar construction materials, but no history of processed uranium operations (or any other radioactive materials use for that matter). The background exposure rates range from

8.6 to 10.2 µR/h, with an average of 9.4 µR/h. While the exposure rates in Building 5 are not much different than those in the background reference area, the difference in exposure rates will be assumed due to the processed uranium contamination in the overhead roof. The direct radiation dose can be calculated assuming a building occupancy of 2000 h/y.

$$\text{Dose}_{\text{bldg occ}} = \left(1.4\frac{\mu R}{h}\right)\left(2000\frac{h}{y}\right)\left(1\frac{mrem}{mR}\right)\left(\frac{1mR}{1000\mu R}\right) = 2.8\,mrem$$

Therefore, if the roof panels are left in place, the estimated dose to a future office worker in Building 5 is 2.8 mrem in 1 y.

5.5.1.2 Dose to Demolition Worker

The future demolition worker, unlike the office worker, is expected to have intimate contact with the roof panels. Specifically, not only must the direct radiation pathway be assessed, but during the removal of the roof panels it is likely that the worker will stir up the contaminated dust, thus requiring that the inhalation and ingestion pathways be considered as well.

5.5.1.2.1 Inhalation Dose   The inhalation pathway will be addressed first. The inhalation dose received by the demolition worker is dependent upon the airborne concentration of uranium that results from the roof panel removal. This in turn depends on the uranium concentration in the dust and the mass loading factor or resuspension of dust in the air. The mass loading factor depends on a number of variables, such as the nature of the dust material (density, particle size) and the activity being performed (e.g., debris removal, floor scabbling, nuclear weapons blast). RESRAD uses a value of 200 µg/m³ for the mass loading factor, which will be used for this analysis.

The airborne concentration is calculated via the following equation:

$$C_{air} = (C_{dust})(\text{mass loading factor})\left(\frac{10^{-6}\,g}{1\mu g}\right)$$

At this point it is probably a good idea to express the complete nature of the processed uranium contamination. On the basis of alpha spectrometry analyses, the uranium isotopic fractions were determined to be 0.485 for U-238, 0.493 for U-234, and 0.022 for U-235—these are the natural isotopic ratios of uranium, unaltered by enrichment or depletion processes.

Hence, the concentrations of U-234 and U-235 in the dust can be calculated based on the concentration of U-238 in the dust, 62 pCi/g. By taking advantage of the known isotopic ratios of uranium, the U-234 concentration is 63 pCi/g and the U-235 concentration is 2.8 pCi/g. The U-238 concentration in air can be calculated as

$$C_{air, U-238} = \left(62\frac{pCi}{g}\right)\left(200\frac{\mu g}{m^3}\right)\left(\frac{10^{-6} g}{1\mu g}\right) = 1.24E-2\frac{pCi}{m^3}$$

In a similar fashion, the U-234 and U-235 concentrations are determined to be 1.26E–2 and 5.6E–4 pCi/m³, respectively.

The inhalation dose can be calculated from the airborne concentration. The additional parameters required for this calculation include the breathing rate and dust exposure time. A fairly heavy breathing rate of 20 m³/8 h (or 2.5 m³/h) will be assumed for this example. Compare this value to the breathing rate for a light activity of 1.2 m³/h provided for reference man in ICRP Publication 23 (ICRP 1975).

The dust exposure time can be estimated by assuming that if it takes 40 min to remove each roof panel, so for 160 roof panels, the total exposure time is given by 40 min/panel × 1 h/60 min × 160 panels = 106.7 h. The inhalation dose is calculated as

$$\text{Dose}_{inhalation} = (C_{air})(\text{breathing rate})(\text{time})(\text{DCF})$$

where the DCF is the dose conversion factor for inhalation found on Table 2.1 of Federal Guidance Report No. 11 (EPA 1988). The inhalation DCFs for U-238, U-234, and U-235 are 0.12, 0.13, and 0.12 mrem/pCi, respectively. Thus, the inhalation dose from the U-238 concentration in air is calculated as

$$\text{Dose}_{inh, U-238} = \left(1.24E-2\frac{pCi}{m^3}\right)\left(\frac{20 m^3}{8h}\right)(106.7h)\left(0.12\frac{mrem}{pCi}\right)$$

$$= 0.38 \, mrem$$

The inhalation doses for U-234 and U-235 are 0.42 and 0.017 mrem, respectively. This yields a total uranium inhalation dose of 0.82 mrem.

Now, the sum of these three doses from the uranium isotopes would provide the total inhalation dose from the roof panel removal, as long as the DCFs for U-238 and U-235 account for Th-234 and Pa-234 m, and

Th-231, respectively. Federal Guidance Report No. 11 was reviewed to determine whether or not the DCFs for the uranium isotopes included the contribution from the beta-emitting progeny; unfortunately, no statements could be found indicating that the progeny were included with the parents. Furthermore, the fact that DCFs are provided for Th-231 and Th-234 offers evidence that each of these radionuclides needs to be considered separately. There was no DCF entry for Pa-234 m, likely because of its short half-life (although it is precisely its short half-life that ensures that it is always in equilibrium with its parents Th-234 and U-238).

Reluctantly, the next step is to calculate the inhalation doses from each of these beta-emitting progeny. The airborne concentrations of Th-234 and Pa-234 m are of course the same as that of U-238, while the Th-231 airborne concentration is the same as its parent U-235. The DCFs for the thorium isotopes are 8.8E–7 and 3.5E–5 mrem/pCi, respectively, for Th-231 and Th-234. Because these DCFs are more than four orders of magnitude less than those for the uranium isotopes, there is no need to calculate the inhalation doses for these thorium isotopes. This makes intuitive sense because the alpha-emitting uranium isotopes are expected to deliver significantly more internal dose than the beta emitters. The fact that Pa-234 m DCF is not provided is somewhat troubling, though it too would likely have a DCF much smaller than the uranium DCFs. So as it turns out, the total inhalation dose from the removal of roof panels is the sum of the three uranium isotopes, or 0.82 mrem.

5.5.1.2.2 Ingestion Dose   The ingestion dose received by the demolition worker is dependent upon the dust concentration, which was calculated previously—62 pCi/g for U-238, 63 pCi/g for U-234, and 2.8 pCi/g for U-235. The time for exposure is the same time used for the inhalation dose—106.7 h of roof panel removal. The ingestion dose is calculated by

$$\text{Dose}_{\text{ingestion}} = (C_{\text{dust}})(\text{ingestion rate})(\text{time})(\text{DCF})(1 \text{ d/8 h})(1 \text{ g/1000 mg})$$

where the ingestion rate assumes that some of the dust is accidently eaten. RESRAD provides guidance on the incidental ingestion rate, specifying a value of 100 mg/d (ANL 1993b). It is uncertain how much of the demolition worker's daily allowance of iron, magnesium, or selenium would be provided in this dust intake. Jokes aside, incidental ingestion can readily occur if workers are not diligent to wash their hands prior to eating during a break time or lunch.

The DCF for ingestion can be obtained from Table 2.2 of Federal Guidance Report No. 11 (EPA 1988). The ingestion DCFs for U-238, U-234, and U-235 are 2.5E−4, 2.8E−4, and 2.7E−4 mrem/pCi, respectively. The ingestion dose from the U-238 dust concentration is calculated as

$$
\text{Dose}_{\text{ing, U-238}} = \left(62\frac{\text{pCi}}{\text{g}}\right)\left(\frac{100\,\text{mg}}{\text{day}}\right)(106.7\,\text{h})\left(2.5\text{E}{-}4\frac{\text{mrem}}{\text{pCi}}\right)
$$
$$
\times\left(\frac{1\,\text{day}}{8\,\text{h}}\right)\left(\frac{1\,\text{g}}{1000\,\text{mg}}\right) = 2.1\text{E}{-}2\,\text{mrem}
$$

The ingestion doses for U-234 and U-235 are 2.4E−2 and 1.0E−3 mrem, respectively. This yields a total uranium ingestion dose of 4.6E−2 mrem (about 5% of the inhalation dose). Finally, because the ingestion DCFs for the thorium isotopes are more than a factor of 10 less than those for the uranium isotopes, it was decided that the ingestion doses for Th-234 and Th-231 were inconsequential. As before, a Pa-234 m DCF was not available.

5.5.1.2.3 External Radiation Dose   The external radiation dose is the final pathway to be considered. Contributions to external dose will be considered both from the surface activity on the interior of the roof panels and from the dust trapped between the roofing panels. The surface activity on the interior roof panels will be considered first.

The average surface activity from processed uranium was determined to be 16,000 dpm/100 cm². The GM detector that was used to make the surface activity measurements was calibrated specifically to processed uranium—this calibration is described in Chapter 8, and results in the determination of an efficiency that considers the radiation emitted from each of the uranium isotopes and their immediate progeny. Therefore, the total uranium activity of 16,000 dpm/100 cm² can be multiplied by the specific fractions that each of the uranium isotopes are present in processed natural uranium. When this is done, the surface activity source term can be expressed as

U-238	7760 dpm/100 cm²
U-234	7888 dpm/100 cm²
U-235	352 dpm/100 cm²

Federal Guidance Report No. 12, Table III.3 (EPA 1993) provides dose coefficients for exposure to contaminated ground surfaces. These DCFs will be assumed sufficiently similar to the exposure expected from the interior roof panels. FGR-12 is pretty clear on how it handles progeny contributions—"the values for each radionuclide do not include any contribution to dose from radioactive decay products formed in the spontaneous nuclear transformation of the nuclide." Unlike FGR-11, this report does provide external DCFs for Pa-234 m. The DCF for the progeny for the external radiation pathway are generally more important than their uranium parents, and therefore must be included in the dose calculations. The DCFs (in mrem/y per μCi/cm^2) are as follows:

U-238	644
Th-234	9720
Pa-234 m	17,900
U-234	874
U-235	173,000
Th-231	21,600

The external radiation doses from each radionuclide are calculated by the product of the source term activity, DCF, exposure time (106.7 h), and conversion factors. Remember that the source terms for Th-234 and Pa-234 m are equal to the U-238 source term, and the Th-231 source term equals that for U-235. To illustrate, the external dose for Pa-234 m is given by

$$\text{Dose} = \left(7760\frac{\text{dpm}}{100\,\text{cm}^2}\right)(106.7\,\text{h})\left(17,900\frac{\text{mrem/y}}{\mu\text{Ci/cm}^2}\right)$$
$$\times\left(\frac{1\,\text{y}}{8760\,\text{h}}\right)\left(\frac{1\,\mu\text{Ci}}{2.22\text{E}6\,\text{dpm}}\right) = 7.62\text{E}{-}3\,\text{mrem}$$

The following doses were obtained for each of the radionuclides as follows:

U-238	2.74E–4 mrem
Th-234	4.14E–3 mrem
Pa-234 m	7.62E–3 mrem
U-234	3.78E–4 mrem
U-235	3.34E–3 mrem
Th-231	4.17E–4 mrem

The total external dose from the surface activity on the interior of the roof panels is 1.62E–2 mrem.

Finally, the external radiation dose from the dust is evaluated. This is accomplished quite similar to what was just done for the surface activity. Federal Guidance Report No. 12, Table III.4 (EPA 1993) provides dose coefficients for exposure to contamination to a depth of 1 cm. This was considered a reasonable approximation to the dust thickness. The DCFs (in mrem/y per pCi/g) are as follows:

U-238	8.26E–5
Th-234	8.52E–3
Pa-234 m	1.78E–2
U-234	1.89E–4
U-235	1.77E–1
Th-231	1.32E–2

The external radiation doses from each radionuclide are calculated by the product of the source term activity in pCi/g, DCF, exposure time (106.7 h), and conversion factors. As before, the source terms for Th-234 and Pa-234 m are equal to the U-238 source term, and the Th-231 source term equals that for U-235. To illustrate, the external dose for U-235 is given by

$$\text{Dose}_{\text{ext, U-235}} = (2.8\,\text{pCi/g})(106.7\,\text{h})\left(0.177\frac{\text{mrem/y}}{\text{pCi/g}}\right)\left(\frac{1\,\text{y}}{8760\,\text{h}}\right)$$

$$= 6.04\text{E–3}\,\text{mrem}$$

The following doses were obtained for each of the radionuclides:

U-238	6.24E–5 mrem
Th-234	6.43E–3 mrem
Pa-234 m	1.34E–2 mrem
U-234	1.45E–4 mrem
U-235	6.04E–3 mrem
Th-231	4.50E–4 mrem

The total external dose from the dust trapped between the roofing panels and the purlins is 2.65E–2 mrem. The total dose from the external radiation pathway is the sum of these two doses (from surface activity and the dust) and is 4.27E–2 mrem.

A recap of the dose to the demolition worker from all three pathways is as follows: 0.82 mrem from inhalation pathway, 4.6E–2 from ingestion pathway, and 4.27E–2 mrem from external radiation pathway. This yields a total dose to the demolition worker of 0.91 mrem.

So what can we conclude from this hazard assessment? Do the doses justify leaving the roof panels in place? Remember, the desired outcome was to demonstrate that removing the roof panels did not result in significant dose reduction to justify the financial costs or safety risks to workers. Well the dose to the building occupant was estimated as 2.8 mrem/y, while the dose to the demolition worker was 0.91 mrem over the course of the roof panel removal. But the greater risk to the demolition worker is in gaining access to these overhead areas. Given that the estimate of dose to the future building occupant is minimal, it is likely that a decision would be made to keep the roof panels in place.

Note: One can make the argument that since the building will ultimately be demolished, there will eventually be a demolition worker scenario. The difference being that when demolition finally happens (perhaps many years later), the workers will not be aware of the contamination, and will not take any radiological precautions. Does this change the decision?

Some tips for hazard assessments: Always clearly state your assumptions and the bases for them and provide current references for the parameters used. Also, whenever using DCFs for radionuclides that may have short-lived progeny, it is imperative to determine whether their progeny are included. For the case of the FGR-11 report, it was somewhat intuitive that progeny were not included, but fortunately, the FGR-12 report explicitly stated that progeny were not included. This is in contrast to the common modeling codes today, where radionuclides such as U-238 or Ra-226 include a number of their progeny.

5.5.2 Hazard Assessment for Contaminated Underground Pipes at Sealed Source Facility

Between 1974 and 1995, the Calibration Products Company (CPC) produced low-activity encapsulated sources of radioactive material for use in classroom educational projects, instrument calibration, and consumer products. A variety of radionuclides were used in the production of sealed sources, but the only ones having a half-life greater than 60 days were

Co-60 and Cs-137. In April 1995, licensed activities were terminated and decommissioning began in earnest. Identification and removal of contaminated equipment and facilities was initiated; currently, decontamination activities have been nearly completed and preparations are being made for the final status survey. However, one major decommissioning decision remains: how to dispose an underground pipe that had transported radioactive liquids from the liquid waste hold-up tank (HUT). Significant project savings could be realized if the pipe could be surveyed and released in place. The plan is to prepare a hazard assessment demonstrating that leaving the pipe in place is a justifiable practice. The hazard assessment will consider that sometime in the future the pipe may be removed without consideration of radiological controls, so it is of interest to calculate this potential future dose. The regulator will be making the final determination on whether the pipe should be removed or not prior to terminating the license.

The 25-cm-diameter underground pipe was estimated to be 12 m long, with depths below grade varying from 0.5 m at the building (where it was previously connected to the HUT) to more than 1.5 m, before it emerged above the creek. The liquid waste HUT itself has been removed during the earlier D&D efforts. Limited characterization surveys had indicated the presence of Co-60 and Cs-137 in the pipe, but at very low levels. Following the HUT removal, along with some of the surrounding soils at the tank–pipe interface, the underground pipe was more fully characterized.

Both ends of the pipe were surveyed with radiation detectors, allowing approximately 2 m on each end of the pipe to be assessed. Additionally, sediment samples from within the pipe were collected where material was available for collection. (Note: The sediment layer was no more than 8 cm at its deepest point, and it is estimated that the pipe was about 20% filled with sediment overall.) Residual radioactivity concentrations within the pipe sediment sample were 48 pCi/g for Co-60 and 115 pCi/g for Cs-137 based on gamma spectrometry analysis. It will be assumed that these sediment concentrations are representative of the entire underground pipe length.

Considering that the underground pipe is too deep to deliver an external dose to an individual directly above the pipe, the only reasonable scenario for the residual radioactivity to deliver dose involves excavation and removal of sediment from the pipe. Therefore, the scenario considered in this hazard assessment is that of a future construction worker removing the underground pipe and being exposed to the residual radioactivity in the pipe. The total calculated dose is based on the future construction worker excavating the pipe, dismantling the pipe at each joint, tamping each section of pipe to dislodge sediment, and loitering in the area of the dislodged sediment pile. Before the inhalation, ingestion, and external radiation doses can be calculated, the radioactive source term (sediment) must be derived. The total volume of the pipe can be calculated as follows:

$$\text{Pipe volume} = \pi\, r^2\, h = \pi \times 12.5 \text{ cm}^2 \times 1200 \text{ cm} = 5.89\text{E}5 \text{ cm}^3$$

The total amount of sediment in the pipe is based on the pipe being 20% full of sediment, which yields a sediment volume of 1.18E5 cm³. The total effective dose equivalent was calculated by summing the dose that would result from inhalation, ingestion, and external radiation exposure from the Co-60 and Cs-137 concentrations in the sediment. Calculations for each dose pathway are provided next.

5.5.2.1 Inhalation Dose

The inhalation dose received by the future construction worker is dependent upon the airborne concentration of Co-60 and Cs-137 that results from tamping out the pipe segments. The dislodged sediment is assumed to be accumulated into a pile located near the construction worker's activities. The inhalation dose of course depends on the radionuclide concentration in the sediment and the mass loading factor. We will use the same mass loading factor as in the first hazard assessment—that is, the RESRAD value of 200 µg/m³ will be used for this analysis.

The airborne concentration is calculated via the following equation:

$$C_{air} = (C_{sediment})(\text{mass loading factor})\left(\frac{10^{-6}\,\text{g}}{1\,\mu\text{g}}\right)$$

The Co-60 concentration in air can be calculated as

$$C_{air,Co-60} = (48\,\text{pCi/g})(200\,\mu\text{g/m}^3)\left(\frac{10^{-6}\,\text{g}}{1\,\mu\text{g}}\right) = 9.6\text{E}-3\,\text{pCi/m}^3$$

In a similar fashion, the Cs-137 airborne concentration is determined to be 2.3E−2 pCi/m³.

The inhalation dose is then calculated directly from the airborne concentration. As before, the additional parameters required for this calculation include the breathing rate and airborne concentration exposure time. A breathing rate of 20 m³/8 h will be assumed for this future construction worker. It will also be assumed that the total job duration of excavating the underground pipe is 10 days (working about 8 h each day), or 80 h. The inhalation dose is calculated as

$$\text{Dose}_{\text{inhalation}} = (C_{\text{air}})(\text{breathing rate})(\text{time})(\text{DCF})$$

where the DCF is the dose conversion factor for inhalation found on Table 2.1 of Federal Guidance Report No. 11 (EPA 1988). The inhalation DCFs for Co-60 (Class Y) and Cs-137 are 2.19E−4 and 3.19E−5 mrem/pCi, respectively. DCFs for Ba-137 m, short-lived progeny of Cs-137, were not provided. The inhalation dose from the Co-60 concentration in air is calculated as

$$\text{Dose}_{\text{inh, Co-60}} = (9.6\text{E}{-}3\,\text{pCi/m}^3)\left(\frac{20\,\text{m}^3}{8\text{h}}\right)(80\text{h})(2.19\text{E}{-}4\frac{\text{mrem}}{\text{pCi}})$$

$$= 4.2\text{E}{-}4\,\text{mrem}$$

The inhalation dose calculated in a similar manner for Cs-137 is 1.5E−4 mrem. The total inhalation dose for both radionuclides is 5.7E−4 mrem.

5.5.2.2 Ingestion Dose

The ingestion dose received by the future construction worker is also dependent upon the sediment concentration of 48 pCi/g Co-60 and 115 pCi/g Cs-137. The same exposure time (80 h) is used for the inhalation dose, and the equation for ingestion dose is given by

$$\text{Dose}_{\text{ingestion}} = (C_{\text{dust}})(\text{ingestion rate})(\text{time})(\text{DCF})(1\ \text{d}/8\ \text{h})(1\ \text{g}/1000\ \text{mg})$$

where the ingestion rate assumes that some of the dust is accidently eaten. We will use the RESRAD guidance on the incidental ingestion rate, which specifies a value of 100 mg/d (ANL 1993b). The DCF for ingestion can be obtained from Table 2.2 of Federal Guidance Report No. 11 (EPA 1988). The ingestion DCFs for Co-60 and Cs-137 (again, Ba-137 m was not provided) are 2.69E–5 and 5.00E–5 mrem/pCi, respectively. The ingestion dose from the Co-60 sediment concentration is calculated as

$$
\begin{aligned}
\text{Dose}_{\text{ing,Co-60}} &= (48\,\text{pCi/g})\left(\frac{100\,\text{mg}}{\text{day}}\right)(80\,\text{h})\left(2.69\text{E}-5\frac{\text{mrem}}{\text{pCi}}\right) \\
&\times \left(\frac{1\,\text{day}}{8\,\text{h}}\right)\left(\frac{1\,\text{g}}{1000\,\text{mg}}\right) = 1.29\text{E}-3\,\text{mrem}
\end{aligned}
$$

The ingestion dose for Cs-137 is 5.75E–3 mrem. This yields a total ingestion dose of 7.0E–3 mrem.

5.5.2.3 External Radiation Dose

The external radiation dose from the sediment pile will be the final pathway considered. The external exposure dose calculations were determined using the Microshield™ modeling code. The sediment pile was modeled as a cylinder measuring 20 cm high with a radius of 42 cm—these dimensions account for 1.11E5 cm³ of the total sediment volume (1.18E5 cm³), the difference being attributed to the sediment that remained trapped in the pipe. The radionuclide inventory was calculated as 7.68E–5 µCi/cm³ for Co-60 and 1.84E–4 µCi/cm³ for both Cs-137 and its progeny, Ba-137 m. These concentrations were calculated by multiplying the radionuclide concentration in pCi/g by the sediment density of 1.6 g/cm³ and a conversion factor.

The dose point was selected at a distance of 50 cm from the pile and at 50 cm above the ground, to represent the location of the construction worker from the pile. The resulting exposure rate at this point was 1.06E–2 mR/h. For the assumed 80-h excavation job, the external exposure was calculated as 0.85 mR. Because the conversion from mR to mrem is essentially unity, the external radiation dose is estimated at 0.85 mrem.

To summarize the dose to the future construction worker from all three pathways: 5.7E–4 mrem from the inhalation pathway, 7.0E–3 from the ingestion pathway, and 0.85 mrem from the external radiation pathway. This yields a total dose to the future construction worker of 0.86 mrem.

So, what can we conclude from this hazard assessment? Well, if the underground pipe remains in place, the dose to a member of the general public is negligible (immeasurable) due to the depth of the pipe. The exposure scenario considered in this hazard assessment was the dose to a future construction worker who removed the pipe. The estimated dose to this worker was only 0.86 mrem. Considering the estimated dose to the future construction worker, leaving this material in place does not pose an unreasonable risk to the future construction worker (assumed maximally exposed individual). Therefore, this hazard assessment supports the decision to allow the underground pipe to remain in place—the remediation costs to excavate the pipe do not seem to warrant the averted dose.

QUESTIONS AND PROBLEMS

1. Given the following code output for RESRAD, calculate the radio-nuclide $DCGL_W$ that corresponds to 25 mrem/y. The output for the particular radionuclide was 0.034 mrem/y for an input activity of 1 pCi/g.

2. State four significant differences between RESRAD and DandD modeling codes.

3. Given the following results from repeated RESRAD runs, calculate the area factors for contaminant areas of 300, 100, 40, 10, and 1 m².

Source Area (m²)	Dose Rate (mrem/y)
2000	7.432E−1
300	6.945E−1
100	3.233E−1
40	7.216E−2
10	4.352E−2
1	5.841E−3

4. Discuss three important considerations or tips when performing hazard assessments.

5. What is the essence (major steps) of translating a dose criterion to a measurable quantity? Which step has the greatest uncertainty?

6. When using RESRAD and DandD models for decay series, how does each model handle Th-232 series?

7. Determine the $DCGL_W$ for Co-60 as a function of contaminated area sizes: 2000, 5000, 7500, and 10,000 m². Perform the same calculation for C-14. What can you conclude about each radionuclide's sensitivity to contaminated area size changes?

CHAPTER 6

Preliminary Survey Design Concerns and Application of DCGLs

IMPORTANT OBJECTIVES OF THE scoping and characterization surveys include identifying site contaminants, determining relative ratios among the contaminants, and establishing DCGLs and conditions for the contaminants that satisfy the requirements of the responsible agency. Chapter 5 discussed how DCGLs for individual radionuclides can be determined using exposure pathway modeling. This can be a very detailed and complex process, often requiring knowledge of various modeling codes and may even require a good deal of site-specific information to obtain "realistic" DCGLs. This chapter discusses how these individual DCGLs can be combined and applied when more than one radionuclide is present at a site. Of course, as many can attest, the typical D&D site often has more than one radionuclide contaminant present.

One of the first issues facing the engineer in planning for the final survey is to state how compliance with the release criteria will be demonstrated. It is not sufficient to simply reference the source of the release criteria (e.g., 10 CFR Part 20 Subpart E or related Federal Register notice), but to explain how the referenced release criteria will be implemented. That is, once DCGLs are calculated for each radionuclide, it is necessary to state how the DCGLs will be applied for one or more radionuclides. Obviously, if n radionuclides are present at a facility, having each radionuclide present at a concentration equal to its particular $DCGL_W$ would result in the release criterion being exceeded by n times. (Note: This is precisely the rationale behind the use of the sum-of-the-ratios frequently employed at D&D sites.) It is therefore necessary to modify the DCGLs to account for multiple radionuclides. The choices include the use of surrogate measurements, gross activity DCGLs, or the unity rule—all of which will be discussed in this chapter.

To modify the DCGLs for multiple radionuclides, it is essential to know the contaminants at the site, as well as the relative ratios of these contaminants, if a relative ratio indeed exists. (The meaning of "relative" or "consistent" ratio will be discussed later in this chapter.) In fact, the objectives of the HSA and preliminary surveys include, in part, identifying the site contaminants, determining relative ratios among the multiple contaminants, and establishing DCGLs (if developed from pathway modeling and using site-specific parameters). Identification of radionuclide contaminants at the site is generally performed through laboratory analyses—for example, alpha and gamma spectrometry. For instance, the radionuclide mixture of contamination at a reactor facility may be assessed by collecting characterization samples from representative plant locations, and performing gamma spectroscopy analyses to determine the relative fractions of activation and fission products present. Radionuclide analyses are not only used to determine the relative ratios among the identified contaminants, but also provide information on the isotopic ratios and percent equilibrium status for common contaminants such as uranium and thorium decay series. Refer to Chapter 16 for a detailed description of survey design considerations for uranium and thorium decay series.

6.1 DIRECT APPLICATION OF DCGLs

In the situation where only one radionuclide is present, the DCGL may be applied directly to survey data to demonstrate compliance. This involves assessing the surface-activity levels or volumetric concentrations of the radionuclide and comparing measured values to the appropriate DCGL. For example, consider a site that fabricated radioactive thermal generators (RTGs) using Pu-238 sources, and that was the only radionuclide used throughout its operational lifetime. The default DCGL for Pu-238 on building surfaces and in the soil may be obtained from the responsible agency or generated using an appropriate modeling code, which may include use of site-specific modeling parameters. Surface-activity measurements and/or soil samples are then directly compared using the sign or WRS statistical tests to the surface- and volume-activity concentration DCGLs for Pu-238 to demonstrate compliance. Though straightforward, this approach is not possible when more than one radionuclide is present.

Hopefully not to confuse matters, but rather to clarify the direct application of DCGLs, it is important to consider the case of decay series. That is, it is possible to use the direct application of DCGLs for simple parent–daughter relationships or even decays series, provided that the DCGL used for the parent accounts for all of the progeny involved. Confusing? Consider the case of Sr-90 surface contamination. It is certainly a single radionuclide, but it is almost always present with its short-lived Y-90 daughter (64 h half-life)—thus, each disintegration of Sr-90 results in two betas being emitted. Further, the DCGL for Sr-90 in dpm/100 cm^2 will usually include, and account for, the presence of Y-90. Therefore, although this situation truly consists of two radionuclides, the direct application of DCGLs is possible because the parent DCGL (Sr-90) also accounts for the daughter.

One final technical point to consider with this example is how to consider the double beta emission from Sr-90 and its progeny Y-90. That is, to calculate the total efficiency it is necessary to evaluate the specific beta radiations emitted per decay of Sr-90. Let us assume that we use a gas proportional detector for the measurements of SrY-90, and that the total efficiency for the Sr-90 beta (~545 keV) is 25% and the efficiency for Y-90 (2.27 MeV) is 45%. Since both beta emissions are present with each decay of Sr-90, the estimated total efficiency for the pair in secular equilibrium is simply the sum of each individual total efficiency because both have a radiation yield of 100%, which results in a total efficiency of 70%. The total efficiency derived in this example would be applied to the DCGL for Sr-90 that is based on Sr-90 being in equilibrium with Y-90. So, the calculated net counts on the gas proportional detector would be divided by a total efficiency of 70% to yield the Sr-90 surface activity in dpm/100 cm^2.

To recap at this point, the direct application of DCGLs is possible for sites with single radionuclides, as well as for those situations where the parent DCGL accounts for progeny. It is very important to understand whether or not particular progeny radiations are included in the parent DCGL. In the above example with Sr-90, it is easy to believe that Y-90 would always be included in the Sr-90 DCGL due to its short half-life. But just the same, it is critical to confirm this assumption. Modeling codes usually state a daughter's half-life cutoff value for deciding to include particular progeny with the parent. A quick check of RESRAD for the handling of U-238 finds that the code assumes that its two short-lived beta progeny, Th-234 and Pa-234 m, are included with U-238. One final caution, not all codes handle these radionuclides in the same fashion—that is, all modeling codes may not include the two beta progeny of U-238 in the dose factor for U-238. It is imperative that one has a definitive understanding of what, if any, progeny are included in the parent DCGL.

When more than one radionuclide is present (which is usually the case), we need to provide a way of accounting for the multiple radionuclides present in a manner that does not exceed the release criteria.

6.2 USE OF DCGLs FOR SITES WITH MULTIPLE RADIONUCLIDES

Each radionuclide DCGL corresponds to the release criterion—that is, the regulatory limit in terms of dose or risk. And as discussed previously for multiple radionuclides, allowing each radionuclide to be present at its DCGL concentration would result in the release criterion being exceeded. As an example, if a facility had three contaminants of concern (e.g., Cs-137, Co-60, and Am-241), permitting the concentration levels of each radionuclide to approach its $DCGL_W$ would result in the total dose exceeding the release criterion by a factor of 3. To remedy this situation, the FSS design needs to consider, and account for, the presence of multiple radionuclides contributing to the total dose. One technique for adjusting the DCGLs is to modify the radionuclide source term entered during exposure pathway modeling to include the multiple radionuclides. To do this, each of the multiple radionuclides must be entered at its fractional amount of the total mix. For example, let us assume that the contaminants are present at relative amounts given by 20%, 28%, and 52% for Cs-137, Co-60, and Am-241, respectively. These precise fractions for each radionuclide can be entered into the dose model. We can illustrate this approach using the D&D model.

The following activities were entered into the residential scenario for D&D, Version 1.0: 0.20 pCi/g for Cs-137, 0.28 pCi/g for Co-60, and 0.52 pCi/g for Am-241. Hence, a total source term of 1 pCi/g was entered into the code. The resulting peak dose was 13.8 mrem, and it occurred at 5 years. The DCGL for this gross activity can be calculated by dividing the peak dose into the release criterion of 25 mrem/y—that yields a value of 1.8 pCi/g. This DCGL is actually the sum of the three radionuclide concentrations for the specific radionuclide ratios entered. That is, the individual DCGLs based on these inputs are as follows:

Cs-137 $(0.20) \times (1.8 \text{ pCi/g}) = 0.36 \text{ pCi/g}$

Co-60 $(0.28) \times (1.8 \text{ pCi/g}) = 0.50 \text{ pCi/g}$

Am-241 $(0.52) \times (1.8 \text{ pCi/g}) = 0.94 \text{ pCi/g}$

Note that this approach assumes that the radionuclide mixture is well-known and is relatively "fixed" for the particular area (i.e., survey unit) being modeled. While this approach is certainly feasible, it is not likely to be precise because each of these contaminants is easily measured using gamma spectrometry. Furthermore, because each of these radionuclides is easily measured, there is no reason to rely on the assumption that particular radionuclide fractions exist. That is, the more typical approach to demonstrating compliance would be to measure each radionuclide and demonstrate compliance using the unity rule (which is described in a later section in this chapter). However, before we leave this example, it is worthwhile to point out the use of the gross activity DCGL to solve this example.

The gross activity DCGL is calculated using the following equation:

$$\text{Gross activity DCGL} = \frac{1}{((f_1/\text{DCGL}_1) + (f_2/\text{DCGL}_2) + \cdots + (f_n/\text{DCGL}_n))}$$

(6.1)

where f_i is the radionuclide fraction for each radionuclide present.

To use this equation in our current example, we need to calculate the individual DCGLs for each of the contaminants. This was accomplished by running D&D, Version 1.0 and determining the individual radionuclide DCGLs based on a release criterion of 25 mrem/y. The results were 0.88 pCi/g for Cs-137, 3.64 pCi/g for Co-60, and 1.8 pCi/g for Am-241. Next, these values were substituted into the gross activity DCGL equation:

$$\text{Gross activity DCGL} = \frac{1}{((0.2/0.88) + (0.28/3.64) + (0.52/1.8))}$$
$$= 1.69 \text{ pCi/g}$$

This gross activity DCGL result is somewhat less than the 1.8 pCi/g value calculated when each radionuclide was entered at the same fractions. The difference is likely, due to the time at which the peak dose occurs when all the radionuclides are entered together (5 years), and the time for the peak dose to be reached for the individual radionuclide DCGLs: 6 years for Cs-137, 1 year for Co-60, and 1 year for Am-241. At five years, the Co-60 concentration has undergone approximately one half-life, and therefore, the gross DCGL when the radionuclides are entered together is 1.80 pCi/g, as compared to the more limiting 1.69 pCi/g when each radionuclide DCGL is entered into the gross activity DCGL equation. In other words, the calculational approach calculates the gross DCGL using the individual DCGLs of each radionuclide determined at the peak dose for each radionuclide—but the peak doses for each radionuclide occur at different times, thus rendering the 1.69 pCi/g gross DCGL technically inaccurate. Again, the reader should note that this example is more of academic interest than of practical value—that is, the likely FSS design in this instance would be to employ the unity rule. Now let us consider some other methods for handling multiple radionuclides.

First, surrogate measurements describe another method for adjusting the DCGL to account for multiple radionuclides. Other methods include the development of a gross activity DCGL (primarily for surface activity) and use of the unity rule (for soil concentrations) to adjust the individual radionuclide DCGLs. The unity rule is satisfied by summing the ratios of the concentration of each radionuclide to its respective DCGL and ensuring that the sum does not exceed 1 (unity). Each of these techniques is taken in turn in this section.

However, there is one final consideration that should be addressed before we describe the DCGL modification techniques in this chapter. Once the DCGLs have been modified using one of the approaches to account for multiple radionuclides, it is important to consider the effect that the modified DCGL has on instrument selection. Specifically, the data quality objectives process for instrument selection should include an evaluation of the instrument sensitivity relative to the modified DCGL. For instance, the detection capability (i.e., MDC) of the selected field survey instrument or laboratory analysis for each radionuclide must be less than the predetermined percentage of its $DCGL_W$ (MARSSIM recommends 50% or lower). For example, assume that C-14 is one of several site contaminants and its $DCGL_W$ is 20 pCi/g, and further, that its analytical MDC is 8 pCi/g—so the MDC is 40% of the $DCGL_W$. Now consider the situation where the modified $DCGL_W$ for C-14 is reduced to 10 pCi/g due to the presence of other radionuclides. Note that this may be the result of employing the gross activity DCGL as described above, or from using

C-14 as a surrogate for another radionuclide (perhaps for tritium). Now the MDC for the laboratory analysis of C-14 is 80% of the $DCGL_W$ and therefore no longer satisfies the DQOs for laboratory analysis. Therefore, it will likely be necessary to improve the sensitivity of the laboratory analytical technique to reduce the MDC for C-14. The MARSSIM user is encouraged to be aware of the potential impact that the modification of DCGLs has on instrument and laboratory analysis selection.

The data quality objectives process for instrument selection should include an evaluation of the instrument sensitivity (MDC) relative to the modified $DCGL_W$ (e.g., modified to account for multiple radionuclides); it is important to consider the effect that the modified $DCGL_W$ has on instrument selection.

6.2.1 Use of Surrogate Measurements

The use of surrogate measurements is an important FSS design tool. Because most D&D sites have more than one contaminant, the survey designer must assess the best way to account for the multiple radionuclides present. This section presents the application of surrogate measurements for soil concentrations, surface activities, and exposure rates.

6.2.1.1 Surrogates for Soil Concentrations

For sites with multiple contaminants, it may be possible to measure just one of the contaminants and demonstrate compliance for one or more of the contaminants present. Both time and cost can be saved if the analysis of one radionuclide is simpler and less expensive than the analysis of the other. This may be the case where the radionuclide is more easily detected than another (e.g., gamma versus alpha emissions) or when analysis of a radionuclide is cheaper than for another (e.g., gamma spectrometry for Cs-137 versus wet chemistry techniques for Sr-90). For example, using the measured Am-241 concentration as a surrogate for Pu-239 reduces the analysis costs by not having to perform expensive alpha spectroscopy analyses for Pu-239 on every soil sample.

A sufficient number of measurements should be made to establish a "consistent" ratio between the surrogate radionuclide (i.e., the measured radionuclide) and the inferred radionuclide. The number of measurements needed to determine the ratio is selected using DQO process based on the chemical, physical, and radiological characteristics of the radionuclides and the site. It is expected that site process knowledge would offer a basis for the relative ratio to exist in the first place. If consistent radionuclide ratios cannot be identified from HSA based on the existing information, it is recommended that one of the objectives of scoping or characterization be a determination of the ratios. (Note: Once a ratio has been determined and implemented during the FSS, MARSSIM recommends that 10% of the measurements include analyses for all radionuclides of concern to confirm the ratios used.)

> The DQO Process should be used to ensure that a sufficient number of measurements have been made to establish a consistent ratio between the surrogate radionuclide and the inferred radionuclide. The regulator must "buy-in" to the selected surrogate ratio.

We consider a simple example where eight samples are collected in a survey unit to establish the Am-241 to Pu-239 ratio. This ratio exists because in the production of Pu-239 additional isotopes of plutonium are also produced. One such isotope, Pu-241, decays rather quickly to Am-241, providing a basis for a ratio of Am-241 to Pu-239 to exist. The soil samples are analyzed for Am-241 by gamma spectrometry and Pu-239 by alpha spectrometry to determine if a consistent ratio can be justified. The ratios of Pu-239 to Am-241 (i.e., Pu-239/Am-241) were as follows: 8.4, 6.5, 7.2, 11.5, 8.7, 6.5, 6.8, and 10.2.

An assessment of this example data set results in an average surrogate ratio of 8.22, with a standard deviation of 1.85. There are various approaches that may be used to apply a surrogate ratio from these data—but each must consider the variability and level of uncertainty in the data and include frequent consultation with the responsible regulatory agency. According to MARSSIM, one may consider the variability in the surrogate ratio by selecting the 95% upper bound of the surrogate ratio (to yield a conservative value of Pu-239 from the measured Am-241). This is calculated by adding two sigma (3.7) to the average, to yield 11.9. Similarly, one may make the case that simply selecting the most conservative value from the data set (11.5) should suffice. The DQO process should be used to assess the use of surrogates. The reader should note that the use of Am-241 as a surrogate for transuranics was used at the Big Rock Point D&D Project in Charlevoix, MI. The authors concluded "samples indicate that the ratio of Am-241 to the other transuranics is relatively constant and can be relied upon to determine levels of other transuranics by scaling" (English et al. 1994, p. 257).

The most direct benefit of using the surrogate approach is the reduced cost of not having to perform costly wet chemistry analyses on each sample.

This benefit should be considered relative to the difficulty in establishing the surrogate ratio, as well as the potential consequence of unnecessary investigations that result from the error in using a "conservative" surrogate ratio. Selecting a conservative surrogate ratio ensures that potential doses from individual radionuclides are not underestimated.

Appendix I in the MARSSIM provides the general equation for multiple radionuclides that can be assessed with one radionuclide. In general, it is difficult to extend the use of surrogates beyond one inferred radionuclide for each surrogate radionuclide. This is due to the difficulty of demonstrating that a consistent ratio exists between two radionuclides, not to mention the increasing difficulty of obtaining consistent ratios for more than two radionuclides. However, that being said, there are some circumstances where a surrogate radionuclide may be used for more than one inferred radionuclide. Consider the use of U-238 as a surrogate for total uranium, which consists of U-234 and U-235, in addition to the U-238. Also consider the use of Am-241 as a surrogate for not only Pu-239, but for all of the plutonium isotopes that may be present. That is, the analysis of Am-241 by gamma spectrometry may be used to infer Pu-238, Pu-239, Pu-240, and Pu-241. While this is possible, and may be supported by process knowledge, the justification for demonstrating consistent ratios will probably be more demanding.

Finally, let us look at a situation that often occurs at power reactor decommissioning projects. A residue sample is collected from a floor drain in the reactor building that is expected to be contaminated due to the history of spills in this area. The sample is processed and analyzed in the laboratory via gamma spectrometry initially, followed by radiochemical analyses for Sr-90 and Ni-63, and transuranics (e.g., Am-241, Np-237, and plutonium isotopes). The results are provided to the FSS planning engineer for interpretation:

Cs-137	15 pCi
Co-60	3 pCi
Fe-55	65 pCi
Ni-63	120 pCi
Sr-90	2 pCi
Np-237	0.08 pCi
Am-241	0.005 pCi
All Pu	<0.008 pCi

The experienced survey planners will quickly categorize the easily detectable radionuclides in this mixture from the hard-to-detect radionuclides. Usually the hard-to-detect radionuclides (Fe-55, Ni-63 and most transuranics) are also grouped with the expensive-to-detect radionuclides—Sr-90 in this case. The easy-to-detect radionuclides (Cs-137, Co-60, and perhaps Am-241) are classified as such because of the relative ease to detect and quantify them using gamma spectrometry.

We must mention a couple of important points before we go further. First, remember that this is only ONE sample from a specific area of the facility. Using the radionuclide ratios that are derived from one sample, and extrapolating on other site areas is generally not recommended. It is usually the burden of the survey planner to prove that the radionuclide ratios used to develop the surrogate approach are representative for the area that the surrogates are being used. Does it make sense that the radionuclide fractions from one sample can apply to all site areas, even though it is probable that the radionuclide ratios will vary from one survey unit to the next?

Second, it may be tempting to simply select one radionuclide in the mixture, let us say Cs-137, and relate all of the radionuclides to Cs-137. Not only is this problematic for the reason mentioned in the preceding paragraph, but also because the use of surrogates presupposes that a physical mechanism exists for two or more radionuclides to be correlated to a degree. So a good deal of thought must be given as to why a relative ratio is expected to exist before one starts the calculational process of computing relative ratios.

Here is an example to hopefully clarify this discussion. Both Cs-137 and Sr-90 are fission products, and therefore related by their respective fission yields. Now they also have similar half-lives, yet their chemical properties are dissimilar. This is sufficient basis to move beyond the process history basis to the empirical basis (collecting and analyzing samples) for establishing surrogates. That is, before using one radionuclide as a surrogate for another radionuclides, the underlying process knowledge that provides a basis for expecting two or more radionuclides to have reasonably consistent ratios must be satisfied. Another consideration is that the radionuclide involved should have reasonably long half-lives (perhaps several years or longer), otherwise radioactive decay is another source of ratio variability.

Furthermore, it is observed that the use of Co-60 as a surrogate for Ni-63 might be more appropriate considering their similar modes of production (both activation products). So in this example, the survey design might include a number of surrogate radionuclides and single radionuclides, combined in the unity rule (see Section 6.2.3). Bottom line, the use of surrogates requires careful study of the underlying bases for expecting radionuclides to have a consistent ratio, followed by the collection of a number of representative samples to support the technical justification to use one radionuclide as a surrogate for another.

Once an appropriate surrogate ratio has been estimated, it is necessary to consider how compliance will be demonstrated using these surrogate measurements. That is, it is necessary to modify the DCGL of the measured radionuclide to account for the inferred radionuclide.

The following illustrates how the DCGL for the measured radionuclide is modified ($DCGL_{meas,mod}$) to account for the inferred radionuclide. The derivation of the modified DCGL equation is simply based on the unity rule. The variables used include $DCGL_{meas}$ to indicate the radionuclide that will be measured, while $DCGL_{infer}$ will represent the radionuclide that is inferred from the measured radionuclide. The basic premise is to calculate the modified DCGL for the measured radionuclide that provides the same ratio of contamination level-to-DCGL as one would calculate if the sum of the fractions for each contaminant were computed. This is probably better stated through the use of an equation:

$$\frac{C_{meas}}{DCGL_{meas,mod}} = \frac{C_{meas}}{DCGL_{meas}} + \frac{C_{infer}}{DCGL_{infer}}$$

Given the above relationship, one then must solve for $DCGL_{meas,mod}$. The first step is to determine a common denominator:

$$\frac{C_{meas}}{DCGL_{meas,mod}} = \frac{C_{meas}DCGL_{infer} + C_{infer}DCGL_{meas}}{(DCGL_{meas})(DCGL_{infer})}$$

Next, the equation is inverted and the C_{meas} term is factored out and canceled from both sides of the equation:

$$DCGL_{meas,mod} = \frac{(DCGL_{meas})(DCGL_{infer})}{[DCGL_{infer} + (C_{infer}/C_{meas})DCGL_{meas}]}$$

Finally, the equation can be rewritten to explicitly show the fraction by which the DCGL for the measured radionuclide is reduced to account for the inferred radionuclide:

$$DCGL_{meas,mod} = (DCGL_{meas})\left(\frac{DCGL_{infer}}{(C_{infer}/C_{meas})DCGL_{meas} + DCGL_{infer}}\right) \quad (6.2)$$

where C_{infer}/C_{meas} is the surrogate ratio for the inferred to the measured radionuclide.

Continuing with our example above for Am-241 and Pu-239, the modified DCGL for Am-241 is calculated according to the equation we just derived:

$$DCGL_{Am,mod} = (DCGL_{Am})\left(\frac{DCGL_{Pu}}{(C_{Pu}/C_{Am})DCGL_{Am} + DCGL_{Pu}}\right)$$

where C_{Pu}/C_{Am} is the surrogate ratio for Pu-239 to Am-241.

Assume that the $DCGL_{Pu}$ is 2 pCi/g and $DCGL_{Am}$ is 5 pCi/g, and that the surrogate ratio is 11.9 (as derived previously). The modified $DCGL_W$ for Am-241 ($DCGL_{Am,mod}$) becomes

$$DCGL_{Am,mod} = (5)\left(\frac{2}{(11.9 \times 5) + 2}\right) = 0.16\,pCi/g$$

This modified DCGL for Am-241 is then used for FSS design purposes. Note that in some instances, the surrogate radionuclide, which may be easily detectable at its DCGL level when by itself, can become extremely challenging to measure at its modified DCGL level. This is the case with the above example for Am-241. By itself, Am-241 is easily measured at 5 pCi/g using gamma spectrometry; however, trying to measure Am-241 at 0.16 pCi/g is quite a different proposition. An even more extreme situation can arise for Cs-137 when it is used as a surrogate.

The DCGL for Cs-137 based on recent screening models is approximately 10 pCi/g, and by all accounts, an easy radionuclide to measure at that level. In fact, given that typical background levels for Cs-137 are a fraction of 1 pCi/g, the surveyor would likely decide to use the sign test, rather than take credit for the Cs-137 present in background. However, the survey designer must be cautious when using Cs-137 as a surrogate, in that the $DCGL_{Cs,mod}$ can be reduced to a level that not only may be more difficult to measure, but also may make it unlikely to "eat" the Cs-137 level in the background. That is, the WRS test might be selected instead of the sign test in this situation. One must think through all possible situations when considering the use of surrogates.

Let us consider another example involving the use of surrogates. Perhaps surprisingly, this particular example may not even seem like an example of surrogate use. The contaminant is 2.5% enriched uranium in soil, perhaps a soil area located outdoors at a fuel fabrication facility. This example will demonstrate how U-238 can be used as a surrogate for the total uranium concentration. Remember that total uranium consists of the sum of U-238, U-235, and U-234, with U-235 present at 2.5% by weight. The first calculation performed is to assess the isotopic fractions of 2.5% enriched uranium. The assumption used is that the U-234:U-235 ratio is 21:1. The percent U-235 enrichment is calculated by dividing the U-234, U-235, and U-238 activity concentrations by their respective specific activities, and determining the ratio of the U-235 isotopic weight to the total uranium weight. Specifically,

$$\% \, EU = \frac{U\text{-}235/2.14E6}{U\text{-}235/2.14E6 + U\text{-}238/3.33E5 + U\text{-}234/6.19E9} \quad (6.3)$$

where U-235, U-238, and U-234 are the activity concentrations in pCi/g, and the numerical values are the respective specific activities for each isotope. Actually, for this equation to yield the % EU, the result needs to be multiplied by 100.

The above equation is used to determine the isotopic fractions of uranium by inputting 0.025 as the value for % EU (i.e., 2.5%), 1 pCi/g for U-235, and 21 pCi/g for U-234, and solving for the U-238 activity concentration. Thus, solving for the U-238 activity concentration yields 6.1 pCi/g.

Therefore, the isotopic fractions of 2.5% uranium can be determined by dividing each activity concentration by the total uranium activity of 28.1 pCi/g. This results in the following uranium isotopic fractions:

U-238	0.217
U-235	0.036
U-234	0.747

The interesting point with this example is that these uranium ratios are quite consistent as long as the material is exactly 2.5% enriched uranium. In the previous paragraphs of this section, we had stressed the importance of carefully examining the underlying bases for expecting the radionuclides to have a consistent ratio. Well for a given enrichment of uranium, the isotopic fractions of uranium will be nearly constant. Hence, we should be in good shape as far as using U-238 as a surrogate for 2.5% enriched uranium. While it is probably obvious, it is worth stating the primary reason for designing the survey based on measurements of U-238. Simply, for uranium in soil, U-238 can be measured via gamma spectrometry, while U-234 requires alpha spectrometry.

At this point in our example we should state the DCGLs to be used for uranium. Let us assume that the NRC-provided DCGLs apply in this example—that is, 14 pCi/g for U-238, 8 pCi/g for U-235, and 13 pCi/g for U-234. The problem at hand is to determine the modified DCGL for U-238 that accounts for the total uranium concentration. This can be accomplished using the general surrogate equation provided in MARSSIM Appendix I. Specifically, equation I-14 on MARSSIM page I-32 can be used to calculate the modified DCGL for U-238.

$$DCGL_{U\text{-}238,mod} = \frac{1}{((1/D_1) + (R_2/D_2) + (R_3/D_3))}$$

where D_1 is the DCGL$_W$ for U-238 by itself, D_2 is the DCGL$_W$ for the second radionuclide (U-238 is the first radionuclide) that is being inferred by U-238. R_2 is the ratio of the concentration of U-235 to that of U-238 (or 0.164), and R_3 is the ratio of the concentration of U-234 to that of U-238 (or 3.44). Therefore, DCGL$_{U\text{-}238,mod}$ can be calculated using the isotopic ratios of 2.5% enriched uranium as follows:

$$DCGL_{U\text{-}238,mod} = \frac{1}{((1/14) + (0.164/8) + (3.44/13))} = 2.8 \text{ pCi/g}$$

To demonstrate compliance, the statistical tests employed in the FSS would be performed using a DCGL$_W$ of 2.8 pCi/g for U-238. Alternatively, the same answer results by using Equation 6.1 to calculate the gross uranium activity DCGL, and then multiplying that value by the U-238 fraction (0.217). Of course, one could opt to use alpha spectrometry and measure all three uranium isotopes and demonstrate compliance using the unity rule.

The MARSSIM advises that the potential for shifts or variations in the radionuclide ratios means that the surrogate method should be used with caution. Physical or chemical differences between the radionuclides may produce different migration rates, causing the radionuclides to separate and changing the radionuclide ratios. Remediation activities have a reasonable potential to alter the surrogate ratio established prior to remediation. It is recommended that when the ratio is established prior to remediation, additional post-remediation samples be collected to ensure that the data used to establish the ratio are still appropriate and representative of the existing site condition. Again, remember that the MARSSIM recommends that 10% of the soil samples collected during the FSS be analyzed for both the measured and inferred radionuclides to validate the surrogate ratio used. If these additional post-remediation samples are not consistent with the pre-remediation data, surrogate ratios should be reestablished, or possibly abandoned in favor of the unity rule.

6.2.1.2 Surrogates for Surface Activity

Compliance with surface-activity DCGLs for radionuclides of a decay series (e.g., thorium, uranium, radium) that emit both alpha and beta radiation, may be demonstrated by assessing alpha, beta, or both radiations. However, relying on the use of alpha surface contamination measurements often proves problematic due to the highly variable level of alpha attenuation by rough, porous, and dusty surfaces. Beta measurements typically provide a more accurate assessment of thorium and uranium contamination on most building surfaces because surface conditions cause significantly less attenuation of beta particles than alpha particles. Beta measurements, therefore, may provide a more accurate determination of surface activity than can be achieved by alpha measurements.

The relationship of beta and alpha emissions from decay series or various enrichments of uranium should be considered when determining the surface activity for comparison with the $DCGL_W$ values. When the initial member of a decay series has a long half-life, the radioactivity associated with the subsequent members of the series will increase—at a rate determined by the individual half-lives—until all members of the decay series are present at activity levels equal to the activity of the parent (secular equilibrium).

Consider the MARSSIM example where the average surface-activity $DCGL_W$ for natural thorium is 600 dpm/100 cm², and that all of the progeny are in secular equilibrium—that is, for each disintegration of Th-232 there are six alpha and four beta particles emitted in the thorium decay series. Note that in this example it is assumed that the surface-activity $DCGL_W$ of 600 dpm/100 cm² applies to the total activity from all members of the decay chain (it is critical to identify exactly what the DCGL refers to).

In this situation, the corresponding alpha activity $DCGL_W$ should be adjusted to 360 dpm/100 cm^2 and the corresponding beta activity $DCGL_W$ to 240 dpm/100 cm^2, in order to be equivalent to 600 dpm/100 cm^2 of natural thorium surface activity.

To demonstrate compliance with the beta activity $DCGL_W$ for the natural thorium example, beta measurements (in cpm) must be converted into activity using a weighted beta efficiency that accounts for the energy and radiation yield of each beta particle. For decay series that have not achieved secular equilibrium, the relative activities between the different members of the decay chain can be determined as previously discussed for surrogate ratios. We will further discuss the gross activity of the DCGLs for surface activity, but first one more example of using beta measurements in place of alpha measurements will be provided here.

Consider the case of depleted uranium contamination on building surfaces. Assuming that the uranium depletion yields only 0.3% U-235 by weight, as compared to the natural U-235 weight percent of 0.7%, the following ratios are obtained:

U-238	0.701
U-235	0.014
U-234	0.285

Now let us assume that the $DCGL_W$ for each uranium isotope is 2000 dpm/100 cm^2. Because each uranium isotope is an alpha-emitter we can plan to make alpha-only measurements to demonstrate compliance with the release criteria. It does not matter which uranium isotope produced the alpha count, because our assumption is that each isotope has the same $DCGL_W$ and we can further assume that each of the alphas emitted are close enough in energy to yield the same efficiency. No problem with this survey design, other than the fact that alpha measurements are not always reliable due to variable attenuation in the field. So how could beta measurements, as opposed to alpha measurements, be used to demonstrate compliance?

It is first necessary to understand the decay characteristics of each uranium isotope present (Chapter 16 covers this in detail); U-238 has two beta-emitting progeny (Th-234 and Pa-234 m), U-235 has Th-231, and U-234 has none. Each of the three beta-emitting progeny are assumed to be in secular equilibrium with their uranium parent, and therefore are present at their parent's isotopic fraction—Th-234 and Pa-234 m both at 0.701 and Th-231 at 0.014. Consequently, the total beta fraction is 1.42, which compares to the alpha fraction of 1. In other words, for this depleted uranium, the beta to alpha ratio is 1.42:1. Finally, if we decide to make only beta measurements to demonstrate compliance with the uranium release criteria, the $DCGL_W$ for beta is increased by a factor of 1.42–2830 dpm/100 cm², and the total beta efficiency is weighted according to the fractional amounts that each beta-emitter is present. Two notes are necessary at this point. First, if our example involved enriched uranium, then the beta to alpha ratio would be less than 1 and the beta $DCGL_W$ would be lower than that for alpha. Second, while instructional, this approach to handling decay series has been somewhat improved by the techniques described in Chapters 8 and 10 that make use of the ISO-7503 approach and consider each radionuclide's instrument and surface efficiency.

6.2.1.3 Surrogates for Exposure Rate

Another example for the use of surrogates involves the measurement of exposure rates—in place of surface- or volume-activity concentrations— for radionuclides that deliver the majority of their dose through the direct radiation pathway. That is, instead of demonstrating compliance with soil or surface contamination DCGLs that are derived from the direct radiation pathway, compliance is demonstrated by direct measurement of exposure rates. To implement this surrogate method, HSA documentation should provide reasonable assurance that no radioactive materials are buried at the site and that radioactive materials have not seeped into the soil or groundwater. This surrogate approach is also attractive because exposure rate measurements generally exhibit less sensitivity to local variations in soil concentrations, as compared to discrete soil samples.

Appendix C in NUREG-1500 provides data on those radionuclides that deliver greater than 90% of their dose via the direct radiation pathway (NRC 1994b). Common radionuclides fitting this bill include Co-60, Cs-134, Eu-152, Eu-154, and Fe-59. It is interesting to note that Cs-137 was not included in Appendix C. This seemed a bit peculiar given that Cs-137 is often characterized by its significant gamma emission (recognizing that it is actually from Ba-137 m). RESRAD Version 5.95 was run with Cs-137 as the soil contaminant to assess the fractional dose attributed to the direct radiation pathway. The results were consistent with NUREG-1500; the direct radiation dose was 83.5%, with the plant and meat pathways contributing 7.6% and 6.6%, respectively.

Not only is this approach limited to certain radionuclides, but also as mentioned above, it is critical that the soil contamination is on the surface, not covered by clean soil or other material. Once again, RESRAD can be called upon to better illustrate the problem if the soil contaminant is not at the surface. The direct radiation dose provided by RESRAD provides an indication of how exposure rate measurements would be impacted in the field. To illustrate this point, RESRAD was run assuming that the contaminated layer was 15 cm thick, for cover thicknesses of 0, 10, and 25 cm. The contaminants assessed included Co-60 and Cs-134. As shown in Table 6.1, significant reductions in the direct radiation dose result when a soil cover is added.

The degree to which exposure rate measurements might replace soil sampling is directly related to the level of confidence the regulator has that the soil contamination is on the surface. While HSA data may go a long way toward justifying the use of this approach, it might not be possible to overcome the inherent uncertainty, however small, in the HSA data. Therefore, while one might very well make a strong technical case for the use of exposure rates as a surrogate for soil sampling, the regulatory approval for this surrogate approach rests squarely on the regulator's comfort zone.

6.2.2 Gross Activity DCGLs for Surface Activity

When applying the gross activity DCGL or unity rule, only significant radionuclides remaining at the time of the FSS need to be considered—that is, some radionuclides present at small fractions of the total inventory may be excluded from the gross activity DCGL or unity rule. Of course, this point needs to be agreed upon by the responsible regulatory agency. This determination requires the analysis of representative samples, usually collected during site characterization, to determine the radionuclide composition. For example, radionuclides that would contribute greater than a predetermined percentage (perhaps 10%) of the total radiation dose from all contaminants or which are present at concentrations that exceed

TABLE 6.1   Direct Radiation Dose Reduction as a Function of Soil Cover Thickness

Radionuclide	Soil Cover Thickness (cm)	Direct Radiation Dose (mrem/y)
Co-60	0	7.5
	10	2.2
	25	0.4
Cs-134	0	4.2
	10	1.0
	25	0.1

the predetermined percentage of their respective DCGLs need be considered as significant contaminants. Again, the only conclusive guidance offered in this matter is that whatever approach is used to define "significant radionuclides," the regulator must be in agreement.

Surface contamination DCGLs apply to the total (fixed plus removable) surface activity present on a surface. For cases where the surface contamination is due entirely to one radionuclide (not the likely situation), the $DCGL_W$ for that radionuclide is used for comparison to measurement data, called the direct application of DCGLs. For situations where multiple radionuclides, each with its own $DCGL_W$, are present, a gross activity DCGL can be developed. This approach enables field measurement of gross activity, rather than determination of individual radionuclide activity for comparison to the DCGL via the unity rule. That is, the alternative would be to use the unity rule and calculate the sum-of-the-ratios value at each measurement location, which of course requires radionuclide-specific data at each location. Can you imagine how resource intensive it would be to determine the radionuclide composition at each direct measurement location?! Luckily, we can use the gross activity DCGL to demonstrate compliance in this case.

The gross activity DCGL for surfaces with multiple radionuclides is calculated as follows:

1. Determine the relative fraction (f) of the total activity, contributed by the radionuclide. This may be accomplished by collecting representative samples and performing radionuclide-specific analyses—the DQO process should be used to determine how many samples are needed.

2. Obtain the $DCGL_W$ for each significant radionuclide present—those radionuclides remaining at the time of the FSS, that would contribute greater than the predetermined percentage of the total radiation dose from all contaminants or which are present at concentrations that exceed the predetermined percentage of their respective DCGLs, need be considered as significant contaminants. This clearly needs to be worked out between the responsible regulatory agency and the MARSSIM user.

   Note: A precedent for "significant" may be the guidance in draft NUREG/CR-5849 (p. A-1), "For sites with multiple radionuclides, only those radionuclides remaining at the time of license termination, which would contribute greater than 10% of the total radiation dose from all contaminants or which are present at concentrations which exceed 10% of their respective guideline values, need be considered as significant contaminants." More recently, NUREG-1757 (p. 3-5) provided similar guidance:

Licensees may eliminate insignificant radionuclides and exposure pathways from further detailed consideration. However, the dose from the insignificant radionuclides and pathways must be accounted for in demonstrating compliance with the applicable dose criteria. Insignificant means no greater than 10% of applicable dose criterion.

Ten percent is an aggregate limit; total dose contributions of all radionuclides and all exposure pathways considered insignificant should not exceed the 10% limitation (NRC 2006a).

3. Substitute the values of f and DCGL in the following equation:

Gross activity DCGL

$$= \frac{1}{(f_1/\mathrm{DCGL}_1) + (f_2/\mathrm{DCGL}_2) + \cdots + (f_n/\mathrm{DCGL}_n)}$$

The gross activity DCGL is then used for FSS design purposes.

Note that the above equation may not work for sites that exhibit surface contamination from multiple radionuclides that have unknown or highly variable concentrations of radionuclides throughout the site. In these situations, the best approach may be to select the most conservative surface contamination DCGL from the mixture of radionuclides that are present— and assume that all of the detector's response is from this one radionuclide.

A sample calculation for the determination of gross surface-activity DCGL is provided. Assume that 45% of the total surface activity was contributed by C-14 with a DCGL of 35,000 dpm/100 cm^2; 30% by Cs-137 with a DCGL of 18,000 dpm/100 cm^2; and 25% by Am-241 with a DCGL of 320 dpm/100 cm^2.

$$\text{Gross activity DCGL} = \frac{1}{(0.45/35,000) + (0.30/18,000) + (0.25/320)}$$
$$= 1230\,\mathrm{dpm}/100^2$$

Remember that the total efficiency of the detector must be weighted by the relative fraction that each radionuclide is present. In some cases, the total efficiency for a radionuclide may be close to zero, or in fact zero, for some hard-to-detect radionuclides. In this example, let us assume that a gas proportional detector is used to make FSS measurements, and that the total efficiency for the radionuclides present is 0.12 for C-14, 0.32 for Cs-137, and 0.06 for Am-241. The weighted total efficiency is calculated:

$$\text{Weighted total efficiency} = (0.45 \times 0.12) + (0.30 \times 0.32) + (0.25 \times 0.06)$$
$$= 0.165$$

This total efficiency would then be used to determine the instrument's MDC to ensure that it is sufficiently sensitive relative to the gross activity DCGL.

Some D&D sites have a mixture of possibly a dozen or more radionuclides potentially contaminating building surfaces. Certainly nuclear power plants undergoing decommissioning fit this bill. The typical reactor radionuclide mixture consists of fission products (e.g., Cs-137, Sr-90), activation products (e.g., Co-60, Fe-55, Ni-63), and a very small amount of largely alpha-emitting transuranics (e.g., plutonium isotopes, Am-241, Np-237). It is very important to consider how these multiple contaminants are treated in the FSS design. For example, the desired approach may be to calculate the gross activity DCGL once an appropriate radionuclide mixture has been developed. This may be easier said than done given the variability in the radionuclide mixture.

The gross activity DCGL can be significantly impacted by the transuranics potentially present. Specifically, even though the transuranics are usually present at very small fractional amounts, their low DCGLs tend to bring the overall gross activity DCGL down (try Problem #4 at the end of this chapter). It is also worth noting the difficulty in determining their fractional abundance using laboratory analytical procedures, given their low abundance. Additionally, the variability in the radionuclide mixture, particularly the transuranic fraction, as a function of plant area greatly influences the gross activity DCGL—making it difficult to defend the representativeness of the gross activity DCGL given the fluctuations of the radionuclide mixture. One solution to handling the transuranics is to measure them separately with alpha-only surface-activity measurements. This would invoke the unity rule since one type of surface-activity measurement would deal with the activation and fission products via the gross beta activity DCGL, and a second surface-activity measurement would account for only the transuranics by comparing with the gross alpha DCGL. Of course, the DQO process should be used to determine the most advantageous approach for the radionuclide mixture potentially present. Perhaps only a small number of plant areas even have a reasonable potential for transuranic contamination, so many plant areas would not have to consider transuranics whatsoever.

6.2.3 Use of the Unity Rule

The theme of this chapter continues—for sites where more than one radionuclide exists in a survey unit, the individual DCGLs for each radionuclide cannot be used directly to assess compliance. As previously discussed, one option to address the existence of multiple radionuclides in a survey unit is to use the surrogate approach. Another option is the unity rule, which is performed by determining the ratio between the concentration of each significant radionuclide in the mixture and its DCGL. Note: As discussed for the gross activity DCGL, only those radionuclides remaining at the time of the FSS, which would contribute greater than the predetermined percentage of the total radiation dose from all contaminants or which are present at concentrations that exceed the predetermined percentage of their respective DCGLs, need be considered as significant contaminants. Again, regulatory approval of the predetermined percentage is absolutely critical—do not charge ahead in this regard without getting the regulator on board. Finally, the sum of the ratios for all (significant) radionuclides in the mixture should not exceed 1 (unity).

> The unity rule must be used whenever more than one measurement is made at a particular location, including multiple radionuclides in a soil sample or alpha and beta surface activity measurements at the same location.

For soil contamination, it is likely that specific radionuclides, rather than gross activity, will be measured for demonstrating compliance. The unity rule is used when multiple radionuclides exhibit unknown or variable relative concentrations throughout the site, and it is not feasible to use surrogate measurements. In this sense, the unity rule may be considered the default approach for multiple radionuclides in soil. The unity rule may be applied to radionuclides whether they are present in the background or not. For radionuclides that are present in the natural background, the two-sample WRS test should be used to determine if residual soil contamination exceeds the release criterion. To account for multiple radionuclides that are present in the background, the sum of each radionuclide concentration divided by its $DCGL_W$ is determined in both the survey unit and background reference area. This sum-of-the-ratios is termed the "weighted sum":

$$\text{Weighted sum} = \frac{C_1}{DCGL_1} + \frac{C_2}{DCGL_2} + \cdots + \frac{C_n}{DCGL_n}$$

where C is the radionuclide concentration.

It is also necessary to determine the standard deviation of the above sum for each survey unit and the background reference area. When using the unity rule, the $DCGL_W$ is 1 (unity) and it is necessary to normalize the standard deviation, as shown below:

$$\sigma = \sqrt{\left(\frac{\sigma_{C_1}}{DCGL_1}\right)^2 + \left(\frac{\sigma_{C_2}}{DCGL_2}\right)^2 + \cdots + \left(\frac{\sigma_{C_n}}{DCGL_n}\right)^2}$$

An example is provided for clarification. Consider an outdoor survey unit contaminated with both radium and thorium, and that the DCGLs for these materials specifically apply to Ra-226 and Th-232, both in equilibrium with their progeny. Assume that the DCGL for Ra-226 is 2.8 pCi/g and that for Th-232 is 1.4 pCi/g, and that the estimated standard deviations in the survey unit are 1.2 pCi/g for Ra-226 and 0.5 pCi/g for Th-232—12 soil samples were collected during the characterization and analyzed using gamma spectrometry to obtain the estimates for the contaminant's variability. The characterization survey results also indicated that the concentrations of the Ra-226 and Th-232 were not correlated in the survey unit, and thus the surrogate approach is not feasible. The weighted sums of the measured concentrations of Ra-226 and Th-232, relative to their respective DCGLs, are calculated for each soil sample location in both the survey unit and the background reference area:

$$\text{Weighted sum} = \frac{C_{Ra\text{-}226}}{DCGL_{Ra\text{-}226}} + \frac{C_{Th\text{-}232}}{DCGL_{Th\text{-}232}}$$

The WRS test (see Chapter 13) would then be performed on the weighted sums from both the survey unit and the reference area. That is, the DCGL (i.e., unity) is added to the weighted sum for each sample results from the background reference area.

However, first the normalized standard deviation must be determined for the weighted sums:

$$\sigma = \sqrt{\left(\frac{\sigma_{Ra\text{-}226}}{DCGL_{Ra\text{-}226}}\right)^2 + \left(\frac{\sigma_{Th\text{-}232}}{DCGL_{Th\text{-}232}}\right)^2} = \sqrt{\left(\frac{1.2}{2.8}\right)^2 + \left(\frac{0.5}{1.4}\right)^2} = 0.56$$

Thus, the standard deviation of the weighted sum is used for FSS design purposes for determining the sample size, via the relative shift. Refer to the example in Section 13.2.2.2 for how the unity rule is applied to the WRS test for uranium and thorium.

The same approach applies for radionuclides that are not present in the background. In this instance, the one-sample sign test is used in the place of the two-sample WRS test. When the contamination at a site is comprised of radionuclides both present in the background and not in the background, the two-sample WRS test should be used, simply substituting zero for the background reference area concentration of the contaminant not present in the background. More fully developed FSS designs employing the unity rule are illustrated in Chapter 13.

QUESTIONS AND PROBLEMS

1. What are the limitations imposed by the direct application of DCGLs? In other words, what conditions must prevail for the direct application of DCGLs to be used?

2. Describe the instrument selection concern that may arise when an individual DCGL is reduced due to the presence of multiple radionuclides. [Hint: The MDC divided the $DCGL_W$ should be less than 50%].

3. State the possible strategies that may be employed to handle multiple radionuclides.

4. Given the following power reactor radionuclide mixture and individual DCGLs:

Cs-137	15%	22,000 dpm/100 cm²
Co-60	27.8%	7000 dpm/100 cm²
Sr-90	10%	8000 dpm/100 cm²
Fe-55	28%	200,000 dpm/100 cm²
Ni-63	18%	85,000 dpm/100 cm²
Pu-239	1.2%	40 dpm/100 cm²

Calculate the gross activity DCGL for this mixture. Calculate the gross activity if the Co-60 fraction was lowered to 25.8%, and the Pu-239 fraction was increased to 3.2%. Would you consider the gross activity DCGL sensitive to small changes in the Pu-239 fractional abundance?

5. Co-60 will be used as a surrogate for Fe-55. The DCGLs for Co-60 and Fe-55 are 3.2 and 125 pCi/g, respectively. Given that the surrogate ratio of Fe-55 to Co-60 is 2.4, what is the modified DCGL for Co-60 used as a surrogate?

6. Using a surrogate radionuclide for more than one inferred radionuclide is difficult in the sense of establishing relative ratios between more than one set of radionuclides (getting consistent ratios for one set of radionuclides is difficult enough). What are the conditions where it may be possible for one surrogate to be used for two or more radionuclides?

7. The site contaminants in the soil at a gaseous diffusion facility include uranium isotopes at natural isotopic ratios (i.e., 0.485 for U-238, 0.493 for U-234, and 0.022 for U-235) and Tc-99. Recent characterization results indicate the radioactive mix is 70% Tc-99 and 30% for the total of uranium isotopes. The DCGLs are as follows:

Tc-99	20 pCi/g
U-238	8 pCi/g
U-234	7 pCi/g
U-235	5 pCi/g

What is the DCGL$_W$ for U-238 used as a surrogate for the uranium isotopes? What is the DCGL$_W$ for U-238 used as a surrogate for both the uranium isotopes and Tc-99? Which of these approaches should the health physicist recommend for measuring and demonstrating compliance for these site contaminants? (Hint: While having U-238 account for Tc-99 lowers analytical costs, at the same time it reduces the U-238 DCGL which will drive up sample size.)

CHAPTER 7

Background Determination and Background Reference Areas

IN THIS CHAPTER, WE will discuss the importance of background measurements in the context of FSSs. MARSSIM provides comprehensive guidance on issues relating to the background and its determination. In this chapter, we will supplement the MARSSIM guidance with additional details on the background measurements and considerations for identifying the background reference areas. Section 7.2 deals with issues relating to surface material backgrounds, while Sections 7.3 and 7.4 consider radionuclide concentrations in background soils and how the scenario B FSS design can be applied. Scenario B represents a fundamental change in how the null hypothesis is stated MARSSIM (scenario A)—that is, the scenario B null hypothesis is that the residual radioactivity in the survey unit is indistinguishable from the background. Finally, it should be noted that the actual performance of the background measurements are covered in Chapter 10.

The point was made in Chapter 4 that most release criteria and guidelines are stated in terms of acceptable residual radioactivity levels above the background. For example, the 35 pCi/g guideline for depleted uranium in the NRC's 1981 Branch Technical Position (NRC 1981b) does not include the background concentration of uranium in the soil. Similarly, the pre-approved authorized limits for radium-226 and radium-228 in soil are explicitly in excess of background levels as stated in DOE O 458.1 (DOE 2011). The well-known surface activity guidelines (Chapter 4) in terms of dpm/100 cm² do not include the naturally occurring activity in the surface material being measured. The only exception that comes to mind is the EPA's 4 mrem/y limit on groundwater—for example, if the background concentration of radionuclides in groundwater delivers 1.5 mrem/y, the residual radioactivity above the background cannot result in a dose greater than 2.5 mrem/y. Furthermore, the dose-based decommissioning rule clearly indicates that sites seeking unrestricted release must demonstrate that residual radioactivity distinguishable from the background (i.e., clearly above the background) must be less than 25 mrem/y. The point of all of this is that the assessment of background levels are necessary because decommissioning release criteria for surface activity and soil contamination are presented in terms of radiation or radioactivity levels above ambient levels at a D&D site.

The background reference areas provide a location for background measurements which are used for comparisons with survey data. The conventional approach of determining background levels—by collecting a number of samples some distance from the site boundary—may no longer be appropriate. Furthermore, some may be tempted to review the literature on background soil concentrations and radiation levels and cite these background values for comparison to their own survey measurements. These background documents might come in handy as a sanity check on site background levels, but make no mistake, it is necessary to have the raw background measurement data to perform the statistical tests described in MARSSIM. Two especially useful references on the background include NUREG-1501, Background as a Residual Radioactivity Criterion for Decommissioning (NRC 1994a) and NCRP Report 94, Exposure of the Population in the United States and Canada from Natural Background Radiation (NCRP 1987).

7.1 BACKGROUND REFERENCE AREAS AND RELATED CONSIDERATIONS

The considerations for selecting background reference areas can be expressed as follows: (1) the background location should be representative of the survey unit location (assuming that the survey unit was never contaminated), and (2) the background location should be nonimpacted from site operations. This implies that background reference areas should reasonably be close to the D&D site (to be representative), and generally upstream and upwind of site operations. The second point is that candidate reference areas must not be impacted or tainted by licensed operations. There are several sources of information concerning the selection of background reference areas—NUREG/CR-5849, NUREG-1501, NUREG-1757, and of course the MARSSIM. And NUREG-1757, vol. 2 (2006) notes that the background survey takes an added importance since the licensee may decide to use a statistical test (i.e., WRS test) that compares survey unit data to a background reference area to demonstrate compliance with the release criteria. These references all recommend that the background reference area for soils should have the same radiological, geological, hydrogeological, and biological characteristics as the survey unit itself. Biological similarities? Reportedly, earthworm activity can distribute radioactivity quite effectively—so perhaps a little thought should be given to biological characteristics in reference areas as compared to survey units.

For surface material background reference areas, the selected surface material backgrounds should be determined in areas of similar construction but without a history of radioactive material use. Because different building materials (poured concrete, concrete block, metal, wood, etc.) can have very different background levels, average background counts are determined for each material encountered in the surveyed area, at a background reference location of similar construction. Construction materials may exhibit significant variability based on the origin of the materials used in the construction of the material. Concrete blocks which may look quite similar may exhibit very different activities based on variations in naturally occurring material.

It may be necessary for the D&D site to have a number of background reference areas. An example may be those sites that exhibit varying geology and soil types—for example, the background concentrations of Cs-137 are higher in wooded areas than in open fields, therefore requiring a D&D site to have a minimum of two reference areas. Similarly, Cs-137 fallout is near surface for undisturbed soil, while Cs-137 concentrations are not expected below several centimeters. This situation may result in the D&D contractor using a Cs-137 background for surface soil areas, but not for soils that are some depth beneath the surface. Another complication is due to the fact that many areas have been landscaped or backfilled at one time or another. This continues to add to the background variability, and the difficulty in identifying representative background reference areas.

It may be necessary to select a reference area that is part of an impacted area. While background reference areas should be selected from nonimpacted areas, sometimes this is just not possible. In these circumstances it is necessary to use class 3 areas for surface material backgrounds. Justification for this practice is the responsibility of the D&D contractor. When making the case for this practice, evidence may include HSA and preliminary survey data, particularly analytical laboratory results that can demonstrate that any radiation present is from natural sources. This confirmation can include *in situ* gamma spectrometry measurements or laboratory analyses. Additionally, with experience, the surveyor already has an idea of the typical background response to certain surface materials. How do the measurements compare to the generally expected background measurements based on experience? Another consideration is the relative importance of background levels as compared to the $DCGL_W$. If the anticipated background levels are small compared to the $DCGL_W$, then it may be possible to be conservative and not take credit for the background at all.

Sometimes the survey unit itself can serve as its own background reference area. While MARSSIM recognizes the possibility for a survey unit to serve as its own reference area, this strategy will be used in limited instances and always with regulatory approval. Figure 7.1 shows a histogram of Th-232 concentrations in a class 1 survey unit. A strong argument

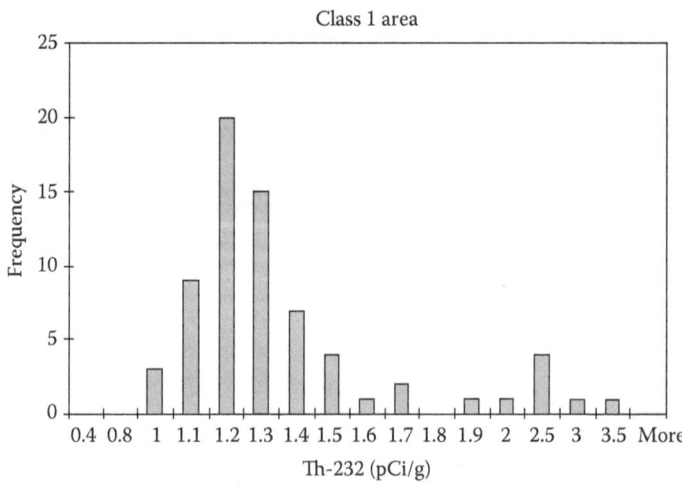

FIGURE 7.1   Histogram of Th-232 concentrations in a class 1 survey unit.

can be made that the Gaussian-shaped distribution on the left in Figure 7.1 is the most representative background reference area that might be found for this survey unit.

Another interesting background consideration is the fundamentally different way that MARSSIM deals with background measurements. That is, many D&D professionals are familiar with the conventional survey approach of subtracting background levels from gross measurements of surface activity. The same technique is routinely performed for laboratory analyses of soil—that is, background concentrations of radionuclides were subtracted from survey unit measurements. The background measurements are compared to the survey unit measurements when using the MARSSIM statistical approach, rather than subtracting the background from the survey unit measurements. A driver for this background comparison approach was that, in some instances, it was not possible to demonstrate compliance on a sample-by-sample basis due to the variability of the background. In these situations, compliance can be demonstrated by comparing two distributions of radionuclide concentrations in the soil, and assessing whether the mean difference is less than the release criterion. This is the essence of the MARSSIM two-sample WRS test approach.

> In the MARSSIM WRS test, the population of measurements in the survey unit is compared to the population of measurements in the background reference area.

One final consideration is the indistinguishability from the background survey design approach. This approach is called scenario B and is based on the NRC license termination rule stated in paragraph 20.1402 of 10 CFR Part 20 Subpart E: "A site will be considered acceptable for unrestricted use if the residual radioactivity that is distinguishable from background radiation results in a TEDE to an average member of the critical group that does not exceed 25 mrem (0.25 mSv) per year, including that from groundwater sources of drinking water, and the residual radioactivity has been reduced to levels that are as low as reasonably achievable (ALARA)" (NRC 1997). Scenario B is covered in Sections 7.3 and 7.4.

7.2 SURFACE MATERIAL BACKGROUNDS

Surface material backgrounds are necessary in calculating appropriate surface activity levels. The conventional approach for calculating the surface activity requires an assessment of both the gross radiation levels on surfaces in the survey unit and the background radiation levels from appropriate surface materials. Specific details on the performance of surface material background measurements are covered in Chapter 10. A well-known background for the detector and the specific surface type is subtracted from the gross surface activity measurement. The background level should account for both the ambient exposure rate background (from natural gamma radiation) and the surface material background. Therefore, the background response of the detector is the sum of the ambient exposure rate background (commonly referred to as instrument background) and the contribution from naturally occurring radioactivity in the surface material being assessed.

An additional complication arises due to fluctuating radon levels. The background measurements performed during the early morning hours may be influenced to a greater degree by radon and its progeny due to less mixing of radon in the atmosphere, and more so, less mixing of the building air due to it being closed overnight. The background measurements made at the same location later in the day may be markedly less, a good indication of the temporal fluctuations of radon. Therefore, one needs to be cognizant that the background radiation levels for materials that have no significant naturally occurring radioactivity component (metals, wood, dry wall, etc.), are driven by the ambient gamma radiation background and radon levels—and that these considerations must be made when selecting appropriate background measurement locations.

Note, in most cases, it is the instrument background that produces the majority of the overall detector background response. The detector's operating voltage and threshold settings, and window thickness represent detector factors that can influence background levels. Nevertheless, the background levels for surface activity measurements do vary because of the presence of naturally occurring radioactive materials in building surfaces, and the possible shielding effect that these construction materials can provide. As mentioned above, the background levels are influenced by the presence of naturally occurring radioactivity in construction materials. Masonry brick and ceramic tiles, for example, frequently contain elevated levels of naturally occurring Th-232, U-238, and K-40. Table 7.1 provides typical background levels for both alpha and beta radiation on a number of common surface materials and survey instruments (reproduced from NUREG-1507).

TABLE 7.1   Background Count Rate for Various Materials

	Background Count Rate (cpm)[a]					
	Gas Proportional					
Surface Material	α Only	β Only	α + β	GM	ZnS	NaI
Ambient[b]	1.00 ± 0.45[c]	349 ± 12	331.6 ± 6.0	47.6 ± 2.6	1.00 ± 0.32	4702 ± 16
Brick	6.00 ± 0.45	567.2 ± 7.0	573.2 ± 6.4	81.8 ± 2.3	1.80 ± 0.73	5167 ± 23
Ceramic block	15.0 ± 1.1	792 ± 11	770.2 ± 6.4	107.6 ± 3.8	8.0 ± 1.1	5657 ± 38
Ceramic tile	12.6 ± 0.24	647 ± 14	648 ± 16	100.8 ± 2.7	7.20 ± 0.66	4649 ± 37
Concrete block	2.60 ± 0.81	344.0 ± 6.2	325.0 ± 6.0	52.0 ± 2.5	1.80 ± 0.49	4733 ± 27
Drywall	2.60 ± 0.75	325.2 ± 8.0	301.8 ± 7.0	40.4 ± 3.0	2.40 ± 0.24	4436 ± 38
Floor tile	4.00 ± 0.71	308.4 ± 6.2	296.6 ± 6.4	43.2 ± 3.6	2.20 + 0.58	4710 ± 13
Linoleum	2.60 ± 0.98	346.0 ± 8.3	335.4 ± 7.5	51.2 ± 2.8	1.00 ± 0.45	4751 ± 27
Carbon steel	2.40 ± 0.68	322.6 ± 8.7	303.4 ± 3.4	47.2 ± 3.3	1.00 ± 0.54	4248 ± 38
Treated wood	0.80 ± 0.37	319.4 ± 8.7	295.2 ± 7.9	37.6 ± 1.7	1.20 ± 0.20	4714 ± 40
Untreated wood	1.20 ± 0.37	338.6 ± 9.4	279.0 ± 5.7	44.6 ± 2.9	1.40 ± 0.51	4623 ± 34

Source: From NUREG-1507.

[a] Background count rates determined from mean of five 1-min counts.

[b] Ambient background determined at the same location as for all measurements, but without the surface material present.

[c] Uncertainties represent the standard error in the mean count rate, based only on counting statistics.

As will be discussed in the next chapter, the survey instruments frequently used for alpha measurements of surface activity are gas proportional and zinc sulfide scintillation detectors. The background count rate range for these detectors is typically of the order of 0–3 cpm for many surfaces. Therefore, the differences in levels of naturally occurring radioactive materials in building surfaces may not be as important for alpha measurements as it is for beta measurements, but it is important to be cognizant of the building surface types that do exhibit alpha backgrounds greater than the typical alpha background range. Also, remember that elevated radon and radon progeny levels can skew the alpha background count rate.

7.3 SCENARIO B SURVEY DESIGN: INDISTINGUISHABLE FROM THE BACKGROUND

The scenario B survey design discussion might have easily fit in Chapter 13 with the other MARSSIM final survey designs and strategies. The reason for keeping it in Chapter 7 is that scenario B is inexorably linked to the background—that is, it is selected because of the concern for variability in the background on the survey design, and cannot be used unless this variability is proven. The reader is hereby cautioned that the concepts discussed in Sections 7.3 and 7.4 presuppose an underlying knowledge of MARSSIM survey designs. If this is not the case, Sections 7.3 and 7.4 may be difficult to follow, and it is recommended that you skip these sections for now. For those reading on, it is recommended that you work through these sections in this chapter with at least one finger holding open Chapter 13. Furthermore, Chapter 13 of NUREG-1505 should also be open as you study this section.

The fundamental objective for performing an FSS is to demonstrate that the release criterion has been met. A number of MARSSIM survey designs are provided in Chapter 13 to illustrate various FSS design strategies. In each case, the statistical survey design is based on one particular statement of the null hypothesis, specifically, that residual radioactivity in the survey unit exceeds the release criterion. This null hypothesis statement is referred to as scenario A and is the exclusive null hypothesis statement used in MARSSIM. However, when the radionuclide of concern is present in the background, and its $DCGL_W$ is relatively close to the background concentration level for this radionuclide, and the background level is variable, then using scenario A can be disastrous (i.e., very high sample sizes and high probability of failing clean survey units). A potential remedy in these difficult situations is the scenario B survey design.

Scenario B refers to the revised manner in which the null hypothesis is stated. For those wondering, it is still considered a MARSSIM survey design. Instead of stating that the survey unit is "dirty," as is the case for scenario A, the null hypothesis for scenario B is that the measurements in the survey unit are indistinguishable from those in the background reference area. More formally, we can write the scenario B null hypothesis as:

H_0: The difference in the median concentration of residual radioactivity in the survey unit and in the reference area is less than the LBGR; versus the alternative hypothesis.

H_a: The difference in the median concentration of residual radioactivity in the survey unit and in the reference area is greater than the $DCGL_W$.

Note that scenario B is very well-described in NUREG-1505, A Nonparametric Statistical Methodology for the Design and Analysis of Final Status Decommissioning Surveys (NRC 1998a), and to a lesser degree in the NRC's NUREG-1757, vol. 2 (NRC 2006a). The latter document states that scenario B is expected only for a small number of facilities, and should be used in special cases when the DCGL is small compared to measurement and/or the background variability.

There are three major steps in designing the scenario B FSS. The first step is to demonstrate that the D&D site in question has sufficient background variability. This can be viewed as a requirement for using scenario B. That is, in order to take advantage of the scenario B survey design, the site must qualify by exhibiting a certain level of the background variability. The Kruskal–Wallis (K–W) nonparametric test is performed to demonstrate that sufficient variability exists. The second step is to determine a concentration level that is deemed indistinguishable from the background. This concentration value is basically some number of standard deviations above the overall background mean, and is called the lower bound on the gray region (LBGR). Concentrations greater than the LBGR are considered distinguishable, those less than the LBGR are considered indistinguishable from the background. The final step is to determine the sample size for the WRS test and the Quantile test.

Once the scenario B FSS has been designed and implemented, the collected survey data are reduced. If the WRS test null hypothesis is rejected, then the survey unit fails; if the null hypothesis is not rejected (i.e., accepted), then we perform the Quantile test. The Quantile test is performed to assess the level of spotty contamination that may be present. It is also important to note that if the scenario B null hypothesis is not rejected (which is the desired outcome), then a retrospective power curve is necessary to ensure that the survey unit was not released simply because the FSS was not powerful (sensitive) enough to detect that residual radioactivity which exceeded the $DCGL_W$. Retrospective power curves are discussed in Chapter 14 of this book.

We will now begin our example of how scenario B can be applied to a D&D site that is potentially contaminated with Th-232. Initial characterization samples from a number of potential background reference areas indicated that the Th-232 concentrations in soil were somewhat variable. The D&D contractor identified four different background reference areas and requested that 16 soil samples be randomly collected from each area. Concerning the number of samples from each reference area, NUREG-1505 recommends that 10–20 samples be collected from each area, and that four reference areas should be selected. Table 7.2 shows the Th-232 concentrations from each of the four background reference areas.

TABLE 7.2   Th-232 Concentrations (pCi/g) from
Four Background Reference Areas

Area 1	Area 2	Area 3	Area 4
1.23	1.35	0.66	1.22
1.04	0.64	0.93	1.22
1.01	1.29	1.12	1.03
1.02	1.11	0.88	1.23
0.84	1.44	1.22	1.25
0.86	1.16	1.26	0.96
1.23	1.16	1.37	1.36
1.15	0.84	1.06	1.30
1.39	0.64	0.98	1.53
1.21	0.65	0.80	1.29
1.31	1.07	0.95	1.17
1.17	0.83	0.98	1.08
1.08	1.02	1.00	1.28
0.62	1.24	1.48	1.37
0.82	1.25	1.47	1.02
0.88	1.25	1.45	1.14

The K–W test must be performed on the Th-232 concentrations in Table 7.2 to demonstrate that there is sufficient variability in the data to justify the use of scenario B. The null hypothesis for the K–W test can be stated as follows:

H_0: No significant variability exists between the four background reference areas; versus the alternative hypothesis.

H_a: Significant variability exists between the four background reference areas.

Therefore, the D&D contractor desires to reject this null hypothesis in favor of the alternative hypothesis, to conclude that there is significant variability in the background. As with any hypothesis testing approach (see Chapter 12), it is necessary to determine a test statistic from the data, and then compare the test statistic to a critical value. For the K–W test, the test statistic is denoted as K, and the critical value is labeled as K_c. To reject the null hypothesis, K must be greater than K_c, allowing the D&D contractor to conclude that the background variability was sufficient to warrant the use of scenario B. Let us now proceed with the calculation of the test statistic (K) from our data in Table 7.2.

First, the mean and standard deviations of the Th-232 concentrations in each reference area are determined. Second, the individual measurements (total of 64) are pooled and ranked from low to high, ensuring that the origin of each sample (i.e., its particular reference area) is maintained. Finally, the sum of the ranks (R_j) for each of the four reference areas are determined. These calculations are shown in Table 7.3.

We note that, when ranking, if several measurements are tied, they are all assigned the average rank of that group of tied measurements. Although there appears to be tied data in Table 7.2, there are no ties since the ranking was performed on the raw data that has three decimal places (only data with two decimal places are shown).

The K–W test statistic, K, can be calculated from the data in Table 7.3 using the following equation:

$$K = \frac{12}{N(N+1)}\left(\sum \frac{R_i^2}{n_i}\right) - 3(N+1),$$

TABLE 7.3   K–W Test on Th-232 Concentrations in Soil

	Background Th-232 Concentrations					Ranks			
	Area 1	Area 2	Area 3	Area 4		Area 1	Area 2	Area 3	Area 4
	1.23	1.35	0.66	1.22		42	55	5	39
	1.04	0.64	0.93	1.22		25	3	14	40
	1.01	1.29	1.12	1.03		20	52	31	24
	1.02	1.11	0.88	1.23		21	30	12	44
	0.84	1.44	1.22	1.25		9	60	41	46
	0.86	1.16	1.26	0.96		11	35	49	16
	1.23	1.16	1.37	1.36		43	34	58	56
	1.15	0.84	1.06	1.30		33	10	26	53
	1.39	0.64	0.98	1.53		59	2	17	64
	1.21	0.65	0.80	1.29		38	4	6	51
	1.31	1.07	0.95	1.17		54	27	15	36
	1.17	0.83	0.98	1.08		37	8	18	29
	1.08	1.02	1.00	1.28		28	23	19	50
	0.62	1.24	1.48	1.37		1	45	63	57
	0.82	1.25	1.47	1.02		7	48	62	22
	0.88	1.25	1.45	1.14		13	47	61	32
Mean	1.05	1.06	1.10	1.21	Sum (R_j)	441	483	497	659
Sigma	0.208	0.261	0.250	0.146	Total		2080		

where N is the total number of background measurements (64), n_i is the number of measurements in each reference area (16), and R_i is the sum of the ranks for each of the four reference areas. Therefore, K is calculated:

$$K = \frac{12}{64(64+1)}\left(\frac{441^2}{16} + \frac{483^2}{16} + \frac{497^2}{16} + \frac{659^2}{16}\right) - 3(64+1) = 4.95$$

Now, we must determine the critical value of the K–W test, which is based on the type I error selected. First note that the type I error (incorrect rejection) means that we would reject the null hypothesis, and falsely conclude that there was significant variability between the different reference areas. Therefore, the larger the type I error, the more tolerant we are being in allowing the null hypothesis to be falsely rejected. Remember, scenario B is stopped in its tracks if we cannot prove via the K–W test that there is significant variability in the background data. Recognizing the importance of the type I error (commonly denoted α), NUREG-1505 recommends that the type I error be 0.1, while the NRC's NUREG-1757 suggests a type I error of 0.2.

Table 13.1 in NUREG-1505 provides critical values for the K–W test as a function of the type I error and the degrees of freedom $(k-1)$, where k is the number of individual reference areas. Therefore, for $k-1$ equal to 3 in our example, K_c is equal to 6.3 for a type I error of 0.1, and 4.6 for a type I error of 0.2. Remember that the null hypothesis is rejected only when the test statistic (K) exceeds the critical value (K_c). For a type I error of 0.1, we cannot reject the null hypothesis (i.e., 4.95 is not greater than 6.3). For a type I error of 0.2, we can indeed reject the null hypothesis, since 4.95 is greater than 4.6. This is a perfect illustration of the paramount importance of selecting the type I error—if a value of 0.1 was selected for the type I error, scenario B would not be possible because the K–W test did not conclude that there was significant variability in the background. In that situation, the D&D contractor would be forced to use scenario A, or push for the type I error to be increased to 0.2. Let us suppose that the regulator approved a type I error of 0.2, and therefore we have demonstrated significant background variability.

Now that the K–W test has justified that the background variability was significant at a type I error of 0.2, we continue with scenario B by determining the radionuclide concentration that is indistinguishable from the background. This concentration will be termed the LBGR for survey design purposes, and will be determined as a function of the overall background variability. We will be analyzing the background variance both within, and among, the four background reference areas. In fact, the analysis of variance steps outlined in NUREG-1505, Section 14.3 will be performed on our data set to determine the component of variance, ω^2, that forms the basis of the LBGR. In fact, NUREG-1505 points out that many analysis of variance (ANOVA) software packages can perform these calculations. Suffice to say, even a simple spreadsheet greatly expedites the necessary calculations to determine ω^2, as evident in Table 7.4.

While not showing the specific nature of each of the NUREG-1505 ANOVA calculations, important interim results will be provided. The first calculation was to determine the mean square within the reference areas, s_w^2—which resulted in 4.90E-2. The next calculation was for the mean square between reference areas, s_b^2—which yielded 8.63E-2. The term n_0 is simply equal to 16, since 16 measurements were collected from each of the four reference areas. Finally, we can calculate ω^2:

$$\omega^2 = \frac{(s_b^2 - s_w^2)}{n_0} = \frac{(8.63E-2) - (4.90E-2)}{16} = 2.33E-3$$

TABLE 7.4   Interim Calculations Used to Determine LBGR via ω^2 for Scenario B

	Background Th-232 Concentrations				Measurements Squared			
	Area 1	Area 2	Area 3	Area 4				
1	1.23	1.35	0.66	1.22	1.506	1.831	0.438	1.479
2	1.04	0.64	0.93	1.22	1.077	0.415	0.865	1.486
3	1.01	1.29	1.12	1.03	1.020	1.656	1.243	1.061
4	1.02	1.11	0.88	1.23	1.032	1.234	0.773	1.515
5	0.84	1.44	1.22	1.25	0.707	2.062	1.488	1.550
6	0.86	1.16	1.26	0.96	0.742	1.348	1.598	0.926
7	1.23	1.16	1.37	1.36	1.510	1.341	1.874	1.836
8	1.15	0.84	1.06	1.30	1.325	0.714	1.130	1.698
9	1.39	0.64	0.98	1.53	1.935	0.404	0.967	2.332
10	1.21	0.65	0.80	1.29	1.464	0.429	0.643	1.651
11	1.31	1.07	0.95	1.17	1.724	1.141	0.893	1.367
12	1.17	0.83	0.98	1.08	1.371	0.694	0.969	1.171
13	1.08	1.02	1.00	1.28	1.169	1.038	0.998	1.631
14	0.62	1.24	1.48	1.37	0.387	1.530	2.187	1.869
15	0.82	1.25	1.47	1.02	0.674	1.563	2.161	1.036
16	0.88	1.25	1.45	1.14	0.777	1.560	2.091	1.300
Sum	16.86	16.94	17.61	19.43	18.420	18.960	20.318	23.906
Mean	1.05	1.06	1.10	1.21				
Mean squared	1.111	1.121	1.211	1.474				

Taking the square root of ω^2, the ω for Th-232 in the background is 0.048 pCi/g. The LBGR concentration is defined as a multiple of ω, usually taken to be three (3). Note that there is a fair amount of subjectivity involved in deciding what level is distinguishable from the background. Following guidance in NUREG-1505, the LBGR is equal to 3ω or 0.145 pCi/g. Remember that this is the net Th-232 concentration that is considered distinguishable. This means that differences between the survey unit and the background reference area smaller than 0.145 pCi/g Th-232 are not considered distinguishable from the background. Consequently, net Th-232 concentrations greater than the LBGR (0.145 pCi/g) are indeed considered to be distinguishable from the background.

Thus, we have performed the first two of the three steps for scenario B—justifying that the background is sufficiently variable, and determining a concentration that is distinguishable from the background (LBGR). The third step is to determine the WRS test sample size for scenario B. This will be performed in the next section as we compare the scenarios A and B survey designs. Note that all of the background reference measurements collected in support of the K–W test do indeed count toward the reference area measurement needs for the WRS and Quantile tests. If additional background reference measurements are necessary, they should be collected randomly, ensuring that all individual reference areas are equally likely to be sampled.

7.4 SCENARIO A VERSUS SCENARIO B FSS DESIGNS

By now you may be wondering what the bottom line difference is between scenarios A and B. Here it is, bottom line—the difference in survey designs can be astounding; implementation of scenario B is very likely to be well-worth the added complexity described in the previous section. At the risk of discussing scenario A survey design before it is formally introduced—it is covered in Chapter 13—we can address this difference in survey strategies.

First, let us assume that an FSS is being designed for a particular class 2 survey unit that is potentially contaminated with Th-232. The D&D project engineer is interested in whether the standard MARSSIM survey design (scenario A) or the scenario B survey design should be used. The project engineer decides to calculate prospective power curves for both survey designs to aid in making this decision. The regulator has approved a Th-232 $DCGL_W$ of 0.8 pCi/g. We will first design the survey using scenario A.

Scenario A

A number of soil samples were collected from the survey unit during characterization. The Th-232 results were reported as 1.36 ± 0.40 (1σ). The project engineer believes that the most representative background reference area for this survey unit is reference area 2. The Th-232 concentrations from this reference area are shown in Table 7.2. Specifically, the Th-232 results are reported as 1.06 ± 0.26 (1σ). Therefore, the best estimate of the net median concentration in the survey unit is 1.36 – 1.06, or 0.3 pCi/g. Given the uncertainty in the Th-232 concentrations, the project engineer selects the LBGR (remember this is for scenario A) at 0.4 pCi/g Th-232, somewhat higher than the expected median concentration. The larger of the two standard deviations from the reference area or survey unit is used in the survey design; so the standard deviation in the survey unit (0.40 pCi/g) is selected. Other DQOs for this survey design include type I and type II errors of 0.05.

The MARSSIM COMPASS code will be used to calculate the sample size and construct the power curve. This scenario A survey design required 32 soil samples. The probability that the survey unit passes as function of Th-232 concentration is shown in Figure 7.2. To facilitate the upcoming comparison with scenario B, let us determine the probability of passing the survey unit at specific net Th-232 concentrations using Figure 7.2:

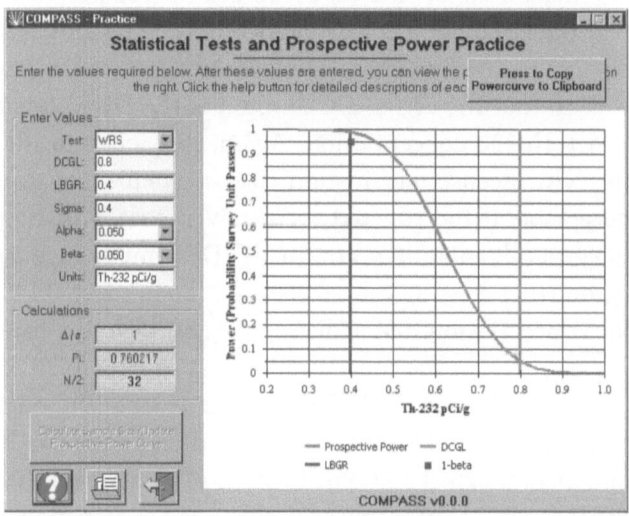

FIGURE 7.2 Probability that survey unit passes versus Th-232 concentration for scenario A.

Th-232 (pCi/g)	Probability of Passing Survey Unit
0.145	~100%
0.5	89%
0.6	60%
0.7	25%
0.8	5% (type I error at DCGL$_W$)
0.945	~0%

Scenario B

The survey unit characteristics described in scenario A are summarized here. The Th-232 results were reported as 1.36 ± 0.40 (1σ). The first difference we confront is the nature of the background reference area. In scenario A the project engineer determined that reference area 2 was the most representative background reference area, and planned the survey design using that background reference area. For scenario B, the idea is to use all 64 background measurements that comprise the four individual reference areas. The overall mean background level of Th-232 from the 64 samples was 1.11 ± 0.22 (1σ). This points out that the standard deviation from the survey unit is greater than that from the background reference area, and will be used in the statistical design, as it was in scenario A.

Similarly, the best estimate of the net median concentration in the survey unit is 1.36 – 1.11 pCi/g, or 0.25 pCi/g. However, we do not set the LBGR at the median concentration as we did in scenario A, rather, we use the LBGR (3ω) of 0.145 pCi/g as defined in the previous section. The agreed upon DCGL$_W$ (0.8 pCi/g) is actually the width of the gray region in scenario B, so that the actual DCGL limit is the DCGL$_W$ plus 3ω, or 0.945 pCi/g net Th-232 concentration. Again, this refers to the manner in which the NRC license termination rule is stated—that unrestricted release is acceptable if the residual radioactivity that is distinguishable from the background radiation results in a TEDE to an average member of the critical group that does not exceed 25 mrem. So in effect, 0.145 pCi/g of Th-232 is the concentration that is distinguishable from the background, and 0.8 pCi/g of Th-232 is the concentration that equates to 25 mrem/y.

Finally, we can calculate the WRS test sample size. The relative shift is calculated by (0.945 – 0.145)/0.4, which equals 2. Other DQOs for this survey design include type I and type II errors of 0.05. Given the reversed null hypothesis from that of scenario A, the type I error is now the D&D contractor's error of concern, while the type II error is the regulator's. The overall type I error of 0.05 must be divided by two, since the WRS test must be followed by the Quantile test (refer to NUREG-1505) using the same data set. Again, the MARSSIM COMPASS code will be used to calculate the sample size and construct the power curve. This scenario B survey design required only 15 soil samples. The probability that the survey unit passes as a function of Th-232 concentration is shown in Figure 7.3.

Some explanatory notes are necessary in Figure 7.3. First, the probability that the survey unit passes is not the power $(1 - \beta)$, but actually the β error. This type II "error" represents the fact that we accept the scenario B null hypothesis that the survey unit is indistinguishable from the background, and is shown as a function of the Th-232 net concentration in Figure 7.3. Strictly speaking, the type II error does not exist until the concentration exceeds that concentration given by $DCGL_W$ plus 3ω, where the error is set at 0.025. The type I error is set at the LBGR, and represents the error in concluding that the survey unit is distinguishable from the background concentration by more than the $DCGL_W$ level (0.8 pCi/g) when it is not. This type I error progressively increases as the concentration approaches the concentration given by $DCGL_W$ plus 3ω. The concentration between the LBGR (0.145 pCi/g) and the $DCGL_W$ plus 3ω (0.945 pCi/g) is called the gray region, and only type I errors (those impacting the D&D contractor) can occur in the gray region.

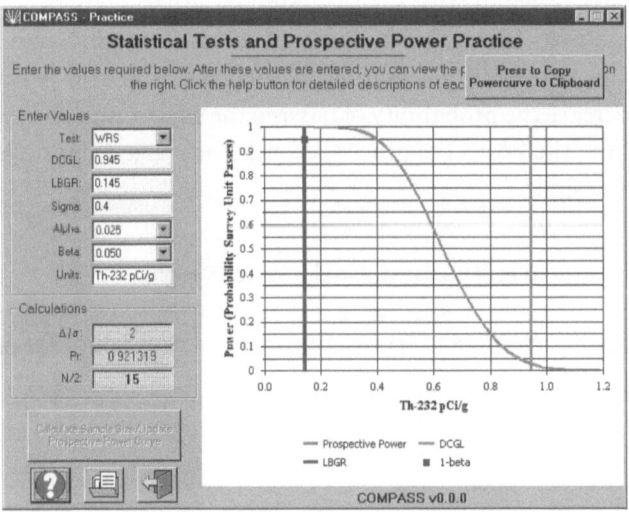

FIGURE 7.3 Probability that survey unit passes versus Th-232 concentration for scenario B.

Assuming that the scenario B survey design is implemented, there are some points to consider. First, it is important to note that the WRS test is performed somewhat differently for scenario B than it is for scenario A (see Chapter 14). For instance, the survey unit measurements are adjusted by subtracting the LBGR from each measurement, rather than the scenario A adjustment on the background measurements. Also, once all the data have been pooled and ranked, it is the adjusted survey unit measurements that are summed to determine the scenario B test statistic. The null hypothesis is rejected if the test statistic exceeds the critical value, meaning that the survey unit measurements are distinguishable from the background measurements by more than the $DCGL_W$. In this case remediation is the likely outcome. If the null hypothesis is not rejected, then a retrospective power analysis is performed. The D&D contractor constructs a retrospective power curve to ensure that the statistical design has sufficient statistical power. Retrospective power curves are covered in Chapters 14 and 18. Provided that the retrospective power curve indicates a powerful survey design, the Quantile test (refer to Chapter 7 of NUREG-1505) is then performed with the survey data to check for spotty contamination in the survey unit that may be missed by the WRS test.

Finally, getting back to the D&D project engineer's decision on whether to use the standard MARSSIM survey design (scenario A) or the scenario B survey design, the project engineer has constructed Table 7.5 to allow a comparison between the two survey designs. The first point of comparison is that the scenario A requires 32 soil samples while scenario B only needs 15. Perhaps most noteworthy is that of the increased scenario A sample size does not increase the probability of passing the survey unit shown in Table 7.5. Therefore, the project engineer recommends that scenario B be used for the FSS design.

TABLE 7.5   Comparison of Scenarios A and B Survey Designs

Th-232 (pCi/g)	Scenario A Probability of Passing SU (%)	Scenario B Probability of Passing SU (%)
0.145	~100	~100
0.5	89	81
0.6	60	58
0.7	25	35
0.8	5	15
0.945	~0	2.5
Sample size	32	15

This example demonstrates the attractiveness of the scenario B survey design. It becomes more appealing relative to scenario A as the background variability increases and the $DCGL_W$ decreases. These conditions lead to high sample sizes required by scenario A, as well as an unreasonably high risk of failing a survey unit that is not contaminated at all. That is, the survey unit might differ from the background reference area only due to variations in the background concentrations. Hence, scenario B was primarily established to minimize the risk of having the D&D contractor remediate background concentrations of radioactivity.

QUESTIONS AND PROBLEMS

1. Given the following series of 1-min measurements with a gas proportional detector on dry wall, calculate the mean and standard deviation:

   252, 237, 254, 251, 231, 248, 267, 255

   When calculating the mean background, what nonparametric statistical test is indicated? (Hint: Refer to Section 13.3.)

2. Given the following measurements from a reference area, are two reference areas necessary or is one sufficient? Cs-137 concentrations (pCi/g) in soil were as follows: 0.4, 0.7, 0.2, 0.5, 0.9, 0.4, 1.1, 0.7, 0.6, and 0.7. What additional information would assist in this decision? What if the $DCGL_W$ is 3 pCi/g? What if it is 0.8 pCi/g?

3. Outline the steps necessary to implement the scenario B survey design at a D&D site. Under what circumstances is it most likely that scenario B will provide a better survey design than scenario A? Do you think that scenario B survey design would be an effective survey design for building surfaces? Why or why not?

CHAPTER 8

Survey Instrumentation Selection and Calibration

THIS CHAPTER PROVIDES AN INTRODUCTION to the various survey instruments that can be used to perform decommissioning surveys. The focus is on those survey instruments that are used to perform the most common decommissioning survey activities—surface activity measurements, exposure rate measurements, and soil analyses.

The discussion of surface activity measurements requires a close coordination with Chapter 10. The determination of surface activity level at a particular location depends on two efficiency parameters—the instrument efficiency and the surface efficiency. The instrument efficiency, which depends on the operating characteristics of the survey instrument, will be covered in this chapter, while the surface efficiency is more appropriately discussed in Chapter 10. It may be helpful to look ahead to Section 8.4 to become familiarized with the general equation for surface activity.

A number of useful guidance documents address the issue of survey instrumentation calibration. The standard calibration guidance for decommissioning survey instrumentation comes from ANSI N323A (1997), ASTM E 1893 (1998), and ISO-7503 (1988a). ANSI N323A is the primary reference for calibration issues, as it provides a thorough coverage on portable radiation survey instrumentation test and calibration. Essentially, ANSI N323A requires that survey instrumentation be calibrated for the radiation types and energies to be measured and for the environmental conditions expected. This concise statement says it all when it comes to calibrating survey instrumentation.

8.1 CALIBRATION FOR SURFACE ACTIVITY MEASUREMENT INSTRUMENTS

The successful calibration of survey instruments begins with the selection of an appropriate calibration source. Unfortunately, many believe that as long as the source is "NIST-traceable," an appropriate calibration source has been selected. While traceability is certainly critical, it is by no means the only consideration. In this section, we will explore a number of considerations when selecting a calibration source for survey instruments used to perform surface activity measurements.

Calibration sources should be traceable to the National Institute of Standards and Technology (NIST) to ensure the accuracy of the source activity. According to ISO-8769 (1988c), calibration sources should be completely stated in terms of radiation type, quantity, and amount of backscattered or absorbed radiation, and for surface activity measurement calibrations, it is important that these sources be specified in terms of both activity and surface emission rate. Additionally, ISO-8769 provides guidance on the performance criteria for the uniformity of the surface emission rate across the calibration source. The calibration source activity should be of sufficient strength to provide adequate counting statistics, but at the same time should not be more than 100 or 1000 times the surface emission rates anticipated during the decommissioning surveys. Also, the radionuclide half-life should be at least several years long to reduce the potential for improper decay corrections on the source activity (ANSI 1997).

The radiation energies of the calibration source should approximate the radiation energy of the contaminant(s) to be surveyed. Furthermore, the calibration source should exhibit the same backscatter and self-absorption characteristics as the surface being measured. This is essential for the accurate determination of surface activity levels. However, this is rarely the case at most decommissioning sites, and strengthens the argument for using ISO-7503, which recommends separate assessments of instrument and surface efficiency during calibration.

The selection of an alpha calibration source is straightforward because most alpha emitters emit alpha radiation in the 4–5 MeV range, and detectors do not exhibit substantially different efficiencies over this energy region. Common alpha radiation calibration sources include Th-230, Am-241, and Pu-239. The selection of an appropriate beta calibration source is a bit more involved because beta emissions are polyenergetic—each beta emitter has a characteristic maximum beta energy emission and average energy. Because of the multiple beta emissions potentially present at D&D sites, it is necessary to determine the average beta energy for the decay series progeny present, or for any other mixture of radionuclides that may be present. Common beta radiation calibration sources include C-14, Tc-99, Tl-204, Cl-36, and SrY-90.

For radioactive materials or decay series that emit both alpha and beta radiation, it may be necessary to consider the detector's response to each of the alpha and beta emissions present. This method is addressed in Section 8.4 (see the case study "Determining Instrument Efficiency for 4% Enriched Uranium"), and requires that the decay scheme of the contamination be completely understood in terms of radiation type, energy, and abundance. The detector's efficiency for each radiation emission can be determined by calibrating to a representative radionuclide and by empirical relationship. For example, the instrument efficiency for alpha and beta emissions may be assessed during calibration by using appropriate alpha or beta calibration sources. Additionally, instrument efficiencies for beta emissions that may not be determined during calibration due to the lack of appropriate calibration standard beta energy, may be calculated empirically through an efficiency versus beta energy calibration curve (Section 8.4). Table 8.1 provides example calibration data for a Ludlum 43-68 gas proportional detector in terms of instrument efficiencies for C-14, Tc-99, Tl-204, and Sr/Y-90.

These data can be used to generate instrument efficiency versus energy curve (using either the average beta energy or the maximum beta energy) as shown below.

Let us assume that we are interested in the instrument efficiency for an average beta energy of 220 keV. Because this energy is not one of the calibration energies, we can use the empirical equation to determine the instrument efficiency. Since the beta energy of 220 keV is fairly close to the average energy of Tl-204 (244 keV), we would expect the instrument efficiency to be close to, and somewhat less than, 0.36. Using the average

TABLE 8.1 Calibration Data for Generating Efficiency versus Beta Energy Calibration Curve

Radionuclide	Average Beta Energy	Maximum Beta Energy	Instrument Efficiency (keV)
C-14	49.7	156.5	0.04
Tc-99	84.6	293.5	0.24
Tl-204	244	763.4	0.36
Sr/Y-90	565	1413	0.50

beta energy empirical equation derived from the calibration data, and substituting in 220 keV:

$$\text{Instrument efficiency} = 0.1747 \ln(220) - 0.5962 = 0.346$$

Therefore, the instrument efficiency is 0.35 for an average beta energy of 220 keV.

As an aside, a similar approach to obtain an efficiency versus energy curve is used for gamma spectrometry. Efficiencies are determined for a dozen or more gamma radiation energies in a mixed-energy gamma standard for a particular geometry. The energy range is typically on the order of 50 keV to more than 2000 keV; the wider the range in energy, the larger the number of radionuclides whose efficiency must be determined. Efficiency calibration data are often converted to a log-log scale and fit to a several-term polynomial function using regression analysis. Once the efficiency versus energy curve has been generated, it is important to check the resulting fitted curves.

8.2 OVERVIEW OF SURVEY INSTRUMENTATION AND PHILOSOPHY OF INSTRUMENT SELECTION

The selection of the proper survey instrumentation is critical to ensure that the data obtained are appropriate for the particular decommissioning survey. To optimize the instrument selection process, the surveyor must understand the operating characteristics and limitations of each survey instrument. Fundamentally, the survey instrument is selected based on the characteristics of the radiation being emitted from the radionuclides of concern at the D&D site. Obviously, an alpha detector would be chosen for contaminants that emit alpha radiation, while a thin-windowed gas proportional detector would be a fine choice for low-energy beta emitters such as Ni-63 or C-14. Secondary considerations in instrument selection include anticipated survey conditions, such as surface conditions, temperature, ambient gamma exposure rates, and radon levels. Also important is the expected performance of the survey instrument in the field. How rugged is the instrument? Can it be repaired in the field? How well does it perform during cold temperatures? Is it overly sensitive to physical shock?

Suffice to say, there are a number of factors that influence survey instrument selection—for example, cost, sensitivity, field ruggedness, performance under temperature, and pressure extremes are just a few. The physical probe area (detector size) is also important. When the surface area to be scanned is large, the obvious choice is the detector that covers the greatest area, that is, the largest probe area. This usually means the floor monitor—which, by the way, can be detached from the cart and used to scan wall surfaces as well. Perhaps even better is the large surface contamination monitor (SCM) from Shonka Research Associates, that can provide a detector width on the order of meters. However, some survey applications require accessibility into tight spaces. In these instances, small detectors (e.g., GM or ZnS detectors) are valued because they can provide access to floor drains, piping, or ductwork that otherwise could not be surveyed with larger detectors.

In fact, survey instrument selection lends itself quite nicely to the data quality objectives process. That is, once a given detector has been proven to meet the minimum requirements of the survey task, there is certainly a balance between desired detector performance—for example, efficiency, durability, automated documentation capabilities—and instrument cost. But clearly, the driving issues in instrument selection are the type of measurement data required (e.g., surface activity, gamma radiation level, soil concentration), the nature of the contaminants being assessed, and the sensitivity of the instrument. The MARSSIM recommends that the MDC for static measurements of surface activity be no greater than 50% of the $DCGL_W$. Therefore, it is essential that the survey designer has a complete knowledge of the nature of the contaminants at their D&D site, and particularly, an understanding of contaminant's decay characteristics—for example, half-life, type of radiation emitted, energy, and radiation yield. To facilitate this understanding, Appendix A includes radionuclide and natural decay series characteristics for common radionuclides encountered at D&D sites.

A challenge in selecting survey instrumentation is that many radiation detectors respond to more than one type of radiation. To counter this problem, survey instruments can be designed or modified such that only the radiations of concern are measured. For example, alpha-absorbers can be applied to detectors to eliminate the detector's response to alpha radiation when only the beta response is desired in a mixed radiation field. Consider the case where the contaminants of concern include both alpha and beta emitters, or where the radioactive material is a natural decay series that emits both alpha and beta radiation. The surveyor must decide if separate measurements of alpha and beta radiation are best to satisfy the survey design, or if the detector can be operated in the alpha plus beta mode. An important consideration in whether to use the gas proportional detector in an alpha-only and beta-only mode, versus an alpha plus beta mode, is (1) the MDC achievable in each mode, and (2) whether or not it is possible to establish relative ratios for the alpha and beta emitters potentially present, since the alpha plus beta measurements would be compared to a gross activity DCGL that combined the two radiations.

CASE STUDY SURFACE ACTIVITY MEASUREMENTS WHEN
ALPHA AND BETA RADIATION ARE PRESENT

The following cases are considered: (1) when separate measurements of alpha-only and beta-only measurements are necessary, and (2) when alpha plus beta measurements are appropriate.

CASE 1: SEPARATE MEASUREMENTS OF ALPHA-ONLY AND BETA-ONLY MEASUREMENTS

This is usually performed when it is not possible to establish defensible ratios of alpha to beta radiation, for instance, consider a site that has both Cs-137 and Am-241 contaminants. Furthermore, let us assume that based on the process knowledge for the site, there is no reason to believe that these contaminants have a relatively constant ratio with one another. Thus, it is necessary to use alpha measurements to assess the Am-241 surface contamination and beta measurements to assess the Cs-137 contamination. Note that alpha-only measurements can be performed either with a gas proportional detector operated in the alpha-only mode or by using a ZnS scintillation detector. Similarly, beta-only measurements can be performed either with a gas proportional detector operated in the beta-only mode or by using a GM detector. These instrument types are discussed more fully in the next section.

CASE 2: DETECTOR OPERATED IN THE ALPHA PLUS BETA MODE

This case includes the situation where it is possible to justify consistent ratios for two or more radionuclides that emit both alpha and beta radiation. This case also applies to the decay series with a known equilibrium status (e.g., natural thorium) and to processed, enriched, or depleted uranium (DU). In this case, the gas proportional detector would be operated in the alpha plus beta mode and the result compared to a gross activity DCGL.

More likely than not, case 1 would be used for individual radionuclides that emit either alpha or beta radiation since the alpha-to-beta ratio is generally unknown, while case 2 would be chosen for decay series radionuclides because it is easier to justify relative ratios. Refer to the MARSSIM survey design strategies and examples provided in Chapter 13.

8.3 SURVEY INSTRUMENTATION FOR SURFACE ACTIVITY MEASUREMENTS AND SCANNING

In the next three sections, we will discuss an overview of field survey instruments, a number of conventionally used survey instruments at D&D sites, and a brief introduction to advanced survey instrumentation. These discussions focus on the survey instrumentation that is used for assessing surface activity, both in the static measurement mode and by scanning.

8.3.1 Overview of Field Survey Instruments

Before we discuss the specific survey instrument types, it is beneficial to spend some time covering the basics of survey instrumentation. First, note that most field survey instrumentation consists of a detector (or probe) and a meter, connected by a cable. The survey meter houses the power supply that supplies the operating voltage to the detector. The survey meter also displays the results of the survey measurement—using a needle indicator on an analog scale and/or a digital display that indicates the number of counts for fixed count time (scaler). A survey meter that has both capabilities is commonly referred to as a ratemeter/scaler. In either case, the meter is equipped with an audible output (clicks or high-pitched chirps). For surface activity assessments, it is generally the rule that scaler counts, rather than an indicator needle, used for the determination of surface activity. During scanning, the indicator needle is periodically observed as the surveyor listens to the audible output of the meter, via a headset.

The detector is the part of the survey instrument that is actually responding to the ionizing radiation. Detectors used for decommissioning surveys are generally gas-filled (GM or gas proportional detectors for instance) or scintillation (NaI or ZnS) detectors. Many detectors have a window thickness that may attenuate the incident radiations to some degree, particularly alpha radiation. This can be a desirable feature, such as when it is necessary to discriminate between alpha and beta radiation with a gas proportional detector. Detectors also come in different sizes— such as varying physical probe areas. Section 8.3.2 discusses some of the more common detector types used during decommissioning surveys.

The survey cable connects the meter to the detector, and as such, is a critical component in the overall operability of the survey instrumentation. At the same time, recognize that survey cables can be a source of survey instrument failure. A quick diagnostic for survey cable failure is to move the cable around and note if there are a number of spurious counts. On a related matter, and while not strictly an example of instrument failure, it is important to recognize that using longer cables can cause differing instrument responses than standard-sized cables used for calibration. In 1993, Ludlum provided a technical note to its users that cautioned that changing coaxial cable lengths between the detector and survey meter can change the overall response of the detector.*

Ludlum explains in the technical note that scintillator or gas proportional detectors are more likely to be affected by cable length changes than GM detectors due to the detector pulse output varying in amplitude with radiation energy levels. That is, increasing or decreasing the coaxial cable lengths from a GM detector to a survey meter is more forgiving when compared to gas proportional and scintillation detectors. Unlike the gas proportional or scintillation detector, increasing the coaxial cable length from 1 to 2 m would not require recalibration or detector operating voltage adjustment to compensate for the increase in cable capacitance. The large pulse amplitudes, typically 1–6 V, produced by GM detectors allow longer cable lengths to be substituted without reducing the signal below the survey meter's threshold level. The detector operating voltage (HV) can be used to compensate for gain/losses or increases due to the change in cable length. Additionally, the threshold (input sensitivity) can be lowered to achieve the cable length compensation for an increase in cable length. Changes in the operating voltage or threshold will usually require recalibration of the survey instrument; if possible, it may be best to simply calibrate a dedicated survey meter/detector with a long cable if the conditions are likely to warrant its use.

A relatively recent enhancement in the use of survey instruments is the availability of data logging. This is particularly useful when many measurements—perhaps hundreds to thousands of measurements—are planned. In this situation, the characterization or final status survey can be greatly facilitated by data logging. While it can be packaged in a number of ways, it essentially allows the survey results to be automatically recorded and stored in a microprocessor. Furthermore, this survey result is correlated to a specific location, generally through the use of a bar code sticker that is affixed to a specific location in the survey unit. Data logging finds tremendous application with the large D&D projects that have substantial interior surface areas—perhaps power reactor or gaseous diffusion plant decommissioning projects come to mind.

Many of the same survey instruments discussed in the next section can be used for both scanning and surface activity measurements. A major

---

* The Ludlum Report, vol. VIII, nos. 2 and 4, June and December 1993.

difference is the MDCs that are attained for same instrument/detector pair depending on whether the instrument is used for scanning or surface activity measurements. Chapter 9 provides extensive discussion on MDC concepts, for both static and scan MDCs.

8.3.2 Conventional Survey Instrument Types

This section discusses some of the more common radiation detectors used for surface activity measurements and scanning. It certainly represents the current arsenal of survey instruments found on most D&D projects today. But at the same time, by no means should this section be construed as a complete list of decommissioning survey instruments, nor will it make anyone a survey instrument specialist. Perhaps the reader should consult a textbook that details the operation of radiation detectors. Well-regarded offerings include Knoll's *Radiation Detection and Measurement*, 4th edition (2010), Turner's *Atoms, Radiation, and Radiation Protection*, 3rd edition (2007), Cember and Johnson's *Introduction to Health Physics*, 4th edition (2009), and NCRP's *A Handbook of Radioactivity Measurements Procedures*, 2nd edition (1985). Topics to review should include gas-filled detectors that can be operated in the current mode or pulse mode and scintillation detectors. For a first-rate, one-page description on these and other detector types, refer to Appendix H of MARSSIM.

In the following discussion of survey instrument types, the format will consist of identifying the detector's survey applications, including both its strong and weak points. A very brief description of instrument operation will provide a basic theory of operation.

8.3.2.1 ZnS Detector: Alpha Measurements

The ZnS scintillation detector is used for both alpha surface activity measurements and surface scans for alpha radiation. The alpha scintillator is based on the principle that alpha particles will produce light (scintillation) in a phosphor. The phosphor may be a thin sheet of plastic scintillator or zinc sulfide embedded in a transparent tape. The thickness of ZnS detectors is limited to about 10 mg/cm^2 because poor light transmission of thicker layers will decrease detector efficiency. A very thin aluminized Mylar foil is wrapped around this to exclude light, while allowing the alpha particles to penetrate. Because of their low penetrating power, alpha particles are stopped in the thin scintillator and deposit nearly all of their energy there.

ZnS phosphors are relatively insensitive to beta or gamma radiations because beta particles and gamma photons will lose only a small fraction of their energy in the thin scintillator, and therefore produce much smaller pulses. If the discriminator level is set to accept the larger pulses due to alpha radiation, and reject the smaller pulses produced by beta and gamma radiation, the instrument can be used for detecting alpha contamination, even when beta–gamma contamination may be present.

The physical probe area is usually between 50 and 100 cm²—thus, this survey instrument is not ideally suited for floor scans. The detector is fairly rugged, but does suffer from light leaks if the Mylar window is punctured during surveys. These light leaks can be diagnosed by holding the detector close to a light source and listening for counts above background, which would indicate a light leak. Although a light leak renders the detector inoperable, the silver lining in this instrument failure is that it is an obvious failure. Some failures can be subtle and are quite problematic, such as a slow rate of P-10 gas leaking from a gas proportional detector.

The bottom line is that the ZnS is an attractive survey instrument because it does not require an external counting gas supply, and it is able to discern only alpha radiation in a mixed radiation field. In addition to light leaks, another disadvantage is that the detector's efficiency varies by roughly 10% across the probe face—therefore, the recommendation is to calibrate the detector to a calibration source that is at least as large as the physical probe area (ISO 1988c).

8.3.2.2 GM Detector: Beta Measurements

Maybe the most recognizable field survey instrument at the decommissioning sites is the pancake GM detector. This survey instrument consists of the GM detector coupled to a survey meter, and is deployed in the field. This detector—characterized by a mica window thickness of approximately 1.7 mg/cm², and probe area of about 20 cm²—continues to enjoy widespread use for a number of reasons. It can respond to alpha, beta, and gamma radiation, albeit by varying degrees, small and can fit in tight spaces, and is a sealed gas detector (no P-10 gas and the requisite DOT shipping papers are needed). The sealed gas is composed of an inert gas, which is a mixture of argon, helium, neon, and a halogen-quenching gas. Another benefit is that the GM detector is quite rugged, and even if a GM tube is punctured while surveying, the detector can be easily fitted with a replacement tube in the field.

The GM detector is primarily used to perform surface activity measurements, but also finds use for surface scanning, particularly for the clearance of materials and when difficult-to-access locations (pipe interiors, small ducts, etc.) are encountered. The GM could be used to scan floor surfaces, but do not expect this to make you popular with the survey tech nicians (remember the size of the GM detector—20 cm²). Although, for covering larger areas, detectors can be purchased that consist of multiple GM tubes housed within a single detector in a diamond-shaped array. This provides a greater effective physical probe area, which is especially helpful for scanning.

The GM is frequently stated as a beta–gamma detector, but it should be recognized that the beta efficiency is much greater than the efficiency to gamma radiation. For example, the beta efficiency (in counts/disintegra-tion) may be about 0.20 for Co-60 betas, while it will likely be less than 1% for the Co-60 gamma radiation. Further, the GM detector can respond to alpha radiation but is highly dependent upon the attenuation produced by the surface conditions. (Note: Alpha efficiencies with a GM detector can be as high as 0.08 counts/disintegration for a clean, nonattenuating calibration source surface.)

The GM can also be shielded (an example is the Ludlum 44-10) with a small amount of shielding to reduce detector background. This applica-tion may come in handy when the ambient gamma levels result in high GM background levels. This application is useful during characterization, to determine where additional remediation is necessary in a high back-ground field.

8.3.2.3 Plastic Detector: Beta and Gamma Measurements

Plastic scintillators can be used for both beta and gamma measurements, with the thickness of the scintillator being the key factor upon which radi-ation type is primarily detected; that is, if beta response is desired, without a significant gamma response, then a scintillator thickness of 0.25 mm might be used. Specifically, the scintillator thickness is determined by the range of the most penetrating particle to be observed. For very low ener-gies, 0.25 mm thickness can provide near 100% efficiency for charged par-ticles penetrating the window, but almost no gamma response. Thicker scintillators may be used if gamma sensitivity is desired or is not a problem.

Thus, plastic scintillators are useful for detecting beta radiation over a wide energy range, and they are also sensitive to alphas and gammas in most cases, depending of course on the window material used (less for alpha detection) and the scintillator thickness (more for gamma detec-tion). The Mylar window thickness is usually sufficient to allow some alpha response, but can be increased so that the detector does not respond at all to alpha particles. In fact, aluminum thicknesses can be added to stop unwanted alpha particles from penetrating the detector window.

As with the ZnS scintillation detector, plastic scintillators are also prone to light leaks, and can be checked by holding the detector up to a light source. Sealing against light is provided by a light tight reflector. The specifications for a commercially available beta scintillator are provided below:

Ludlum Model 44-116 Beta Scintillator

Scintillator: 0.25-mm-thick plastic scintillation material (PVT)

Window: Typically 1.2 mg/cm² aluminized Mylar

Window area: Active—125 cm² (e.g., physical probe area)

Open—100 cm²

Efficiency (4π geometry): Typically 15% for Tc-99; 22% for S-90/Y-90; 22% for Cl-36

Background: Typically 30–35 cpm/μR/h (e.g., in a 10 μR/h, the background would be 300–350 cpm)

Nonuniformity: Less than 10%

The beta efficiency for plastic scintillators is less than that for gas proportional detectors, but an advantage is that no counting gas supply is required with the plastic scintillator.

8.3.2.4 Dual Phosphor Detectors: Alpha and Beta Measurements

Dual phosphor, or phoswich, detectors are composed of two detector materials (scintillators) that are "sandwiched" together. A common example of a phoswich detector used for the measurement of alpha and beta surface activity consists of a plastic scintillator for beta detection coated with ZnS for alpha detection. Thus, the surveyor can assess both alpha and beta surface activity at a particular location. That is, by simply throwing a toggle or switch on the meter, the surveyor can record both the alpha cpm and the beta cpm.

The potential drawbacks are slightly reduced beta efficiency since the betas have to travel through the ZnS prior to interacting with the plastic scintillator. For example, the Tc-99 instrument efficiency with a gas proportional detector (0.8 mg/cm²) is about 40%, while the instrument efficiency for the dual phosphor detector may only be about 25% for Tc-99. However, an even more disturbing drawback may be the cross-talk that occurs between the alpha and beta channels. While this cross-talk can go both ways, the alpha-to-beta cross-talk is more prevalent, sometimes reaching 10% (i.e., 10% of the alpha response are interpreted as beta counts). Nonetheless, this detector type can be a very good choice for decommissioning sites that require both alpha and beta measurements. Examples of dual phosphor detectors include the Bicron AB50 (50 cm^2) and the Ludlum 43-89 (100 cm^2).

8.3.2.5 Gas Proportional Detector: Alpha or Beta Measurements

Gas proportional detectors are gas-filled detectors that can be set up to measure alpha, beta, or alpha plus beta radiation. The counting gas used in these detectors is P-10 (10% methane, 90% argon). It is common to operate the detector in a continuous gas flow mode. That is, once the detector is purged and checked out with a radioactive source, P-10 gas is passed through the detector at a rate of about 40 cm^3/min. It is also possible to operate the gas proportional detector with a static gas charge—that is, disconnected from the gas supply—but this must be done with caution.

The potential problem in operating the gas proportional detector on a static charge is that the P-10 gas may leak from the detector. Even though the manufacturer states that a static gas charge can last for a substantial duration (e.g., one claims more than 15 h), many times the gas seal is not what it once was and gas leaks out. The big problem is that this is not easily detectable—such as a light leak on a scintillator—but rather, the efficiency of the detector progressively gets worse, negatively affecting the data quality due to reduced efficiency. Notwithstanding the problem of operating on a static gas charge, sometimes it is necessary, such as reaching upper walls and ceiling or other locations at a significant distance from the P-10 gas supply. It is important to keep close tabs on the detector background while operating on a static charge. If it drops to a noticeable degree, maybe 5–10% below the expected background, it is necessary to repurge the detector. Increasing the frequency of source checks to ensure proper functioning is also a good idea.

Detector sizes (i.e., physical probe areas) generally range from about 100 to 600 cm^2, with the larger sizes finding extensive application as floor monitors. The gas proportional detector window material is typically aluminized Mylar, with a standard thickness of about 0.8 mg/cm^2. It is precisely the thickness of the Mylar window that permits discrimination of the alpha radiation, if desired. Furthermore, thinner Mylar windows (0.4 mg/cm^2) can be used for the detection of low-energy beta emissions, such as C-14 or even Ni-63. According to the Ludlum instrument catalog,

4π detection efficiencies for the gas proportional detector with a 0.8 mg/cm^2 Mylar window are quoted as follows: 20% for Pu-239; 30% for Tc-99; 15% for C-14; and less than 1% for gamma radiation. Based on ORAU measurements on Tc-99 calibration sources, the 4π detection efficiencies as a function of Mylar window thickness are about 0.28, 0.25, and 0.16 counts per disintegration, respectively, for 0.4, 0.8, and 3.8 mg/cm^2 thicknesses.

The standard alpha operating voltage is typically between 1000–1500 V. With this voltage setting, only alpha pulses produce enough ionization to allow the radiation to be detected; beta pulses do not exceed the detection threshold. This is reported as the alpha-only mode (more on the mode of operation and instrument selection strategy follows below). The background level is less than 5 cpm (though it is material dependent; see Chapter 7) when operating on the alpha-only plateau region.

The gas proportional detector can also be operated in the beta-only mode. This occurs on the beta plateau, which is typically 1700–1800 V. Now, if alpha radiation is present, it too will be detected—hence, the name "alpha plus beta mode." If it is desirable to discriminate against the alphas, then a thicker window is used (3.8 mg/cm^2); this is termed the beta-only mode. The background level on the beta plateau is typically 400 cpm or less, again dependent on the surface material being surveyed.

The mode of operation for a gas proportional detector is an important consideration when selecting survey instrumentations (refer to the case study "Surface Activity Measurements When Alpha and Beta Radiation Are Present"). This is particularly important when both alpha and beta emitters are present. The alpha-only mode is established by discriminating against the beta radiation that may be present. This is accomplished by using an operating voltage that puts the detector on the alpha plateau. (It may be beneficial for the reader to consult a textbook on gas-filled radiation detectors if additional information on plateau curves is desired.) At the alpha plateau voltage, only the alpha radiation produces sufficient ionization energy to result in detector response. Thus, by keeping the operating voltage on the alpha plateau—which is several hundred volts less than the alpha plus beta plateau—only alpha radiation is detected. The standard window thickness of 0.8 mg/cm^2 will permit alpha radiations to enter the window.

To measure alpha and beta radiations, the gas proportional detector is operated on the alpha plus beta plateau—which is usually around 1700–1800 V. Therefore, as long as the alpha radiations can make it through the detector window, the gas proportional detector is operating in the alpha plus beta mode. This is the preferred mode when the detector is used for surface scanning.

Finally, the gas proportional detector can be used to measure only beta radiation. This is accomplished by keeping the operating voltage on the alpha plus beta plateau, but using a thicker window to attenuate the alpha radiation. The amount of window thickness should be selected to eliminate the detector's response to alpha radiation, while at the same time maximizing the detector's response to beta radiation. The standard window thickness used for this purpose is 3.8 mg/cm^2 which can be obtained from Ludlum.

One of the more well-known uses of the gas proportional detector is its use as a floor monitor. The floor monitor uses a large detector (573 cm^2) deployed on a cart that maintains a minimal spacing between the detector and surface being scanned. Again, the floor monitor can be operated in the alpha-only mode, beta-only mode, or alpha plus beta mode—depending on the contaminants of potential concern.

One final note about gas proportional detectors is to pay special attention to windowless gas proportional detectors. Several instrument manufacturers market these models for the measurement of tritium surface activity levels, the basic idea being that the extremely low-energy beta from tritium just might produce a count in the detector provided that the window does not get in the way. In theory, windowless gas proportional detectors can accurately assess tritium surface activity levels, but in practice, at least in this author's experience, they are not reliable and use a lot of P-10 counting gas. A common problem is trying to maintain a seal between the detector and the surface, and keeping dirt and static charges from producing false readings. To be fair, some D&D professionals have reported success with these windowless detectors.

8.3.2.6 NaI Scintillation Detectors: Gamma Radiation

NaI scintillation detectors are very sensitive to gamma radiation, and therefore make them the ideal choice for locating elevated radiation from gamma emitters in soil. As would be expected, these detectors are almost always used in the scanning mode, especially for gamma scans over land areas. Various sizes of NaI crystals are available for radiation detectors. Common detector sizes include 1″ × 1″, 1.25″ × 1.5″, and 2″ × 2″ NaI detectors. The larger the detector, the more sensitive the detector generally becomes and greater the background level. Thus, the issue of detector size can be an important consideration in detector selection. On occasion, NaI scintillation detectors, particularly low-energy gamma detectors, have been used to measure surface contamination from low-energy photon emitters such as I-129, I-125, and others.

A special NaI crystal used to detect low-energy gamma radiation and x-rays is the FIDLER (field instrument for the detection of low-energy radiation). The FIDLER consists of a 5″ diameter by 1/16″ thick crystal that finds extensive use for the detection of plutonium x-rays (17 keV x-rays) and Am-241 gamma radiation (60 keV). It has even been used successfully at sites contaminated with DU, which has a significant component of low-energy gamma emissions from Th-234 (U-238 progeny) and Th-230. It is usually operated with an energy window setting that corresponds to the photon energy of interest, rather than a gross counting mode.

8.3.3 Advanced Survey Instrument Types

A number of advanced survey instruments are gaining wide acceptance in the decommissioning industry. These survey instruments are quite specialized and are relatively expensive compared to the commercially available instrumentation discussed in the previous section. For indoor survey applications, examples of advanced survey instruments include the SCM for scanning large floor surfaces, and the Trimble Total Station for capturing geospatial data necessary for mapping radiological contamination indoors. Innovative outdoor survey applications include *in situ* gamma spectrometry (ISGS) and gamma radiation scanning systems that consist of an array of NaI scintillation detectors mounted on various platforms (e.g., truck, utility vehicle, mule, human) coupled to a global positioning system (GPS). Although these innovations have been developed beyond the design phase and are deployed at a number of D&D sites, a number of advanced survey instruments are indeed in the early stages of development.

While high-purity germanium (HPGe) detectors continue to be best-in-class in terms of energy resolution, they are heavy and relatively expensive due to the need for cryogenic cooling. Cadmium zinc telluride (CdZnTe, or just CZT) has become a leading semiconductor detector for gamma spectroscopy at room temperature. The CZT detector is an attractive survey instrument because it does not require a supply of liquid nitrogen for cooling. Owing to its high Z composition, this detector is finding increasing application in the field identification of gamma-emitting radionuclides. While CZT detectors have excellent resolution characteristics, it is expensive to grow CZT crystals large enough to provide good collection efficiency. The energy resolution of CZT at the Co-57 122 keV gamma line is 1.5 keV—much closer to the resolution of HPGe (0.4–0.5 keV) than to the NaI detector (about 12 keV). However, the CZT detector efficiencies are notably lower than HPGe and NaI detectors due to its smaller detector active thicknesses. The FLIR Radiation Raider is one example of a hand-held survey instrument that uses an array of CZT detectors to identify gamma-emitting radionuclides in the environment. While this detector is portable, CZT detectors are not expected to be used for quantifying radionuclide concentrations.

Lanthanum bromide ($LaBr_3(Ce)$) detectors, which are becoming increasingly popular, are inorganic scintillation gamma radiation detectors that operate at room temperature and exhibit improved resolution (by roughly a factor of 2) and temperature stability compared to NaI(Tl) detectors. Typical crystal sizes for spectroscopy systems range from 1×1 in. to 3×3 in. Even though lanthanum bromide detectors do not approach the outstanding resolution of HPGe detectors, their efficiency is about 1.3 times that of NaI(Tl) for the same crystal volume. While lanthanum bromide detectors are capable of finding energy peaks more quickly due to their superior efficiency compared to a comparably sized NaI(Tl) detector, $LaBr_3$ detectors exhibit internal radioactivity that reduces its spectral resolution at energies below 100 keV. Stay tuned for emerging applications of $LaBr_3$ on decommissioning projects, particularly for quickly identifying gamma-emitting radionuclides in the field.

The next two sections provide an introduction of some of the state-of-the-art survey technologies that are currently being deployed at a number of D&D sites.

8.3.3.1 In Situ *Gamma Spectrometers: Gamma Radiation Measurements*

ISGS is often touted as having the potential to reduce, if not completely eliminate, the need for soil samples. The usefulness of ISGS is likely neither as good as its advocates claim nor as poor as its critics contend. In this section, we will try to draw the lines more clearly on exactly where the utility of ISGS lies. With the state of the art ever increasing, it is not farfetched to consider that advances in technology will soon render the application of ISGS even more useful than that described herein.

A brief history of the field of ISGS may be appropriate at this point. First, the field of ISGS for assessing radionuclide concentrations in soil has been around for about 30 years. In fact, when the scientists at the Environmental Measurements Laboratory (EML) began this technique the state-of-the-art detector was a NaI. The technology continued to advance along with the development of GeLi detectors with better resolution—but had to be maintained at liquid nitrogen (LN_2) temperatures at *all* times—and over the past decade, the detector used for ISGS is the HPGe detector. Recent advances in computing allowed instrument manufacturers to use Monte Carlo N-Particle (MCNP) modeling to develop efficiencies for various geometries. Canberra's ISOCS (*in situ* object counting system) system is very popular, with the manufacturer providing built-in efficiencies for a number of commonly encountered geometries at D&D sites. It seems that the ISOCS works well as an object counter precisely because it provides control over the geometry (a particular container size, a drum, etc.). However, there is no control over soil contaminant geometry and distribution in realistic D&D situations; therefore, using ISOCS becomes a bit more challenging.

The ISGS system usually consists of a semiconductor detector, electronics for pulse amplification and pulse height analysis, computer system for data collection and analysis, and a portable cryostat to cool the detector. HPGe detectors continue to be the semiconductor of choice, although advances in technology may soon result in the use of other semiconductor detectors (e.g., CZT detectors discussed earlier) that can function at room temperature. Another practical innovation has been the development of rugged, multiattitude LN_2 cryostats that have allowed the ISGS systems to be deployed in rather harsh (real) D&D environments.

The overall efficiency and sensitivity of an ISGS system for quantifying volumetric contamination depends on the intrinsic detector efficiency, radionuclide gamma ray energy and abundance (yield), background, count time, and geometry. The intrinsic detector efficiency depends on gamma energy, and is experimentally determined for various photon energies. At low energies, gamma radiation is often absorbed outside the detector (e.g., attenuation by the radioactive source), and in the detector casing or cap. As the gamma energy increases, the efficiency rises rapidly due to the abrupt reduction in the attenuation mentioned above. A maximum intrinsic efficiency is reached for an energy value that depends on the detector and source characteristics. Above a few hundred keV, the efficiency decreases monotonically (Knoll 2010).

ISGS systems are often deployed using shielding and collimation to reduce background levels. High background levels decrease the sensitivity of the ISGS, particularly at lower energies of concern due to the Compton continuum contributions. For example, measuring the 60 keV gamma line from Am-241 can be impacted by the presence of Co-60 radioactivity levels due to the increased Compton continuum created by the higher-energy

Co-60 photons. To reduce the impact of background, lead shielding and collimation is often used. While collimation lowers the overall efficiency of the ISGS system by effectively shielding the contamination from the detector, the sensitivity of the ISGS measurement is increased. The net effect is to improve the signal-to-noise ratio for the specific gamma energy of interest.

So, what are the specific decommissioning applications that offer the most promise for ISGS? Several applications for ISGS are briefly outlined below, each with a subjective, qualitative rating on the expected success of that particular application. A useful reference in this regard is Dehmel and Schneider (2003). Note that the application of ISGS for the clearance of solid materials is covered in Chapter 15.

> *Determining radionuclide concentrations in soil.* Clearly, determining radionuclide concentrations in soil is a major reason why these systems are purchased, which of course coincides with this also being a major selling point used by the manufacturers. Currently, this application is deemed to have a fair chance at success. Soil samples continue to be the accepted methodology for demonstrating compliance, and until regulators are comfortable with the large averaging areas provided by the ISGS units, it is unlikely that this state of affairs will change in the next few years. The principal drawback is that in many instances it is difficult to model the distribution of contamination in soil with a sufficient degree of confidence.

One strategy that does seem to work well for validating the depth of contamination is to compare the level of attenuation exhibited by two gamma lines at sufficiently different energies. An important condition of course is that at least two appropriate gamma energies are present in the contamination. One example might be the depth determination of U-238 using its immediate progeny—the very low 63 keV gamma from Th-234 and the high-energy gamma (1001 keV) from Pa-234 m. The concentrations for each of these radionuclides will be quantified using *in situ* analysis based on an estimated depth distribution from initial soil sampling results. Accounting for statistical uncertainties in the concentration calculation, if the concentration of Th-234 (63 keV) is less than Pa-234 m (1001 keV), then the contamination is deeper than that modeled, while if the Th-234 is greater than the Pa-234 m, then we can conclude that the contamination is shallower than that modeled.

A related application, or more correctly a subset application, is the ISGS's ability in determining the average radionuclide concentration or for large soil areas. In this regard, the ISGS's rated success is excellent. In fact, if there were no regulatory limits on averaging areas (i.e., one could average the contamination over hundreds of square meters) then the *in situ* system is infinitely better than soil sampling (again, with the caveat of proper calibration). So, the balance seems to be in the survey design that takes advantage of the inherent *in situ* ability to average the gamma fluence over large soil areas, while not overlooking the contribution from small areas of elevated contamination, which are of dosimetric concern. To summarize, the ISGS system can provide a very accurate assessment of the incident gamma fluence at the detector but the difficulty lies in interpreting the soil concentrations and distributions that give rise to this gamma fluence.

Providing "negative data" for certain site areas. This application of the *in situ* system is expected to offer an excellent success of confirming the absence of contamination in land areas. Essentially, the *in situ* system is set up to count for a sufficient time to achieve the desired MDC. As long as the MDCs are appropriately selected, the user will be able to confidently state that contamination is not present from specific radionuclides. This application can be expected to be widely used in class 3 survey units.

Identifying radionuclide contaminants. Again, this *in situ* application is anticipated to offer excellent success, not only in identifying the specific radionuclides but also in determining the relative ratios of multiple radionuclides in the field (of course, if they are gamma emitters). This application would also benefit the user at uranium sites, where the relative concentrations of U-238 and U-235 can be used to estimate the enrichment levels.

Scanning. This *in situ* application is rated as "good." While the *in situ* unit provides large area coverage when it is not collimated, it suffers from not being able to specifically identify the location of potential hot spots. This application may prove more useful if used in conjunction with conventional gamma scans using a NaI scintillation detector.

Use as field counting lab. This application has already proven to be an excellent use of the *in situ* system, though technically, can this really be called "*in situ*"? But nonetheless, using the *in situ* system to analyze soil samples in standard geometries helps to keep laboratory analysis costs to a minimum, and greatly expedites the turnaround time from sample collection to results. There is also a connection to the first point above. One way to obtain information on the distribution of the contamination is to collect soil samples at various depths and surface locations. This information can then be used to model the contamination profile.

Monitoring progress of remediation. The success of the *in situ* system is good in this application. It can be used *in situ* to measure soil concentrations, but maybe using action limits that are somewhat less than the actual soil DCGLs. This would address the above concern of the potential inaccuracies of the system, while still being compliant with the DQOs of the remedial action support survey.

An additional consideration is how ISGS might be worked into the MARSSIM survey approach. First, it is important to recognize that each of the applications discussed above certainly relate, to some degree, to the performance of MARSSIM surveys. The most direct way that ISGS can be included in the MARSSIM survey approach is for it to replace conventional soil sampling. Specifically, each *in situ* measurement would be used in the nonparametric statistical tests to demonstrate compliance with release criteria. The reader is referred to Chapter 13, Section 13.6 for a general discussion on how *in situ* gamma spectrometer measurements may be used in a MARSSIM survey design. Pertinent references for the reader interested in more detail on ISGS include ICRU 53 (1994) and NUREG-1506 (NRC 1995a).

CASE STUDY   EVALUATING THE ISGS FOR DETECTING
             DISCRETE PARTICLES IN SOIL

The *in situ* gamma spectrometer response for a "discrete particle" (or hot spot) was assessed in an ORAU study (Chapman et al. 2006). The objective of the research was to determine what level of radioactivity in a discrete particle will trigger a positive measurement result (and investigation), distinguishable from the background. Or, more simply, what is the discrete particle detectability using an ISGS? Discrete particles were defined as small localized volume of soil that contains one or several particles of radioactivity that is significantly more radioactive than the average low-level concentration of the surrounding soil.

ORAU performed ISGS measurements with reference source materials of Cs-137, Co-60, and natural thorium. A 38% efficient HPGe detector was positioned at 1 and 2 m heights above the ground; sources were placed on the surface and subsurface at 7.5 and 15 cm and in addition, at radii of 0 (on axis), 1, and 2 m. The amount of activity selected for each source was calculated to be somewhat less than the equivalent volumetric average concentrations, associated with nominal DCGLs published in the literature (see Chapter 4).

The primary feature of this study was to calculate the discrete particle activity, when located at various radial and depth locations that would result in further investigation. This discrete particle activity represents the "hot spot MDA" for the specified test conditions. Two pieces of information are required to calculate the hot spot MDA as defined in this study: (1) the minimum detectable (MD) counts in the photopeak region from background spectra, and (2) the hot spot efficiency for a specific source geometry. Hot spot MDAs were then calculated by dividing the MD counts obtained from the background spectra by the particular detector efficiency.

For example, with no source present, the MD counts at the 1 m detector height for the Co-60 1173 keV gamma line, was 32.2 counts based on the Nuclide MDA Report, which provided 0.1467 µCi/unit in the photopeak region. This value was multiplied by the efficiency (6.60E-6), conversion factor (2.22E6 dpm/µCi), and 15-min live time.

For the 1 m detector height, a Co-60 test source (1.2 µCi) was positioned at a 1 m radial location and surface position (zero depth). The Co-60 source produced a net count in the photopeak region (1173 keV) of 117 counts. The detector efficiency for this particular discrete particle geometry is given by

$$\text{Detector efficiency} = \frac{117 \text{ counts}}{1.2 \text{ µCi}} = 97.5 \text{ counts/µCi}$$

The hot spot MDA is calculated by dividing the MD counts by the detector efficiency:

$$\text{Hot spot MDA} = \frac{32.2 \text{ counts}}{97.5 \text{ counts/µCi}} = 0.33 \text{ µCi}$$

Hot spot MDAs are provided for the various experimental configurations of the discrete source activity (Chapman et al. 2006). The results indicate that when the discrete source is directly beneath the detector at 1 m height, it is possible to detect 0.02–0.04 µCi of Co-60 or Cs-137. When the detector height is increased to 2 m, the hot spot MDA for each radionuclide increases to 0.08–0.16 µCi. This is due to the detector's greater field of view at the 2 m height and the corresponding $1/r^2$ decrease in geometric efficiency. Furthermore, as the discrete source is moved from directly beneath the detector to 1 m radius, the hot spot MDAs increase by a factor of 5–8 for both Co-60 and Cs-137. Finally, it is interesting to note that increasing the depth of the discrete source burial from the surface to 15 cm has less of an impact on the hot spot MDA than moving it to the 1 m radial location.

In summary, based on this study, it is generally possible to detect 1 μCi (and many times less than 1 μCi) of a discrete particle using ISGS for Co-60, Cs-137, and Th-232, for a number of experimental conditions that include depths up to 15 cm and radial locations out to 2 m.

8.3.3.2 Automated Scanning and Measurement Systems

Advances in survey instrumentation have reached the point where it is now possible to scan an area and have a microprocessor produce an automated report. Too good to be true? Perhaps it is too expensive for some D&D budgets, but certainly the capability exists. These systems can track both the position and the output of the radiation survey instrumentation, for example, the instrumentation that links the radiation measurement with its location, such as the SCM built by Shonka Research Associates, Inc (SRA). This floor monitor uses a position-sensitive proportional counter that allows one detector to act as the equivalent of hundreds of individual detectors—which results in the collection of tremendous amounts of data. Process software saves the survey data and correlates the data as a function of survey location.

The SRA SCM has been primarily used to scan a large surface area within a building, such as the massive gaseous diffusion buildings in Oak Ridge, Tennessee. It consists of a cart-mounted SCM and a survey information management system (SIMS). Figure 8.1 shows the SCM being used to scan floor surfaces in an experiment designed to evaluate its performance to assess alpha contamination. As stated previously, its position-sensitive proportional counter can be configured in nominal 1 and 2 m detector lengths and used to scan surfaces in a series of 1 or 2 m wide

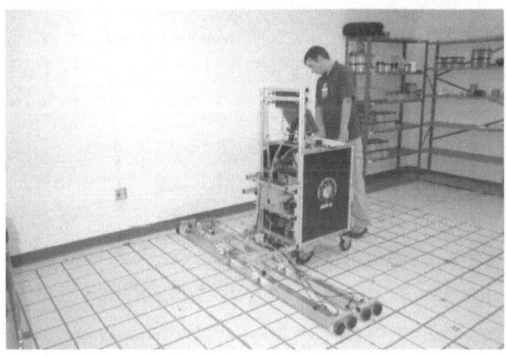

FIGURE 8.1   SRA surface contamination monitor.

strips. The locations of these measurements are referenced to a surface grid using a wheel encoder. Once calibrated, the system automatically records the detector output and location information in the surveyed area.

Outdoor radiation surveys are routinely performed using GPS and geographic information system (GIS) technologies to characterize, map, and report radiological conditions. Common tools such as Trimble GPS units provide accurate positioning to correlate the location and magnitude of the instrument response at user-defined intervals (often 1 s intervals). GIS integrates GPS location data and radiation survey data to display color-coded maps of radiation levels in the survey unit.

Another survey system that tracks the position and output of detectors is the ultrasonic ranging and data system (USRADS) unit. This technology consists of ultrasonic and radiotransmitter/receiver that use the radio and ultrasonic transmissions from the field surveyor to determine surveyor location and the instrument response at that location (ITRC 2006). Data are collected and processed, providing a documented survey that correlates the location and magnitude of instrument response at 1 s intervals.

Many of these systems can accommodate detection devices other than radiation detectors. For example, survey platforms can be equipped with ground-penetrating radar (GPR) to gather information about subsurface soil anomalies (e.g., unexploded ordnance). GPR can be linked to GPS/GIS to offer a more detailed geophysical survey that integrates hydrologic and geologic data. Subsurface soil layers can be evaluated using GPR and their geographic position with highly accurate GPS to create 3-D subsurface models. Thus, no matter the detector (NaI, HPGe, or GPR), these survey systems allow large areas to be surveyed (scanned) rapidly, and offer automatic acquisition of survey data. And it largely eliminates the issue of human factors, as it pertains to scan MDCs in Chapter 9.

Conveyor survey monitors (CSMs), discussed in Chapter 15, provide an effective approach for quickly scanning materials such as soil, copper chop, and smaller pieces of concrete rubble. CSMs operate by moving bulk materials past radiation detectors (e.g., plastic scintillators) using a conveyor system while automatically storing and analyzing the resulting detector signals via a data acquisition system. During calibration, investigation levels can be programmed into the software to stop the conveyor at specified radiation levels that correspond to release criteria.

To summarize, many of these advanced survey systems allow both the radiation level and the location to be tracked using mapping software. The survey data can be analyzed and formatted into a report of the radiological conditions, with measurements exceeding release criteria highlighted. The bottom line is that these automated scanning and measurement systems may replace the need for statistical survey designs provided that the instruments can be accurately calibrated and provide reproducible results. This begs the question: why collect a sample of the population when you can have the entire population?

8.3.4 Environmental Effects on Survey Instrument Operation

It is interesting that most planning for decommissioning survey activities tends not to consider the real conditions encountered at many D&D sites. At issue is the nature of the D&D site where survey measurements are made, and the environmental conditions that impact these measurements. Perhaps there are serious building concerns—is the roof leaking? Can the building be heated to keep the instruments properly functioning? What about surface conditions such as scabbled, dirty, and damp surfaces? [Note: These surface effects are covered in Chapter 10 in the discussion of the surface efficiency (ε_s).] At the same time, the environmental effects of temperature and pressure on instrument performance should be duly considered. There have been a number of assessments of instrument performance following the procedures in ANSI N42.17 (1989), such as the comprehensive report by Swinth and Hickey that evaluated the performance of health physics instrumentation subject to various environmental conditions. Some of the more common instrument effects that crop up on decommissioning surveys are highlighted in this section.

Most of the time, measurements of surface activity are performed indoors, at room temperature. In these situations, the temperature response of the survey instrument is largely a nonfactor. However, when measurements are being made outside—release of materials, paved lots, building exteriors—or within old buildings that are without heat or electricity (e.g., many FUSRAP sites), temperature effects can be the paramount concern. Swinth and Hickey studied the effects of temperature variations on a variety of health physics instrumentation. Both alpha scintillators and gas proportional detectors were among the instrument types tested. The researchers performed the evaluation by obtaining measurements at 20°C (ambient room temperature) and at 10° increments over a range of −20°C to 50°C. The instrument was allowed to stabilize for at least 30 min at each temperature prior to making a measurement. The performance criterion used for assessing failure due to temperature response was defined as a change in response of over 15% from the response at 20°C. The results were somewhat surprising.

Considering a temperature range of 0–40°C, *all* of the gas proportional detectors failed the temperature test. They noted that failures occurred about equally at both ends of the temperature range, 0 and 40°C—although the manufacturer indicated that meter errors are greater at the lower temperatures (this had also been ORAU's experience). They also noted that the instrument manufacturer specifies that the gas proportional instruments may require the resetting of the high-voltage control to achieve <10% meter reading error over the temperature range of −40°C to 60°C. This is because the voltage plateau of the detector shifts as a function of temperature.

Interestingly, the researchers noted that the GM detectors performed exceedingly well on the temperature tests—none out of 23 failed the 0–40°C temperature test. Conversely, alpha scintillators did not fare so well—two of five alpha scintillators failed. The researchers concluded that in some cases voltage adjustments can compensate for inadequate instrument response.

ORAU has adopted an approach for assessing temperature response, particularly for gas proportional detectors. For cold weather temperatures, defined as less than 32°F (0°C), the survey instrument will be calibrated at the cold temperature. This is a result of experimentation to quantify the temperature at which it is likely that the detectors no longer fall within the QC source check-out range. Figure 8.2 clearly shows how the voltage plateau shifts to the right with decreasing temperature. The problem with this shift is that the gas proportional detector may have a markedly lower efficiency than it did when calibrated at room temperature. That is, suppose that the voltage plateau was determined at 70°F (21°C), then from Figure 8.2 it can be seen that the operating voltage on the plateau is about 1700 V. Now, if the temperature drops to 30°F (−1.1°C), the high voltage determined during calibration (1700 V) does not keep the gas proportional detector operating on the plateau, which requires a voltage of about 1775 V. The net effect is that the detector calibrated at room temperature and used to make survey measurements at 30°F (−1.1°C) will suffer from a significant under response. The magnitude of this effect is less pronounced, but quite evident at temperatures from 30°F (−1.1°C) to 50°F (10°C). Temperatures slightly cooler than room temperature, perhaps as low as 50°F (10°C), are not impacted to any appreciable degree.

Another significant environmental effect on the survey measurements is from radon and its progeny. Radon levels within a building can exhibit significant diurnal variations, usually with the highest radon levels present

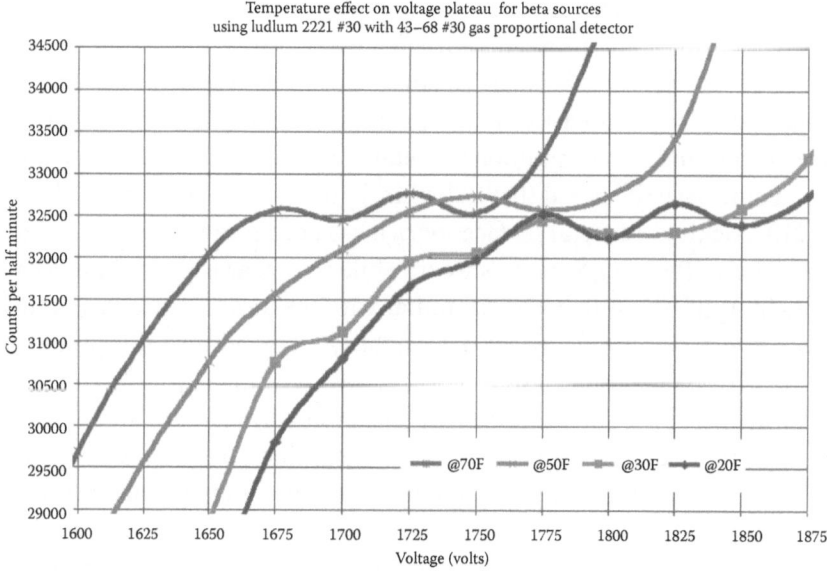

FIGURE 8.2   Temperature effect on voltage plateau for gas proportional detector. (Courtesy of ORAU.)

in the morning hours. Alpha surface activity measurements can be greatly impacted by radon progeny that have accumulated on building surfaces, perhaps being responsible for an increase in the alpha background from 1–2 cpm to 4–7 cpm. An easy way to determine if the increased alpha count rate is from radon progeny is to cover the location with paper for 30 min, and then perform a follow-up measurement. If the alpha count rate has been reduced, it is probably a result of the radon progeny decay.

Not only can radon and its progeny wreak havoc with background levels and their variability, but can also result in a certain radon progeny (Po-210) being mistaken as alpha contamination. This situation can be described as the Po-210 interference in measuring alpha surface contamination, and is particularly troublesome at transuranic (TRU) sites due to the historically low TRU surface activity limits (i.e., 100 dpm/100 cm²). Furthermore, this plate-out of radon progeny has been most noticeable on metal surfaces. In particular, the interference from Po-210 alpha radiation has been identified during the course of performing final status surveys of outdoor metal structures, especially on roofs. This problem has been confirmed at three DOE sites that have prominent alpha contaminants, including Rocky Flats (principal contaminant is Pu-239), Mound

Laboratories (principal contaminant is Pu-238), and the K-25 site in Oak Ridge (principal contaminant is enriched uranium).

The Po-210 deposition phenomenon seems to be readily observable by the presence of elevated fixed alpha contamination primarily on galvanized metal surfaces, in addition to metal that is rusty, oxidized, or weathered. Some have theorized that a chemical phenomenon is responsible for Po-210 "sticking" to metal surfaces once it has been produced by its longer-lived parent, Pb-210. (Note: It is worthwhile to review the uranium decay series in Appendix A to become familiar with the progeny involved in this phenomenon). A number of actions can be taken to address this problem.

The first item is to recognize that this phenomenon exists. This means that when you encounter unexpected alpha "contamination," do not be so quick to rule out the presence of Po-210, especially if measurements are made on metal surfaces. Second, confirm that the elevated alpha readings are from Po-210, and not plutonium or uranium. Metal "coupon" samples can be collected and analyzed by alpha spectrometry to identify the alpha energy present. It might also be possible to make separate measurements of gross beta and gross alpha surface activity—if it is Po-210, then the Bi-210 beta should also present, which can be compared to the two plutonium sites that have no significant beta contamination. The DQO approach should be used to address this problem. The solution may be to work with the regulator to develop an acceptable approach that might involve both field measurements of surface activity and limited number of coupon samples.

8.4 DETERMINATION OF INSTRUMENT EFFICIENCY FOR SURFACE ACTIVITY MEASUREMENTS

We will begin this discussion with a review of the conventional equation used to calculate the surface activity level in units of dpm/100 cm²:

$$\text{Surface activity} = \frac{R_{S+B} - R_B}{\varepsilon_T \times (\text{probe area}/100\,\text{cm}^2)}$$

where

R_{S+B} is the gross count rate of the measurement in cpm

R_B is the background count rate in cpm

ε_T is the 4π detector efficiency in counts per disintegration

The conventional equation uses an efficiency term that represents the 4π geometry and relates the number of counts measured to the number of disintegrations (i.e., efficiency in units of counts per disintegration). As with any instrument calibration procedure, it is important that the calibration source be representative of the conditions under which the detector will be used. That is, the total efficiency should accurately represent the surface type and condition—accounting for the overlying material present, dirty or scabbled surfaces, and so on. This may be achieved provided that the calibration source exhibits characteristics similar to the surface contamination, that is, radiation energy, backscatter effects, source geometry, and self-absorption.

However, it is not reasonable to always expect that the calibration source is representative of the field conditions where measurements are performed. To address this reality, the MARSSIM has adopted the ISO-7503 approach. The ISO-7503 guidance for surface activity measurements divides the total 4π efficiency into two components, instrument efficiency and surface efficiency (ISO 1988a):

$$A_s = \frac{R_{S+B} - R_B}{\varepsilon_i \varepsilon_s W}$$

where

R_{S+B} is the gross count rate of the measurement in cpm
R_B is the background count rate in cpm
ε_i is the instrument efficiency (unitless)
ε_s is the surface efficiency (unitless)
W is the area of the detector window (cm^2)

We will revisit this surface activity equation in Chapter 10. However, it is important to address that component of the overall efficiency factor that deals with the selection of instrumentation—called the instrument efficiency, ε_i. Consequently, determining the instrument efficiency is an important aspect of calibration for surface activity measurements. The instrument efficiency is formally defined as the ratio of the counts recorded by the instrument and the number of particles of a given type above a given energy emerging from the front face of the source per unit time (ISO 1988a). Hence, this term relates the number of counts recorded by the detector to the 2π emission rate from the surface.

The instrument efficiency is determined during calibration by dividing the net counts obtained with the detector over the calibration source by the surface emission rate of the calibration source. The surface emission rate is the 2π particle fluence that embodies both the absorption and scattering processes that affect the radiation emitted from the source. The fact that many of the radiations striking the detector may be backscattered is already accounted for by the 2π emission rate. That is, the surface emission rate is expressed in units of alphas per minute or betas per minute. The calculation of instrument efficiency can be expressed as

$$\varepsilon_i = \frac{R_{S+B} - R_B}{q_{2\pi,sc}}$$

where $q_{2\pi,sc}$ is the surface emission rate of the calibration source area that is subtended by the physical probe area of the detector.

The calibration source manufacturer usually quotes both the activity and surface emission rate of the source. ISO-8769 recommends that calibration sources be fabricated with an area of at least 150 cm^2 (ISO 1988c). The standard further states that survey instruments should be calibrated using sources that are larger than the physical probe area. For example, if the physical probe area is 126 cm^2, and the calibration source is 150 cm^2, then the quoted surface emission rate of the source must be corrected for the area that is subtended by the probe area.

Consider the following illustrative example. Surface activity measurements will be made with a ZnS detector that has a physical probe area of 74 cm^2. The calibration source area is 150 cm^2 and the surface emission rate of the Pu-239 source is 27,600 alphas per minute. The detector is placed on the source for 1 min and the gross counts are 4819 cpm. Assuming a background of 2 cpm, the instrument efficiency is calculated as follows:

$$\varepsilon_i = \frac{4819 - 2}{27,600 \times (74/150)} = 0.35$$

It is imperative to remember that the instrument efficiency is only a 2π efficiency, and as such, should always be multiplied by the surface efficiency before the surface activity is calculated. Otherwise, the calculated surface activity will be underestimated by approximately a factor of two, depending of course on the actual value of the surface efficiency. Again, the surface efficiency is covered in Chapter 10 since it is not impacted by the survey instrument used.

Table 8.2 provides a number of instrument efficiencies obtained by ORAU for gas proportional, GM, ZnS, and phoswich detectors. Instrument

TABLE 8.2 Instrument Efficiencies for Common Calibration Radionuclides

Radionuclide (Average Beta Energy)	Gas Proportional (Alpha + Beta)	GM	Gas Proportional (Alpha-Only)	ZnS
C-14 (49.4 keV)	0.21	0.10	—	—
Tc-99 (84.6 keV)	0.39	0.21	—	—
Tl-204 (244 keV)	0.50	0.36	—	—
SrY-90 (563 keV)	0.59	0.51	—	—
Th-230	—	—	0.40	0.33

Source: Courtesy of ORAU.

efficiency depends on the type and energy of radiation, and increases with increasing energy.

Let us consider a couple of examples to illustrate the determination of the instrument efficiency for multiple radionuclides. Assume that the gas proportional detector operated in the alpha plus beta mode will be used for beta measurements of SrY-90 and Tc-99. Based on historical site assessment data and supplemented with characterization data, the relative ratios of SrY-90 and Tc-99 were 60% and 40%, respectively. The weighted instrument efficiency can be determined using the individual instrument efficiencies from Table 8.2:

$$\text{Weighted } \varepsilon_i = (0.60) \times (0.59) + (0.40) \times (0.39) = 0.51$$

Thus, the instrument efficiency for this radionuclide mixture is 0.51. Remember that the surface efficiency must be determined and multiplied by the instrument efficiency to yield the total efficiency. While the surface efficiency will be covered in Chapter 10, it is worthwhile to discuss the way in which the surface efficiency can be weighted. Let us assume that the individual surface efficiencies are 0.5 for SrY-90 and 0.25 for Tc-99. First of all, we might note that it is more useful to calculate the weighted total efficiency than to calculate either of the intermediaries—that is, weighted instrument or surface efficiency. The weighted total efficiency is the value that would be used to convert net count rate to surface activity. It is calculated by multiplying the radionuclide fraction for each radionuclide by its instrument and surface efficiency as follows, then summing up the results for each radionuclide:

$$(0.60)(0.59)(0.5) = 0.177$$
$$(0.40)(0.39)(0.25) = 0.039$$
$$\text{Weighted total efficiency} = 0.216$$

The weighted surface efficiency can be calculated by dividing the weighted instrument efficiency into this weighted total efficiency (0.216/0.51). This yields a weighted surface efficiency of 0.424.

The same weighted surface efficiency result can be obtained in a more direct fashion by weighting the individual surface efficiencies by the fraction of the weighted instrument efficiency that each radionuclide represents. This is probably better explained using an equation:

$$\text{Weighted } \varepsilon_s = (0.50) \times \left(\frac{(0.60)(0.59)}{(0.51)} \right) + (0.25) \times \left(\frac{(0.40)(0.39)}{(0.51)} \right) = 0.424$$

Let us consider a second example for determining the weighted instrument efficiency. Suppose that a GM detector is being calibrated to measure Co-60 and Ni-63 on reactor fuel building wall surfaces. Part 61 analyses on waste from this building indicate that the radionuclide mixture is 30% Co-60 and 70% Ni-63. The individual instrument efficiencies are 0.22 for Co-60 and 0.03 for Ni-63. The weighted instrument efficiency is determined as

$$\text{Weighted } \varepsilon_i = (0.30) \times (0.22) + (0.70) \times (0.03) = 0.087$$

In some instances, one or more of the radionuclides may not be detectable at all. Suppose that the Ni-63 was replaced by H-3, which is not detectable with the GM detector (ε_i equals 0). In this case, 70% of the weighted instrument efficiency is zero, which results in a weighted instrument efficiency of 0.066, entirely from the Co-60 component.

The decay series emit both alpha and beta radiation and require additional considerations. It may be feasible for either alpha or beta activity, or both, to be measured for determining the surface activity level. Because alpha radiation may be attenuated to a variable degree when the surface is porous, dirty or scabbled, the measurement of alpha radiation may not be a reliable indicator of the true surface activity levels. Such surface conditions usually cause significantly less attenuation of beta radiation. As previously discussed in Chapter 6, a common practice has been to use beta measurements as a surrogate for alpha measurements to demonstrate compliance with surface activity guidelines expressed in alpha activity.

When applying beta measurements to assess compliance with the decay series surface activity DCGL, consideration should be given to the radionuclide, and its energy, used to calibrate the detector. For example, SrY-90, a high-energy beta emitter, is often used to calibrate the detector for surface activity measurements of uranium. That is, a SrY-90 calibration source is assumed to be sufficiently representative of the beta emissions—specifically the high-energy beta emission from Pa-234 m— from the uranium surface contamination, and therefore, it is assumed that the total efficiency using a SrY-90 source is representative of the uranium contamination.

Additionally, the relationship of beta to alpha emissions must be determined (e.g., by alpha spectroscopy analyses), so that the surface contamination DCGL specified in terms of alpha surface activity may then be expressed in terms of beta surface activity. Because most detectors used for surface activity assessment can respond to alpha, beta, and gamma radiations to varying degrees, using a single radionuclide—or even one in equilibrium with another radionuclide, SrY-90—for calibration may not be representative of the complex decay scheme of the uranium decay series (see Appendix A). The most representative calibration process would use a source prepared from the radioactive material (e.g., uranium) that is being measured in the field, and would then determine the detector's response directly from this specific calibration source material. An alternative approach is addressed in the following case study.

CASE STUDY   DETERMINING INSTRUMENT EFFICIENCY
             FOR 4% ENRICHED URANIUM

Consider a case where the contaminant on building surfaces is 4% enriched uranium—for example, 4% U-235 by weight. The approach used to determine the instrument efficiency is to consider the detector's response to each of the alpha and beta emissions in the decay of low enriched uranium. This method requires that the decay scheme of the contamination be completely understood in terms of radiation type, energy, and abundance. Table 8.3 illustrates this approach for 4% enriched uranium, as measured by a 126 cm^2 gas proportional detector in alpha plus beta mode with a standard 0.8 mg/cm^2 Mylar window thickness. The alpha fractions of U-234, U-235, and U-238 have been determined for 4% enriched uranium using alpha spectrometry analysis results (Table 8.3). The detector's instrument efficiency for each radiation emission should be determined by experiment and/or empirical relationship. For example, the detector's response to the alpha emissions of U-234, U-235, and U-238 may be assessed experimentally with a Th-230 calibration source, the Th-231 beta energies from the U-235 series may be determined using a Tc-99 calibration source. Recall that instrument

TABLE 8.3   Instrument Efficiency for Low Enriched Uranium (4%) Using a 126 cm²
Gas Proportional Detector with a 0.8 mg/cm² Window in Alpha Plus Beta Mode

Radionuclide	Radiation/ Average Energy (MeV)	Alpha Fraction	Radiation Yield (%)	Instrument Efficiency	Weighted ε_i
U-238	Alpha/4.2	0.150	100	0.40	0.06
Th-234	Beta/0.0435	0.150	100	0.19	0.028
Pa-234 m	Beta/0.819	0.150	100	0.69	0.104
U-234	Alpha/4.7	0.809	100	0.40	0.328
U-235	Alpha/4.4	0.041	100	0.40	0.016
Th-231	Beta/0.0764	0.041	100	0.36	0.015
			Total weighted instrument efficiency		0.55

efficiencies for beta emissions that may not be determined during calibration due to the lack of appropriate calibration standard beta energy may be calculated empirically through an efficiency versus beta energy calibration curve (Section 8.1).

As shown in Table 8.3, the majority of the gas proportional detector's response is due to U-234 and its large alpha fraction. The weighted instrument efficiency for this detector for low enriched uranium is 0.55. This weighted instrument efficiency must be multiplied by an appropriately weighted surface efficiency. This will be shown in a corresponding case study in Chapter 10.

The measurement of the Ra-226 series is further complicated by the phenomenon of radon emanation. That is, to calculate the instrument efficiency, it is necessary to evaluate the number of alpha and beta radiations emitted per decay of Ra-226. However, this is influenced by the equilibrium radon and its progeny. To illustrate, assume that complete equilibrium exists throughout the Ra-226 decay series. It is then possible to calculate the total weighted instrument efficiency for a GM detector for this equilibrium situation (Table 8.4).

The estimated instrument efficiency for the entire decay series (nine radionuclides) assuming equilibrium is 1.87, which averages to about 0.21 per radionuclide. However, considering that complete equilibrium of radon and its progeny may not exist of the particular surface of interest, the fraction of progeny present must be assumed or measured. It may be possible to collect an *in situ* gamma spectrum on representative surfaces to determine the relative ratios of progeny present, or it may be appropriate to use indoor equilibrium factors of radon and its progeny. In either case, once the fractional abundance of progeny has been assessed, the instrument efficiency can be modified to account for the actual mixture of radionuclides present. Remember that in performing these evaluations, erring on the side of conservatism will likely be better received by regulators and technical reviewers.

The instrument efficiency alone cannot be used to convert net count rate measurements to surface activity; it is necessary to determine the surface efficiency for this decay series. Note the instrument efficiency assigned to

TABLE 8.4   Instrument Efficiency for Radium-226 Assuming Complete Equilibrium of Radon and Its Progeny Using a GM Detector

Radionuclide	Radiation/ Average Energy (MeV)	Progeny Equilibrium Fraction	Radiation Yield (%)	Instrument Efficiency	Weighted ε_i
Ra-226	Alpha/4.8	1	100	0.10	0.10
Rn-222	Alpha/5.5	1	100	0.10	0.10
Po-218	Alpha/6.0	1	100	0.10	0.10
Pb-214	Beta/0.219	1	100	0.33	0.33
Bi-214	Beta/0.632	1	100	0.59	0.59
Po-214	Alpha/7.69	1	100	0.10	0.10
Pb-210	Beta/0.006	1	100	0	0
Bi-210	Beta/0.389	1	100	0.45	0.45
Po-210	Alpha/5.5	1	100	0.10	0.10
			Total weighted instrument efficiency		1.87

the alpha emitters of 0.10; when these values are multiplied by a nominal surface efficiency of 0.25 for alpha emitters (refer to ISO-7503 in Chapter 10), the total alpha efficiency for the GM detector is estimated as 2.5%, which is strongly dependent on surface conditions. The main contributors to detector response are the beta emitters from Pb-214, Bi-214, and Bi-210.

Let us suppose that the overall weighted surface efficiency is 0.30. The total efficiency in this case would be 1.87 times 0.30, or 0.56. While this value may seem high, remember that it accounts for nine radionuclides in equilibrium with Ra-226. The important point is to ensure that the way the total efficiency is calculated is consistent with the DCGL$_W$ for Ra-226. That is, the total efficiency of 0.56 would apply to the DCGL$_W$ for Ra-226 that was based on it being in equilibrium with all of its progeny. In this situation, the DCGL$_W$ for Ra-226 is probably going to be relatively low because it assumes that all the progeny are present. At the same time, this is consistent with a relatively high efficiency, again, because all of the progeny are assumed to be present.

8.5 SURVEY INSTRUMENTATION FOR EXPOSURE RATE MEASUREMENTS

Exposure rate measurements are not expected to be a major factor during the performance of decommissioning surveys, especially not for final status surveys since DCGLs are expressed in terms of building surface activity or soil concentrations. Historically, exposure rates were measured during the final status survey precisely because there was a specific exposure rate guideline that needed to be satisfied (e.g., 5, 10, or 20 μR/h). With the move to dose-based release criteria, it does not seem likely that these measurements will be required by regulators. That being said, exposure rate measurements can be used as a surrogate for surface activity measurements (activated materials) or for surface soil contamination. As previously discussed in Chapter 6, Appendix C in NUREG-1500 gave the green light for this approach, providing data on those radionuclides that deliver greater than 90% of their dose via the direct radiation pathway (NRC 1994b).

8.5.1 Pressurized Ionization Chambers

The pressurized ionization chamber (PIC) is widely recognized as the industry standard for the measurement of exposure rates, especially for low-level exposure rates at μR levels. The PIC exhibits a reasonable flat energy response across a wide gamma energy range. A major drawback of the PIC is evident during its field deployment—it is heavy, and maybe even worse, it is bulky. The PIC is composed of two major components—the pressurized gas detector that measures the exposure rate and the electronics package that reports the data—connected by a thick "umbilical cord" used for data and power transmission.

Oftentimes, NaI scintillation detectors are cross-calibrated to PICs to determine the cpm per μR/h for a particular energy (or contaminant mixture of gamma energies). This allows the NaI scintillation detector to be used for exposure rate measurements in the field, which is much easier than lugging the PIC. This same procedure has also been used to cross-calibrate energy-compensated GM detectors to assess exposure rate levels.

8.5.2 Micro-R and Micro-Rem Meters

Perhaps the most common micro-R meter is the Ludlum Model 19. This detector features a 1″ × 1″ NaI detector that is completely housed within the meter box and reads out in μR/h. The micro-R meter is therefore energy dependent because it uses NaI detector. Specifically, it exhibits a substantial overresponse (e.g., 2–10 times the true exposure when calibrated to Cs-137) at an energy range of 50–300 keV, and underresponds to energies greater than 700 keV. Nonetheless, it is very sensitive to gamma radiation—for example, it has a reported gamma efficiency of 200,000 cpm per mR/h for Cs-137—providing a very good indicator of above-background radiation levels that may be present.

One attribute that makes the micro-R meter so popular is that it travels well—no cables are needed; one just needs to ensure that the batteries are fresh. It is a good instrument to choose for scoping surveys when qualitative measurements are sufficient. However, if reliable exposure rate measurements are required, you probably need an instrument with a flatter energy response, such as a micro-rem meter. An example of micro-rem meter is the Bicron micro-rem meter. This survey instrument has been a popular alternative to the PIC, for the assessment of exposure rates at micro-R levels. It is not difficult to provide a justification for the practice of reporting the Bicron measurements in μR/h, as opposed to the calibrated value on the meter face (e.g., μrem/h).

The Bicron micro-rem meter is a tissue-equivalent organic scintillator that responds to incident gamma radiation and reads out directly in dose equivalent units of μrem/h—supposedly of value for the situation where the guideline was stated specifically in terms of μrem/h, as opposed to the more common μR/h. While this instrument readout is close to the exposure rate in μR/h, it is important to recognize and understand the resultant error in this approximation.

It can be readily shown that 1 R is roughly equal to 0.877 rads in air. It is important to note at this point that the unit Roentgen is only defined for exposure in air. The following relationship may be used to relate the dose in air to other media, such as tissue:

$$D_{\text{tissue}} = D_{\text{air}} \frac{(\mu_{en}/\rho)_{\text{tissue}}}{(\mu_{en}/\rho)_{\text{air}}}$$

This relationship between the dose in tissue to the dose in air can then be established for a range of gamma energies. Table 8.5 illustrates the ratio of mass energy absorption coefficients for air and tissue (striated muscle).

From Table 8.5 it can be seen that the ratio of the mass energy absorption coefficients for tissue and air is virtually constant—ratio is

TABLE 8.5   Ratio of Mass Energy Absorption Coefficients for Tissue and Air

Photon Energy keV	$(\mu_{en}/\rho)_{\text{air}}$ (cm²/g)	$(\mu_{en}/\rho)_{\text{tissue}}$ (cm²/g)	$(\mu_{en}/\rho)_{\text{tissue}}/(\mu_{en}/\rho)_{\text{air}}$
100	0.0234	0.0256	1.09
200	0.0268	0.0294	1.10
400	0.0295	0.0325	1.10
600	0.0295	0.0325	1.10
1000	0.0278	0.0306	1.10
1500	0.0254	0.0280	1.10
2000	0.0234	0.0257	1.10

approximately 1.1—over the gamma energy range typically encountered during environmental surveys. Because 1 R is equivalent to 0.877 rads in air, it is possible to establish the relationship between 1 R and the tissue dose in rads:

$$D_{\text{tissue}} = 0.877 \times 1.1 = 0.965 \text{ rads}$$

The last step in determining the necessary identity between R and rem in tissue is to recognize that gamma radiation has a quality factor of 1 (1 rem per rad); thus, 1 R equals 0.965 rem in tissue. Therefore, 1 rem in tissue is equal to 1.036 R, and the error is slightly nonconservative because when the Bicron meter is reading 1 μrem/h it is measuring 1.036 μR/h. Consequently, it is recommended that no correction be made to convert the μrem/h reading to μR/h due to the relatively large fluctuations in reading the Bicron micro-rem meter's analog scale. Hence, it can be stated that the conversion from μrem/h to μR/h is essentially unity.

One final point to make about the micro-rem meter concerns the instrument read-out on an analog meter. The lowest scale on the instrument usually allows the measurement of exposure rates up to a couple tens of μR/h (e.g., the Bicron meter goes up to 20 μR/h on the lowest scale). At this level, the analog meter needle may fluctuate to a fair degree. So it is reasonable to ask how one should record the fluctuating exposure rate at a particular location. Do you record the result of the highest reading during these fluctuations? Do you try to determine an "average" reading after observing the bouncing needle for several seconds? One possible solution that provides an unbiased reading is to record 10 instantaneous observations. This can be done by covering the meter face with your hand, and then uncovering the meter face for an instant and reporting the observed needle reading. This "peek-a-boo" method is repeated for 10 measurements, and the average of these 10 instantaneous readings provides an unbiased exposure rate measurement at that particular survey location.

8.6 LABORATORY INSTRUMENTATION

Surveys in support of decommissioning frequently involve sampling and laboratory analyses on collected media samples. The purpose of this section is to briefly introduce the common laboratory instruments used to perform these analyses. A number of reference documents exist in this regard, including Appendix H in MARSSIM, provides a concise operational overview of several laboratory instruments, and of course, the

Multi-Agency Radiological Laboratory Analytical Protocols manual cannot go unmentioned.

The MARLAP is a three-volume guidance manual prepared for planning, implementation, and assessment of projects that require the laboratory analysis of radionuclides (NRC 2004). According to its abstract, MARLAP's objective is to "provide guidance and a framework for project planners, managers, and laboratory personnel to ensure that radioanalytical laboratory data will meet a project's or program's data requirements." The material covered in MARLAP could easily fill a two-semester advanced course on the generation and analysis of radioanalytical data. Though somewhat rigorous, MARLAP is a useful reference that addresses a variety of data collection activities, including many germane to decommissioning—for example, site characterization, final status survey compliance demonstration, and waste management activities.

The most common laboratory analyses performed during decommissioning surveys are smear counting and soil analyses. Hence, laboratory instrumentation typically includes low-background gas proportional counters for smear counting and HPGe detectors for gamma spectrometry of soil samples. Other laboratory techniques are routinely performed on soil and other environmental media. For example, the preparation of environmental samples essentially involves the concentration of radioactive material from the form it was sampled in the field to a much smaller quantity. That is, while hundreds of grams of soil may be sampled for Sr-90, following the preparation and wet chemistry separations to isolate the strontium, the actual sample that is counted is likely less than 1 g. The radioactive material in the sample is concentrated in this way to increase the sensitivity of the analysis and to ensure the sample is in a more suitable counting geometry.

The principal methodology for counting alpha emitters in a laboratory is alpha spectroscopy. This technique uses a high-resolution silicon diode surface barrier detector. The resolution achieved is typically 10–20 keV, which makes alpha spectroscopy possible. For some radionuclides, the alpha energies are too close to resolve, such as the case for Pu-239 and Pu-240, which is often reported as Pu-239/240. A complication that can arise in the laboratory is that of a poor deposition on the counting planchet, which causes some degree of alpha attenuation (self-absorption). This blurs the resolution and makes it difficult to quantify the alpha emitters. Sometimes a complete reanalysis is warranted; other times it may be sufficient to simply redeposit the sample prior to counting.

For quantifying gamma emitters, the HPGe detector is the industry standard. It is usually calibrated to analyze samples in a variety of geometries, the most common of which may be the 1 or 2 L Marinelli geometry. Software that assists in detector calibration and energy peak determination is standard. Chapter 10 provides a discussion on the interpretation of gamma spectra.

The NaI detector can also be used, but of course it suffers in terms of resolution as compared to the HPGe. One application where the NaI detector is more than sufficient would be the analysis of single radionuclides, or a few that are well separated in energy. In this case, it is unnecessary to have high resolution because the radionuclides being analyzed are well known. An example is the Ra-226 counting at wind-blown tailing areas at uranium mills. It essentially becomes a question of instrument DQOs—if they can be satisfied by the inexpensive-to-operate NaI (e.g., liquid nitrogen is not required), then follow your DQOs; it always knows.

For radionuclides that emit only low-energy beta radiation, such as H-3, S-35, C-14, and Ni-63, the laboratory analysis of choice is liquid scintillation counting (LSC). Gross beta counting using a gas proportional counter, as was discussed for smears, can be used once the sample has been separated, and counted for example, for Sr-90. In general, it is difficult to resolve beta spectra with LSC, although for up to a few radionuclides present in a sample, limited success has been demonstrated.

QUESTIONS AND PROBLEMS

1. For the following survey needs and radiological conditions at a D&D site, describe which type of detector(s) would be selected:

   - Scans inside a small floor drain opening to a depth of 20 cm for a beta emitter

   - Alpha scans of a large floor area

   - Beta measurements of surface activity in the presence of an alpha emitter

   - Surface activity measurements in a survey unit contaminated with Co-60, Ni-63, Cs-137, and Am-241

   - Surface activity assessment in a survey unit contaminated with H-3, C-14, and Co-60

2. For hard-to-reach locations, a common decommissioning survey trick is to attach a longer cable between the meter and the detector. What may need to be done to ensure that the increased cable capacitance does not change the overall response of the detector? Which two of the following three detector types are most affected by increased cable length—GM, gas proportional, and scintillation detectors?

3. What is the major problem with operating the gas proportional detector on a static gas purge? How can this affect the quality of the survey data?

4. What are the three modes that the gas proportional detector can be operated? Briefly describe how each mode is achieved.

5. Calculate the instrument efficiency for a GM detector given the following information. The calibration sheet for the 180 cm² Tl-204 source states that the activity is 23,500 dpm and the surface emission rate is 15,000 betas per minute. The GM was placed on the source for 2 min and 1080 counts were observed. The detector background is assumed to be 60 cpm. (Hint: the physical probe area for the GM is 20 cm².)

6. State the relative advantages of using the Bicron micro-rem meter and the PIC to measure exposure rates.

7. Calculate the weighted instrument efficiency for the thorium series (assume it is in secular equilibrium throughout) with a gas proportional detector in alpha plus beta mode.

8. Name three environmental phenomena that can significantly influence the performance and survey results of radiation detectors. Describe solutions for overcoming each of these potential environmental pitfalls.

CHAPTER 9

Detection Sensitivity

Static and Scan MDCs

A S DISCUSSED IN THE previous chapter, a major factor influencing survey instrument selection is detection sensitivity. The detection sensitivity of an instrument or measurement procedure is given the term minimum detectable concentration (MDC). The value of the MDC for field survey instruments and laboratory analyses in relation to the DCGLs is necessary, in particular, to assess compliance with the MARSSIM recommendation that detection sensitivities for direct measurements of surface activity and laboratory analyses of soil samples be less than 50% of the appropriate DCGL$_W$. The MDC of field survey instruments relative to the derived concentration guideline levels is a major factor affecting the planning and design of final status surveys.

Fundamentally, MARSSIM users with potentially contaminated building surfaces and land areas are required to demonstrate that residual radioactivity at their sites meet the applicable release criteria. This is performed by making measurements of surface activity or laboratory analyses for radionuclide concentrations in soil samples, in conjunction with scanning for elevated activity in building surfaces and land areas. Thus, detection limits for field survey instrumentation are an important criterion in the selection of appropriate instrumentation and measurement procedures. For the most part, detection limits need to be determined to evaluate whether a particular instrument and measurement procedure is capable of detecting residual activity at the regulatory release criteria (i.e., DCGLs). Before any measurements are performed, the survey instrument and measurement procedures to be used must be shown to possess sufficient detection capabilities relative to the surface activity DCGLs; that is, the detection limit of the survey instrument must be less than the appropriate surface activity $DCGL_W$. Two statistical concepts necessary for the calculation of MDC—the critical level and the detection limit—are discussed in the next section.

9.1 CRITICAL LEVEL AND DETECTION LIMIT

Detection sensitivity can be expressed in terms of the net counts that are detectable above background levels, and then translating these "minimum detectable net counts" in terms of surface activity or radionuclide concentrations in soil. It is this second step that can be significantly affected by field conditions encountered at decommissioning sites. The MDC (or minimum detectable activity for those who are more comfortable with MDA) corresponds to the smallest activity concentration measurement that can be achieved with a specified survey instrument and type of measurement procedure. Note that the detection sensitivity depends not only on the survey instrument but also on how the instrument is to be used, such as specifying the count time and surface-to-detector geometry. Specifically, the MDC depends not only on the particular instrument characteristics (e.g., instrument efficiency, background, and count time) but also on the factors involved in the survey measurement process, which include surface type, source-to-detector geometry, and surface efficiency (backscatter and self-absorption).

NUREG-1507, titled "Minimum Detectable Concentrations with Typical Radiation Survey Instruments for Various Contaminants and Field Conditions" (NRC 1998b), provides experimental data that were obtained for parameters that affected the instrument efficiency (e.g., source-to-detector geometry, distance from the source), and for parameters that affected surface efficiency (e.g., surface type and coverings, such as oil, paint, and dust). Taken together, these data illustrate the composite effect that the survey instruments and field conditions have on detectability.

The measurement of residual radioactivity during surveys in support of decommissioning often involves the measurement of residual radioactivity at near-background levels, especially for naturally occurring uranium, thorium, and radium decay series and other alpha-emitting radionuclides, whose release criteria are only marginally above the background. Thus, it is important to assess the minimum amount of radioactivity that may be detected, especially for radionuclides whose release criteria are close to background levels. In general, the MDC is the minimum activity concentration on a surface or within a soil volume, which an instrument is expected to detect most of the time. It has been a convention in the health physics field to define the activity concentration that is detected "most of the time" as that activity expected to be detected with 95% confidence (or 95% of the time). It is important to note, however, that this activity concentration, or the MDC, is determined *a priori*, that is, before survey measurements are conducted. This is necessary to have sufficient confidence that the survey instrument selected has sufficient DQOs—that is, can detect residual radioactivity at or below the DCGL$_W$ level.

As generally defined, the detection limit, which may be a count or count rate, is independent of field conditions such as scabbled, oxidized (rusty), wet, or dusty surfaces. Furthermore, the detection limit is independent of the radionuclide distribution in soil. Rather, the detection limit is based on the number of counts and does not necessarily equate to measured activity under field or laboratory conditions. These field conditions do, however, affect the instrument's "detection sensitivity" or MDC. Therefore, it is important to recognize this subtle difference and note that the terms MDC and detection limit cannot be used interchangeably.

> The detection limit is based on the number of counts and does not necessarily equate to measured activity; rather, the MDC is the detection limit that accounts for field or laboratory conditions.

The underlying statistical framework for the determination of the MDC is critical to a sound understanding of what is meant by "detection sensitivity." Because this is such an important topic to many decommissioning health physicists, the derivation of the MDC concepts provided below will spare few details. Again, note the logical progression to obtain the MDC: first determine the critical level, upon which the detection limit may be derived, and finally the MDC is calculated from the detection limit.

The MDC concepts are derived from statistical hypothesis testing (refer to Chapter 12 for a discussion of hypothesis testing), in which a decision is made on the presence of activity. Specifically, a choice is made between the null hypothesis (H_0) and the alternative hypothesis (H_a). The null hypothesis can be stated as "no net activity is present in the sample," while the alternative hypothesis states that the observed counts are greater than the background, and hence, that net activity (i.e., residual contamination) is present. These statements are written as

H$_0$: No net activity is present in the sample

H$_a$: Net activity is present in the sample

The first step in setting up the detection sensitivity hypothesis testing framework, is to consider an appropriate blank or background distribution for the medium (building surface, soil, smear, etc.) to be evaluated. Currie (1968) defines the blank as the signal resulting from a sample that is identical, in principle, to the sample of interest, except that the residual activity is absent. This determination must be made under the same geometry and counting conditions as used for the sample.

Consider the distribution of counts obtained from measurements of the blank, which is characterized by a population mean (μ_B) and standard deviation (σ_B). Now consider the measurement of a sample that is known to be free of residual activity. This zero-activity (background) sample has a mean count (C_B) and standard deviation (s_B). The net count (and, subsequently, residual activity) may be determined by subtracting the blank counts from the sample counts:

$$C_B - \mu_B = 0,$$

which results in a zero-mean net count probability distribution that is approximately normally distributed (see Figure 9.1). The standard deviation of this distribution, σ_0, is obtained by propagating the individual errors associated with both the blank (σ_B) and the zero-activity samples (s_B). That is

$$\sigma_0 = \sqrt{\sigma_B^2 + s_B^2}. \tag{9.1}$$

A net count level may then be determined from this distribution that can be used as a decision tool to decide when the activity is present. This critical level, L_C, is that net count in a zero-mean count distribution having a probability, denoted by α, of being exceeded. It is a common practice to set α equal to 0.05 and to accept a 5% probability of incorrectly concluding that activity is present when it is not (false positive). If the observed net count is less than the critical level, the surveyor correctly concludes

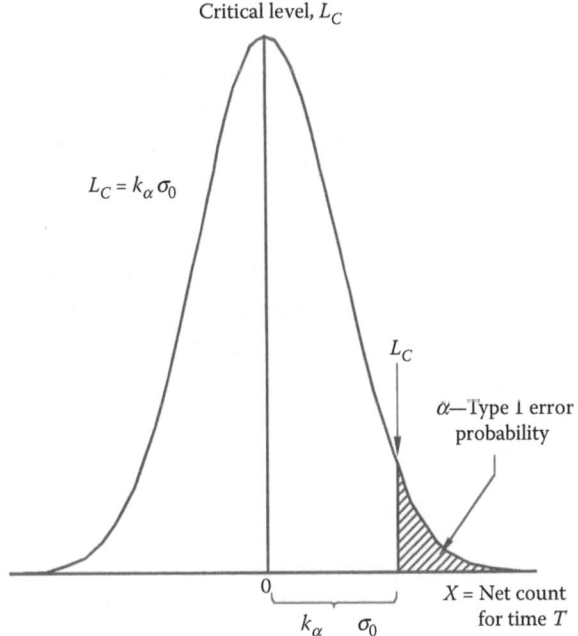

FIGURE 9.1   Zero-mean count frequency distribution used to establish the critical level.

that no net activity is present. When the net count exceeds L_C, the null hypothesis is rejected in favor of its alternative, and the surveyor falsely concludes that net activity is present in the blank sample. It should also be noted that the critical level, L_C, is equivalent to a given probability (e.g., 5%) of committing a type I error (false-positive detection). The expression for L_C is generally given as

$$L_C = k_\alpha \sigma_0. \qquad (9.2)$$

As stated previously, the usual choice for α is 0.05, and the corresponding value for k_α is 1.645.

This value for k_α actually relates to a 90% confidence level, meaning that each tail of the distribution is 5%. In the determination of the critical level, we are only concerned with the upper bound tail, that is, where only false positives occur (concluding that activity is present when it is just at background).

> *Caution:* The type I (α) and type II (β) decision errors used in the derivation of MDC concepts in this chapter are different from the meaning of type I (α) and type II (β) decision errors used in MARSSIM; the meaning of type I (α) and type II (β) decision errors should always be defined in the context of how the null hypothesis is stated.

For an appropriate blank counted under the same conditions as the sample (e.g., same counting geometry and background levels), the assumption may be made that the standard deviations of the blank and zero-activity sample are equal (i.e., σ_B equals s_B). Thus, the critical level may be expressed as

$$L_C = 1.645\sqrt{2s_B^2} = 2.33 s_B. \tag{9.3}$$

Note that the L_C value determined above is in terms of net counts, and as such, the L_C value should be added to the background count if comparisons are to be made to the directly observable instrument gross count.

Returning for a moment to the issue of false positives, one can discuss situations where values other than α equal to 0.05 are warranted. It is first necessary to understand specifically what happens when a false positive occurs, that is, what is the cost of the false positive. Does it mean that additional expensive sample analyses are necessary? Does it convey unwarranted angst and worry for a person informed of a "positive" bioassay result? Or does it simply mean pausing for a few additional seconds during a surface scan? Obviously, the greater the cost of the false positive, the more reluctant one is to accept false positives. Sometimes, an alpha error of 0.05 is too high (e.g., if expensive analyses are the result), other times it may be too low (e.g., when scanning requires additional pauses). This is precisely what is meant by using the DQO process to set the acceptable false-positive rate based on the context of the situation.

The detection limit, L_D, is defined to be the number of mean net counts obtained from samples for which the observed net counts are almost always certain to exceed the critical level. It is important to recognize that L_D is the mean of a net count distribution. Let us assume that we have a sample with an activity level that produces the number of net counts that equals the detection limit. The net counts (C_S) obtained from this sample can be determined by subtracting the blank from the gross count—that is,

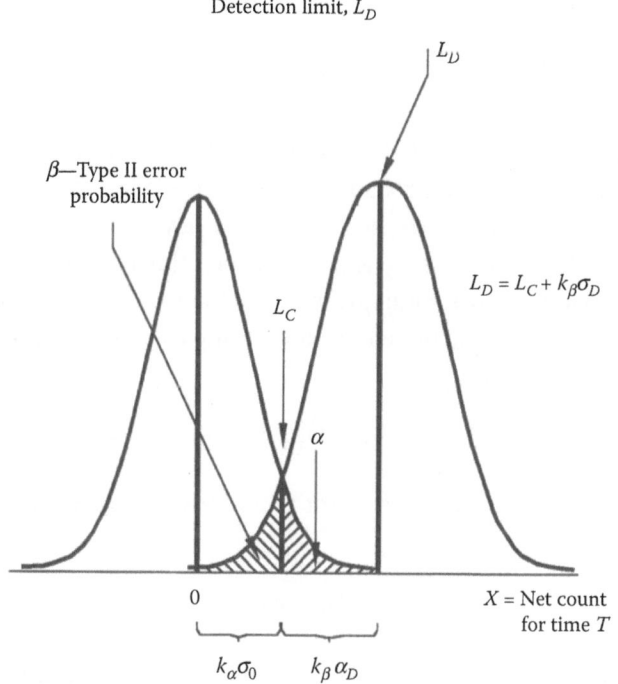

FIGURE 9.2   Net count distributions that identify the setting of the detection limit.

$C_S = C_S + B - \mu_B$. So, in effect, the detection limit is simply the net counts (C_S) that have been defined to "almost always exceed the critical level." Therefore, we must clearly state what is meant by "almost always exceed the critical level."

The detection limit is positioned far enough above zero so that there is a probability, denoted by β, that the L_D will result in a signal less than L_C (see Figure 9.2). As with the α error, it is common practice to set β equal to 0.05 and to accept a 5% probability of incorrectly concluding that no activity is present, when it is indeed present (type II error). That is, the surveyor has already agreed to conclude that no net activity is present for an observed net count that is less than the critical level; however, an amount of residual activity that would yield a mean net count of L_D is expected to produce a net count less than the critical level 5% of the time. This is equivalent to missing residual activity when it was present. Note that L_C is often referred to as the *a posteriori* decision—an experimental result is concluded to be detected (i.e., above background) when net counts exceed L_C.

The expression for L_D is generally given as

$$L_D = L_C + k_\beta \sigma_D \tag{9.4}$$

where k_β is the abscissa of the standardized normal distribution corresponding to a one-tailed probability level of $1 - \beta$ for detecting the presence of net activity, and σ_D is the standard deviation of the net sample count (C_S), when C_S equals L_D. Again, consider the measurement of a sample that provides a gross count given by C_{S+B}, at the detection level. The net sample count, C_S, is calculated by subtracting the mean blank count (μ_B) from the gross count (C_{S+B}):

$$C_S = C_{S+B} - \mu_B.$$

Recognizing that the gross count is composed of two components, we can write

$$C_S = (C_S + C_B) - \mu_B.$$

Then, the detection limit may be written as follows, recognizing that C_S equals L_D:

$$L_D = C_S + (C_B - \mu_B).$$

The standard deviation of the net sample, σ_D, is obtained by propagating the error in the gross count and from the background count when the two are subtracted to obtain L_D. As previously derived, the standard deviation of this zero-net count distribution, σ_0, is obtained by propagating the uncertainties associated with both the blank (C_B) and the zero-activity samples (μ_B), and recognizing the equality between the mean and variance for Poisson distributions:

$$\sigma_D = \sqrt{(\sigma_{C_S}^2 + \sigma_0^2)} = \sqrt{(C_S + \sigma_0^2)} = \sqrt{(L_D + \sigma_0^2)}. \tag{9.5}$$

This expression for σ_D may be substituted into the above equation and solved for L_D.

The detection limit will now be solved explicitly using a number of algebraic steps and simplifying assumptions. We can start by substituting the expression (Equation 9.5) for σ_D into the detection limit expression (Equation 9.4):

$$L_D = L_C + k_\beta \sqrt{(L_D + \sigma_0^2)},$$

and recognizing that σ_0^2 from Equation 9.2 can be set equal to $(L_C/k_\alpha)^2$, and rearranging terms

$$\frac{L_D - L_C}{k_\beta} = \sqrt{L_D + \frac{L_C^2}{k_\alpha^2}}.$$

At this point, we can square both sides of the equation and multiply through by k_β^2:

$$L_D^2 - 2L_D L_C + L_C^2 = k_\beta^2 L_D + L_C^2 \frac{k_\beta^2}{k_\alpha^2}.$$

Again, we can rearrange terms to get the L_D terms on the left, and then factor out an L_D:

$$L_D^2 - [2L_C + k_\beta^2]L_D = L_C^2 \frac{k_\beta^2}{k_\alpha^2} - L_C^2.$$

We are now ready to complete the square on the left side of the equation (only left side of the equation is shown):

$$\left[L_D^2 - (2L_C + k_\beta^2)L_D + \left(\frac{2L_C + k_\beta^2}{2} \right)^2 \right] - \left(\frac{2L_C + k_\beta^2}{2} \right)^2.$$

At this point, the right side of the equation can be shown with the left side:

$$\left(L_D - \left(\frac{2L_C + k_\beta^2}{2} \right) \right)^2 = \frac{k_\beta^2}{k_\alpha^2} L_C^2 - L_C^2 + \left(\frac{2L_C + k_\beta^2}{2} \right)^2.$$

Completing the square on the right side of the equation and multiplying through by $4k_\alpha^2$

$$4k_\alpha^2 \left(L_D - \left(\frac{2L_C + k_\beta^2}{2} \right) \right)^2 = 4k_\beta^2 L_C^2 + 4k_\alpha^2 k_\beta^2 L_C + k_\alpha^2 k_\beta^4.$$

Next, we can factor out the $k_\alpha^2 k_\beta^4$ term from the expression on the right side of the equation, divide both sides of the equation by $4k_\alpha^2$, and take the square root of both sides:

$$L_D - \left(\frac{2L_C + k_\beta^2}{2} \right) = \frac{k_\beta^2}{2} \sqrt{1 + \frac{4L_C}{k_\beta^2} + \frac{4L_C^2}{k_\alpha^2 k_\beta^2}}.$$

We can now expand the left side of the equation and rearrange terms to isolate L_D:

$$L_D = L_C + \frac{k_\beta^2}{2} + \frac{k_\beta^2}{2} \sqrt{1 + \frac{4L_C}{k_\beta^2} + \frac{4L_C^2}{k_\alpha^2 k_\beta^2}}.$$

Substituting the expression for L_C from Equation 9.2 and factoring out $k_\beta^2 / 2$:

$$L_D = k_\alpha \sigma_0 + \frac{k_\beta^2}{2} \left(1 + \sqrt{1 + \frac{4k_\alpha \sigma_0}{k_\beta^2} + \frac{4k_\alpha^2 \sigma_0^2}{k_\alpha^2 k_\beta^2}} \right).$$

At this point in the derivation, we will focus only on the term under the radical, canceling out the k_α^2 and completing the square:

$$\sqrt{\left(1 + \frac{4\sigma_0}{k_\beta} + \frac{4\sigma_0^2}{k_\beta^2} \right) + \left(\frac{4k_\alpha \sigma_0}{k_\beta^2} - \frac{4\sigma_0}{k_\beta} \right)}.$$

Simplifying these terms and factoring out a term on the right side:

$$\sqrt{\left(1 + \frac{2\sigma_0}{k_\beta} \right)^2 + \frac{4\sigma_0}{k_\beta} \left(\frac{k_\alpha}{k_\beta} - 1 \right)}.$$

At this point, we can rebuild the complete equation for the detection limit:

$$L_D = k_\alpha \sigma_0 + \frac{k_\beta^2}{2} \left(1 + \sqrt{\left(1 + \frac{2\sigma_0}{k_\beta} \right)^2 + \frac{4\sigma_0}{k_\beta} \left(\frac{k_\alpha}{k_\beta} - 1 \right)} \right). \qquad (9.6)$$

We will pause at this point in the derivation of the MDC to comment on a couple of more interesting aspects of this general detection limit equation. It is doubtful that many have committed this detection limit equation to memory. Equation 9.6 is noteworthy in the sense that no simplifying assumptions have been made. This is not the case for the more common detection limit expressions, where the assumptions are frequently taken for granted, or possibly not even recognized. For instance, nearly all expressions for MDC assume equal values of the type I and type II decision errors, and further, they are usually both expressed as 0.05. In Equation 9.6, not only can the decision errors be something other than 0.05, but the type I and type II errors can also be different from one another. Perhaps a circumstance will arise where a higher false-positive (type I) error rate is deemed acceptable. For example, maybe the cost of false positives are relatively minor, so a decision is made to allow a type I error of 20%. In this case, k_α is 0.842, rather than the standard 1.645 for a 5% type I decision error.

The detection limit equation in Equation 9.6 is also generic in the sense that the standard deviation (σ_0) for the zero-activity net distribution still depends on the background measurement method. This standard deviation is usually determined from one of two general approaches. The first approach is to estimate σ_0 from the background counting rate (R_B) and the background (T_B) and sample (T_{S+B}) count times:

$$\sigma_0 = \sqrt{R_B T_{S+B}\left(1 + \frac{T_{S+B}}{T_B}\right)}.$$

Note that this equation can be derived by propagating the error in the net counting rate, $R_{S|B}$, when the gross counting rate is equal to the background counting rate (i.e., $R_{S+B} = R_B$). Chapter 17 provides an overview of the propagation of errors.

Another option is to assume paired measurements of the blank and sample, which basically implies that the standard deviations of the blank and sample are equal (i.e., σ_B equals s_B). This can be justified, at least to a degree, since the number of net counts that define the detection limit should be relatively close to the background; this supports the assumption that the two standard deviations are equal. In this case, $\sigma_0^2 = 2s_B^2$, and $\sigma_0 = \sqrt{2}s_B$. The background standard deviation (s_B), which is an experimental estimate of the blank standard deviation (σ_B), can be determined from repeated measurements. Each background measurement should be counted for the same counting time interval as the sample.

Let us illustrate each of these two options with examples. Suppose that a 10-min measurement was performed of the background, and the resulting background counting rate was 68 cpm. The sample count rate is 1 min. The standard deviation is calculated as

$$\sigma_0 = \sqrt{68 \text{ cpm}(1 \text{ min})\left(1 + \frac{1}{10}\right)} = 8.65 \text{ counts}.$$

Alternatively, let us assume that the background count rate was determined by 10 repetitive measurements, each one for a 1-min count time. The sample count time will also be 1 min for these paired measurements. The following background measurement data (in counts for 1 min) were collected: 74, 73, 61, 70, 66, 60, 69, 72, 66, and 68. The mean is 67.9 counts and the standard deviation (s_B) is 4.75 counts. Therefore, the background standard deviation, σ_0, is calculated as

$$\sigma_0 = \sqrt{2} \times (4.75) = 6.72 \text{ counts}.$$

To summarize, Equation 9.6 is a very useful general expression of the detection limit. It allows the flexibility to calculate the detection limit for a variety of type I and type II decision errors, as well as allowing for different methods to calculate the standard deviation (σ_0) for the zero-activity net distribution.

Before we continue our derivation of the MDC equation, it is beneficial to consider when it might be advantageous to use a different type I or type II error. The DQO process should be employed to determine if type I or type II decision errors other than 0.05 are warranted. For example, it may be necessary to determine the activity level that is detectable 99% of the time, if it is essential that a certain activity level not be missed (translates to a type II error of 0.01). Conversely, a detection rate of only 67% (i.e., type II error rate of 0.33) may be acceptable in particular situations. Obviously, lowering the type II error rate increases the detection limit—for example, suppose that 50 net counts are detectable 95% of the time, then it might take 66 net counts to be detected 99% of the time—while increasing the type II error rate decreases the detection limit.

Sometimes, it is wondered why the detection limit cannot be established at the critical level. Indeed, many miscues have occurred because of this confusion. The allure to doing this would be the artificially small L_D that results, and being able to quote a superb, though misleading, detection sensitivity. Consider what it means to have the L_D set at the L_C—50% of the L_D distribution would be less than the critical level (a type II error of 50%). In other words, stating the detection limit that is calculated at the critical level means that there would be a 50% chance of concluding that no activity is present, when it was present. This is clearly an unacceptable practice—no better than flipping a coin.

Returning to the derivation of the MDC, one of the first simplifications we can entertain is to set the type I and type II decision errors equal. Setting $k_\alpha = k_\beta = k$, and the right term under the radical goes to zero, Equation 9.6 becomes

$$L_D = k\sigma_0 + \frac{k^2}{2}\left(1 + \sqrt{\left(1 + \frac{2\sigma_0}{k}\right)^2}\right).$$

Performing simple algebra, the detection limit can be written as

$$L_D = k\sigma_0 + k^2\left(1 + \frac{\sigma_0}{k}\right) = 2k\sigma_0 + k^2. \tag{9.7}$$

It is interesting to note at this point that from Equation 9.2 we can write

$$L_D = 2L_C + k^2.$$

Finally, we can simplify Equation 9.7 by selecting the type I and type II errors equal to 0.05 ($k = 1.645$), and recognizing that for the case of paired observations of the background and sample, $\sigma_0 = \sqrt{2}s_B$. Therefore, the detection limit may be expressed in a recognizable form as

$$L_D = 2.71 + 4.65s_B.$$

The assumption that the standard deviation of the count (σ_D) is approximately equal to that of the background greatly simplified the expression for L_D, and is usually valid for total counts greater than 70 for each sample and blank count (Brodsky 1992). (This requirement may be difficult to satisfy for alpha counting, where the background levels during the counting interval are usually less than 10.) Brodsky has also examined this expression and determined that in the limit of very low background counts, s_B would be zero and the constant 2.71, which is an artifact of the derivation, should be 3 based on a Poisson count distribution (Brodsky and Gallaghar 1991). Thus, the expression for the detection limit becomes

$$L_D = 3 + 4.65 s_B. \tag{9.8}$$

The detection limit given by Equation 9.8 may be stated as the net count having a 95% probability of being detected when a sample contains an activity at L_D, which translates to a 5% probability of falsely interpreting sample activity as activity due to the background (type II error). The MDC can now be calculated directly from the detection limit.

9.2 STATIC MDC

The MDC—which follows directly from the detection limit concepts—is a level of radioactivity, either on a surface or within a volume of material (soil), that is practically achievable by an overall measurement process (EPA 1980). The expression for MDC may be given as

$$\mathrm{MDC} = \frac{3 + 4.65 s_B}{KT} \tag{9.9}$$

where K is a proportionality constant that relates the detector response to the activity level in a sample for a given set of measurement conditions and T is the counting time. This factor typically encompasses the detector efficiency, self-absorption factors, and probe area corrections.

The standard deviation of the background, s_B, can be determined by repetitive measurements of the background, as shown in the previous section. It is a common decommissioning survey practice to perform a number of background measurements to provide a better estimate of the background value that must be subtracted from each gross count when using the Sign test for surface activity assessments. Therefore, by performing multiple measurements of the background, one can calculate the background standard deviation directly, as well as improve the accuracy of the background value.

Another common expression for MDC is based on the Poisson esti-mate of background standard deviation. That is, the square root of the background mean count ($\sqrt{C_B}$) is equal to the background standard devia-tion (σ_B). The MDC equation becomes

$$\text{MDC} = \frac{3 + 4.65\sqrt{C_B}}{KT} \tag{9.10}$$

where C_B is the background count in time, T, for paired observations of the sample and blank. The quantities encompassed by the proportionality constant, K, such as the instrument efficiency, surface efficiency, and probe geometry, should also be average, "well-known" values for the instrument. For making assessments of MDC for surface activity measurements, the MDC is given in units of disintegrations per minute per 100 square centi-meters (dpm/100 cm²).

> Note: Static MDC equations are appropriate for (1) paired observations (i.e., same count times) of the sample and blank, or (2) different count times of sample and blank.

For cases in which the background and sample are counted for different time intervals, the MDC becomes (Strom and Stansbury 1992)

$$\text{MDC} = \frac{3 + 3.29\sqrt{R_B T_{S+B}\left(1 + \dfrac{T_{S+B}}{T_B}\right)}}{KT_{S+B}} \tag{9.11}$$

where R_B is the background counting rate, and T_{S+B} and T_B are the sample and background counting times, respectively.

CASE STUDY   EXAMPLES OF STATIC MDC CALCULATIONS

Let us first consider a static MDC example using Equation 9.10 for paired mea-surements. Assume that a gas proportional detector is used to make surface activity measurements for Co-60. Suppose that a 1-min background count was performed, and C_B was equal to 386 counts. The quantities encom-passed by the proportionality constant, K, are as follows: $\varepsilon_i = 0.36$, $\varepsilon_s = 0.42$, and physical probe area is 126 cm². The MDC is calculated as

$$MDC = \frac{3 + 4.65\sqrt{386}}{(0.42)(0.36)(126/100)} = 495 \text{ dpm/100 cm}^2.$$

Note that the primary purpose in calculating the MDC is to determine, if the survey instrumentation and its procedure for use (i.e., count time) are sufficiently sensitive to detect the DCGL$_W$. MARSSIM has defined "sufficiently sensitive" as 50% of the DCGL$_W$. So, what is the recommended course of action if the DCGL$_W$ for Co-60 in this example was 800 dpm/100 cm²? The short answer may be to count the background longer—to lower the static MDC.

Using the same example data for the paired MDC equation, we can calculate the MDC for the longer count time using Equation 9.11. In this case, assume that the background count time is 10 min and the sample count time is 1 min, and we will assume that the background count rate is 386 cpm:

$$MDC = \frac{3 + 3.29\sqrt{386(1)(1 + 1/10)}}{(1)(0.42)(0.36)(126/100)} = 372 \text{ dpm/100 cm}^2.$$

Owing to the longer background count, the static MDC is now less than 50% of the DCGL$_W$. (Note: This example is not intended to convey that static MDCs greater than 50% of the DCGL$_W$ are not acceptable, rather, it illustrates an easy technique for lowering the static MDC a tad.)

Now let us consider an example where the static MDC is based on multiple measurements of the background. Suppose that a GM detector has been calibrated to measure depleted uranium on dry wall surfaces in a processing facility. Assume that the weighted instrument efficiency was 0.45 and the weighted surface efficiency was 0.31. A dry wall reference area was identified and the background count rate was determined by 10 random measurements on dry wall, each one for a 1-min count time. The sample count time will also be 1-min for these paired measurements. The following background measurement data (in counts for 1-min) were collected: 51, 64, 60, 61, 53, 51, 59, 55, 57, and 60. The mean is 57.1 counts and the standard deviation (s_B) is 4.46 counts. The MDC for the GM can be determined using Equation 9.9:

$$MDC = \frac{3 + 4.65s_B}{KT} = \frac{3 + 4.65(4.46)}{(0.45)(0.31)(20/100)} = 850 \text{ dpm/100 cm}^2.$$

Notice that the background standard deviation will usually be smaller for multiple measurements of the background (using Equation 9.9), than for a single measurement of the background (using Equation 9.10). For example, if we had used Equation 9.10 to calculate the GM detector MDC, then the background standard deviation for a single measurement of 57 counts would be the square root of 57, or 7.5, counts. This is about 70% higher than the background standard deviation of 4.46 counts obtained from multiple measurements of the background; correspondingly, the MDC for the single background measurement result of 57 counts is 1370 dpm/100 cm².

9.3 SCAN MDC

Scanning is performed to identify elevated levels of direct radiation in the survey unit and is therefore an integral part of the final status survey. According to the *Multiagency Radiation Survey and Site Investigation Manual* (NRC 2000a), the number of data points required in the design of a class 1 survey unit depends on the minimum concentration that is detectable (scan MDC) using typical scanning instruments—for example, gas proportional floor monitors and hand-held detectors, GM detectors, and NaI scintillation detectors. The MDC for a scan survey depends on many of the same factors that influence the detection of contamination under static conditions—for example, the level of the background, the type of potential contamination, the intrinsic characteristics of the detector, and the desired level of confidence (type I and type II decision errors). Additional factors related to the surveyor come into play; for example, the scan MDC is influenced by the surveyor's ability to move the detector over a surface at a prescribed rate. However, other important factors pertaining to the surveyor's abilities and decision processes typically are not taken into account.

9.3.1 Signal Detection Theory for Scanning

The concepts and methods of signal detection theory (Green and Swets 1988) provide a means for characterizing the performance of surveyors using scanning instrumentation to detect radioactive contamination. The theory applies statistical decision techniques to the detection of signals in noise (background) by human observers, and is described in NUREG/CR-6364 (Brown and Abelquist 1998). If the distributions underlying a detection decision can be specified (similar to Figures 9.1 and 9.2 used for static MDC), the theory can define the performance (or detectability) expected based on Poisson statistics. Signal detection theory is valuable because it allows the basic relationships among important parameters—for example, background rate and length of observation interval—to be anticipated, and it provides a standard of performance against which to compare a surveyor's performance. The observation interval is the period that the detector can respond to the contamination as it is moved over the surface—it depends on the scan speed and the size of the hot spot. To emphasize, the observation interval depends on the *a priori* size of the hot spot, and how quickly the detector is moved over the surface. As a general rule, the observation interval is on the order of seconds, perhaps 0.5–2 s represents a reasonable range.

Specifying achievable performance during scanning depends on the limits applied to type I and type II errors. The null hypothesis for the scanning decision is the same as for the static MDC case, "no net activity is present in the sample," while the alternative hypothesis is that the observed counts are greater than the background, and hence, that net activity is present. In scanning, the error rates are embodied in the criterion established by the observer for deciding, based on an instrument's output, that contamination is present. According to the signal detection theory, the *a priori* probabilities of the events and the values and costs associated with the outcomes influence the placement of the criterion. Simply stated, human factors plays an important role in determining the scan MDC.

The detection of a signal in a background of noise is determined not only by the magnitude of the signal relative to the background but also by the observer's willingness to report that a signal is present, that is, the criterion for responding "yes, a signal is present." The importance of this concept for assessing decision performance is that different observers (or even the same observer at different times) may differ in the proportion of signals correctly detected, and yet be equally sensitive to the signal. Signal detection theory provides an index of sensitivity—which is independent of the observer's criterion—called d', which represents the number of standard normal deviates between the background and signal means of normally distributed activity. Table 6.5 in MARSSIM shows the values of d' associated with various true-positive $(1 - \beta)$ and false-positive (α) rates.

For the ideal observer construct—that is, not influenced by human factors—the number of source counts required for a specified level of performance can be calculated using Poisson statistics by multiplying the square root of the number of background counts by the detectability value associated with the desired performance (as reflected in d'):

$$s_i = d'\sqrt{b_i} \qquad (9.12)$$

where s_i and b_i are, respectively, the number of source and background counts in the observation interval i and the value of d' is selected based on the required true-positive $(1-\beta)$ and false-positive rates (α). Essentially, s_i is the minimum detectable count rate in the observation interval.

For example, consider the increment in a 960 cpm background that would be detectable by an ideal observer using a 1-s observation, assuming a target performance of a 90% correct detection rate and false-positive rate of 10%; therefore, $\alpha = \beta = 0.10$. From MARSSIM Table 6.5, the value of d' for the specified level of performance is roughly 2.5; the number of background counts in a 1-s interval is 16 (i.e., 960 cpm divided by 60 s/min). Therefore, the estimated detectable net source counts for an ideal observer is 2.5 multiplied by 4 (the square root of 16) or 10 counts in a 1-s interval; this is equivalent to 600 cpm or 1560 cpm gross.

9.3.2 Decision Processes of the Surveyor (Human Factors)

What is meant by the statement that the scan MDC is influenced by human factors? One way we can answer this question is to study what is actually happening while a surveyor performs a scan survey. As will become apparent in the following discussion, a major difference between the static MDC and scan MDC is that the static MDC is largely devoid of a human decision on whether activity is present or not. Rather, a scaler count exceeding the static MDC is assumed to be detectable at the specified type I and type II error rates. Scanning, on the other hand, requires the surveyor to compare the audible gross count rate to the expected background count, and make a decision on whether detectable activity is present. Thus, scanning is inherently linked to human factors based on the underlying processes that define scanning.

Scanning consists of two components—continuous monitoring and stationary sampling. In the first component, characterized by continuous movement of the detector over the surface, the surveyor has only a brief "look" at potential sources. This observation interval is determined by the speed of the scan and the size of the hot spot. The surveyor's criterion for deciding that a signal is present at this stage would be expected to be liberal, since it is important not to miss potential contamination. That is, the only "cost" of a false positive at this initial scan stage is a little time involved in pausing over the suspected hot spot.

The second component occurs only after a positive response was made during the first stage. The surveyor interrupts the scanning and holds the detector over the potential contamination for a period, while comparing the instrument's output signal to the background counting rate. Because of the longer observation interval, sensitivity is relatively high compared to the first stage. For a positive decision during the second stage, the criterion should be more strict, since the cost of a "yes" is to spend considerably more time taking a static surface activity measurement or collecting a soil sample. If the observation interval is sufficiently long, an acceptable rate of source detection can be maintained, despite applying the more stringent criterion.

> The two stages of scanning are continuous monitoring and stationary sampling. It is advantageous to pause frequently during the first scanning stage—it increases the overall detectability of the scan, but with no "cost" other than spending a little time pausing over the suspected hot spot.

The implication of this view of the scanning task is that to perform within specified limits on type I and type II error rates, surveyors adopt the criteria at each stage of the scanning process to keep the joint performance within limits. For example, consider the requirement that α and β are less than 0.10. Surveyors performing as described above might choose a criterion for the continuous stage of the process, which gave a true-positive rate of 0.95, at the cost of a 0.55 false-positive rate. For the stationary stage, a more stringent criterion would be chosen that would significantly reduce the number of final decisions that were false positives, while maintaining a high proportion of correct detections. For example, an optimal criterion for the stationary stage might result in a true-positive rate of 0.95 and a false-positive rate of just 0.20. Assuming a simple multiplicative model of the application of these criteria, the resulting performance is roughly equal to the requirement specified above—that is, continuous-stage true-positive rate of 0.95 times 0.95 for stationary stage yields a overall correct detection rate of 0.90.

Admittedly, it is one thing to simply specify a particular false-positive rate, it is an entirely different matter to ensure that surveyors are scanning in a fashion that results in a specified false-positive error rate. So, how is this done? To answer in one word—training. The surveyor's performance is evaluated and the desired performance is achieved by training the surveyor to operate in a manner that results in the desired false-positive rate—this will likely require empirical evaluations of surveyor performance.

Consider again a 960 cpm background and a 1-s observation, the latter representing the residence time of the source during continuous scanning. Assuming that the surveyor chooses the criteria as described above (i.e., to achieve a true-positive rate of 0.95 with a false-positive rate of 0.55), the value of d' used to estimate the detectable increment in counting rate is roughly 1.5 (obtained from MARSSIM Table 6.5). The detectable increment for the continuous monitoring stage, then, is 6 counts, or 360 cpm. The increment for the second stage is estimated similarly, except that a longer interval is assumed, since it represents the length of time for which the surveyor pauses. For a 4-s pause, the number of background counts in the interval is 64; the value of d' corresponding to the second-stage performance specified above (i.e., a true-positive rate of 0.95 with a false-positive rate of 0.20) is roughly 2.5. The detectable increment at this stage is 2.5 multiplied by 8 (the square root of 64) or 20 counts; this is equivalent to 300 cpm.

For estimating the detectable increment (s_i), the larger of the two values is more appropriate; typically, this is the value associated with the continuous (first) stage, owing to the relatively short interval upon which the decision is made. However, if the interval associated with the second stage (the pause) is not long enough, increments correctly identified at the first stage will not be reliably confirmed in the second. For example, if the hypothetical surveyor pauses for only 1 s (rather than four) upon identifying an increased count rate, the detectable increment is not 300 cpm but instead it is 600 cpm, and now the second stage represents the larger value.

Field studies and computer simulations were conducted to determine whether surveyors behaved as described above; results are described in detail in NUREG/CR-6364 (Brown and Abelquist 1998). From the above discussion, given the performance goal (i.e., the desired true-positive rate and false-positive limit) and observation interval, it is possible to arrive at a minimum increment in counting rate detectable by the ideal observer for any given background rate. However, surveyors are not perfect counting devices, they are also influenced by human factors. Therefore, it becomes necessary to consider the efficiency of the surveyor, p, relative to the ideal observer. Taking surveyor efficiency into account, the estimate of the detectable number of source counts in the interval becomes

$$s_{i,\text{surveyor}} = \frac{d'\sqrt{b_i}}{\sqrt{p}}.$$

An estimate of the proportion p was based on the simulation experiments reported in Brown and Abelquist (1998); although there was considerable variation in the survey or efficiency, a value of 0.5 was selected.

To adjust an estimated minimum detectable counting rate (MDCR) (calculated as described previously) to reflect an assumed efficiency, the counting rate is divided by the square root of the efficiency. Continuing with the previous example, the detectable counting rate for the continuous (first) stage, taking the surveyor's efficiency into account, is calculated by dividing 360 cpm by the square root of 0.5; this is roughly equal to 500 cpm.

So, what are some examples of human factors that can impact the scan MDC? Well, think about the following questions. How close is it to lunch, or to the end of the day? How long has the surveyor been scanning? After about an hour it becomes very difficult for most surveyors to stay focused on the scanning task, especially when scanning walls and ceilings. It is a good idea to regularly work in other survey activities for the surveyor to break up the monotony of scanning. Is the surveyor preoccupied with something other than the sound of radiation clicks in the headset? These questions provide insight into what is meant by human factors; it is difficult to accurately account for them in an equation.

Another example of the human factors impact on scan performance is the surveyor's *a priori* expectation of contamination being present in a survey unit. That is, a surveyor may readily expect to find contamination while scanning a class 1 survey unit, so any clicks above background are immediately assumed to be indication of a hot spot. Conversely, if the surveyor is scanning in a class 3 land area survey unit, the surveyor's expectation for contamination might be phrased like, "Why am I scanning out here in the middle of no where? What a waste of my time, there's nothing here." So, if the same number of clicks register on the meter as in the class 1 survey unit example, the surveyor is more likely to chalk it up to a background fluctuation, rather than believing there is activity in a class 3 survey unit.

Finally, we might consider the supervisor's direction to the surveyor. If the supervisor indicates that it is very important not to miss any contamination, perhaps because the regulator is planning to perform a confirmatory survey, then the surveyor will perform the scan very deliberatively, not being overly concerned with committing a number of false positives. Conversely, if the supervisor tells the surveyor to only flag locations that truly have contamination, then the surveyor is going to think twice about flagging a location, minimizing the number of false positives because of the cost the supervisor has placed on them.

9.3.3 Scan MDCs for Structure Surfaces

The MARSSIM survey design for determining the number of data points for areas of elevated activity depends on the scan MDC for the selected survey instrumentation. Because of their low background levels (typically only a few cpm), the approach defined in Sections 9.3.1 and 9.3.2 is not generally applicable to scan MDCs for alpha contaminants. In this section, we will treat the scan MDCs on building structures separately for alpha and beta radiation.

9.3.3.1 Scan MDCs on Structure Surfaces for Alpha Radiation

Scanning for alpha emitters must be derived differently than scanning for beta and gamma emitters. For the most part, the background response of most alpha detectors is very close to zero, typically registering no more than 2 or 3 cpm. An important consideration in the performance of alpha scans is that surfaces that are dirty, wet, or scabbled can significantly affect the detection efficiency and therefore alter the expected MDC for the scan. The use of reasonable detection efficiency values instead of optimistic values is highly recommended.

According to the recently released ASTM standard on the Selection and Use of Portable Radiological Survey Instruments for Performing *In Situ* Radiological Assessments in Support of Decommissioning (1998), for typical alpha background levels (1–2 cpm), a "two to three fold increase in audible signal is required to be recognizable." The standard provides the following equation for the alpha scan MDC, which the standard terms the minimum surface sensitivity (MSS):

$$MSS = \frac{3B_0}{\varepsilon_0 (A_d/100)},$$

where B_0 is given as the background count rate in cpm, ε_0 is the detector efficiency in counts per disintegration, and A_d is the window area of the detector probe in cm².

Unfortunately, it seems that this equation is not technically justifiable for alpha radiation. The whole basis for the MSS is embedded in the assumption that an alpha radiation level three times background is detectable. Empirical evidence has generally shown that it takes substantially more than just three times the alpha background level to be detectable. Consider an example to illustrate this point.

Assume that the scan will be performed using a gas proportional detector (126 cm² physical probe area) and that the alpha background is 1 cpm, with an alpha efficiency of 10% (total efficiency). The MSS using the above equation yields

$$MSS = \frac{(3) \times (1 \text{ cpm})}{0.10(126/100)} = 24 \text{ dpm/100 cm}^2.$$

Health physics technicians who have logged a number of hours scanning can attest that detecting 30 dpm/100 cm² of alpha contamination with a scan is simply not possible. Rather, their experiences of the detectable alpha contamination levels while scanning are much closer to 200 or 300 dpm/100 cm². This empirical evidence can be derived by noting the locations that are identified as elevated while scanning, and then determining the alpha contamination levels that triggered the scan hit by taking a direct measurement of surface activity. One final comparison of the MSS approach for alpha scan MDC is to calculate the static MDC for alpha measurements using this detector. All parameters being equal, the static MDC of an instrument is less than its scan MDC, due to the better statistics obtained with a longer, static count. However, when the static MDC is calculated for a 1-min count using this survey instrument, it turns out to be greater than its scan MDC:

$$\text{MDC} = \frac{3 + 4.65\sqrt{1}}{(0.1) \times (126/100)} = 60 \text{ dpm/100 cm}^2.$$

The scan MDC with its brief observation interval cannot be more sensitive than a direction measurement of surface activity. So, what is one to do?

One avenue for the consideration of alpha scan MDCs is presented in the MARSSIM. MARSSIM offers the following guidance: "Since the time a contaminated area is under the probe varies and the background count rate of some alpha instruments is less than 1 cpm, it is not practical to determine a fixed MDC for scanning. Instead, it is more useful to determine the probability of detecting an area of contamination at a predetermined DCGL for given scan rates." Generally, given a known scan rate, a background count rate of zero, and a level of surface contamination, the probability of detecting a single count while passing over the contaminated area is given in MARSSIM, Appendix J as

$$P(n \geq 1) = 1 - e^{\frac{-G\varepsilon d}{60v}} = 1 - e^{\frac{-G\varepsilon t}{60}}$$

where
   $P(n \geq 1)$ = probability of observing a single count
   G = hot spot activity (dpm)
   ε = detector efficiency (4π)
   d = width of the detector in the direction of the scan (cm)
   v = scan speed (cm/s)
   t = residence time of detector over activity (s)

Appendix J in MARSSIM provides a complete derivation and discussion of this formula.

The scan process is composed of two stages: continuous monitoring and stationary sampling (pausing). During the continuous monitoring stage, the surveyor listens to the number of clicks. Because the background is on the order of 0–3 cpm, a single count gives the surveyor sufficient cause to stop and investigate further—by pausing for an additional number of seconds. (Note: this is analogous to the beta/gamma derivation that defines the minimum count rate above background, MDCR, covered in the next section.) Therefore, the scan MDC for alpha contamination is determined based on the continuous monitoring stage.

Before we continue, a word about the residence time (or observation interval) of the detector over the activity is warranted. Of course, prior to performing the alpha scan, the surveyor does not know the size of the hot spots that may be present. If the hot spots are fairly large in area, for example, greater than several hundred square centimeters, the detector will have a reasonably long residence time. On the other hand, if the hot spot is truly more like a spot, let us say on the order of 10–50 cm², the residence time may be no more than 1 or 2 s, at best. For these reasons, it seems reasonable to simply postulate a certain size hot spot for the purpose of these scan MDC calculations. The postulated hot spot size for surface contamination on building surfaces is 100 cm²—which may be considered to be a 10 cm by 10 cm area. Hence, the residence time is dependent upon the scan rate. For example, a scan rate of 5 cm/s yields a residence time of 2 s.

Obviously, slower scan rates provide greater residence times, but there are bounds to how slow one can scan. For instance, one cannot continue to increase the observation interval (by slowing down) and expect to reduce the scan MDC without limit. Human factors suggests that if a hot spot was not detected during 5 or 6 s of scanning over an elevated area, another several seconds added to the observation interval does not improve the scan sensitivity.

The probability of detecting given levels of alpha surface contamination can be calculated by the use of Poisson summation statistics. If we define the scan MDC at a certain Poisson probability of being detected, then one can calculate the minimum alpha activity that can be detected by solving for G (now defined as the alpha scan MDC):

$$\alpha \text{ scan MDC} = \frac{[-\ln(1 - p(n \geq 1))]60}{\varepsilon_i \varepsilon_s t} \tag{9.13}$$

where *t* is calculated based on the scan rate but should not be greater than several seconds.

In the application of this equation, one must recognize that a point of diminishing returns exists for how large *t* can be. That is, once the probe is over the source for 5 s (as may be the case for distributed sources), increasing the residence time to several seconds will not significantly reduce the scan MDC. Finally, the use of *a* as 100 cm² hot spot size allows the calculation of alpha scan MDC in units of dpm/100 cm². Care must be exercised to calculate the alpha instrument efficiency using a 100 cm² calibration source. The next section on beta scan MDCs provides a more exhaustive discussion on scanning instrument efficiencies.

> The scan MDC cannot be reduced without limit simply by scanning slower and slower to increase the observation interval.

As an example, consider evaluating the scan MDC for Pu-239 on a stainless-steel benchtop in a laboratory. Assume that the scan speed (2.5 cm/s) is such that a residence time of 4 s is maintained over the contamination. The instrument efficiency is assumed to be 0.44, and the surface efficiency according to ISO-7503 is 0.25. Therefore, the scan MDC is based on a 90% probability of detecting 1 count:

$$\alpha \text{ scan MDC} = \frac{[-\ln(1-0.9)]60}{(0.44)(0.25)(4)} = 310 \text{ dpm}/100 \text{ cm}^2.$$

Once a count is recorded and the surveyor stops; the surveyor should wait for a sufficient period of time such that the first stage of scanning is the limiting concern in the derivation of the scan MDC. This can be accomplished by pausing for a period of time such that if the activity at the MDC is present, it would be readily apparent. How long should one need to pause? *Answer*: Not long, no more than a few seconds at 310 dpm would be necessary to produce additional counts discernable above background. Why? For an alpha activity of 310 dpm and stated efficiencies of 0.44 and 0.25, pausing for 4 s after the initial count is registered will yield more than two alpha counts:

$$(310 \text{ dpm})(0.44)(0.25)(4 \text{ s})(1 \text{ min}/60 \text{ s}) = 2.3 \text{ counts}.$$

So, if no additional counts are observed, the initial count is either at background levels or less than scan MDC.

For backgrounds greater than zero, for example, 1–3 cpm, the calculational approach and scan MDC result are still valid; however, it is at the expense of an increased false-positive rate. That is, the surveyor will be more likely to mistake background as contamination. For background count rates on the order of 5–10 cpm, a single count should not cause a surveyor to investigate further, primarily because there would be an inordinate amount of false positives. A counting period long enough to establish that a single count indicates an elevated contamination level would be prohibitively inefficient. For these types of instruments, the surveyor usually will need to get at least two counts while passing over the source area before stopping for further investigation. A similar approach can be used to obtain the Poisson probabilities for this situation.

Experiments relevant to alpha scanning sensitivity have been reported using a semiempirical approach (Goles et al. 1991). MDCs were defined as that activity that could be detected 67% of the time under standard survey conditions. The instruments evaluated included, for alpha detection, 50 cm^2 portable alpha monitor, 100 cm^2 large area scintillation monitor, and a 100 cm^2 gas proportional counter. The test procedure involved maintaining a scan rate of 5 cm/s, with a scan height held at 0.64 cm. Alpha sources were 2.54 cm diameter, electroplated sources. Residence times were on the order of 2 s. The MDC for alpha activity was defined as the amount of activity that produces 1 count as the detector passes over the surface (alpha background was considered to be zero), based on whether it could be detected 67% of the time. Goles reported detection at 300 dpm with the gas proportional detector, and further provided a range from 200 to 350 dpm for both the gas proportional detector and the scintillation monitor over a range of alpha energies.

The above data are provided for a relatively small areal source (~5 cm^2). Concerning distributed sources, Goles stated that due to the increased residence time over the extended source (~100 cm^2), which has a lower surface concentration, the MDCs are found to be generally the same as if all surface contamination was concentrated in a single point.

Using the available Goles experimental data, we can evaluate the alpha scan MDC using the above equation. The scan speed is 5 cm/s and the detector dimensions (~10 cm) provide a residence time of about 2 s over the alpha source. The total efficiency is assumed to be 11% and the scan MDC is based on a 67% probability of detecting 1 count

$$\alpha \text{ scan MDC} = \frac{[-\ln(1 - 0.67)]60}{(0.11)(2)} = 300 \text{ dpm}$$

which is remarkably consistent with the range of alpha scan MDCs reported by Goles.

9.3.3.2 Scan MDCs on Structure Surfaces for Beta Radiation

The scan MDC for beta radiation on structure surfaces is based on the signal detection theory for scanning presented in Sections 9.3.1 and 9.3.2. The scan MDC may be determined by applying conversion factors that account for detector and surface characteristics and the surveyor's efficiency to the MDCR. The MDCR is calculated by converting the minimum detectable counts in the observation interval, s_i, to cpm. The MDCR accounts for the background level, performance criteria (d'), and observation interval—which depend on the size of the hot spot. In this context, the size of the hot spot relates to the area of detection defined by the detector-to-source geometry (for instance, a 2 mm² point source of activity may produce an effective hot spot area of over 100 cm²). Therefore, the greater the effective area of the contamination and the slower the scan rate, the greater the observation interval. Because the dimensions of potential hot spots in the field cannot be known *a priori*, it is necessary to postulate a certain area, and then to select a scan rate that provides a sufficient observation interval to determine a scan MDC. The scan MDC for structure surfaces may be calculated as

$$\text{Scan MDC} = \frac{\text{MDCR}}{\sqrt{p\varepsilon_i\varepsilon_s}} \tag{9.14}$$

where

ε_i is the instrument efficiency
ε_s is the surface efficiency

Notice that there is no physical probe area correction in the above equation. This is in contrast to the static MDC equation, as well as the calculation of surface activity. The reason is that to calculate scan MDC we must postulate a certain size hot spot. As long as this postulated hot spot measures 100 cm² the scan MDC will be determined in the appropriate units (dpm/100 cm²). Care must also be taken in determining the instrument efficiency—specifically, ε_i should be determined using a 100 cm² calibration source.

An example may help clarify this point. A GM detector with a physical probe area of 20 cm^2 will be used to first scan equipment surfaces and then to make direct measurements of surface activity. For surface activity calibration, the instrument efficiency can be calculated during calibration by dividing the net counts by the 2π emission rate, where the emission rate was corrected by the calibration source area subtended by the physical probe area of the GM. That is, if the calibration source area was 100 cm^2, the emission rate quoted by the calibration source manufacturer would be reduced by 20/100 to determine the instrument efficiency.

However, when the GM is to be used for scans, the instrument efficiency must be determined for the instrument's response to a full 100 cm^2 area. This can be achieved by dividing the net counts by the 2π emission rate, not corrected by the detector's physical probe area. This represents the detector's efficiency to an entire 100 cm^2 surface area—which was assumed to be the hot spot size of concern. Hence, as long as 100 cm^2 is used as the size of the postulated hot spot and the instrument efficiency is calculated for the same area, there is no need for a probe area correction. OK, so what happens if the calibration source area is greater than 100 cm^2—for example, 150 cm^2? In this instance, the 2π emission rate that is specified for the full 150 cm^2 area is reduced by the ratio 100/150. This ensures that the scanning instrument efficiency for the GM detector is normalized to a 100 cm^2 area.

Before we leave this discussion, one final point should be made concerning detectors with physical probe areas larger than 100 cm^2, such as floor monitors and some hand-held detectors. In Chapter 8, the physical probe area of the gas proportional floor monitor was stated to be about 600 cm^2. To determine scanning instrument efficiencies for the floor monitor, the detector response should also be calibrated to a 100 cm^2 surface area. But what happens if the calibration source area is 150 cm^2? In this circumstance, we would not modify the 2π emission rate, simply because of the detector-to-source geometry involved. That is, regardless of the calibration source area—100 or 150 cm^2—the floor monitor physical probe area completely covers the calibration source size, and the instrument efficiency for scanning is best determined using the full 2π emission rate from the calibration source. In a similar fashion, the 126-cm^2 gas proportional detector would ideally be calibrated to a 100 cm^2 area. However, if the calibration source area is 150 cm^2, the 2π emission rate of the source would not be modified, again because of the detector-to-source geometry involved and the fact that the detector would likely respond to radiations from the full 150 cm^2 source.

TABLE 9.1 Instrument Efficiencies for GM Detector Used for Scanning

Radionuclide	Average Energy (keV)	Calibration Source Size (cm²)	2π Emission Rate (cpm)	Net Count Rate (cpm)	Instrument[a,b] Efficiency (ε_i)
C-14	49.4	150	255,000	4663	0.027
Tc-99	84.6	150	14,400	506	0.052
Tl-204	244	150	191,985	10,368	0.081
SrY-90	563	150	302,878	17,582	0.087
Ru-106 (Rh-106)	1410	150	102,624	10,517	0.153

Source: Courtesy of ORAU.

[a] Instrument efficiency calculated by dividing the net cpm by the emission rate, normalized to a 100 cm² area.

[b] The detector-to-surface spacing was nominally 1–2 cm, and was maintained by the survey technician holding the probe over the surface to simulate the spacing maintained during scanning.

TABLE 9.2 Instrument Efficiencies for Floor Monitor Used for Scanning

Radionuclide	Average Energy (keV)	Calibration Source Size (cm²)	2π Emission Rate (cpm)	Net Count Rate (cpm)	Instrument[a] Efficiency (ε_i)
C-14	49.4	150	255,000	45,575	0.18
Tc-99	84.6	150	14,400	3489	0.24
Tl-204	244	150	191,985	74,549	0.39
SrY-90	563	150	302,878	127,759	0.42
Ru-106 (Rh-106)	1410	150	102,624	69,609	0.68

Source: Courtesy of ORAU.

[a] Instrument efficiency calculated by dividing the net cpm by the emission rate—no probe area corrections performed because of the large physical probe area of floor monitor (~600 cm²).

The point of the discussion in the preceding paragraphs on scanning instrument efficiency can be summarized as follows: In consideration of the physical probe area of the detector and the calibration source area, calculate the instrument efficiency for scanning that best approximates the detector's response to an *a priori* 100 cm² hot spot. Tables 9.1 through 9.3 provide instrument efficiencies for scanning for common survey instruments.

As an example, consider evaluating the scan MDC for C-14 on a stainless-steel benchtop in a biomedical laboratory. The scan MDC will be determined for a background level of 300 cpm and a 1-s interval using a

TABLE 9.3   Instrument Efficiencies for Gas Proportional Detector (0.8 mg/cm^2) Used for Scanning

Radionuclide	Average Energy (keV)	Calibration Source Size (cm^2)	2π Emission Rate (cpm)	Net Count Rate (cpm)	Instrument[a,b] Efficiency (ε_i)
C-14	49.4	150	255,000	72,046	0.28
Tc-99	84.6	150	14,400	4375	0.30
Tl-204	244	150	191,985	88,264	0.46
SrY-90	563	150	302,878	139,201	0.46
Ru-106 (Rh-106)	1410	150	102,624	74,193	0.72

Source: Courtesy of ORAU.
[a] Instrument efficiency calculated by dividing the net cpm by the emission rate (no probe area corrections performed).
[b] The detector-to-surface spacing was nominally 1–2 cm, and was maintained by the survey technician holding the probe over the surface to simulate the spacing maintained during scanning.

hand-held gas proportional detector (126 cm^2 probe area). For a specified level of performance at the first scanning stage of 95% true-positive rate and 60% false-positive rate (and assuming the second-stage pause is sufficiently long to ensure that the first stage is the more limiting), $d' = 1.38$ (from MARSSIM Table 6.5) and the MDCR is calculated as follows:

$$b_i = (300 \text{ cpm})(1 \text{ s})(1 \text{ min}/60 \text{ s}) = 5 \text{ counts,}$$
$$s_i = (1.38)(5)^{1/2} = 3.1 \text{ counts}$$

$$\text{MDCR} = (3.1 \text{ counts})[(60 \text{ s/min})/(1 \text{ s})] = 185 \text{ cpm.}$$

Using a surveyor efficiency of 0.5, and assuming the instrument and surface efficiencies of 0.28 (refer to Table 9.3) and 0.25 (from ISO-7503 recommendations), respectively, the scan MDC is calculated as

$$\text{Scan MDC} = \frac{185}{\sqrt{0.5}(0.28)(0.25)} = 3740 \text{ dpm/100 cm}^2.$$

What if we increased the background level to 380 cpm? The scan MDC is calculated in the same manner, keeping the other parameters constant:

$$b_i = (380 \text{ cpm})(1 \text{ s})(1 \text{ min}/60 \text{ s}) = 6.3 \text{ counts},$$
$$s_i = (1.38)(6.3)^{1/2} = 3.5 \text{ counts}$$

$$MDCR = (3.5 \text{ counts})[(60 \text{ s/min})/(1 \text{ s})] = 208 \text{ cpm}.$$

The scan MDC is now calculated as

$$\text{Scan MDC} = \frac{208}{\sqrt{0.5}(0.28)(0.25)} = 4210 \text{ dpm}/100 \text{ cm}^2.$$

Therefore, increasing the background by about 25% (from 300 to 380 cpm) results in the scan MDC being increased by 12%. This is expected because the MDCR is proportional to the square root of background counts in the observation interval.

Finally, what if the false-positive rate of 60% is deemed to be too high? Let us assume that we want the scan MDC for a false-positive rate of 25% and true-positive rate of 95%. MARSSIM Table 6.5 provides a d' that equals 2.32. The scan MDC for a background level of 380 cpm is calculated to be 7080 dpm/100 cm^2, up from 4210 dpm/100 cm^2. This makes intuitive sense; if one is not willing to pause as frequently during the scan—which is precisely the impact of changing from a false-positive rate of 60% to 25%—it is going to require a greater contamination level to be detectable 95% of the time. Hopefully, this example points out the importance of the false-positive rate in determining the scan MDC.

These straightforward examples illustrate the scan MDC range as a function of background level and acceptable performance in terms of desired false-positive rate. The scan MDC for the gas proportional detector calibrated to C-14 varied from 3740 to 7080 dpm/100 cm^2, as the background level was varied from 300 to 380 cpm, and the false-positive rate was lowered from 60% to 25%. NUREG-1507 (NRC 1998b) provides additional examples for calculating the scan MDC.

Let us consider one more example before leaving this section. As our final example, assume that the same stainless-steel benchtop in the biomedical laboratory will be scanned using a GM detector. As before, the contaminant is C-14. This time the scan MDC will be determined for a background level of 65 cpm and a 1-s interval. The specified level of performance at the first scanning stage will be 95% true-positive rate and 60% false-positive rate, which yields a $d' = 1.38$ (again refer to MARSSIM Table 6.5). The MDCR is calculated as follows:

$$b_i = (65 \text{ cpm})(1 \text{ s})(1 \text{ min}/60 \text{ s}) = 1.08 \text{ counts,}$$
$$s_i = (1.38)(1.08)^{1/2} = 1.43 \text{ counts}$$

$$\text{MDCR} = (1.43 \text{ counts})[(60 \text{ s/min})/(1 \text{ s})] = 86 \text{ cpm.}$$

Using a surveyor efficiency of 0.5, and assuming instrument and surface efficiencies of 0.027 (refer to Table 9.1) and 0.25 (from ISO-7503 recommendations), respectively, the scan MDC is calculated as

$$\text{Scan MDC} = \frac{86}{\sqrt{0.5(0.027)(0.25)}} = 18,000 \text{ dpm}/100 \text{ cm}^2.$$

Obviously, the larger scan MDC for the GM detector is driven by its low instrument efficiency for C-14.

9.3.4 Scan MDCs for Land Areas

Assessing gamma scan MDCs using NaI scintillation detectors for soil areas not only depends on the minimum detectable count rate but also on the energy and level of gamma radiation from a postulated hot spot area. That is, in addition to the MDCR and detector's response characteristics, the scan MDC (in pCi/g) for land areas is based on the areal extent of the hot spot, its depth in the soil, density and elemental composition of the soil, and the radionuclide characteristics—that is, energy and radiation yield of gamma emissions. If constant parameters are used for each of the above variables, except the specific radionuclide being assessed, the scan MDC can be reduced to a function of the radionuclide alone.

An overview is given of the approach used to determine scan MDCs for soil. The MDCR may be calculated by selecting a given level of performance and scan speed, and determining the background level of the NaI scintillation detector. The surveyor's efficiency is selected (usually taken to be 0.5), and then the surveyor's MDCR, which accounts for the surveyor efficiency, $\text{MDCR}_{surveyor}$, is related to the radionuclide concentration in soil. This correlation requires two steps—first, the relationship between the NaI detector's net counting rate to net exposure rate (cpm per μR/h) as a function of energy is established; second, the relationship between the radionuclide contamination in soil and exposure rate is determined.

> The scan MDC for land areas requires two fundamental steps once MDCR is determined: (1) establish the relationship between the NaI detector count rate and the exposure rate, and (2) establish the relationship between the exposure rate and the radionuclide concentration in soil.

For a particular gamma energy, the relationship of the NaI scintillation detector's counting rate (cpm) and exposure rate may be determined analytically. The approach for establishing the cpm per μR/h relationship is discussed in NUREG-1507; results for two sizes of NaI scintillation detectors are provided in Table 9.4.* Once this relationship is established, the MDCR$_{surveyor}$ (in cpm) of the NaI scintillation detector can be related to the minimum detectable net exposure rate. It is necessary to convert

TABLE 9.4   NaI Scintillation Detector cpm per μR/h Values

Gamma Energy (keV)	2″ × 2″ NaI Detector	1.5″ × 1.25″ NaI Detector
20	2200	990
30	5160	2320
40	8880	3990
50	11,800	5320
60	13,000	5830
80	12,000	5410
100	9840	4420
150	6040	2710
200	4230	1890
300	2520	1070
400	1700	700
500	1270	510
600	1010	390
662	900	350
800	710	270
1000	540	200
1500	350	130
2000	260	100
3000	180	70

---

* For gamma energies not provided in Table 9.4, it is recommended that the data be fit to polynomial and power functions. The 20 to 60 keV data and 60 to 100 keV data are best fit to separate polynomial functions (3rd order and 2nd order, respectively), while the 100 to 3000 keV data can be best fit to a power function.

the MDCR to the minimum detectable exposure rate because the modeling code used for this work (Microshield™) relates soil concentrations to exposure rates above the soil surface. This minimum detectable exposure rate is used to determine the minimum detectable radionuclide concentration (i.e., the scan MDC) by modeling a specified postulated hot spot. Modeling can be performed using commercially available codes such as Microshield, or even better, more powerful codes that make use of Monte Carlo techniques.

One such code is the MCNP (Monte Carlo N-Particle) code maintained by the Los Alamos National Laboratory. MCNP, which can be run on personal computers, can be used to model the photons emanating from the radionuclide source term. The code accounts for a number of possible photon interactions, including Compton scattering, photoelectric absorption, and bremsstrahlung. Basically, MCNP solves the problem by simulating photon histories for a specified geometry, while codes like Microshield use a deterministic method to describe the average photon behavior. One shortcoming with the Microshield modeling code is that it considers only primary gamma energies when evaluating the buildup from scattered photons. The NaI detector's response will be greater during field applications compared to the calculated response because the detector is more efficient at detecting scattered photons of lower energy. This situation is expected to generate a conservative determination of the detector's response and the resulting assessment of scan MDC. Nevertheless, it is recommended that future work in this area makes use of the MCNP code.

Microshield was used for the determination of scan MDCs in this text. Soil concentrations within the postulated hot spot are modeled, which determines the net exposure rate produced by the input radionuclide concentration at a typical detector distance above the surface during scanning. This height above the soil surface is usually taken to be about 10 cm, which represents the nominal height of the NaI scintillation detector as it is swung in a pendulum fashion over the soil. Factors that are considered in the modeling include the radionuclide(s), concentration of radionuclide, areal dimensions of hot spot, depth of hot spot, location of dose point (i.e., NaI scintillation detector height above the surface), and density and composition of soil.

Modeling analyses were conducted by selecting a radionuclide (or radioactive material decay series) and then varying its concentration. Also, the areal dimension of the cylindrical hot spot was 0.25 m^2 (radius of about 0.28 m); it was uniformly contaminated to a depth of 15 cm, the dose point was 10 cm above the surface, and the density of soil was 1.6 g/cm³. The hot spot size is a key factor in the determination of the scan MDC, but yet it cannot possibly be known prior to conducting actual scans. The nominal size selected (0.25 m²) represents what may be considered a reasonable size at many D&D sites, but the reader should recognize that this value does not have a strong technical basis; rather it was selected based on survey experience at D&D sites. Accordingly, these variables should be changed to reflect site-specific conditions and needs of the DQO process. The objective was to determine the radionuclide concentration that was correlated to the minimum detectable net exposure rate.

The scan MDC for Cs-137 using a 1.5″ × 1.25″ NaI scintillation detector is considered as an example. The assumptions are that the background level is 3600 cpm, the desired level of performance for the first stage is 95% correct detections and 60% false-positive rate, and the detector is paused long enough at the second stage so that the MDCR is determined from the first stage. A scan rate of 0.5 m/s provides an observation interval of 1 s (based on a hot spot diameter of about 0.56 m). The MDCR for these conditions was calculated previously as 640 cpm, therefore, assuming a surveyor efficiency of 0.5 yields an MDCRsurveyor of 900 cpm, that is, $640/(0.5)^{1/2}$.

The corresponding minimum detectable exposure rate is determined for the detector and radionuclide. The manufacturer of this particular 1.5″ × 1.25″ NaI scintillation detector quotes a counting-rate-to-exposure-rate ratio for Cs-137 of 350 cpm per μR/h, which is assumed to account for the 662 keV gamma emission from its short-lived progeny, Ba-137m. The minimum detectable exposure rate is calculated as

$$\frac{900 \text{ cpm}}{350 \text{ cpm}/(\mu R/h)} = 2.57 \ \mu R/h.$$

Both Cs-137 and its short-lived progeny, Ba-137m, were chosen from the Microshield library. Note that many health physicists assume that the decay of Cs-137 results in a 662 keV gamma emission directly, but in reality it is the Ba-137m that is formed from the decay of Cs-137 that yields this gamma emission. Therefore, it is extremely important that the user of the Microshield be knowledgeable of how the code handles progeny relationships—and in this case, both Cs-137 and Ba-137m need to be input from the library. The source activity for each radionuclide and other modeling parameters were entered into the modeling code. The source activity was selected based on an arbitrary concentration of 5 pCi/g, and converted to the appropriate units for input into the modeling code

$$(5\ \text{pCi/g})(1.6\ \text{g/cm}^3)(1\ \mu\text{Ci/10}^6\ \text{pCi}) = 8 \times 10^{-6}\ \mu\text{Ci/cm}^3.$$

The modeling code determined an exposure rate of 1.31 μR/h, which accounts for buildup, although it uses the primary photon energy for scattered fluence rate. Finally, the radionuclide concentrations of Cs-137 and Ba-137m necessary to yield the minimum detectable exposure rate 2.57 μR/h may be calculated as

$$\text{Scan MDC} = (5\ \text{pCi/g})\frac{2.57\ \mu\text{R/h}}{1.31\ \mu\text{R/h}} = 9.8\ \text{pCi/g}.$$

It must be emphasized that while a single value for the scan MDC can be calculated for a given radionuclide, other scan estimates may be equally justifiable depending on the values chosen for the various factors, including the MDCR (which depends on background level, acceptable performance criteria, and observation interval), the surveyor's efficiency, detector parameters, and the modeling conditions of the contamination. Determining the scan MDC for radioactive materials—like uranium and thorium—must consider the gamma radiation emitted from the entire decay series. Table 9.5 provides scan MDCs for common radiological contaminants, many of which were published in NUREG-1507 (NRC 1998b).

Now that we have discussed the technical approach for determining the scan MDC, let us take a closer look at the steps involved, particularly the steps involved with the Microshield code. Specifically, we will consider the procedure necessary to calculate the scan MDC for Am-241 using a 1.5″ × 1.25″ NaI scintillation detector. Let us assume that the background for this detector is 4000 cpm and that the observation interval is 1 s for a scan rate of 0.5 m/s. The specified level of performance at the first scanning stage will be 95% true-positive rate and 60% false-positive rate, which yields a $d' = 1.38$. The MDCR is calculated as follows:

$$b_i = (4000\ \text{cpm})(1\ \text{s})(1\ \text{min/60 s}) = 66.7\ \text{counts},$$
$$s_i = (1.38)(66.7)^{1/2} = 11.3\ \text{counts}$$

$$\text{MDCR} = (11.3\ \text{counts})[(60\ \text{s/min})/(1\ \text{s})] = 676\ \text{cpm}.$$

TABLE 9.5   NaI Scintillation Detector Scan MDCs for Common
Radiological Contaminants[a]

| Radionuclide/Radioactive Material | Scan MDC (pCi/g) | |
	1.5″ × 1.25″ NaI Detector	2″ × 2″ NaI Detector
Am-241	45	32
Co-60	5.8	3.4
Cs-137	10.4	6.4
Cs-134	4.0	2.5
Eu-152	5.8	3.7
I-129	120	85
Fe-59	10.4	6.3
Th-230	3000	2100
Ra-226 (in equilibrium with progeny)	4.5	2.8
Th-232 decay series (sum of all radionuclides in thorium decay series, in equilibrium)	28	18
Th-232 alone (in equilibrium with progeny in decay series)	2.8	1.8
Depleted uranium[b](0.34% U-235)	80	56
Processed natural uranium[b]	115	80
3% Enriched uranium[b]	140	96
20% Enriched uranium[b]	150	110
50% Enriched uranium[b]	170	120
75% Enriched uranium[b]	190	130

[a] The background level for the 1.5″ × 1.25″ NaI detector was assumed to be
    4000 cpm and 10,000 cpm for the 2″ × 2″ NaI detector. The observation
    interval was 1 s and the level of performance was selected to yield d' of 1.38.
[b] Scan MDC for uranium includes the sum of U-238, U-235, and U-234
    isotopes.

Using a surveyor efficiency of 0.5, we can calculate the $MDCR_{surveyor}$ from $676/(0.5)^{1/2}$, which equals 960 cpm. It is at this point that we can focus on the specifics of the Microshield code (version 4.21):

1. Choose the source geometry: cylindrical volume with end shields is selected from the menu.

2. Enter dose point coordinates: $x = 0$, $y = 25$ cm (15 cm of soil, 10 cm in air), and $z = 0$.

   The cylinder height equals 15 cm with an air gap of 10 cm; the cylinder radius is 28.2 cm. No shield data were entered.

3. Enter material—soil is not an option in Microshield version 4.21, so concrete was used with a density of 1.6 g/cm³.

4. Radionuclide source selected from library of nuclides. Select Am-241 and calculate volumetric concentration that equates to 1 pCi/g:

$$(1\,pCi/g)(1.6\,g/cm^3)(1\,\mu Ci/10^6\,pCi) = 1.6 \times 10^{-6}\,\mu Ci/cm^3.$$

5. At this point, Microshield performs a calculation of the activity for a number of the energy groups. For Am-241, the bulk of the gamma activity should be associated with the 60 keV gamma emission.

6. Run Microshield.

7. Display results for "Photon source; dose point results with buildup." Record total exposure rate with buildup from all energy groups (60 keV should dominate). These results are used to determine the energy-weighted cpm per μR/h. The result for 60 keV was 3.691E − 3 μR/h, no other energies had dose rates that were at least 1% of this value, so the energy-weighted cpm per μR/h was the value for 60 keV (5830 cpm per μR/h), obtained from Table 9.4 for the 1.5″ × 1.25″ NaI scintillation detector.

8. Calculate the minimum detectable exposure rate, based on the MDCR and the energy-weighted cpm per μR/h value:

$$\text{Minimum detectable exposure rate} = \frac{960\,\text{cpm}}{5830\,\text{cpm}/\mu R/h}$$
$$= 0.165\,\mu R/h.$$

9. Calculate the scan MDC, given the dose rate based on the 1 pCi/g of Am-241 input:

$$\text{Scan MDC} = (1\,pCi/g)\frac{0.165\,\mu R/h}{3.691E-3\,\mu R/h} = 44.7\,pCi/g.$$

Therefore, the scan MDC for Am-241 using a 1.5″ × 1.25″ NaI scintillation detector, for the specific parameters input is about 45 pCi/g. How accurate is this value? How would one even begin to determine the accuracy of this scan MDC for Am-241? One approach would be to empirically determine what levels of Am-241 are detectable with NaI scanning. Another way might be to analyze the individual components, or variables, and assess their uncertainty. It may turn out that some of the assumptions used in the above calculation are not that sensitive to the final result.

A sensitivity analysis could be performed to determine how each of the above variables impact the scan MDC for Am-241. These variables include the background level, d' value, observation interval, surveyor efficiency, depth of contamination, dose point, areal size of hot spot, density, and the cpm per μR/h value for the NaI scintillation detector at 60 keV. The idea is to identify the variables that have the greatest impact on the scan MDC. While there are a number of ways that a sensitivity analysis might be performed, the approach described herein was to vary each of the variables by ±20%, and calculate the scan MDC, holding all other variables constant. For a reference point, the scan MDC (44.7 pCi/g) at ±20% ranged from 35.8 to 53.6 pCi/g.

The first variable studied was the background level. The assumed background level was 4000 cpm, so the sensitivity analysis used values at 3200 and 4800 cpm. So, using the same approach as illustrated above, the scan MDC for the background level at 3200 cpm was 39.7 pCi/g, while the scan MDC for the higher background level (4800 cpm) was 48.7 pCi/g. Therefore, the 20% variation in background levels resulted in a scan MDC range of 39.7–48.7 pCi/g.

The second variable studied was d'. Rather than try to change the true-positive rate or false-positive rate individually, the overall d' was modified by 20%. The reader should recognize that at any particular d' value, there are a number of possible true-positive/false-positive combinations. The 20% range on d' was 1.10–1.66 (nominal d' value was 1.38). The d' of 1.10 for instance might represent a performance level of 80% true positives and 40% false positives, while the d' of 1.66 may represent 95% true positives and 50% false positives (refer to MARSSIM Table 6.5). The scan MDC for the d' value at 1.10 was 35.4 pCi/g, while the scan MDC for the d' value at 1.66 was 53.4 pCi/g. Therefore, the 20% variation in d' values in a scan MDC range of 35.4–53.4 pCi/g. This range matches that of scan MDC (44.7 pCi/g) at ±20%; therefore, the d' value is a very sensitive parameter in the determination of scan MDC.

The third parameter evaluated was the observation interval, which of course embodies the scan speed. The nominal observation interval was 1 s; therefore, the 20% variation was 0.8–1.2 s. The scan MDC for the observation interval at 0.8 s was 49.7 pCi/g, while the scan MDC for the longer observation interval (1.2 s) was 40.6 pCi/g. Therefore, the 20% variation in observation intervals resulted in a scan MDC range of 40.6–49.7 pCi/g, quite similar to the background level variation.

The fourth parameter studied was the surveyor efficiency. The default value used in this text is 0.5, so the 20% variation resulted in surveyor efficiencies at 0.4 and 0.6. The scan MDC for the surveyor efficiency at 0.4 was 49.7 pCi/g, while the scan MDC for the higher surveyor efficiency was 40.6 pCi/g. Therefore, the 20% variation in surveyor efficiency was exactly equal to the observation interval scan MDC range of 40.6–49.7 pCi/g. Both these parameters exhibit square root relationships to the scan MDC.

Note that the first four variables studied impact the minimum detectable exposure rate, via the $MDCR_{surveyor}$. Further, each of these four variables has a relatively significant impact of the scan MDC calculation, with the d' value being the greatest. The next four variables are related to the source term geometry. The necessary modifications in the Microshield model were made and revised dose rates were noted. The $MDCR_{surveyor}$ was a constant 960 cpm for this part of the sensitivity analysis.

The first Microshield model parameter evaluated was the depth of Am-241 contamination. The nominal value was 15 cm, so the 20% variation in this variable resulted in contamination depths at 12 and 18 cm. The same input concentration of 1 pCi/g was used at both depths. The scan MDC for the depth at 12 cm was 44.8 pCi/g, while the scan MDC for the 18 cm depth was 44.7 pCi/g. Therefore, the 20% variation in contamination depth was extremely minor, with the scan MDC range of 44.7–44.8 pCi/g. Now the reader should be cautioned that these results are for the low-gamma-energy Am-241; the depth of contamination might be more significant for higher-energy gammas like those from Co-60.

The second parameter considered in the modeling code was the dose point. The default used was 10 cm above the hot spot. The 20% variation resulted in dose points at 8 cm and at 12 cm. The scan MDC for the dose point at 8 cm was 40.5 pCi/g, while the scan MDC for the 12 cm dose point was 49.4 pCi/g. Therefore, the 20% variation in dose point was significant, with the scan MDC range of 40.5–49.4 pCi/g. Again, the specific radionuclide is important in assessing this parameter; higher-energy gamma emissions may not exhibit as great a difference.

The third variable studied was the areal size of the hot spot. The nominal size of the hot spot was 0.25 m², so the 20% variation resulted in a range of 0.2–0.3 m². This size variation was actually input in the code as a change in the radius of the cylinder (default of 28.2 cm), with the 20%

variation at 25.2 and 30.9 cm. Now, because the observation interval was kept constant at 1 s, the effective scan speeds were implicitly assumed to change—0.50 m/s at the smaller diameter and 0.62 m/s at the larger diameter. Thus, while a change in hot spot size certainly changes the observation interval, the impact evaluated here was just on the calculated dose from a changing source term size. The scan MDC for the hot spot size at 0.2 m² was 47.8 pCi/g, while the scan MDC for the 0.3 m² hot spot was 42.6 pCi/g. Therefore, the 20% variation in hot spot size was somewhat significant, with the scan MDC range of 42.6–47.8 pCi/g.

The fourth variable considered was the density. The nominal value of density was 1.6 g/cm³, so the 20% variation was 1.28–1.92 g/cm³. Of course the real assessment should be on how concrete (which was used for these assessments) compares to soil. The scan MDC for the density at 1.28 g/cm³ was 45.5 pCi/g, while the scan MDC for the density at 1.92 g/cm³ was 44.2 pCi/g. Therefore, the 20% variation in density was insignificant, with the scan MDC range of 44.2–45.5 pCi/g. Therefore, considering the second set of variables that were modified in the Microshield model, the variable having the greatest impact was the dose point, while the depth of contamination had a negligible impact.

The final parameter evaluated was the cpm per μR/h value for the NaI scintillation detector at 60 keV, which can be thought of as the detector efficiency. The nominal value was 5830 cpm per μR/h, so the 20% variation is 4664–6996 cpm per μR/h. In reality, the NaI scintillation detector has a specific cpm per μR/h value, so it may be more correct to view this variation as the uncertainty in this parameter value, rather than on some range that the parameter value might have been based on surveyor or contamination characteristics. So, in a sense, this parameter value is different than the first eight variables studied. The scan MDC for the cpm per μR/h value at 4664 was 55.8 pCi/g, while the scan MDC for the cpm per μR/h value at 6996 was 37.2 pCi/g. Therefore, the 20% variation in cpm per μR/h value was one of the most significant, with the scan MDC range of 37.2–55.8 pCi/g.

In summary, one of the key variables to focus on when assessing accuracy of the scan MDC for Am-241 would certainly be the cpm per μR/h value at 60 keV. Other important parameters include the d' value, dose point, background level, surveyor efficiency, and observation interval. The depth of contamination, hot spot size, and density—all have less of an impact on the scan MDC for Am-241. Note that these variables may change in importance as a function of the radionuclide being evaluated.

CASE STUDY   APPROACH TO CALCULATING LAND AREA
SCAN MDCS FOR SURVEY INSTRUMENTS

Suppose that you are tasked with calculating the scan MDC for instruments used to scan land areas not covered in this text. How would you begin to tackle this problem? This case study provides a summary of the critical components necessary to determine the scan MDC. Let us begin with an overview of the key steps in calculating the scan MDC.

The following are the four steps for calculating scan MDC for land areas:

1. Calculate the MDCR (in cpm) for a given background, observation interval, and performance level (type I and type II decision errors).
2. Divide MDCR by surveyor efficiency to yield MDCR$_{surveyor}$.
3. Translate MDCR$_{surveyor}$ to the minimum detectable exposure rate using the relationship of net count rate to net exposure rate for a particular survey instrument.
4. Translate minimum detectable exposure rate to scan MDC using a model such as Microshield or MCNP for specific conditions—for example, contamination distribution, soil density, and detector-to-surface spacing.

Let us suppose that the scanning instrument being considered is a $3'' \times 3''$ NaI scintillation detector. According to steps #1 and #2, the MDCR (and subsequently, the MDCR$_{surveyor}$) should be calculated given specific parameter values for the detector background, observation interval, performance level desired, and surveyor efficiency. Of these parameters, only the background level is impacted by the selection of the $3'' \times 3''$ NaI scintillation detector. The background level can be readily determined in an appropriate reference area.

Step #3 involves determining the relationship of the MDCR$_{surveyor}$ (in cpm) to the minimum detectable exposure rate, as a function of energy. This determination requires input from the survey instrument manufacturer. Specifically, the manufacturer of the $3'' \times 3''$ NaI scintillation detector should be able to quote a counting-rate-to-exposure-rate ratio at a minimum of one particular gamma energy (typically, the energy quoted is the 662 keV from Cs-137 (Ba-137m). This value can be regarded as the efficiency of the detector at one particular energy. From this one energy, the detector's response function to other gamma energies can be calculated based on a knowledge of the detector's characteristics (e.g., attenuation coefficient, density and thickness of the detector material). The calculational approach used for establishing the cpm per µR/h relationship is illustrated on page 10 of Abelquist and Brown (1999). The result of this step is the translation of the minimum detectable count rate to the minimum detectable exposure rate. This is necessary to provide a linkage to the modeling results obtained in step #4.

Step #4 results in the minimum detectable exposure rate being related to the MDC (i.e., the scan MDC). This is accomplished by modeling the radionuclide concentration in a specific geometry—considering the contamination areal and depth distribution and soil density—and yields an exposure rate at a specified dose point. The modeling effort is of course independent of the survey instrument being evaluated. So, the only instrument-specific characteristics necessary to calculate the scan MDC for the 3″ × 3″ NaI scintillation detector (or any detector for that matter) include the background level and the cpm per μR/h relationship for the specific detector.

The MARSAME (Section 7.11) provides an exceptionally detailed example of how the FIDLER's scan MDC is calculated for uranium and thorium generally following these steps.

9.3.5 Scan MDCs for Multiple Contaminants for Structure Surfaces and Land Areas

Now that the technical basis for calculating scan MDCs for individual radionuclides has been established, the logical follow-up question is: What is the scan MDC when multiple radionuclides are present? This section will offer answers to this question of scan MDCs when multiple radionuclides are present on both structure surfaces and land areas.

For multiple contaminants on a structure surface that include beta emitters, the scan MDC equation is identical to that presented previously:

$$\text{Scan MDC} = \frac{\text{MDCR}}{\sqrt{p}\varepsilon_i\varepsilon_s}.$$

While the equation is the same, that is probably not true for the efficiencies. The instrument and surface efficiencies should account for the particular radionuclides present, and their relative ratios. Consider an example where a plastic scintillation detector will be used to scan for Co-60 and Sr-90 on wood surfaces. Characterization data indicate that the mixture of the two radionuclides is 25% Sr-90 and 75% Co-60. Assume that the instrument efficiencies used for scanning are 0.51 for Sr-90 and 0.32 for Co-60. The weighted instrument efficiency is simply given

Weighted instrument efficiency $= (0.75)(0.51) + (0.25)(0.32) = 0.46.$

In a similar fashion, the surface efficiencies for each contaminant—0.5 for Sr-90 and 0.25 for Co-60 based on ISO-7503 guidance—must also be weighted:

$$\text{Weighted surface efficiency} = (0.75)(0.5) + (0.25)(0.25) = 0.44.$$

Finally, the scan MDC can be calculated as a function of the background level (assume 410 cpm), observation interval (1 s), and scanning performance level (for 95% true positives and 25% false positives, d' is 2.32). The MDCR is calculated as follows:

$$b_i = (410 \text{ cpm})(1 \text{ s})(1 \text{ min}/60 \text{ s}) = 6.83 \text{ counts},$$
$$s_i = (2.32)(6.83)^{1/2} = 6.06 \text{ counts},$$

$$\text{MDCR} = (6.06 \text{ counts})[(60 \text{ s/min})/(1 \text{ s})] = 364 \text{ cpm}.$$

Using a surveyor efficiency of 0.5, and the weighted instrument and surface efficiencies calculated above, the scan MDC is calculated as

$$\text{Scan MDC} = \frac{364}{\sqrt{0.5}(0.46)(0.44)} = 2500 \text{ dpm}/100 \text{ cm}^2.$$

Hence, to determine scan MDCs for multiple beta emitters, one simply needs to account for the weighted instrument and surface efficiencies.

Multiple alpha contaminants on building surfaces are treated in a similar manner, possibly even more straightforward than that shown for multiple beta emitters. This is because many of the common alpha contaminants emit radiation in a relatively narrow energy band of 4–5 MeV, which tends to keep the variation in alpha instrument efficiency small as well. In practice, many D&D contractors may choose to simply calibrate their alpha survey instruments to an alpha emitter at the low end of the range, perhaps Th-230, and use a single, somewhat conservative instrument efficiency for all alpha radiation. Therefore, it is possible that even with multiple alpha emitters present, the instrument efficiency remains unchanged. This is always the case for the surface efficiency when multiple alpha emitters are present simply because the ISO-7503 recommends an ε_s of 0.25 for all alphas. Thus, the alpha scan MDC for one alpha emitter or a number of alpha emitters is likely to be the same value.

The situation where the building surfaces scan is performed with the detector set to respond to both alpha and beta radiation, should be addressed for completeness. This case may arise when contamination is present from natural decay series, and the instrument efficiency is determined by weighting the response from each radiation, alpha and beta, in the series (as described in Chapter 8). Even in this example, the scan MDC for building surfaces is calculated in the same manner as shown above, ensuring that the instrument and surface efficiencies are properly weighted to account for the radionuclide mix. While the building surface scan MDC for multiple contaminants is merely a matter of accounting for changes in the efficiencies; unfortunately, determining land area scan MDCs for multiple contaminants becomes more involved.

Perhaps the best way to begin this discussion is to provide an example of how the scan MDC can be determined for multiple radionuclides. Let us suppose that a class 1 survey unit is potentially contaminated with both Am-241 and Co-60. From Table 9.5, we can see that the scan MDCs for a 1.5″ × 1.25″ NaI scintillation detector are 45 and 5.8 pCi/g for Am-241 and Co-60, respectively. Now, the first thing that must be considered is whether or not it is possible to establish a consistent ratio between the two radionuclides. If consistent ratios can be established, then it is simply a matter of modeling the individual radionuclides at their fractional abundances. For example, if the mixture was 30% Am-241 and 70% Co-60, then the source term entered into the model would be based on 0.3 pCi/g Am-241 and 0.7 pCi/g Co-60. The modeling code output would be used to determine both the exposure rate at the dose point for this radionuclide mixture and the energy weighted cpm per μR/h value. Indeed, the same approach is used to determine scan MDCs for decay series—individual series radionuclides are entered into the model at their appropriate fractions considering isotopic abundances (for uranium) and equilibrium conditions (NUREG-1507 offers an example for 3% enriched uranium).

So let us continue with our 30% Am-241 and 70% Co-60 example. The fractional amounts of these radionuclides entered into Microshield were

TABLE 9.6   Microshield Exposure Rate Results for 30% Am-241 and 70% Co-60 Mixture

Energy (keV)	Exposure Rate (μR/h)	cpm/μR/h	Weighted cpm/μR/h
60	1.075E−3	5830	8.79
1173	3.363E−1	177	83.85
1332	3.725E−1	152	79.76
Total	7.1E−1		172.4

4.8E–7 μCi/cm³ for Am-241 and 1.12E–6 μCi/cm³ for Co-60, based on a total of 1 pCi/g for both. The same geometry and parameter values discussed previously for Am-241 were used in this evaluation. Microshield results indicated a total exposure rate at the dose point of 0.71 μR/h, which came from the fractional exposure rates for the gamma energies involved (Table 9.6).

The results in Table 9.6 show that the total weighted cpm/μR/h value for this particular radionuclide mixture is 172.4. The individual cpm/μR/h values for each energy were taken from Table 9.4, with interpolation for the Co-60 energies, for the 1.5″ × 1.25″ NaI scintillation detector. Each energy-weighted cpm/μR/h value was determined by multiplying the cpm/μR/h value by the exposure rate for an energy and then dividing by the total exposure rate. The scan MDC for this mixture of Am-241 and Co-60 can now be calculated. First, the minimum detectable exposure is determined based on the MDCR (960 cpm for previously specified parameters) and the energy-weighted cpm per μR/h value:

$$\text{Minimum detectable exposure rate} = \frac{960 \text{ cpm}}{172.4 \text{ cpm/μR/h}} = 5.57 \text{ μR/h}.$$

The scan MDC immediately follows from this result, based on an input of 1 pCi/g for the Am-241/Co-60 mixture:

$$\text{Scan MDC} = (1 \text{ pCi/g}) \frac{5.57 \text{ μR/h}}{0.71 \text{ μR/h}} = 7.8 \text{ pCi/g}.$$

Therefore, the scan MDC for a 30% Am-241 and 70% Co-60 mixture using a 1.5″ × 1.25″ NaI scintillation detector is 7.8 pCi/g. Note that this value is slightly higher than the scan MDC for Co-60 alone, and good deal less than the scan MDC for Am-241 alone.

The major concern with this approach is that it requires a knowledge of a particular ratio between the two radionuclides. We might be reminded of the same dilemma when trying to use the surrogate approach (Chapter 6) in that justifying a consistent ratio can be difficult, or perhaps impossible. However, in this application, we may not have to be as strict. Excusing the lack of a formal proof, the scan MDC for a mixture of two radionuclides cannot be any better (lower) than the lower individual scan MDC, nor any

TABLE 9.7   Microshield Exposure Rate Results for 80% Am-241 and 20% Co-60 Mixture

Energy (keV)	Exposure Rate (μR/h)	cpm/μR/h	Weighted cpm/μR/h
60	2.867E – 3	5830	81.4
1173	9.609E – 2	177	82.96
1332	1.064E – 1	152	78.89
Total	2.055E – 1		243.2

worse (higher) than the higher individual scan MDC. In our example for Am-241 and Co-60, the scan MDC for any possible combination of these radionuclides cannot be less than 5.8 pCi/g nor greater than 45 pCi/g. So, a conservative approach would be to assume the scan MDC for the mixture is the higher of the two, or 45 pCi/g in this instance. A somewhat less conservative approach, related to the idea of a consistent ratio, is to establish a range of ratios. Perhaps the surveyor has performed many analyses of the ratio between Am-241 and Co-60, and the highest fractional amount of Am-241 ever observed was 80%. Now the average fractional amount may have been 30% Am-241 and 70% Co-60, but this ratio may be highly variable. The benefit to using the 80% Am-241 fractional abundance is that it is not as conservative as assuming 100% Am-241—and further, it may not be as difficult to justify as the highly variable average ratio. So let us run the above example with an 80% Am-241 and 20% Co-60 mixture to determine the scan MDC.

The fractional amounts of these radionuclides for this new mixture were 1.28E–6 μCi/cm^3 for Am-241 and 3.2E–7 μCi/cm^3 for Co-60, based on a total of 1 pCi/g for both. Again, the same geometry and parameter values discussed previously for Am-241 were used in this evaluation. Microshield results indicated a total exposure rate at the dose point of 0.2055 μR/h, which came from the fractional exposure rates for the gamma energies involved (Table 9.7).

The minimum detectable exposure is determined based on the MDCR (960 cpm for previously specified parameters) and the energy-weighted cpm per μR/h value:

$$\text{Minimum detectable exposure rate} = \frac{960 \text{ cpm}}{243.2 \text{ cpm/μR/h}} = 3.95 \text{ μR/h.}$$

The scan MDC for the 80% Am-241 and 20% Co-60 mixture becomes

$$\text{Scan MDC} = (1\,\text{pCi/g})\frac{3.95\,\mu R/h}{0.2055\,\mu R/h} = 19.2\,\text{pCi/g.}$$

Therefore, the scan MDC for an 80% Am-241 and 20% Co-60 is more than 50% less than the scan MDC for Am-241 alone (45 pCi/g).

At this point, an interesting observation can be made. It appears that the scan MDC for multiple radionuclides can be expressed using the following relationship:

$$\text{Scan MDC}_{\text{mult. radionuclides}} = \frac{1}{\dfrac{f_1}{\text{scan MDC}_1} + \dfrac{f_2}{\text{scan MDC}_2} + \cdots + \dfrac{f_n}{\text{scan MDC}_n}}$$

(9.15)

where $f_1, f_2, \ldots, f_n$, are the fractional amounts of each radionuclide and scan MDC_1, scan MDC_2, and so on represents the individual scan MDC for that radionuclide.

It should be noted that Equation 9.15 is based on empirical observation and validation by example, as opposed to being rigorously proven. For example, if we substitute the individual scan MDCs for Am-241 and Co-60 for an 80%/20% mixture for the previous example, the scan MDC is

$$\text{Scan MDC}_{80\%\text{Am.}20\%\text{Co}} = \frac{1}{(0.8/45) + (0.2/5.8)} = 19.1\,\text{pCi/g.}$$

Obviously, this result compares well with the scan MDC of 19.2 pCi/g calculated above.

How does the scan MDC for multiple radionuclides impact the use of the unity rule? Consider an example where the results from previous surveys indicated the presence of Ra-226, Th-230, and processed natural uranium. Assume that modeling was performed and that the applicable DCGLs for the contaminants of concern are 60 pCi/g for Th-230, 3 pCi/g for Ra-226, and 28 pCi/g for processed natural uranium. Further, assume that it has been previously established that U-238 will be an appropriate surrogate for the measurement of processed natural uranium, and that

TABLE 9.8   Analytical Results Used to Determine
Relative Fractions of Radionuclide Mixture

	Sample Result	Relative Fraction
Th-230	190 pCi/g	0.936
Ra-226	7.1 pCi/g	0.035
U-238	5.9 pCi/g	0.029

the $DCGL_W$ for U-238 as a surrogate is 13.5 pCi/g. Owing to multiple contaminants at the site, the unity rule will be used and therefore, the $DCGL_W$ should be normalized to 1 (unity), as described in Chapter 6.

The actual MDCs of scanning techniques were determined for performing gamma scanning with NaI scintillation detectors. The scan MDCs for the 1.5″ × 1.25″ NaI scintillation detector for these contaminants are 3000 pCi/g, 4.5 pCi/g, and 115 pCi/g, respectively, for Th-230, Ra-226, and processed natural uranium. The processed uranium scan MDC was derived previously assuming that it was comprised with a U-238 fraction of 0.492, so the scan MDC for U-238 (as processed natural uranium) is 56.2 pCi/g. To calculate the scan MDC for a specific mixture of these three contaminants, it was necessary to collect a representative sample. The analytical results from this sample, and their relative fractions, are shown in Table 9.8.

The scan MDC for this radionuclide mixture was calculated using Microshield. The three primary radionuclides and their progeny were input into the code at their proper fractions. That is, the radionuclides associated with processed uranium were entered based on their relationship to U-238, and the progeny of Ra-226 were assumed to be in a secular equilibrium. A total of 16 radionuclides were entered to represent this mixture. The minimum detectable exposure is determined based on the MDCR (960 cpm for previously specified parameters) and the energy-weighted cpm per μR/h value:

$$\text{Minimum detectable exposure rate} = \frac{960 \text{ cpm}}{361.7 \text{ cpm/μR/h)}} = 2.65 \text{ μR/h.}$$

The scan MDC for this specific mixture of radionuclides is calculated as

$$\text{Scan MDC} = (1 \text{ pCi g}) \frac{2.65 \text{ μR/h}}{0.02521 \text{ μR/h}} = 105 \text{ pCi/g.}$$

As an aside, Equation 9.15 can also be used to calculate the scan MDC for this mixture

$$\text{Scan MDC}_{\text{U,Th,Ra}} = \frac{1}{(0.936/3000) + (0.035/4.5) + (0.029/56.2)} = 116 \text{ pCi/g}.$$

So, in this instance, the simplified Equation 9.15 produces a result that is about 10% different from the full-blown approach. However, the unquestioned value of the simplified Equation 9.15 is that the scan MDC for a number of combinations of these three radionuclides can be calculated quickly. The scan MDC that is ultimately used for this survey unit design may be based on soil concentration data that exhibit a fair amount of variability among the three radionuclides. Being able to quickly determine a prudently conservative scan MDC for this mixture will greatly expedite survey design matters.

Using the larger scan MDC to be somewhat conservative (which is respectable), it is interesting to partition the 116 pCi/g into its radionuclide components based on the radionuclide fractions, and then to divide these results by the individual radionuclide's scan MDC:

Th-230 (0.936) × (116 pCi/g) = 108.7 pCi/g, and dividing by 3000 pCi/g = 3.6%

Ra-226 (0.035) × (116 pCi/g) = 4.1 pCi/g, and dividing by 4.5 pCi/g = 90.3%

U-238 (0.029) × (116 pCi/g) = 3.4 pCi/g, and dividing by 56.2 pCi/g = 6.1%.

Thus, Ra-226 is the primary radionuclide responsible for the scanning detection of this radionuclide mixture because it is present at more than 90% of its individual scan MDC. The "detectability contributions" from the other two radionuclides are necessary for this mixture to be detectable at 116 pCi/g.

We still need to answer the question of whether the scan MDC for this mixture is sufficiently sensitive, relative to the DCGL. To answer this question, it is necessary to have an area factors table for the specific radionuclide mixture. Chapter 5 discusses how area factors for radionuclide mixtures can be determined. Once appropriate area factors are available, the next step is to determine the required scan MDC. This can be expressed as the DCGL$_W$ times the area factor. The trick is to express the DCGL$_W$ for this mixture in the same way that the scan MDC is expressed. That is, the scan MDC for this mixture (116 pCi/g) represents the gross activity, at the specified radionuclide fractions, that is detectable via scanning. Similarly, the gross activity DCGL is the combination of these radionuclides, at their specified fractions, that equates to the dose criterion. In this case, the gross activity DCGL is

$$\text{Gross activity DCGL} = \frac{1}{(0.936/60) + (0.035/3) + (0.029/13.5)} = 34 \text{ pCi/g.}$$

Therefore, the required scan MDC is equal to the DCGL$_W$ (34 pCi/g) times the area factor, which must be determined from modeling this specific radionuclide mixture. The idea is to ensure that the actual scan MDC is less than the required scan MDC. For this to happen, the area factor needs to be at least 3.4 (116 pCi/g divided by 34 pCi/g). So, for multiple radionuclides in soil, the DCGL$_W$ needs to be expressed in terms of the gross activity when assessing the sufficiency of the scan MDC, and also in terms of unity when applying the statistical tests.

9.3.6 Empirically Determined Scan MDCs

The accuracy of the scan MDCs determined using the approaches described in this chapter can be empirically evaluated in a number of ways. For example, by identifying concentrations in the soil flagged by scanning and subsequently sampled, the soil concentrations at the lower end of the range of results should indicate the scan MDC achieved. That is, during the performance of scanning in an outdoor class 1 survey unit, suppose that five elevated areas are detected and subsequently sampled. The results from these samples range from 3 to 26 pCi/g for a particular radionuclide. An empirical evaluation would indicate that the lower values in the range represent a ballpark estimate of the scan MDC. Of course, the more soil samples that are actually flagged during the scan and subsequently sampled will improve the accuracy of the empirical assessment of scan MDC.

Here is an example from a real D&D site that was contaminated with Co-60. A 1.5″ × 1.25″ NaI scintillation detector was used to perform scans in an impacted land area. The background level for the NaI detector varied from 2000 to 3000 cpm during the course of the scan, largely due to geological variations. The survey technician flagged several locations of elevated radiation and collected soil samples. The results were as follows:

NaI Reading at Sampled Location	Co-60 Concentration (pCi/g)
2000 cpm	0.1 (false positive)
25,000 cpm	25.5
7000 cpm	9.2
18,000 cpm	20.8
2800 cpm	2.1

You might note that the surveyor flagged two areas that were generally consistent with the background levels. This happens occasionally because the detector response briefly indicates that elevated radiation seems to be present, which is sufficient to cause the surveyor to decide that a hot spot has been identified. In this small data set, the two near-background radiation levels that were flagged resulted in one false positive and one "hit" at 2.1 pCi/g. For comparison, the calculated value of the Co-60 scan MDC for this detector (Table 9.5) is 5.8 pCi/g. Among other things, the calculated scan MDC represents the activity concentration that is detected 95% of the time. When you think about it, that is a pretty stringent requirement—given that 3.5 pCi/g of Co-60 may be detected 80% of the time. The point is that the 2.1 pCi/g of Co-60 that was detected should not be considered as the scan MDC, even though it certainly was detected. It may very be that it represents the level that is detectable 50% of the time—with this type of empirical evaluation, is it not possible to determine how many times this 2.1 pCi/g of Co-60 was missed by the scan?

This empirical evaluation may be termed the *a posteriori* assessment of scan MDC. It is performed by keeping track of soil samples and surface activity measurements collected as a result of scans. The accuracy of this approach must be considered in the context of the actual versus modeled conditions. In the above example, not only is the perceived versus actual level of performance (d') responsible for differences between the empirical and calculational scan MDCs, but actual field conditions may also be different from the model. The actual size of the hot spot in the survey unit may be larger or smaller than the modeled hot spot area of 0.25 m², or the hot spot may be deeper or shallower than the 15 cm depth modeled. Furthermore, it is important when interpreting empirical results to consider the presence of multiple contaminants in the soil, which may be the cause for overstating the apparent scan detection of a single radionuclide or radioactive material. For example, if you recall from the previous section, the scan MDC for the radionuclide mixture of Th-230, Ra-226, and

U-238 was determined to be 108.7 pCi/g of Th-230, 4.1 pCi/g of Ra-226, and 3.4 pCi/g of U-238. It would be greatly misleading to conclude that Th-230 or U-238 were even remotely detectable at their respective concentrations, considering that Ra-226 was providing 90% of the calculated detection capability with its 4.1 pCi/g concentration.

Similarly, it is possible to design experiments to assess the scan MDC for survey instruments and technicians. This may be termed the *a priori* experimentation of scan MDC. A number of researchers, as well as D&D professionals, have performed mock-ups of contaminated surfaces to assess scanning detectability. For instance, in a study by Goles et al. (1991) empirical results include 305 net cpm detected in 50 cpm background level, 310 cpm in 250 cpm background, and 450 cpm in 500 cpm background for detection frequencies of 67% (as compared to the often quoted 95%). Olsher et al. (1986) reported that 392–913 alpha dpm was detectable 50% of the time with ZnS scintillation detectors. Thelin obtained experimental data from 25 health physics technicians using a portable scintillation detector (NE Technology probe) for surface activity measurements (Thelin 1994). Eight sources were randomly placed against the inner surface of a box with approximate dimensions $46 \times 36 \times 30$ cm. The source levels ranged from 236 to 1516 net cpm. The technicians were requested to scan the outside of the box and to identify locations that exceeded background radiation levels. The number of sources identified by each technician was evaluated and a hyperbolic function was fit to the experimental data. Thelin reports that at a background count rate of 482 ± 52 cpm at 2σ, the technicians were able to locate and identify source levels of 700 cpm approximately 90% of the time. These examples of *a priori* empirical assessments of scan MDC are valuable as validation to the calculational models, as well as determining the scan capabilities for specific survey technicians.

Advances in survey instrumentation and data management systems have resulted in significant improvements that have direct application to the way scan MDCs are determined. Field gamma spectroscopy systems (i.e., *in situ* units) may be used in the scanning mode to identify and generally quantify levels of gamma activity in soil. It is likely that empirical methods will be necessary to determine the scan MDC for using *in situ* gamma spectrometry units in this manner.

Increasingly, survey instrumentation that links the radiation measurement with its location is being used to scan large surface areas at D&D sites. Examples include the position-sensitive gas proportional floor monitor by Shonka Research Associates, Inc., LARADs, and the USRADS system (refer to Chapter 8 for a discussion of these instruments). In each case, the output of the scan survey may consist of an isopleth of radiation levels. The results may be interpreted by correlating the radiation levels with soil concentrations following a sampling campaign. The "minimum detectable contour," in terms of cpm, would have to be established (estimated) based on a fitting program. It is then possible to determine the minimum detectable radiation level, and its corresponding soil concentration (i.e., its scan MDC). It might also be possible to assess the scan MDC by making use of the surface contour plots generated from GPS-logged data. These methods allow one to have an objective way to demonstrate what levels were detectable with a particular survey methodology.

It has also become common to automate the scan by making use of survey meter features such as the "peak-hold mode." During a user-specified scanning interval, the survey meter can be set to record and "trap" the highest meter reading achieved using conventional detectors. Again, empirical methods will probably be necessary to determine the scan MDC for using survey meters that operate in the peak-hold mode.

QUESTIONS AND PROBLEMS

1. Explain the difference between the critical level and the detection limit. Why is it not advisable to set the detection limit at the critical level?

2. Calculate the static MDC in dpm/100 cm^2 for a gas proportional detector with the following parameters (assumed paired measurements in background and survey unit):

   > Background: 280 counts in 1 min
   > Instrument efficiency (ε_i): 0.34
   > Surface efficiency (ε_s): 0.5
   > Physical probe area: 125 cm^2

3. A survey technician at Rocky Flats is using a gas proportional detector (126 cm^2) operated in the alpha-only mode to make measurements of surface activity. The contaminants of concern are Am-241 and plutonium isotopes. The technician states that with a 2-min count time and background level of one count per minute, the MDC is about 100 dpm/100 cm^2. What is the total efficiency ($\varepsilon_i \times \varepsilon_s$) of this detector given the MDC value quoted by the technician? Is this efficiency realistic?

4. A health physicist is designing a final status survey for a research laboratory that is scheduled to be decommissioned. The principal radionuclides used for laboratory experiments were C-14 and S-35. The State regulator provided the release criteria (DCGLs) for each radionuclide as follows:

$$
\begin{array}{ll}
\text{C-14} & 50{,}000 \text{ dpm/100 cm}^2 \\
\text{S-35} & 78{,}000 \text{ dpm/100 cm}^2
\end{array}
$$

The health physicist plans to use a GM detector (physical probe area is 20 cm^2) for surface activity measurements and wants to determine if a 1-min count time will provide an MDC less than 50% of the DCGL$_W$. Given that the background count for a 1-min count time is 72 counts, and that the instrument efficiencies (ε_i) for C-14 and S-35 are 0.16 and 0.17, respectively, determine if the 1-min count will provide a sufficient MDC for the HP. (Hint: Be sure to consider the decay characteristics of each radionuclide, in particular, the radionuclide half-lives.)

5. For Problem 4, assuming that a 1-min count time will provide a sufficient MDC (i.e., less than 50% of DCGL$_W$), determine the shortest count time possible that can still achieve a sufficient MDC? (Assume that paired measurements will be performed.)

6. A MARSSIM user plans to use a xenon detector for surface activity measurements of I-129 on sealed concrete surfaces. RESRAD-BUILD modeling resulted in a DCGL$_W$ for I-129 of 1000 dpm/100 cm^2. The following xenon detector parameters are provided:

Background: 420 counts in 1 min

Total efficiency (ε_{tot}): 0.12

Physical probe area: 100 cm^2

The MARSSIM user believes that 1-min counts will not provide adequate sensitivity (i.e., 50% of DCGL$_W$); therefore, 5-min background measurements be performed, while 1-min surface activity measurements will be performed in the survey unit. Will the MARSSIM user have a sufficient MDC using this measurement scheme? What is the MDC for 1-min paired measurements?

How would your answers above change if the following revised instrument data were provided?

Xenon gas proportional detector
Indicated use: low-level beta–gamma survey
Window: 6.0 mg/cm² aluminized Mylar
Window area:
  Active—169 cm²
  Open—140 cm²
Efficiency (4π geometry): typically 3%—I-129
Sensitivity: typically 36,000 cpm/mR/h (Co-60 gamma)

7. What impacts are there in counting the background longer than survey unit measurements when using the MARSSIM surface activity assessment? (Hint: consider both the WRS test and the Sign test.)

8. Calculate the scan MDC for a 1.5″ × 1.25″ NaI scintillation detector used for land area scans for I-129 (i.e., confirm the value provided in Table 9.5). Assume that the background level is 4000 cpm, observation interval is 1 s, and d' is 1.38. The Microshield code provided the following data for I-129 for an input of 1 pCi/g:

Energy (keV)	Exposure Rate (μR/h)
29	7.31E–4
30	1.37E–3
34	5.57E–4
40	3.86E–4
Total exposure rate = 3.044E–3	

These data are consistent with the low-energy x-ray and gamma emissions from the decay of I-129. (Hint: It is necessary to determine the energy-weighted cpm/μR/h value from Table 9.4. To facilitate this determination, the data in Table 9.4 were fit to a combination of polynomial and power functions as a function of energy. The cpm/μR/h values based on the fit were 2173, 2330, 2984, and 3971, respectively, for 29, 30, 34, and 40 keV.)

9. A NaI scintillation detector is being used to scan for Am-241. The current scan MDC of 50 pCi/g is based on an acceptable false-positive rate of 20% and a scan speed of 1 m/s. What is the estimated scan MDC if the acceptable false-positive rate is increased to 40% and the scan speed is reduced to 0.5 m/s?

10. Describe the difference between the instrument efficiency used for static measurements and the instrument efficiency used for scan MDC calculations for building surfaces.

11. Calculate the scan MDC for a floor monitor used to scan the turbine building floor surfaces for Co-60. Assume that the scan MDC is to be determined for a background level of 1350 cpm and a 2-s interval. The desired level of performance at the first scanning stage will be 95% true-positive rate and 25% false-positive rate. (Hint: Use Table 9.2 to get the instrument efficiency for scanning.)

12. Calculate the scan MDC for a 1.5″ × 1.25″ NaI scintillation detector to a 5% Am-241 and 95% Co-60 mixture in soil.

CHAPTER 10

Survey Procedures and Measurement Data Interpretation

Many of the survey procedures performed in support of decommissioninag are the same for scoping, characterization, and FSSs. This chapter provides a description of the common survey procedures and techniques that make up these decommissioning surveys. ANSI N13.49, "Performance and documentation of radiological surveys" (2001) is a germane reference. The focus will be on those few survey procedures that account for most of the decommissioning survey effort—surface-activity measurements, surface scans, and gamma spectrometry analyses. Of course, soil sampling is an important survey procedure, but it is well-described in many documents, including the MARSSIM. How many ways can one write: "collect sufficient soil quantity for the analysis, do not cross-contaminate the sample, collect soil to the depth that is consistent with dose modeling, ensure that sample locations for statistics are randomly selected"? Instead, the focus will be on the analysis of soil samples using gamma spectrometry, and some of the problems that can negatively impact the results.

Another common survey technique not discussed in this chapter is smear sampling. Smears have certainly diminished in importance with the issuance of the NRC license termination rule—due to the fact that surface-activity DCGLs are based on total activity, and no longer divided into total and removable surface-activity guidelines as has been the custom (Regulatory Guide 1.86 and DOE Order 5400.5). Nevertheless, smears will continue to be a staple at most D&D sites, especially for hard-to-detect nuclides like tritium and where materials are being cleared. The reader is encouraged to refer to Frame and Abelquist (1999) for a mini-dissertation on smears.

One other survey technique discussed earlier is that of *in situ* gamma spectrometry. A number of its capabilities and applications to decommissioning are outlined in Chapter 8. Undoubtedly, *in situ* gamma spectrometry is becoming increasingly popular at many D&D sites. The attraction to this survey technology is in large measure due to its role in reducing analytical costs, by reducing the need for as many soil samples. Over the next several years this technology will continue to gain a more prominent role in FSSs to demonstrate compliance with soil DCGLs. (Note: Regulatory acceptance of this survey technology is vital to its future success as decommissioning sites.) NUREG-1506 (NRC 1995a) provides an appropriate discussion on *in situ* gamma spectrometry and its decommissioning applications.

10.1 SURFACE-ACTIVITY MEASUREMENTS

Simply stated, surface-activity measurements are performed to quantify surface-activity levels. These measurements represent the fundamental compliance measurement for buildings and structures. It is exceedingly easy to measure surface activity, but quite challenging to do so accurately. The emphasis over the last several years has been to make more accurate measurements of surface activity. This is evident with the release of technical guidance documents essentially devoted to the measurement of surface activity, such as ISO-7503, NUREG-1507, and ASTM (1998). The purpose of the NUREG-1507 study was to provide guidance to licensees for selection and proper use of portable survey instruments. An important concept throughout the development of NUREG-1507 was to provide an awareness of and better understanding of how decommissioning field conditions can affect the performance of survey instruments.

The standard assessment of surface contamination levels is performed by converting the instrument response (in cpm) via a static measurement to surface activity (in dpm/100 cm²) using the total efficiency and other correction factors. The use of total efficiency can have a negative impact on accuracy if the calibration source is not representative of field conditions. But insomuch as the calibration source exhibits characteristics similar to the surface contamination—that is, similar radiation energy, backscatter effects, source geometry, self-absorption—then there is little cause for concern. However, in practice this is hardly the case; more likely, total efficiencies (in counts per disintegration) are determined with a clean, high-backscatter stainless-steel source, and are used to measure contamination on a lower-backscatter surface, such as concrete or wood. Consequently, the efficiency determined under ideal conditions in the laboratory, and used for field applications, is often overestimated due to the use of high-backscatter sources with little or no surface absorption. The study undertaken in NUREG-1507 was designed to provide an estimate of the variation expected when making measurements on different surfaces and under different surface conditions. The primary parameter affected by surface conditions is the surface efficiency.

10.1.1 Surface Efficiency (ε_s)

The MARSSIM and other regulatory agency guidance documents support the use of International Standard ISO-7503 for the determination of surface activity. By separating the total efficiency into instrument and surface efficiency components, ISO-7503 provides the surveyor with a greater ability to consider the actual characteristics of the surface contamination, and consequently, to more accurately determine the surface-activity level. A review of the ISO-7503 approach, as compared to conventional practice for surface-activity assessment, is beneficial here. The expression for total alpha or beta surface activity per unit area as obtained from the ISO-7503 guidance clearly distinguishes between instrument efficiency and surface efficiency:

$$A_s = \frac{R_{S+B} - R_B}{(\varepsilon_i)(\varepsilon_s)W} \tag{10.1}$$

where

R_{S+B} is the gross count rate of the measurement in cpm,
R_B is the background count rate in cpm,
ε_i is the instrument or detector efficiency (unitless),
ε_s is the efficiency of the contaminated surface (unitless), and
W is the area of the detector window (cm²).

The conventional approach to surface-activity assessment uses the total efficiency, which is usually calculated by dividing the net counts obtained on a calibration source by the quoted activity. More formally, the total efficiency is determined by selecting an alpha or beta calibration source with energy emissions representative of the expected surface contamination. The net detector response, in cpm, is divided by the calibration source activity, in dpm, to yield a 4π efficiency that is reported in units of counts per disintegration. As stated previously, the conventional approach is technically sound as long as the calibration source and surface contamination exhibit similar characteristics, that is, radiation energy, backscatter effects, source geometry, and self-absorption—which in many instances is not the case.

The determination of ε_i using the ISO-7503 approach (discussed in Chapter 8) is very similar to the conventional practice, with the exception being that the detector response, in cpm, is divided by the 2π surface emission rate of the calibration source. (Note: The 2π surface emission rate and the source activity are usually stated on the calibration source certification sheet.) This value of ε_i is then multiplied by the appropriate ε_s to determine the 4π total efficiency for a particular surface and condition. Therefore, the only difference between the current practice and the ISO-7503 technique of surface-activity assessments is in the determination of the surface efficiency, ε_s.

The minimum level of effort in estimating the surface efficiency, ε_s, is to simply select suitable values based on the radiation and radiation energy. This is precisely what the ISO-7503 recommends in the absence of experimentally determined values. Specifically, the recommendation is to use a surface efficiency of 0.5 for maximum beta energies exceeding 0.4 MeV (e.g., Cs-137, SrY-90), and to use a surface efficiency of 0.25 for maximum beta energies between 0.15 and 0.4 MeV (e.g., C-14, Tc-99, and Co-60) and for alpha emitters. Admittedly, this is a pretty rough cut at establishing source efficiencies, but it does provide for a more realistic and conservative assessment of surface activity, and importantly, does not require any increase in effort over what is currently done. To summarize at this point, using the ISO-7503 approach provides the flexibility needed to determine a more realistic surface activity without an increase in the level of effort over what has conventionally been performed.

ISO 7503 recommends using a surface efficiency (ε_s) of 0.5 for radionuclides with maximum beta energies exceeding 0.4 MeV (e.g., Cs-137, Pa-234m), and using a surface efficiency of 0.25 for radionuclides with maximum beta energies between 0.15 and 0.4 MeV (e.g., C-14, Co-60) and for alpha emitters.

Appendix O of NUREG-1757, vol. 2 states that the NRC "generally considers the ε_s values described in ISO-7503 guidance for alpha and beta emitters to be acceptable estimates, absent site-specific information ...." (NRC 2006a). For situations where surface contamination measurements are planned on irregular, uneven (curved) surfaces such as scabbled concrete and embedded piping, the NRC recommends that licensees determine appropriate site-specific surface efficiency (ε_s) factors.

The purpose of the NUREG-1507 study was to build upon the ISO-7503 approach, and to provide an estimate of the variation expected when making measurements on different surfaces and under different surface conditions—in essence, to provide a more refined estimate of the surface efficiency. These studies were performed primarily by ORAU at ORISE facilities in Oak Ridge, Tennessee. A measurement hood, constructed of Plexiglas, provided a controlled environment in which to obtain measurements with minimal disturbances from ambient airflow. Experiments were performed within the measurement hood using a detector-source jig to ensure that the detector-to-source geometry was reproducible for all parameters studied. Various field conditions were simulated, under well-controlled and reproducible conditions. Surfaces evaluated included concrete, wood, stainless steel, and carbon steel. Surface conditions studied included both physical surface effects, such as scabbled concrete and wood, and the effects of surface coatings (e.g., dust, paint, water, and oil) on efficiency determinations.

The surface efficiency is primarily impacted by the backscatter and self-absorption characteristics of the material surface. A simple experiment was performed to evaluate the backscatter characteristics of SrY-90 deposited on surfaces commonly encountered during the course of performing decommissioning surveys (NRC 1998b). The total efficiencies were determined by dividing the net count rate by the deposited SrY-90 activity, which was the same for all surfaces. The backscatter factor, calculated by dividing the particular surface material total efficiency by the total efficiency for air, ranged from 1.20 to 1.43 for the gas proportional detector with 0.4 mg/cm² window thickness, and ranged from 1.11 to 1.37 for the detector with 3.8 mg/cm² window thickness. Specifically, the backscatter factor for the 0.4 mg/cm² window thickness detector for stainless steel was 1.43—as compared with 1.20 for wood, 1.24 for drywall, 1.25 for a tile floor, and 1.30 for sealed concrete floor.

Consequently, total efficiencies for surfaces other than stainless steel are overestimated by 10–20% due to the backscatter effect alone. Because calibration sources are typically composed of high Z materials like stainless steel or nickel, this may result in the surface activity for surfaces like wood, drywall, and concrete being underestimated 10–20% if the conventional approach to surface-activity assessments is used. However, if the ISO-7503 approach is used, the backscatter factor is well accounted for by the ISO-7503 rules-of-thumb for surface efficiencies based on the nature and energy of radiation emissions. The bottom line is while the backscatter factor can have a measurable impact on the accuracy of surface-activity measurements, an even greater impact is from the surface material coverings (e.g., rust, dust, paint).

The impact of self-absorption was also extensively studied in NUREG-1507. Experiments were conducted by varying the material thicknesses of dust, oil, paint, and water that were applied to the surfaces of calibration sources to measure attenuation. The effect of self-absorption was calculated as a function of increasing material thicknesses, and the data were fit to exponential distributions. More than 30 tables and figures are provided in NUREG-1507 that describes the self-absorption phenomena for a number of surfaces and radionuclides. These experimental data provide more accurate estimates of the surface efficiency for specified surface types and conditions than the limited guidance in the ISO standard. However, it is expected that the simple rules-of-thumb for surface efficiencies in ISO-7503 will certainly suffice in most instances.

In general, a comparison between the ISO-7503 surface efficiencies and similar data in NUREG-1507 indicates that the ISO rules-of-thumb for surface efficiencies are conservative. In fact, this is most apparent for beta-emitting radionuclides with end point energies approximately between 0.25 and 0.4 MeV. These radionuclides are assigned surface efficiencies of 0.25, when NUREG-1507 research indicates that typical D&D conditions would warrant surface efficiencies much closer to 0.5. Therefore, the survey designer should carefully balance the simplicity of adopting the ISO rules-of-thumb for surface efficiencies, with the degree of conservatism provided.

Finally, we might ask if additional surface efficiency studies at specific D&D sites are warranted. Are there advantages to those D&D contractors with the resources and desire to duplicate, or expand upon, the experiments performed in NUREG-1507? It is certainly worth considering (read "apply the DQO Process"), but it is anticipated that in many instances these licensees, upon assessing the particular surface conditions, can select the appropriate surface efficiency from NUREG-1507. That said, one example of necessary experimentation occurs when the contamination is deeper than the saturation layer—that is, when the surface contamination is really volumetric contamination.

The presence of radionuclide concentrations at depths beneath building surfaces (e.g., wood, concrete, dry wall) can result from both activation and contamination that is transported into the surface. Although there does not seem to be consensus on the definition of "surface" in this context, ISO-7503 does define surface contamination as that residing within the saturation layer thickness. That is, the radiation emitted from the deepest contamination in the surface are attenuated to a degree that they just miss penetrating the surface—thus most of the contamination is shallower than the saturation layer thickness and is available for detection. The thickness of the saturation layer is a function of the radiation type and energy, and surface material. It is of course very thin for alpha emitters, and conversely much thicker for SrY-90 betas. So for radionuclides that are no deeper into the surface than the saturation layer thickness, the calibration used to convert the net count rate to dpm/100 cm^2 is adequately covered by the surface efficiency term (ε_s). However, what can we do about the radionuclides that exist at depths greater than their saturation layer thickness—where surface-activity measurements do not provide a complete accounting of the residual radioactivity present?

Perhaps the first step is to establish whether or not contamination or activation has resulted in the presence of residual radioactivity at depth. This can be performed by reviewing the HSA data that might indicate spill locations where contamination may have penetrated into the surface, or surface materials that might have become activated from exposure to neutron fluence. The HSA information should be complemented with surface characterization where it is likely that residual activity exists deeper than the surface layer. For example, the depth profiling of concrete surfaces, via collection of concrete cores, might indicate that the residual contamination resided only on the surface in the spill area. However, if any residual contamination does penetrate into the material surface, technically defensible assessment of surface activity levels in the field should be performed. The point is that the standard surface efficiency factors may no longer apply.

Let us suppose that concrete is contaminated at depth with Co-60. The technical assessment should account for the self-absorption and backscatter characteristics of the concrete surface being measured. The surface efficiency based on the ISO-7503 rule-of-thumb is equivalent to 0.25, but this value decreases as the contamination is found to be deeper into the surface material. Therefore, ε_s should be determined empirically—prior to the implementation of the FSS—by making surface-activity measurements on the concrete surface, and then carefully removing the concrete for analysis in the laboratory to determine the total activity present. The total efficiency is determined by dividing the measured count rate by the laboratory analysis results for the total activity. Then, the surface efficiency is obtained by dividing the total efficiency by the instrument efficiency for the field survey instrument. A straightforward example may be useful here.

Suppose that a concrete core was analyzed in the laboratory for Co-60 content. Prior to the sample being crushed and homogenized for gamma spectrometry analysis, a GM detector was used to measure the net count rate on the surface—the result was 235 cpm. The gamma spectrometry analysis indicated a total Co-60 activity in the core of 18,600 dpm. The instrument efficiency for the GM for Co-60 is 0.23 (based on interpolation from Table 8.2). The first calculation is to determine the total efficiency of the GM detector based on the measured net count rate divided by the total activity—this results in 235 cpm/18,600 dpm, or 0.013. The surface efficiency based on this concrete core assessment is calculated by dividing the total efficiency (0.013) by the instrument efficiency (0.23), which yields a surface efficiency of 0.055. This is only 20% of the standard ISO-7503 surface efficiency for Co-60. Therefore, when surface-activity measurements are performed during the FSS in areas characterized by Co-60 at this depth, the GM measurement data are corrected by an instrument efficiency of 0.23 and a surface efficiency of 0.055.

Another challenge in determining the surface efficiency is when decay series radionuclides are present. The following case study continues the example started in Chapter 8 for determining the instrument efficiency for 4% enriched uranium.

CASE STUDY   DETERMINING SURFACE EFFICIENCY AND TOTAL
EFFICIENCY FOR 4% ENRICHED URANIUM

Table 8.3 showed how the instrument efficiency calculations for 4% enriched uranium, as measured by a 126-cm² gas proportional detector in alpha plus beta mode with a standard 0.8 mg/cm² Mylar window thickness. The weighted instrument efficiency for this detector for low enriched uranium was 0.55. This weighted instrument efficiency must be multiplied by an appropriately weighted surface efficiency to determine the total efficiency, which will be shown in this case study.

There are actually two approaches that can be used to determine the weighted surface efficiency. The first approach allows the direct determination of the total efficiency for 4% enriched uranium for the particular detector. This is usually the bottom line result that is desired. In this approach, each of the radionuclides are taken in turn, multiplying the radionuclide fraction times the instrument efficiency, and then by the surface efficiency, to determine the total efficiency for that radionuclide (Table 10.1). The surface efficiencies applied to each radionuclide are based on the ISO-7503 rules-of-thumb.

Now the total efficiency of 0.163 for this detector calibrated to 4% enriched uranium is the value of interest. This is the efficiency used to convert the raw data to the DCGL units of dpm/100 cm². From this total efficiency result, we can calculate the weighted surface efficiency by dividing the instrument efficiency (0.55) by the total efficiency (0.163). The weighted surface efficiency is 0.296.

An alternative approach to calculate the weighted surface efficiency is to weigh the individual surface efficiencies for each radionuclide by the instrument efficiency fraction represented by each radionuclide. The calculation of the weighted surface efficiency is best shown by an equation:

$$
\varepsilon_s = \left(\frac{(0.15)(0.40)}{0.55} \times (0.25) \right) + \left(\frac{(0.15)(0.19)}{0.55} \times (0.25) \right)
$$

$$
+ \left(\frac{(0.15)(0.69)}{0.55} \times (0.50) \right) + \left(\frac{(0.809)(0.40)}{0.55} \times (0.25) \right)
$$

$$
+ \left(\frac{(0.041)(0.40)}{0.55} \times (0.25) \right) + \left(\frac{(0.041)(0.36)}{0.55} \times (0.25) \right) = 0.296
$$

The value of being able to explicitly determine the weighted surface efficiency comes into play when the D&D site is contaminated with multiple contaminants. For example, now that the instrument and surface efficiencies have been determined for 4% enriched uranium, suppose that the

TABLE 10.1   Total Efficiency for Low Enriched Uranium (4%) Using a 126-cm² Gas Proportional Detector with a 0.8 mg/cm² Window in Alpha Plus Beta Mode

Radionuclide	Radiation/Average Energy (MeV)	Radionuclide Fraction	Instrument Efficiency	Surface Efficiency	Weighted ε_T
U-238	Alpha/4.2	0.150	0.40	0.25	0.015
Th-234	Beta/0.0435	0.150	0.19	0.25	0.007
Pa-234 m	Beta/0.819	0.150	0.69	0.50	0.052
U-234	Alpha/4.7	0.809	0.40	0.25	0.081
U-235	Alpha/4.4	0.041	0.40	0.25	0.004
Th-231	Beta/0.0764	0.041	0.36	0.25	0.004
			Total efficiency		**0.163**

radionuclide mixture at the site includes 75% enriched uranium (4%) and 25% Tc-99. The total efficiency in this situation would be calculated by weighting both the instrument and surface efficiencies of each contaminant by their relative ratios.

10.1.2 Building Material-Specific Backgrounds

A fundamental survey procedure that relates to the surface-activity assessment given by Equation 10.1 is the determination of the background count rate to subtract from the gross count rate. One of the first considerations is to assess the types of surfaces that are likely to be encountered in the course of making surface-activity measurements in the survey units. Then it is necessary to identify appropriate background reference materials (and locations) to make these background measurements. These background measurements should be performed in areas of similar construction, but without a history of radioactive material use.

A number of variables can affect the determination of the background level, including both detector and environmental factors. Some of these factors are described in Chapter 7. This section focuses on the determination of background levels that are performed on various surface types that might be found in the survey unit. It is not a tremendously difficult task to obtain appropriate background measurements, but at the same time, a fair amount of forethought is necessary.

The background measurements should be performed at a minimum of 10–20 locations on the selected surface type (e.g., wood, concrete, ceramic), depending on the results of the DQO process and how these background measurements will be used. That is, sufficient background measurements should be collected to adequately encompass the fluctuations that occur within the area—for example, variations in radon and gamma radiation levels. This may mean initially assessing background at the same location three or four times during a given day to account for radon fluctuations.

The background measurements should be randomly located—that is, measurements should be performed at a number of different locations on the material to account for material variability as well. The goal is to ensure that true background variability is encompassed in the determination of the background standard deviation, σ_r. Furthermore, the background measurements can be used for determining both the average background (for the sign test application), as well as for obtaining the background reference area distribution for the WRS test—as discussed in Section 13.3. The count time of the background measurement should be chosen so that the MDC for the instrument/detector combination is less than the DCGL for the contaminant(s) of concern (Chapter 9). (Note: A spreadsheet would greatly facilitate the manipulation of these background data.)

Consider a simple example where a GM detector is used to perform 18 background measurements on a reference dry wall location. The 1-minute count rate (in cpm) results were as follows:

56	57	65
63	53	67
57	64	57
55	64	62
68	60	66
65	53	58

The mean background for this data set is 60.6 cpm, and the standard deviation is 4.9 cpm. This background would be subtracted from gross count rates on dry wall materials during the data reduction phase, while the standard deviation ($\sigma_r = 4.9$ cpm) might be used in conjunction with the standard deviation in the survey unit to determine the total standard deviation for using the sign test for surface-activity assessment (Section 13.3.3).

After collecting the background data for the surfaces, it may be possible to combine surfaces which exhibit similar background count rates into the same category and use one material-specific background for surfaces that are placed in the same category (refer to Table 7.1, Background Count Rate for Various Materials). For example, wood surfaces may exhibit count rates that are similar to dry wall, or concrete blocks may have count rates that are comparable to poured concrete floors. Therefore, in certain instances, one background count could be used to represent two or more materials. Again, the DQO process should be used to determine appropriate criteria for grouping surface types that seem to have similar background characteristics.

A couple of other points should be made relative to background measurements. First, similar instrument/detector combinations will likely exhibit background measurements that are statistically alike. This can be seen in the following data which justifies the use of one instrument/detector combinations average background for all the like instrument/detector combinations for a particular surface. Second, it should be noted that even though different areas at a facility may have similar construction components, these materials may have different natural radioactive material concentrations (e.g., concrete blocks, bricks, floor tile, and bathroom tiles) which may lead to nonrepresentative background values being subtracted from the surface-activity measurement data. Therefore, one should take note as to whether the majority or all the calculated numbers are negative or positive—if so, then an inappropriate background may have been used and a new construction material-specific background should be determined.

One final point will be made concerning the potential difficulty in determining background values for surface-activity assessments. This complication can arise as a result of highly variable ambient exposure rates. Specifically, surface-activity assessments can be further involved due to the sometimes significant variation of ambient background levels, in addition to variable surface material backgrounds. In these circumstances shielded measurements can be used to distinguish the ambient background from the surface material background. This approach requires the subtraction of the two background components—ambient background and surface material background—from each gross surface-activity measurement in the survey unit.

This can be accomplished by using a plexiglass or acrylic shield to attenuate the most energetic beta radiation expected in the survey unit—for example, a thickness on the order of 300 mg/cm^2 was used successfully at the Fort St. Vrain decommissioning project. The surveyor first makes measurements of the total or gross radiation level at a particular survey unit location. This is the unshielded measurement and includes contributions from both the ambient gamma exposure rate and beta surface activity. The shielded measurement allows a determination of the gamma-only component (or ambient background) of the total detector's response. Therefore, the previously determined surface material background and ambient background are both subtracted from the gross count rate to yield a net count rate, and ultimately the surface activity. Owing to the complexity, this technique should only be used when the variability of the ambient background poses a real problem.

The following surface-activity measurement checklist summarizes the discussion of this section. It provides an outline of surface-activity assessment using the ISO-7503 approach, in consideration of the appropriate background determination.

Surface-Activity Measurements Checklist—for the Sign Test

Prior to making surface-activity measurements:

1. An instrument/detector combination is selected to make surface-activity measurements appropriate for the contaminant(s) of concern. For multiple radionuclides, the instrument selection must consider the radiation types involved.

2. The instrument/detector is calibrated to an appropriate (considering representative energy and geometry) calibration source—ISO-8769 recommends a distributed source of at least 150 cm².

3. The net instrument response in cpm is divided by the 2π surface emission rate of the calibration source to determine the value of ε_i. The instrument efficiency must consider the radionuclides present and their respective ratios (Chapter 8). This value is recorded on the field survey data sheets.

4. MDC is determined for the instrumentation (see Chapter 9 on surface-activity MDCs), and should be less than 50% of the appropriate DCGL$_W$, per MARSSIM. This requires an estimation of the surface efficiencies, ε_s, for various surface types and conditions expected to be encountered, or simply use the ISO-7503 guidance for surface efficiencies. An estimation of surface material backgrounds is also required.

Surface-activity measurements at the D&D site:

5. Measurements are collected for the predetermined time interval—count time must be long enough to provide satisfactory MDCs—and the integrated counts are recorded on the field survey data sheets.

6. When performing direct measurements on various material surfaces at the site (e.g., concrete blocks, bricks, tile, dry wall, metal), the type of material should be noted on the data sheet and the surveyor should take note as to whether a construction material-specific background for that material needs to be determined (or if it is similar to another material, i.e., material grouping).

7. While the measurement is being performed, the surveyor might also note the surface condition or a noteworthy environmental condition (elevated radon suspected) on the field survey data sheet. For

example, an entry might be "concrete surface, no overlying material"; or "steel surface, with some oxidation (rust)." Consider the need to use a site-specific ε_s based on surface conditions.

The data reduction following the FSS are

8. Surface efficiencies are assigned to each measurement based on experimentally determined data in NUREG-1507, or based on guidance in ISO-7503. It is anticipated that several surface efficiency values will suffice for the many different surface types and conditions encountered in the field. For example, a concrete floor, treated wood, and steel surface with oxidation may all be given an ε_s value of 0.50, while a stainless-steel surface is given a value of 0.65.

9. Surface activities are calculated from the recorded gross counts in the field, correcting for the background, instrument, and surface efficiencies. (The use of spreadsheets greatly enhances the calculation of surface activities.)

10.2 SCANNING BUILDING SURFACES AND LAND AREAS

As mentioned in Chapter 9, scanning is performed to identify elevated levels of direct radiation in the survey unit. This is performed by moving the detector over the building surface or land area and listening to the audible output of the detector. Thus, the surveyor must make a decision on the presence of direct radiation relative to the ambient background level. Once an elevated area is identified, the location is flagged, and a more precise surface-activity measurement is performed for indoor scans, or a soil sample is collected for outdoor scans.

Many of the same instruments used for direct measurements are used for scanning building surfaces, and include gas-proportional floor monitors and hand-held detectors, GM detectors, and ZnS scintillation detectors. Outdoor scans are usually performed using NaI scintillation detectors with various sized NaI crystals. For example, for low-energy gamma emitters, the FIDLER is usually the scanning instrument of choice. The discussion in Chapter 9 on the MDC for a scan survey provides a detailed discussion on how the sensitivity of these different detectors can be determined.

A recent study by King et al. (2012) showed that the scan MDC can be significantly impacted by the surveyor's swing mechanics, such as the detector-to-source distance and the detector trajectory during the scan.

The approach described in the paper was to hold most environmental and surveyor conditions constant, and calculate scan MDCs by varying the radionuclide, hot-spot size and the detector motion (either a pendulum arc or flat, fixed-height trajectory over the surface). The best (lowest) scan MDCs were obtained with a flat, fixed-height trajectory. For example, for a nominal detector height above the surface of 10 cm and a 0.1 m^2 Co-60 hot spot, the 2×2 NaI scan MDC was 592.4 Bq/kg for flat trajectory motion versus 624.5 Bq/kg for pendulum arc trajectory. Typical scan MDC differences for the two detector motion variables ranged from a few percent to more than 20%. So the take home message is that hot-spot detection via scanning is optimized by maintaining the detector-to-source distance as low as practical during the scan.

One of the historic shortcomings associated with surface scanning is how the scan results are documented. Oftentimes, the only documentation provided for scanning results was a statement in the report that scanning was performed and x number of hot spots were identified. The survey data forms for scanning might indicate the scan ranges on survey maps, along the magnitude and location of any identified hot spots. Unfortunately, the accuracy of these data entries are entirely dependent on the surveyor's ability to recall the scan ranges in all portions of the survey unit, and to transcribe this information on the survey map. Fortunately, advances in survey instruments can produce automatic documentation of scan results that are quite accurate (Section 8.3.3).

Scanning has also been used at a number of D&D sites as an indicator of when it is necessary to collect a soil sample. Specifically, if the gamma reading is above a predetermined action level (in cpm), then a soil sample is collected and analyzed. Of course, the success of this approach depends on the radionuclides having gamma emissions, as well as the contamination being on the surface (and not at depth), and most importantly, gaining regulatory approval. Important questions to ask include: What level of characterization was performed in areas used to determine the correlation factor to justify that there are no buried radioactive material? Were borehole samples collected to determine the depth profile of Ra-226 concentration in soil?

One example of where these conditions have been satisfied has been at sites contaminated with windblown tailings (Ra-226) (e.g., former uranium mills). The basic survey approach is to survey the area for gamma radiation using a NaI scintillation detector. The gamma radiation counts are integrated for established time periods or over specified grid block areas, and a correlation between NaI count rate and Ra-226 concentration is established.

Let us consider an example where the D&D contractor seeks to establish an empirical relationship between the NaI detector readings and the radionuclide concentrations of Ra-226 in soil. The D&D contractor has process history as well as a substantial data set which indicates that the Ra-226 contamination is present only on the soil surface. During the survey, the D&D contractor scans over the soil in 20 grid blocks with a NaI detector, recording the accumulated NaI counts in 1-min periods. A representative soil sample is collected from each grid block. Table 10.2 presents the results of this empirical study.

The data can be plotted with the NaI reading as a function of the Ra-226 soil concentration, as shown in Figure 10.1. We can then fit the data to a straight line, which results in the following line equation:

$$y = 460.68x + 3457.3$$

where y is the NaI reading in cpm, and x is the Ra-226 concentration in pCi/g.

TABLE 10.2   NaI Readings versus Ra-226 Concentrations Used to Establish Empirical Relationship

NaI Reading (cpm)	Ra-226 Concentration (pCi/g)
3200	0.8
5600	1.2
4200	2.2
5900	5.6
4400	0.8
8600	6.2
3500	1.1
6800	3.6
12,000	15
6100	4.2
4200	6.3
2800	0.6
4100	3.1
3200	1.2
6000	12
3500	2.2
3800	1.7
4800	2.7
5500	5.2
7800	4.3

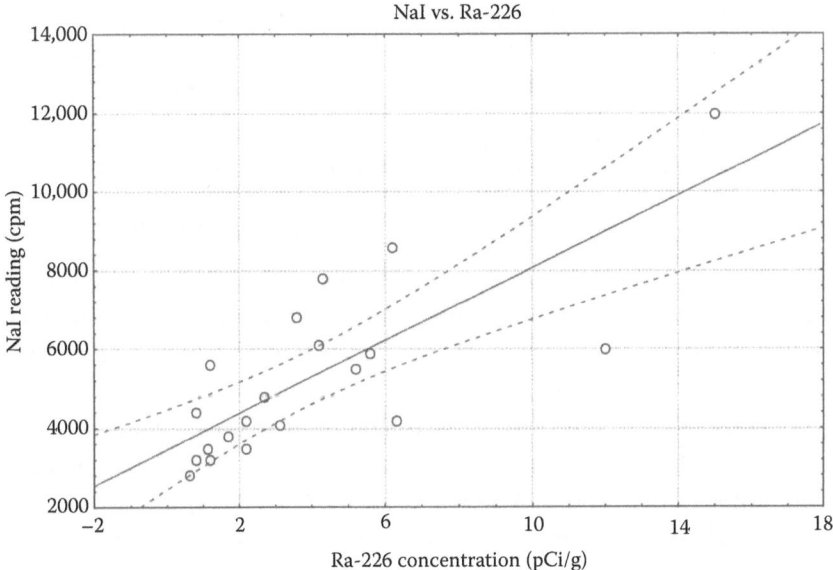

FIGURE 10.1   NaI detector reading (cpm) versus Ra-226 concentration.

Suppose that the D&D contractor wants the action level set at a value that flags a Ra-226 concentration of 6 pCi/g. Substituting 6 pCi/g into the above equation yields an action level of 6220 cpm based on the straight line fit. However, by drawing the best-fit straight line through these data—and determining an action level based on that fit—the action level will underestimate the Ra-226 present roughly 50% of the time. This is not going to satisfy anyone's DQOs, particularly the regulators! It is therefore recommended that a more conservative relationship be used in deriving the action level, such as drawing a line through the data that reduces the probability of underestimating the Ra-226 concentration. For example, one solution might be to draw the line that reduces the probability of underestimating the Ra-226 concentration to only 5%. This is shown as the lower dotted line in Figure 10.1. In this case, the action level based on 6 pCi/g is approximately 5500 cpm according to the graph. Therefore, any reading greater than 5500 cpm would be flagged as an indication that the Ra-226 concentration might exceed 6 pCi/g.

The empirical relationship between the NaI detector readings and the radionuclide concentrations of Ra-226 in soil correction factors should be reevaluated as more and new data becomes available. The effect of uncertainty in the correction factor should be addressed in setting the NaI cpm that triggers a specific action. The scan speed used to establish the NaI cpm-to-pCi/g correction factor should be maintained during the survey, and the effect that scanning faster has on the correction factor may need to be evaluated. Other factors that may impact the correction factor are the depth of the radionuclide and the background variability. If the soil concentration is below the surface to an appreciable degree there will be less detector response as compared to surface contamination.

10.3 GAMMA SPECTROMETRY ANALYSES FOR SOIL

As mentioned at the outset of this chapter, the gamma spectrometry analysis of soil samples is a fundamental decommissioning survey activity. Certainly there are other important analyses that are conducted on soil, such as alpha spectrometry and radionuclide-specific analyses that involve wet chemistry techniques, but a discussion of these are beyond the scope of this chapter. The reader is referred to the Multi-Agency Radiological Laboratory Analytical Protocols (MARLAP) (NRC 2004) for a discussion of these and other analytical procedures for soil. Herein, we will provide a general overview of gamma spectrometry, focusing on the interpretation of gamma spectrometry results.

One important point regarding gamma spectrometry results is the need to report actual values, rather than "<MDC" for radionuclides that are analyzed via gamma spectrometry. While this may pose a problem for some gamma spectrometry analysis packages, it is usually possible to modify the analysis routine to force it to produce a result for a radionuclide, even when it is truly less than the MDC. The critical importance of reporting actual values can be realized when performing the nonparametric statistical tests, which can only support a limited number of "<MDC" values.

10.3.1 Calibration Standards and Geometries for Soil Analyses

By far, the most common laboratory technique performed during decommissioning surveys is gamma spectrometry analyses of soil. High-purity germanium detectors are the state-of-the-art instrumentation for gamma spectrometry. They offer very good energy resolution, and are becoming increasingly more efficient. Basically, the gamma spectrometry analysis of soil is very straightforward. The laboratory establishes counting geometries that will be used for the soil analysis. These might include a 1-L Marinelli, a 0.5-L Marinelli, or smaller 100-g "hockey puck" geometry. Regardless of the geometry used, it must be calibrated using a NIST (National Institute of Standards and Technology) traceable or equivalent standard.

The radionuclide standard is added to the soil matrix and it is uniformly mixed. The fabricated soil standard is transferred to one of the geometries to be used for analyzing soil. A number of calibration standards are fabricated and counted to assess the overall uniformity of the spiked soil. The standard is counted for some period of time to determine the counting efficiencies versus gamma energy for a particular geometry. The standard gamma spectrometry analysis package consists of a peak search by energy and quantifies the radionuclide concentration based on the efficiency versus gamma energy curve.

The accuracy of gamma spectroscopy analysis depends on the soil samples having similar characteristics to the calibration standards used in the analysis. That is, the samples must be of similar composition (e.g., high atomic number [Z] material), mass, density, and moisture content as that of the calibration standard. If the samples are similar to the calibration standard, then the efficiency curve generated from the calibration standard may be used to quantify the radionuclide concentration in the soil samples. It is important to account for the mass of the standard and the amount of high atomic number [Z] materials that can produce significant attenuation of low-energy gammas. If the soil samples being analyzed exhibit a different mass or higher Z material content as compared to the radionuclide standard, then the radionuclide standard is not representative of the soil sample and erroneous results are quite possible. The bottom line is that the calibration standards must be representative of the soil samples being analyzed. Otherwise, the soil sample results should be reported as "semi-quantitative," or even "qualitative."

Let us consider further the case of analyzing a soil sample that contains a significant high Z component. Using a soil calibration standard that contains little or no high Z components to assay a high Z material sample will likely result in an underestimation of the radionuclide concentration in that sample. This is because low-energy gamma radiation is attenuated more in the high Z material sample than it is in the calibration standard. These sample attenuation concerns may be addressed by application of the "direct ratio method" of gamma radiation counting using an external source of gamma radiation. The direct ratio method works by comparing the gamma photopeak energy of interest in the sample to the gamma photopeak in a suitable calibration standard, with both photopeaks corrected for the relative amount of attenuation present in the sample and calibration standard (Abelquist et al. 1996).

10.3.2 Interpreting Gamma Spectrometry Data

Once the FSS has been completed, it is likely that a number of soil samples are awaiting analysis. A couple of weeks go by and the gamma spectrometry results are now at hand. Before we jump right into the data reduction described in Chapter 14, it is important to interpret the data. How many times has a laboratory simply reported everything that showed up in its nuclide library? For example, is it really likely for Be-7 or Ga-67 to be present—given their short half-lives?

Many times the interpretation of gamma spectrometry data is straight-forward. For example, the gamma spectrometry results indicate peaks at 662 and 835 keV, indications that Cs-137 and Mn-54 are present. Table 10.3 provides a listing of the total absorption peaks, that can be used to identify some selected radionuclides. However, complications can arise from radionuclide interferences—that is, from radionuclides present in a sample that have gamma emissions at the same energy (or at similar energies that cannot be resolved by the detector)—that make interpretation of gamma spectrometry results difficult. A classic interference at uranium sites is between Ra-226 (186 keV) and U-235 (185 keV). Substantial levels of Ra-226 at a site have often been incorrectly reported due to interference with U-235.

The following review of gamma spectrometry data is largely based on information gathered by the ORISE laboratory over a number of years of analyzing soil samples at decommissioned sites.

This information helps the D&D project engineer in evaluating the soil sample results, and should be used in conjunction with available HSA data. In many cases the gamma spectrometry data will provide conclusive results, in other cases further analysis by alpha spectrometry or nuclide-specific analysis may be required. Again, the following guidance in deciphering gamma spectrometry results is provided courtesy of ORAU.

Uranium is a contaminant at many D&D sites and therefore it should come as no surprise that it is routinely analyzed by gamma spectrometry. The uranium isotopes that are measured by gamma spectrometry include U-238 and U-235 (U-234 has low-energy photon emission with even a lower radiation yield, so gamma spectrometry is generally not an option at environmental levels). U-238, an alpha emitter, is determined from the peaks of its progeny Th-234. The primary peak is at 63.3 keV and the secondary peak is at 92.6 keV. U-235, also an alpha emitter, is determined from its own gamma emissions—the primary peak is at 185.7 keV and the secondary peak is at 143.8 keV. Unfortunately, there are also a number of natural radionuclides that can interfere with the uranium peaks.

One interfering radionuclide that is frequently present is Ac-228, progeny of the natural Th-232 series. Ac-228 emits a 93.4 keV peak that

TABLE 10.3   Peaks Used for Identification of Selected Radionuclides

Radionuclide	How It is Used in Identification of Radionuclide	Energy of Peak (keV)
U-238 from Th-234	Primary	63.3
	Secondary	92.6
From Pa-234 m	Verification	1001
U-235 in the presence of	Primary	185.7
1–2 pCi/g of Ra-226	Secondary	143.8
	Verification	163.4
U-235 in the presence of higher	Primary	143.8
concentrations of Ra-226	Secondary	163.4
	Verification	205.3
Ra-226 from Pb-214	Primary	351.9
	Secondary	295.2
From Bi-214	Verification	609.3[a]
Th-232 from Ac-228	Primary	911.1
	Secondary	969.1
Th-228 from Pb-212	Primary	238.6[b]
Bi-212, Tl-208	Verification	727.2, 583.1[c]
Cs-137	—[d]	661.6
Co-60	Primary	1173.2
	Secondary	1332.5
Cs-134	Primary	795.8
	Secondary	604.7
Mn-54	—	834.8
Ru-106	Primary	621.8
	Verification	511.9
Am-241	—	59.4
Th-230	—	67.7

Source: Courtesy of ORAU.
[a] Value may be 12–15% low due to summing losses.
[b] Bi-212 has a branching ratio of 64.07% for this gamma.
[c] Tl-208 is produced by the alpha decay of Bi-212 with a 35.93% branching ratio.
[d] — indicates only peak that the radionuclide emits, therefore it is the primary peak used to quantify the radionuclide.

interferes with the 92.6 keV peak of Th-234. Therefore, in the presence of thorium contamination, a better choice for determining the U-238 concentration is the 63.3 keV peak of Th-234. As mentioned above, Ra-226 can interfere with U-235 because it emits a 186.2 keV peak that interferes with the 185.7 keV peak of U-235. Therefore, in the presence of radium contamination, a better choice for determining the U-235 concentration is the 143.8 keV peak of U-235.

CASE STUDY USING THE DQO PROCESS FOR EVALUATING
POTENTIAL INTERFERENCES IN SAMPLES ANALYZED
FOR URANIUM (COURTESY OF ORAU)

The following guidance and investigatory actions are recommended to help resolve potential uranium interferences.

1. Review the data and check to see if interfering contaminants are present. Specifically look at the concentrations of Ra-226 and Ac-228. Natural levels of these radionuclides are generally between 1 and 2 pCi/g. If these values are exceeded, the reviewer should be alerted to the fact that the peaks of U-238 at 92.6 keV and U-235 at 185.7 keV may have additional counts due to interference, and that the results are suspect.

2. After checking for interferences, compare the 63.3 keV and the 92.6 keV activities and uncertainties for statistical agreement. If there is statistical agreement (within 10%) between the activities, either peak activity can be used to quantify U-238. The U-235 concentration can be inferred from the 185.7 peak provided that the levels of Ra-226 remain between 1 and 2 pCi/g.

3. If there is no statistical agreement between the 63.3 and 92.6 keV peaks of Th-234 for U-238, review the software's peak search printout or spreadsheet. For the 92 keV energy peak check to determine that the search routine identified only one peak between the energies of 90 and 94 keV. If there is more than one peak, the nuclide identification program will always calculate the activity from the first peak listed by the peak search routine. At this point, it is possible to reanalyze the spectrum using different Gaussian and peak search sensitivities. If the peak search appears to be correct, go to the next step.

4. If the peaks have been properly identified, next check the resolution via the full-width at half-maximum (FWHM). The range of the FWHM values is 0.9–1.8 keV. FWHM values outside this range are suspect and a request for reanalysis of the spectrum should be made. If the FWHM values are with the range, go to the next step.

5. Check the weight of the sample to ensure it is appropriate based on the weight of the calibration standard. If the sample weight exceeds the recommended limits, the difference between the 63.3 and 92.6 keV lines is probably due to the self-absorption of the 63.3 keV gamma by the sample matrix. The 92.6 keV line is the one of choice.

6. In the presence of higher activity uranium samples, it may be more appropriate to use the 1001 keV peak of Pa-234 m to quantify the sample concentration. This peak has a rather low radiation yield so it does not become viable until the U-238 activity concentration exceeds 10 pCi/g.

7. After reviewing the data, the reviewer can request that further analysis be performed on the spectrum or that a longer count be made on the sample. Longer counts will improve the counting statistics and in some instances this will help with some sample activities that are near guidelines.

QUESTIONS AND PROBLEMS

1. Calculate the weighted total efficiency for a detector calibrated to measure Fe-55 and Co-60 contamination that is present on the floor surfaces in the fuel storage building, where the anticipated mix is 64% Fe-55 and 36% Co-60. The instrument efficiencies for this detector are 0.03 for Fe-55 and 0.38 for Co-60. (Remember that the total efficiency is the product of the instrument efficiency and surface efficiency.)

2. Determine the total efficiency for the D&D site contaminated with 4% enriched uranium and Tc-99. Assume the same gas proportional detector used in the case study for 4% enriched uranium will be used. The radionuclide mixture consists of 60% Tc-99 and 40% enriched uranium. (Table 8.2 provides the instrument efficiency for Tc-99 for this detector.)

3. What are preferred radionuclides and their gamma lines (energies) for the gamma spectrometry analysis of U-238, Th-232, and Ra-226?

4. State the radionuclides and their energies that commonly interfere with the quantification of U-238 and U-235.

5. What is the effect of analyzing a soil sample that weighs more than the recommended limits for that particular soil geometry?

6. What is a possible solution for quantifying radionuclide concentrations in soil, particularly low-energy gamma emitters that are significantly attenuated by high Z material in the sample?

7. What are some conditions that can complicate the correlation of gamma radiation level (cpm) to soil concentration (pCi/g)?

CHAPTER 11

Radiological Hot-Spot Survey Considerations

11.1 INTRODUCTION: CURRENT HOT-SPOT APPROACH

A common assumption in dose modeling is that the contamination is more or less uniformly distributed over a parcel of land or building surface area. In reality however, the contamination distribution is rarely uniform for survey units that exhibit residual radioactivity. In fact, the remediation process itself tends to leave behind a spotty, nonuniform contamination distribution. So while the contamination source term may have been significantly reduced by the cleanup actions, the remaining contamination profile is often characterized as spotty, with isolated pockets of elevated contamination levels called hot spots.

The purpose of this chapter is to present an alternative approach for modeling the receptor dose due to hot spots—to specifically address how environmental pathways and parameters are impacted by hot-spot source terms. The research identifies pathways and parameters that are particularly "hot-spot sensitive"—those pathways and parameters in particular were studied to determine the best way for considering their contribution to receptor dose (Abelquist 2008).

On the basis of experiences in implementing and reviewing FSSs (including MARSSIM surveys) for a number of years, two perspectives on hot spots are clear: (1) the currently accepted hot-spot limits have a weak technical basis, and (2) hot spots are frequently missed during FSS.

This chapter takes a closer look at hot spots associated with decommissioning projects, particularly during FSS. Two topics were considered in a new approach for handling hot spots. First, the fundamental aspect of how acceptable DCGLs for hot spots ($DCGL_{EMC}$) are determined was examined. This work included a detailed look at how hot spots of various sizes actually produce receptor doses for specified environmental and building pathways. The radionuclides evaluated in this work were chosen for modeling due to their varying decay modes, and the fact that they represent a wide range of physical and chemical characteristics that affect environmental transport—for example, deposition, resuspension, volatilization, plant uptake, and solubility. The radionuclides include C-14, Co-60, Sr-90, Tc-99, I-129, Cs-137, Ra-226 (series in equilibrium), Th-232 (series in equilibrium), U-238 (processed uranium), Pu-239, and Am-241. The hot-spot sizes considered for environmental scenarios (outdoors) were 10, 3, 1, 0.5, 0.1, and 0.01 m². Hot-spot sizes considered for the building occupancy scenario (indoors) were 3, 1, 0.5, 0.1, and 0.01 m². The smallest hot-spot size (0.01 m²) may be effectively considered to represent the common averaging area for surface-activity measurements (10 cm × 10 cm, or 100 cm²).

The second aspect of this research was to consider a statistical assessment approach for hot spots that assesses the acceptability of multiple hot spots in the survey unit. This approach does not necessarily depend on the results of the area factor calculations (for $DCGL_{EMC}$) explained above. Rather, it is more of a big picture perspective that places hot spots in the overall context of the contaminant distribution in a survey unit. That is, while the MARSSIM describes a two-pronged approach for separately demonstrating compliance with both the mean contaminant concentration and elevated areas (hot spots), this chapter considers an integrated contaminant distribution concept where compliance is demonstrated for the contaminant distribution as a whole. That is, both the mean and upper percentiles of the contaminant distribution are considered in demonstrating compliance. The value of this approach is twofold. First, it addresses the issue of how to handle hot spots that may exist in the survey unit, but have not been found. Second, it inherently handles multiple hot spots because they are characterized and accounted for in the overall contaminant distribution that is being assessed for compliance with release criteria.

11.2 HOT-SPOT LIMITS IN SOIL

Recall that Chapter 4 provided a historic review of decommissioning guidelines used prior to MARSSIM's development, and included hot-spot guidelines from Regulatory Guide 1.86 that set the hot-spot limit at three times the average limit, and from the DOE 5400.5 where the soil hot-spot limit is calculated by multiplying the average guideline by a factor of $(100/A)^{0.5}$, where A is the area of the hot spot. Chapter 5 described how DCGLs are derived according to the MARSSIM guidance. In particular, Section 5.4 showed how DCGLs and area factors can be calculated using dose modeling codes. Recall that the area factor refers to the amount (factor) of radioactive concentration above the DCGL concentration that may be present in a smaller contamination area and still deliver the same dose (as the DCGL concentration averaged over the entire survey unit). Section 5.4.2 provides a number of illustrative examples of area factors for a few commonly encountered radionuclides that may be worth reviewing prior to working through this chapter.

The proposed approach for determining hot-spot limits is to calculate acceptable hot-spot release criteria based on dose, but not simply by reducing the size of the contaminated area to smaller and smaller hot-spot sizes. Rather, hot-spot release criteria are developed by considering the best estimate of the dose from first principles. An overarching issue that is addressed during implementation of the hot-spot release criteria is the requirement for receptor doses to be ALARA. A practical example of implementing ALARA for hot spots is that once found, the hot spot is remediated (regardless of dose).

A quick review of how MARSSIM handles hot spots may be helpful. Residual radioactivity that appears as small areas of elevated activity (i.e., hot spots) within a larger survey unit area, is usually identified via radiation scanning and then evaluated as individual measurements (i.e., as opposed to nonparametric statistical analysis). The DCGL derived for the individual hot spots is the $DCGL_{EMC}$; it is defined as the DCGL used for the elevated measurement comparison (EMC). The simple relationship between the DCGLs is that the $DCGL_{EMC}$ equals the $DCGL_W$ times the area factor. Thus, the area factor is the magnitude by which the concentration within the hot spot can exceed the $DCGL_W$ while maintaining compliance with the release criterion.

The EMC is intended to flag potential failures in the remediation process by evaluating hot spots that remain at the time of the FSS. Abelquist (1997) provides an example of MARSSIM application of the EMC at a D&D site in Oklahoma. Surface scans and judgmental sampling identified a large hot spot (~20 m²) near the center of a survey unit. The $DCGL_{EMC}$ for the EMC is obtained by multiplying the $DCGL_W$ by the area factor that corresponds to the actual area of the hot-spot concentration. This hot-spot area is deemed acceptable provided that the appropriate $DCGL_{EMC}$ is not exceeded. On the basis of dose modeling, the area factor for the 20 m² hot-spot size is 208, which resulted in a $DCGL_{EMC}$ of 33.3 pCi/g. However, the estimated hot-spot concentration exceeded the $DCGL_{EMC}$—40.6 pCi/g based on two soil samples collected from the 20 m² area. Consequently, this 20 m² hot spot failed the EMC. It should be recognized that any combination of area and radionuclide concentration that exceeds the appropriate $DCGL_{EMC}$ means that the particular hot spot needs to be remediated.

The MARSSIM recommends running RESRAD and RESRAD-BUILD codes at successively smaller contamination areas (e.g., from the RESRAD default of 10,000 m² to 1 m²), and taking the ratio of dose generated by the modeling code for the default area to that generated for the smaller areas studied (NRC 2000a). The dose rate for the smaller contamination area will always be at least as large as that for the default contaminant size. The area factor for a specific contaminant area is simply the dose rate for the smaller contaminant area by the initial dose rate for the default contaminant area. In addition to the contaminant area size, the only other parameter that is changed during the determination of area factors is the length of the contaminant area parallel to the aquifer.

This simplistic approach overlooks the problem that some pathways are not meant to be evaluated at area sizes substantially less than 100 or 1000 m². The assumptions that logically hold for larger land areas and building surface areas—such as the "unlimited reservoir" of contamination for the inhalation pathway—may not support the dose modeling technical basis as the area is reduced to the size of typical hot spots. For ingestion-based agricultural scenarios—plant, meat, milk, and drinking water ingestion—the fraction of food and water originating from these smaller contaminant zones should also be examined. A final point is that the external radiation pathway is modeled by an infinite plane source in RESRAD. The practical effect of reducing the size of the contaminated area from the entire survey unit to a much smaller area is that the receptor is assumed to spend all of their outdoor time directly on a small hot spot (clearly not a realistic assumption). NUREG-1757, vol. 2 (NRC 2006a) recognizes that potential limitations of the current method of determining DCGL values may exist. The NUREG suggests that it is worthwhile to consider alternate risk scenarios when determining acceptable residual radioactivity levels of discrete particles. Simply stated, the dose modeling scenarios used in RESRAD may not be strictly applicable for contaminated areas of 1 m² or smaller—scenarios, pathways, and modeling parameters for nominal hot-spot sizes are questionable, and should be addressed. The next three sections take a closer look at hot spots in soil—for external radiation, inhalation, and ingestion-based environmental pathways.

11.2.1 External Radiation Pathway

The first pathway evaluated is the direct exposure to external radiation from contaminated soil. The receptor dose from a widely distributed source term to the dose from a hot spot of particular size is compared—this ratio of receptor doses allows calculation of the hot-spot limit for that size hot spot. An example case for a hot-spot size equal to 10 m² of Co-60 in a 1000 m² survey unit was evaluated. The actual hot-spot dose determined from first principles was compared to the current practice of obtaining area factors described in MARSSIM, as well as to the result obtained using the MicroShield code (Abelquist 2008).

11.2.1.1 RESRAD Calculation of Area Factor

The RESRAD area factor approach, as described in the MARSSIM, is the conventional approach being used at many decommissioning sites in the United States today. This approach calculates a correction factor that accounts for the difference in the size of the contaminated area, and the resulting change in dose. The area factors are computed by taking the ratio of the dose or risk per unit concentration generated by RESRAD for the assumed contaminated area (survey unit size on the order of 1000–10,000 m²) to that generated for smaller hot-spot sizes.

RESRAD was run assuming that Co-60 contamination was present to a depth of 15 cm over the 1000 m² survey unit. No soil cover was modeled. Unit concentration (1 pCi/g) was input in the modeling code. The default occupancy factor is 0.6, which accounts for an outdoor time fraction of 0.25 plus an indoor time fraction of 0.5 that is weighted by a 70% indoor shielding factor. The resulting dose-to-source ratio (DSR) from the RESRAD run was 7.34 mrem/y per pCi/g. The dose was evaluated by RESRAD to be 7.34 mrem/y at time $t = 0$ years. The ground radiation pathway was responsible for 99.56% of the total dose, while the plant pathway was roughly the remaining about 0.4%. The $DCGL_W$ based on 25 mrem/y can be calculated as follows: 25 mrem/y/(7.34 mrem/y/1 pCi/g), which yields a value of 3.4 pCi/g.

RESRAD was run again to calculate the area factor, and therefore the $DCGL_{EMC}$, for a 10 m² hot spot. The dose from this smaller contaminated area is certainly expected to be less than the dose resulting from the entire survey unit being uniformly contaminated; the dose in this case is 3.21 mrem/y. This time the external ground radiation pathway is responsible for 99.99% of the total dose, with the plant pathway contributing the other 0.01%. The area factor is calculated by dividing the dose from the larger contaminated area (7.34 mrem/y) by the dose due to the smaller hot-spot area (3.21 mrem/y). This ratio is 2.3 and it is the area factor for a 10 m² hot spot of Co-60. The DCGL$_{EMC}$ for the 10 m² Co-60 hot spot is therefore 2.3 times 3.4 pCi/g or 7.8 pCi/g. Hence, the hot-spot limit using this approach is 2.3 times the average guideline.

RESRAD was integral to this research due to the fact that, for all intents and purposes, it is the only modeling code used to obtain area factors needed to derive hot-spot limits. In that context, going back to first principles for some pathways really meant taking a closer look at how the RESRAD code calculated receptor dose, and more specifically, how the receptor dose was related to the size of the contaminated area. An important aspect of this research was to clearly understand how the RESRAD modeling code handles hot spots when calculating receptor dose.

For the external radiation pathway, the environmental transport factor (ETF) is directly impacted by the size of the contaminated area, and therefore it figures prominently in the determination of DSR. One of the factors used to calculate ETF is the radionuclide-specific area factor (or FA). In fact, for the external radiation pathway, FA is solely responsible for determining the area factor.

11.2.1.2 MicroShield Calculation of Area Factor

MicroShield was used to calculate the exposure rate, with buildup, for the case of uniform Co-60 contamination present to a depth of 15 cm over the 1000 m² survey unit. Again, unit concentration in pCi/g was input. The exposure rate result was 2.16E–3 mR/h. The annual dose can be calculated assuming that the same outdoor fraction as used by RESRAD (0.25), and recognizing that 1 mR in air is equivalent to 1 mrem in tissue for gamma emitters:

$$\text{Dose} = (2.16\text{E}{-}3 \text{ mR/h})(8760 \text{ h/y})(0.25) = 4.73 \text{ mrem/y}$$

Once again MicroShield is run to calculate the area factor for a 10 m² hot spot. The receptor is assumed to be located at the center of the hot spot.

The exposure rate in this case is 9.41E–4 mR/h. This result is converted into annual dose as follows:

$$\text{Dose} = (9.41E\text{--}4 \text{ mR/h})(8760 \text{ h/y})(0.25) = 2.06 \text{ mrem/y}$$

As before, the area factor is calculated by dividing the dose from the larger contaminated area (4.73 mrem/y) by the dose due to the smaller hot-spot area (2.06 mrem/y). This ratio is 2.3—the exact same area factor as obtained from the RESRAD code. Therefore, the MicroShield calculation confirms the RESRAD result that the area factor for a 10 m² hot spot of Co-60 is 2.3 times the average guideline. Again, it is important to remember that these results are for the case of the receptor located directly on the hot spot.

Therefore, comparable results are obtained using RESRAD and MicroShield for calculating the receptor dose to a 10 m² hot spot: 3.21 and 2.06 mrem/y. In addition, a hand calculation based on first principles resulted in a receptor dose value (2.36 mrem/y) that was between that determined from RESRAD and MicroShield (Abelquist 2008). The difference between the RESRAD and MicroShield results is about 36%. Differences in the determination of radiation buildup and occupancy factor values are likely causes for this difference. Ultimately this difference is not that important. Rather, the hot-spot area factors are of interest, and they depend on the relative decrease in dose for each method used. The more pressing concern is the unrealistic assumption that the receptor is located directly on the hot spot for all of their time spent outdoors.

11.2.1.3 Receptor Located Some Distance from the Hot Spot

The next step was to evaluate the receptor dose from a 10 m² hot spot when the receptor is located some distance from the hot spot. An arbitrary distance of 6 m was selected to evaluate both RESRAD and MicroShield calculations of the annual receptor dose from a 10 m² Co-60 hot spot. The RESRAD annual dose was 9.51E–2 mrem/y (99.66% from external radiation pathway). As before, this result was based on an outdoor fraction of 0.25, and the receptor was assumed to be located 6 m from the hot spot for 0.25×8760 h/y.

MicroShield (considering buildup) was then used to calculate the receptor dose the same distance from the hot spot. Again, the annual dose was calculated assuming the same outdoor fraction, and recognizing that 1 mR in air is similar to 1 mrem in tissue for gamma emitters:

$$\text{Dose} = (2.74E{-}5 \text{ mR/h})(8760 \text{ h/y})(0.25) = 5.99E{-}2 \text{ mrem/y}$$

The difference between the RESRAD and MicroShield results is about 37%. Differences in the determination of radiation buildup and occupancy factor values are likely causes for this difference. Again, this difference is not that important—rather, the hot-spot area factors are of interest, and they depend on the relative decrease in dose for each method used.

For example, given the receptor dose based on a 6 m distance from the hot spot, the RESRAD area factor is calculated. Specifically, the area factor is calculated by dividing the dose from the 1000 m² contaminated area (7.34 mrem/y) by the dose due to the smaller hot-spot area located 6 m from the receptor (9.51E−2 mrem/y). This ratio is 77, and it represents the area factor for a 10 m² hot spot of Co-60 assuming that the receptor is 6 m from the hot spot. The DCGL$_{EMC}$ for the 10 m² Co-60 hot spot in this case was 77 times 3.4 pCi/g or 262 pCi/g. Again, the important outcome is that the hot-spot limit using this approach is 77 times the average guideline.

The MicroShield area factor is then calculated for the same distance. As before, the area factor is calculated by dividing the dose from the 1000 m² contaminated area (4.73 mrem/y) by the dose due to the smaller hot-spot area located 6 m from the receptor (5.99E−2 mrem/y). The area factor turns out to be 79, which is close to the result determined by RESRAD.

So, the area factor for a 10 m² hot spot is 2.3 when the receptor is located directly on the hot spot, and 77 (or 79) when the receptor is 6 m from the hot spot. What is a technically defensible approach for determining a reasonable receptor-to-hot-spot distance? One approach is to use probabilistic modeling to determine a representative distribution source-to-receptor distances.

11.2.1.4 A Realistic Hot-Spot Dose Assessment

The idea is to use a probabilistic approach for assessing receptor distance from the hot-spot location, and to use that distance in the determination of hot-spot area factor. This approach provides a more realistic assessment of the receptor dose from hot spots by considering the probability of getting dose from these small source terms. The current area factor approach assumes the worst case that receptor has the misfortune of spending all allotted outdoor time perched on the hot spot. This is very unlikely, and should not form the basis for determining hot-spot limits. In that regard, it is important to point out the significant conservatism of assuming that the receptor spends all of their time on the hot spot when outdoors.

A comparison to the inhalation pathway is instructive. RESRAD uses an ETF to calculate inhalation dose to a receptor located at some distance from the source of the airborne contamination. That is, RESRAD does not assume that the receptor is located at the hot-spot location for purposes of inhalation pathway calculations. (Many screening calculations do indeed assume that the receptor is located directly over the hot spot, and assume that the receptor inhales the radioactivity that is resuspended without the benefit of airborne dispersion.) RESRAD uses the transport (e.g., wind) of radioactivity to provide some measure of atmospheric dispersion (dilution) of the airborne radioactivity before it is inhaled by the receptor. So, a parallel assumption for the external radiation pathway would be that the receptor is NOT located directly on the hot spot, but rather some distance from the hot spot. (Note that the RESRAD user does have the flexibility of "moving" the receptor off of the hot spot, but this option is not routinely selected when calculating area factors.)

One possibility as to why RESRAD handles receptor dose from hot spots in this manner is that it is an unintended consequence of the typical receptor-to-source geometry where the receptor is assumed to be located above an infinite plane source. This is the geometry used in Federal Guidance Report No. 12 to obtain the dose coefficients for contaminated soil (EPA 1993), and these dose coefficients are used in the RESRAD code. Hence, the default approach in RESRAD is to position the receptor directly over the source, notwithstanding the RESRAD user's ability to change this receptor-to-source geometry. As long as the source is large (on the order of 10s to 100s of square meters), it matters little where the receptor is located, the external radiation exposure at the receptor location is essentially constant. However, for a small radiation source (i.e., hot-spot size), it no longer makes sense to assume that the receptor is located directly above the source. Conversely, it is more appropriate to assume that the receptor is likely to be some distance from the hot spot over the course of time the receptor spends outdoors.

Furthermore, the NRC has adopted a philosophy of being "prudently conservative" when it comes to dose modeling. Background discussion in Appendix I of NUREG-1757, vol. 2 (NRC 2006a) states the Commission directed NRC staff to address areas of excessive conservatism, and to use a probabilistic approach for calculating the total effective dose equivalent.

This proposed approach is consistent with the NRC's stated philosophy—it will result in a much lower dose from potential hot spots present.

The goal is to generate a distribution of distances (l), and use this variable to calculate a distribution of doses that result when considering that the receptor will usually be located at varying distances from the hot spot. The receptor can be located at any location (x_1, y_1) within the survey unit, and the same goes for the hot spot (x_2, y_2). The distance between the receptor and hot spot is given by

$$l = \sqrt{(y_1 - y_2)^2 + (x_1 - x_2)^2}$$

Assume a class 1 survey unit of 1000 m² with square dimensions of 31.6 × 31.6 m². The minimum distance between the receptor and hot spot is zero (current assumption in practice), and the maximum distance in this case is the diagonal in the survey unit (44.7 m).

Crystal Ball™ was used to generate 1000 trials of random locations for the receptor and hot-spot location. A uniform distribution was assumed for sampling each of the two pairs of coordinates, with a minimum of zero and maximum of 44.7 m. The Crystal Ball output is provided in Table 11.1.

Figure 11.1 shows the Crystal Ball output trials as well as the best fit to these data, which was a beta distribution. The average distance between receptor and hot spot based on this simulation was 16.8 m, with a standard deviation of 7.86 m. The minimum and maximum distances were 0.33 and 38.2 m, respectively. Obviously, the most conservative distance to select would be zero. A reasonably conservative distance might be the 10% percentile value of the distribution—for this simulation the 10% percentile is 6.01 m. That is, only 10% of the expected receptor-to-hot-spot distances are less than 6 m, while 90% are greater than 6 m.

TABLE 11.1   Distribution of Receptor-to-Hot-Spot Distances (m) Using Crystal Ball

Statistic	Forecast Value	Fit Value: Beta Distribution
Mean	16.8	16.8
Median	16.65	16.57
Standard deviation	7.86	7.86
Minimum	0.33	−2.28
Maximum	38.19	39.86

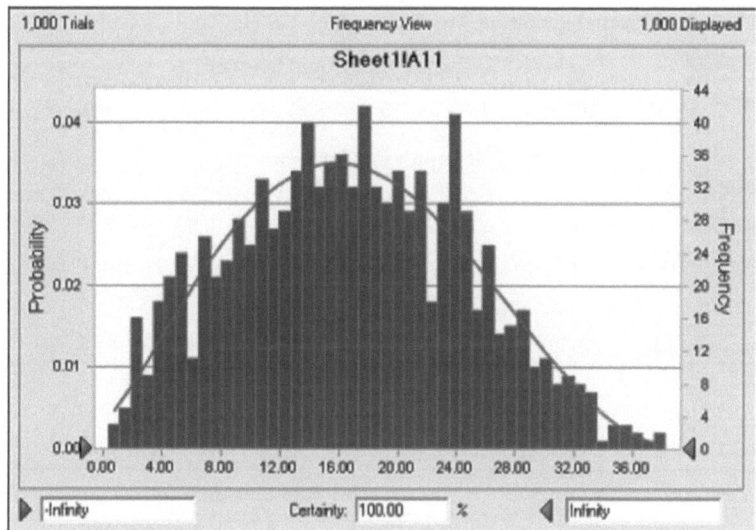

FIGURE 11.1   Output from Crystal Ball simulation of receptor to hot-spot distances.

11.2.1.5 External Radiation Pathway Results

Results from the RESRAD and MicroShield runs for the radionuclides analyzed are shown in Tables 11.2 through 11.9. Each table compares the hot-spot area factors as a function of radionuclide, hot-spot size, and receptor distance from the hot spot for both RESRAD and MicroShield (and the existing MARSSIM AF is provided for reference). The reference survey unit size is 1000 m². The RESRAD and MicroShield area factors are generally comparable for the case where the receptor is located directly on the hot spot. (Note that only MicroShield was used to calculate the dose for the situation where the receptor is located 6 m from the hot spot.)

TABLE 11.2   External Radiation Area Factors for Co-60 Hot Spots in Soil

	Hot-Spot Size (m²)						
	1000	10	3	1	0.5	0.1	0.01
Receptor on Hot Spot							
MARSSIM AF	1	2.1	4.4	9.8	NA	NA	NA
RESRAD AF	1	2.27	4.82	11.3	11.3	11.3	11.3
MicroShield AF	1	2.30	4.73	11.4	21.3	100	990
Receptor 6 m from Hot Spot							
MicroShield AF	1	79.3	250	650	1150	4050	30,000

TABLE 11.3   External Radiation Area Factors for I-129 Hot Spots in Soil

	Hot-Spot Size (m²)						
	1000	10	3	1	0.5	0.1	0.01
	Receptor on Hot Spot						
MARSSIM AF	NA	NA	NA	NA	NA	NA	NA
RESRAD AF	1	1.77	3.50	7.03	7.03	7.03	7.03
MicroShield AF	1	1.76	3.14	6.93	12.6	57.6	575
	Receptor 6 m from Hot Spot						
MicroShield AF	1	122	340	785	1280	3620	13,600

TABLE 11.4   External Radiation Area Factors for Cs-137 Hot Spots in Soil

	Hot-Spot Size (m²)						
	1000	10	3	1	0.5	0.1	0.01
	Receptor on Hot Spot						
MARSSIM AF	1	2.4	5.0	11.0	NA	NA	NA
RESRAD AF	1	2.21	4.67	10.8	10.8	10.8	10.8
MicroShield AF	1	2.18	4.42	10.6	19.8	93.1	918
	Receptor 6 m from Hot Spot						
MicroShield AF	1	83.3	260	672	1170	3930	27,400

TABLE 11.5   External Radiation Area Factors for Ra-226 Hot Spots in Soil

	Hot-Spot Size (m²)						
	1000	10	3	1	0.5	0.1	0.01
	Receptor on Hot Spot						
MARSSIM AF	1	7.8	21.3	54.8	NA	NA	NA
RESRAD AF	1	2.28	4.82	11.2	11.3	11.3	11.3
MicroShield AF	1	2.26	4.63	11.1	20.8	97.8	964
	Receptor 6 m from Hot Spot						
MicroShield AF	1	80.2	252	658	1150	4020	28,900

Further, it should be noted that at the time research was performed, the RESRAD code did not allow the calculation of doses for hot-spot sizes smaller than 1 m² areas. After discussing this RESRAD limitation with Dr. Charley Yu of Argonne National Laboratory, the RESRAD code was modified to handle hot spots less than 1 m². Hence, for areas less than 1 m²,

TABLE 11.6   External Radiation Area Factors for Th-232 Hot Spots in Soil

	Hot-Spot Size (m²)						
	1000	10	3	1	0.5	0.1	0.01
	Receptor on Hot Spot						
MARSSIM AF	1	3.2	6.2	12.5	NA	NA	NA
RESRAD AF	1	2.29	4.85	11.3	11.3	11.3	11.3
MicroShield AF	1	2.31	4.74	11.4	21.3	100	990
	Receptor 6 m from Hot Spot						
MicroShield AF	1	78.6	249	652	1150	4070	29,700

TABLE 11.7   External Radiation Area Factors for U-238 Hot Spots in Soil

	Hot-Spot Size (m²)						
	1000	10	3	1	0.5	0.1	0.01
	Receptor on Hot Spot						
MARSSIM AF	1	11.1	18.3	30.6	NA	NA	NA
RESRAD AF	1	2.13	4.46	10.2	10.2	10.2	10.2
MicroShield AF	1	2.09	4.14	9.78	18.2	85.1	837
	Receptor 6 m from Hot Spot						
MicroShield AF	1	85.5	265	679	1180	3870	23,700

TABLE 11.8   External Radiation Area Factors for Pu-239 Hot Spots in Soil

	Hot-Spot Size (m²)						
	1000	10	3	1	0.5	0.1	0.01
	Receptor on Hot Spot						
MARSSIM AF	NA	NA	NA	NA	NA	NA	NA
RESRAD AF	1	1.98	4.11	9.08	9.09	9.09	9.09
MicroShield AF	1	1.92	3.63	8.41	15.5	72.4	713
	Receptor 6 m from Hot Spot						
MicroShield AF	1	92.2	280	703	1210	3720	19,200

the RESRAD and RESRAD-Offsite models now use either extrapolation or make an assumption that the dose is linearly proportional to the area.

The external radiation pathway produced area factors that do *not* scale directly with the size of the contaminated area. For environmental pathways that are not hot-spot sensitive, the hot-spot dose depends only on

TABLE 11.9   External Radiation Area Factors for Am-241 Hot Spots in Soil

	\multicolumn{7}{c}{Hot-Spot Size (m²)}
	1000	10	3	1	0.5	0.1	0.01
	\multicolumn{7}{c}{Receptor on Hot Spot}
MARSSIM AF	1	96.3	139.7	208.7	NA	NA	NA
RESRAD AF	1	1.61	3.22	6.60	6.60	6.60	6.60
MicroShield AF	1	1.83	3.35	7.50	13.7	63.1	619
	\multicolumn{7}{c}{Receptor 6 m from Hot Spot}
MicroShield AF	1	93.5	283	694	1170	3580	14,400

the inventory (source term) in the contaminated area, not the size of the contaminated area. For non-hot-spot-sensitive pathways, the area factors either scale directly with the size of the contaminated area, or they have even larger area factors (e.g., in the case of water-dependent pathways). The external radiation pathway is hot-spot sensitive, and often results in the smallest (most conservative) area factors.

The consistency of area factors across the radionuclides evaluated was noteworthy. For example, the area factor ranged from roughly 7 to 11 for a 1 m² hot spot, from 12 to 21 for a 0.5 m² hot spot, and 60–100 for a 0.1 m² hot spot. From an application perspective, it might be beneficial to consider establishing the area factors for the external radiation pathway based on the most limiting radionuclide—which was Am-241 or I-129, depending on the model (RESRAD or MicroShield) used to generate the area factor.

The assumption that the receptor might be located 6 m from the hot spot on average had a significant impact on the resulting area factors. These area factors were also consistent across the range of different radionuclides: area factors ranged from 650 to 785 for a 1 m² hot spot and from 1150 to 1280 for a 0.5 m² hot spot.

The obvious conclusion based on the external radiation pathway is that hot-spot doses are much smaller under likely field conditions than assessed under current MARSSIM practice. This is particularly true when making allowance for the receptor to be located 6 m from the hot spot. The area factors for the eight radionuclides evaluated when the receptor was located directly on the hot spot ranged from 6.6 to 11.4 for 1 m² hot spot; and ranged from 650 to 785 when the receptor was located 6 m from the 1 m² hot spot. Thus, allowing the receptor to be on average 6 m from the hot spot over the exposure time results in area factors that are much greater than currently allowed. However, these larger area factors are still more restrictive than those area factors that scale directly with the size of the contaminated area (where the area factor for 1 m² area is 1000).

11.2.2 Inhalation Exposure to Resuspended Soil Pathway

The second pathway evaluated is the inhalation exposure due to resuspended contaminated soil. The receptor dose from a widely distributed source term to the dose from a hot spot of particular size is compared—this ratio of receptor doses allows calculation of the hot-spot limit for that size hot spot. A detailed look at how hot spots of various sizes actually produce receptor doses for the inhalation exposure to resuspended soil is considered in this section. The hot-spot sizes considered are 10, 3, 1, 0.5, 0.1, and 0.01 m². An example case for a hot-spot size equal to 10 m² of Pu-239 in a 1000 m² survey unit was evaluated.

11.2.2.1 RESRAD Area Factor Approach for Inhalation Exposure Pathway
Resuspension is the physical mechanism of re-injecting particulates that have been deposited on the ground from an atmospheric deposition event back into the atmosphere. Once the particulates have been resuspended, they are dispersed as they travel toward the receptor. RESRAD uses an air transport and dispersion model to calculate dispersion coefficients throughout the area of interest for unit releases from each of the resuspension sources. Note that resuspension rates from contaminated soil can increase due to the amount of soil exposed (lack of vegetative cover), size of the area involved, and the resuspension mechanisms (ERG 2004).

The inhalation exposure pathway involves two phenomena to deliver receptor dose: (1) soil contamination becomes airborne, and (2) receptor inhalation of airborne concentration of radionuclides for some duration. The first phenomenon considers the airborne concentration near the source due to resuspension of the contamination, and the second considers the dilution of the airborne concentration as it moves to the receptor location via air dispersion. The *RESRAD User's Manual* (Appendix B) does a nice job describing the dose modeling for the inhalation exposure to resuspended soil pathway (ANL 2001).

Essentially, RESRAD calculates the receptor dose from the inhalation pathway for a uniformly contaminated area (e.g., 1000 m² survey unit). As with the external radiation pathway, the effective dose equivalent limit is converted to a soil concentration by means of dose-to-source ratios (DSRs).

Recall that the DSRs are expressed in terms of three primary factors: dose conversion factors (DCFs), ETFs, and source factors (SFs). For the inhalation exposure to resuspended soil pathway, the dose-to-soil concentration ratio, DSR_i, for the ith radionuclide in mrem/y per pCi/g is given by

$$DSR_i = \sum_j DCF_j \times BRF_{i,j} \times ETF_j \times SF_{i,j}$$

where

DCF_j is the dose conversion factor for the jth radionuclide in mrem per pCi;

$BRF_{i,j}$ is the fraction of the total decay of radionuclide i that results in an ingrowth of radionuclide j;

ETF_j is the environmental transport factor for the jth radionuclide at time, t; and

$SF_{i,j}$ is the source factor that accounts for ingrowth and decay and leaching of the jth radionuclide originating from the transformation of the ith principal radionuclide at time t.

(Note that i and j are index labels for principal radionuclides—i refers to radionuclides that exist initially at time t, and j refers to radionuclides that exist in decay chain of radionuclide i.)

The DCF is the dose-to-exposure ratio—that is, the committed effective dose equivalent to an individual from inhalation exposure of unit radioactivity of the radionuclide present. The DCFs in RESRAD were taken from FGR-11 (EPA 1988). For example, the DCF for Pu-239 is 0.429 mrem/pCi. The SF is essentially a correction factor for the source term that accounts for ingrowth, radioactive decay, and leaching.

The ETF for the inhalation exposure pathway is the ratio of the annual intake of the ith principal radionuclide by dust inhalation to the concentration of that radionuclide in the soil. Of the three primary factors used to determine the DSR for the inhalation exposure pathway, the ETF is impacted by the mass loading of airborne contaminated particles and the size of the contaminated area—called the area factor in RESRAD. The *RESRAD Manual* provides the following equation for the ETF for the inhalation exposure pathway:

$$ETF_i = ASR_i \times FA_i \times FCD_i \times FO_i \times FI_i$$

where

> ASR is the air-to-soil concentration ratio, which also equals the mass loading of airborne contaminated soil particles (RESRAD default is 1E–4 g/m³);
>
> FA is the area factor
>
> FCD is the depth and cover factor
>
> FCD = 1 when contaminated zone thickness exceeds the depth of the soil mixing layer
>
> FO is the occupancy factor
>
> [FO = f_{otd} + (f_{ind} × F_{dust}), where f_{otd} and f_{ind} are outdoor and indoor time fractions, respectively, and F_{dust} is the indoor dust filtration factor]; and
>
> FI is the annual intake of air (8400 m³/y).

RESRAD uses a constant mass loading factor for estimating the airborne concentration near the source. By way of comparison, the NRC's D&D model uses a resuspension factor model to describe the process by which the dust becomes airborne. RESRAD models dilution of the airborne concentration using a zero release height Gaussian plume model. This approach is embodied in the area factor, which depends on the particle size, wind speed, and size of the contaminated area. RESRAD uses least-squares regression to fit the area factor, with the resultant equation shown below:

$$FA = \frac{a}{1 + b(\sqrt{A})^c}$$

where

> A is the size of the contaminated area (m²);
>
> a, b, c are coefficients of least-squares regression that are provided as a function of wind speed.

RESRAD was run for a source term of Pu-239 contamination (1 pCi/g) that was present to a depth of 15 cm over a 1000 m² survey unit. As with the external radiation pathway, no soil cover was assumed. The outdoor time fraction was 0.25, and when combined with an indoor time fraction of 0.5 and dust filtration factor of 0.4, an occupancy factor of 0.45 is obtained for the inhalation pathway. The resulting DSR from the RESRAD run is 0.170 mrem/y per pCi/g (this result considers all environmental pathways). The receptor dose, from all pathways based on 1 pCi/g soil contamination, is therefore 0.170 mrem/y at time $t = 0$ years. The inhalation exposure pathway accounts for 12.65% of the dose—the inhalation pathway was responsible for delivering a dose of 0.022 mrem/y. The soil ingestion and plant pathways were responsible for 56.7% and 30.2% of the receptor dose, respectively. The $DCGL_W$ based on 25 mrem/y can be calculated: 25 mrem/y/(0.170 mrem/y/1 pCi/g) = 147 pCi/g.

RESRAD was then used to calculate the area factor, and therefore the $DCGL_{EMC}$, for a 10 m^2 hot spot. All of the parameters were the same with the exception of the contaminated area size and the length parallel to the aquifer. The dose from this smaller contaminated area was 0.015 mrem/y. Surprisingly, the inhalation exposure pathway was responsible for 89.4% of the total dose, or 0.013 mrem/y; the soil ingestion and plant pathways contributed 6.55% and 3.48%, respectively. It is interesting that the inhalation exposure pathway contribution jumped from 12.65% to nearly 90% as the contaminated area size was reduced from 1000 to 10 m^2.

The area factor is calculated by dividing the dose (from all pathways) from the larger contaminated area (0.170 mrem/y) by the dose due to the smaller hot-spot area (0.015 mrem/y). This ratio is 11.5 and it is the area factor for a 10 m^2 hot spot of Pu-239. The $DCGL_{EMC}$ for the 10 m^2 Pu-239 hot spot is therefore 11.5 times 147 pCi/g, or 1700 pCi/g. Therefore, the hot-spot limit using this approach is 11.5 times the average guideline. However, we are interested in the inhalation pathway.

Focusing exclusively on the inhalation pathway for delivering receptor dose, the area factor based on the inhalation exposure pathway alone is calculated. Recall that the inhalation pathway dose for the 1000 m^2 survey unit was 0.022 mrem/y. The inhalation pathway dose for the 10 m^2 hot spot was 0.013 mrem/y. The area factor based on the inhalation exposure pathway is simply 0.022 mrem/y divided by 0.013 mrem/y, or 1.64. This means that the smaller hot-spot area still results in a sizeable inhalation dose relative to the large 1000 m^2 survey unit.

The inhalation pathway area factor result was checked against the RESRAD area factor (FA). Note that RESRAD uses a default particle size of 1 μm and wind speed of 2 m/s. The linear regression coefficients for these defaults are $a = 1.6819$, $b = 25.5076$, and $c = -0.2278$ (ANL 2001). For a survey unit area of $A = 1000$ m^2, FA is given by

$$FA = \frac{1.6819}{1 + 25.5076(\sqrt{1000})^{-0.2278}} = 0.1333$$

The FA parameter is determined for a hot-spot area $A = 10$ m^2 using the same equation. The result is FA = 0.0816. Next, the ratio of the FA parameters is obtained, 0.1333 divided by 0.0816, or 1.63. This is virtually the same (area factor) result obtained by taking the ratio of the inhalation exposure pathway doses shown above (i.e., 1.64). This means that the difference in receptor dose, as a result of changing contaminated area sizes, is due entirely to the FA parameter.

Using FA equal to 0.0816 for the 10 m^2 hot spot, it is instructive to see how RESRAD calculates the inhalation pathway dose. Recall that the DCF for Pu-239 is 0.429 mrem/pCi. The ETF is calculated as:

$$\text{ETF} = (1\text{E}{-}4 \text{ g/m}^3) \times (0.0816) \times (1) \times (0.45) \times (8400 \text{ m}^3/\text{y})$$
$$= 3.08\text{E}{-}2 \text{ g/y}$$

Now, the inhalation pathway dose from the 10 m^2 hot spot is calculated as:

$$D_{inh} = (0.429 \text{ mrem/pCi}) \times (3.08\text{E}{-}2 \text{ g/y}) \times (1 \text{ pCi/g}) = 0.0132 \text{ mrem/y}$$

This calculation shows how the RESRAD area factor parameter (FA) operates according to the size of the contaminated area (although rather weakly), and confirmed the hot-spot inhalation pathway dose calculation. It is reasonable to conclude that for Pu-239, the inhalation exposure pathway delivers nearly the same dose from a 10 m^2 contaminated area as it does from the soil concentration in a 1000 m^2 area (the difference is only a factor of 1.63). That is, even with 100 times more activity in 1000 m^2 survey unit than in the 10 m^2 hot spot, the inhalation dose from the survey unit is only 1.63 times greater. This seems nonintuitive—and certainly conservative.

One might expect that the inhalation receptor dose would generally scale with the size of the contaminated area, similar to the approach in RESRAD-BUILD. Indeed, the Eastern Research Group (ERG 2004) reports that source term "emission rates might increase, depending on the amounts of soil exposed, the size of the area involved, and the resuspension mechanisms." The next section takes a look at the receptor doses based on first principles.

11.2.2.2 Calculation of Inhalation Pathway Dose Based on First Principles

The inhalation pathway dose was first calculated for a receptor located in a 1000 m^2 survey unit uniformly contaminated with Pu-239 to a depth of 15 cm (no soil cover). The default parameters selected are the same as those used in RESRAD—namely a mass loading factor of 1E–4 g/m³, occupancy factor of 0.45, and an annual intake of air equal to 8400 m³/y. The inhalation dose can be calculated as follows:

$$D_{inh} = (1 \text{ pCi/g}) \times (1E\text{--}4 \text{ g/m}^3) \times (8400 \text{ m}^3/\text{y}) \times (0.429 \text{ mrem/pCi}) \times (0.45)$$
$$= 0.162 \text{ mrem/y}$$

This inhalation dose calculated above is much higher than that determined by RESRAD (by a factor of 7.5) for the same contamination size of 1000 m² (0.022 mrem/y). Indeed, the RESRAD result is 0.133 times the inhalation pathway dose calculated from first principles. The difference is due to the application of an area factor in RESRAD that serves to dilute the airborne concentration that the receptor inhales. The hand calculation makes the conservative assumption that the airborne concentration predicted by multiplying the soil concentration by the mass loading factor is the same concentration breathed by the receptor. This is analogous to simple screening techniques that assume that the airborne concentration at the receptor is equal to the airborne concentration at the point of release (a conservative approach).

Both mechanical disturbances (tilling fields) and wind erosion can generate airborne concentrations. The distance between the point of generation and the receptor location is variable. The Gaussian plume model aspect of FA parameter may adequately consider the receptor distance from the hot spot. That is, the dilution afforded by the FA parameter effectively accounts for the variable receptor to hot-spot distance. The FA parameter in RESRAD adds realism by diluting the airborne concentration that reaches the receptor location. Therefore, the FA factor was used in the calculation of receptor inhalation dose using first principles—the revised inhalation dose for a 1000 m² contaminated area is 0.022 mrem/y. Therefore, applying the FA factor to the hand calculation of inhalation dose for a 1000 m² contaminated area resulted in the same inhalation dose as obtained from RESRAD.

11.2.2.3 Proposal to More Realistically Assess Hot-Spot Dose

As a point of interest, the FA parameter for a 1 m² hot spot for the same conditions is 0.0634, which yields a factor of 2.1 when divided by the FA for the 1000 m² area (0.1333). So, while the FA parameter accounts for the size of the contaminated area, it does so very weakly. This simple example shows that the inhalation dose is only reduced by a factor of 2.1 as the contaminated area is reduced from 1000 to 1 m².

ERG (2004) discusses resuspension of contamination in the context of open field areas that have either unlimited or limited wind erosion potential. An unlimited potential area can be characterized by a smooth field, lacking vegetation, and covered with a thick reservoir of loose sandy soil (unlimited reservoir). Conversely, a limited potential area can be characterized by a heterogeneous field covered with a high density of gravel, rocks, or vegetation. Considering these definitions, it seems that small hot spots would be classified as having limited wind erosion potential—primarily due to the fact that they do not possess an unlimited reservoir of contamination available for resuspension. Once winds begin to resuspend contamination, "the supply of erodible particles is quickly exhausted ...." (ERG 2004, p. 3-7). Therefore, it may be reasonable to conclude that the source term available for inhalation pathway depends more strongly on the contaminated area size than credited by the RESRAD approach.

To increase the effect of contaminated area on the inhalation dose, a simple reduction term defined by dividing the hot-spot area by the survey unit area is proposed. This simply reduces the radionuclide source term available to deliver inhalation dose to the receptor.

Now the inhalation pathway dose to a receptor is calculated from a 10 m² hot spot. As before, assume that the contaminant is Pu-239 to a depth of 15 cm (no soil cover). Assume that the same default parameters as used in RESRAD—namely a mass loading factor of 1E−4 g/m³, occupancy factor of 0.45, and an annual intake of air equal to 8400 m³/y. The FA parameter for a 10 m² hot spot for default conditions described above is 0.0816, and the source term reduction factor is 10/1000, or 0.01. This calculation assumes that the airborne contamination that the receptor breathes is directly proportional to the hot-spot size. The inhalation dose can be calculated as follows:

$$D_{inh} = (1 \text{ pCi/g}) \times (0.0816) \times (0.01) \times (1E{-}4 \text{ g/m}^3) \times (8400 \text{ m}^3/\text{y})$$
$$\times (0.429 \text{ mrem/pCi}) \times (0.45) = 1.32E{-}4 \text{ mrem/y}$$

This receptor dose is less than the 0.0216 mrem/y dose calculated for the 1000 m² survey unit due to the source term reduction factor (0.01) and the ratio of the FA parameters (0.0816/0.133, or 0.613). This calculation assumes that the source term reduction factor effectively accounts for the fact that the receptor inhalation dose delivered from a hot spot reflects the reduced total source term in a hot spot. This source term reduction factor, along with the FA parameter, considers the size of the contaminated area on the determination on receptor inhalation dose. The area factor for a 10 m² Pu-239 hot spot is determined by dividing the 1000 m² dose (0.022 mrem/y) by the 10 m² hot-spot dose (1.32E–4 mrem/y)—which results in an area factor of 163.

11.2.2.4 Inhalation Pathway Results

Tables 11.10 through 11.20 illustrate the hot-spot area factors as a function of radionuclide, hot-spot size, and receptor distance from the hot spot for

TABLE 11.10 Inhalation Pathway Area Factors for C-14 Hot Spots in Soil

	Hot-Spot Size (m²)						
	1000	**10**	**3**	**1**	**0.5**	**0.1**	**0.01**
	Receptor on Hot Spot						
RESRAD area factor	1	9.95	18.1	31.2	43.8	96.3	289
Hand calculation area factor	1	163	621	2100	4540	2.71E4	3.50E5

TABLE 11.11 Inhalation Pathway Area Factors for Co-60 Hot Spots in Soil

	Hot-Spot Size (m²)						
	1000	**10**	**3**	**1**	**0.5**	**0.1**	**0.01**
	Receptor on Hot Spot						
RESRAD area factor	1	1.64	1.86	2.10	2.27	2.71	3.50
Hand calculation area factor	1	163	621	2100	4540	2.71E4	3.50E5

TABLE 11.12 Inhalation Pathway Area Factors for Sr-90 Hot Spots in Soil

	Hot-Spot Size (m²)						
	1000	**10**	**3**	**1**	**0.5**	**0.1**	**0.01**
	Receptor on Hot Spot						
RESRAD area factor	1	1.64	1.86	2.10	2.27	2.71	3.49
Hand calculation area factor	1	163	621	2100	4540	2.71E4	3.50E5

TABLE 11.13   Inhalation Pathway Area Factors for Tc-99 Hot Spots in Soil

	Hot-Spot Size (m²)						
	1000	10	3	1	0.5	0.1	0.01
	Receptor on Hot Spot						
RESRAD area factor	1	1.63	1.86	2.10	2.27	2.71	3.50
Hand calculation area factor	1	163	621	2100	4540	2.71E4	3.50E5

TABLE 11.14   Inhalation Pathway Area Factors for I-129 Hot Spots in Soil

	Hot-Spot Size (m²)						
	1000	10	3	1	0.5	0.1	0.01
	Receptor on Hot Spot						
RESRAD area factor	1	1.63	1.86	2.10	2.27	2.71	3.50
Hand calculation area factor	1	163	621	2100	4540	2.71E4	3.50E5

TABLE 11.15   Inhalation Pathway Area Factors for Cs-137 Hot Spots in Soil

	Hot-Spot Size (m²)						
	1000	10	3	1	0.5	0.1	0.01
	Receptor on Hot Spot						
RESRAD area factor	1	1.64	1.86	2.10	2.27	2.71	3.50
Hand calculation area factor	1	163	621	2100	4540	2.71E4	3.50E5

TABLE 11.16   Inhalation Pathway Area Factors for Ra-226 Hot Spots in Soil

	Hot-Spot Size (m²)						
	1000	10	3	1	0.5	0.1	0.01
	Receptor on Hot Spot						
RESRAD area factor	1	1.63	1.86	2.10	2.27	2.71	3.50
Hand calculation area factor	1	163	621	2100	4540	2.71E4	3.50E5

TABLE 11.17 Inhalation Pathway Area Factors for Th-232 Hot Spots in Soil

	Hot-Spot Size (m²)						
	1000	10	3	1	0.5	0.1	0.01
	Receptor on Hot Spot						
RESRAD area factor	1	1.63	1.86	2.10	2.27	2.71	3.50
Hand calculation area factor	1	163	621	2100	4540	2.71E4	3.50E5

TABLE 11.18 Inhalation Pathway Area Factors for U-238 Hot Spots in Soil

	Hot-Spot Size (m²)						
	1000	10	3	1	0.5	0.1	0.01
	Receptor on Hot Spot						
RESRAD area factor	1	1.63	1.86	2.10	2.27	2.71	3.50
Hand calculation area factor	1	163	621	2100	4540	2.71E4	3.50E5

TABLE 11.19 Inhalation Pathway Area Factors for Pu-239 Hot Spots in Soil

	Hot-Spot Size (m²)						
	1000	10	3	1	0.5	0.1	0.01
	Receptor on Hot Spot						
RESRAD area factor	1	1.64	1.86	2.10	2.27	2.71	3.50
Hand calculation area factor	1	163	621	2100	4540	2.71E4	3.50E5

TABLE 11.20 Inhalation Pathway Area Factors for Am-241 Hot Spots in Soil

	Hot-Spot Size (m²)						
	1000	10	3	1	0.5	0.1	0.01
	Receptor on Hot Spot						
RESRAD area factor	1	1.64	1.86	2.10	2.27	2.71	3.50
Hand calculation area factor	1	163	621	2100	4540	2.71E4	3.50E5

both RESRAD and the hand calculation. The area factors calculated using the RESRAD code were consistent for all of the radionuclides studied with the exception of C-14. With the exception for C-14, these area factors had a very small range, from 1.64 to 3.49 for hot spots ranging in size from 10 to 0.01 m². The area factors calculated using the hand calculations were significantly larger. These area factors were consistent for all 11 radionuclides considered, ranging from a low of 163 for a 10 m² hot spot, to 3.50E5 for a 0.01 m² hot spot.

11.2.2.5 Inhalation Pathway Conclusions

As with the external radiation pathway, the conclusion is that hot-spot doses are much smaller under likely field conditions than assessed using the RESRAD modeling code. Area factors calculated based on first principles (hand calculation) are much greater than currently calculated using the RESRAD code. This is due to the approach of dividing the hot-spot area by the survey unit area—which simply reduces the radionuclide source term available to deliver inhalation dose to the receptor. The inhalation pathway may or may not be considered "hot-spot sensitive" depending on whether the RESRAD or hand calculation approach is used to generate area factors—using the hand calculation approach, this pathway is not hot-spot sensitive. Furthermore, for the inhalation pathway, the area factor calculated does not depend on the radionuclide, and when compared to the external pathway, it is much less limiting. As shown in Table 11.21, the external radiation pathway (using Co-60 results as representative of the external radiation pathway), results in smaller area factors than the inhalation pathway. This is true even for the case when the receptor is 6 m from the hot spot.

TABLE 11.21 Comparison of External Radiation (Co-60) and Inhalation Area Factors in Soil

	Hot-Spot Size (m²)						
	1000	10	3	1	0.5	0.1	0.01
	Receptor on Hot Spot						
MicroShield AF (External radiation, Co-60)	1	2.30	4.73	11.4	21.3	100	990
	Receptor 6 m from Hot Spot						
MicroShield AF (External radiation, Co-60)	1	79.3	250	650	1150	4050	3.00E4
Hand calculation (Inhalation)	1	163	621	2100	4540	2.71E4	3.50E5

Again, recall the consistency of area factors for the external radiation pathway across the radionuclides evaluated. For the case when the receptor was on the hot spot, the area factor ranged from roughly 7 to 11 for a 1 m² hot spot, from 12 to 21 for a 0.5 m² hot spot, and 60 to 100 for a 0.1 m² hot spot. When the receptor was located 6 m from the hot spot, area factors ranged from 650 to 785 for a 1 m² hot spot and from 1150 to 1280 for a 0.5 m² hot spot. Therefore, it is reasonable to conclude that the external radiation pathway, for all radionuclides studied, is more appropriate to use for the determination of area factors as compared to the inhalation pathway. We will now show that the external radiation pathway is also conservative relative to the ingestion-based environmental pathways.

11.2.3 Ingestion-Based Environmental Pathways

Ingestion-based environmental pathways account for potential exposure to contamination in soil and other environmental media from eating food grown on site and drinking water from the site. Example pathways include direct ingestion of soil, ingestion of drinking water from a groundwater source, ingestion of plant products grown in contaminated soil, and ingestion of animal products grown on site (i.e., after animals ingest contaminated drinking water, plant products, and soil). The ingestion-based pathways can be further categorized by water-dependent and water-independent pathways. For example, water-independent pathways include the direct ingestion of soil and ingestion of plant products grown in contaminated soil. Water-dependent pathways include ingestion of drinking water, ingestion of plant products irrigated with contaminated groundwater.

The objective was to evaluate how hot spots impact receptor dose via different environmental pathways. For several ingestion-based pathways, RESRAD was evaluated to determine how the modeling code handles the dose calculation for hot spots (Abelquist 2008). Results for three of the ingestion-based pathways are considered in this section.

11.2.3.1 Direction Ingestion of Soil

The direct ingestion of soil exposure pathway involves the ingestion of contamination by future site occupants. This is a water-independent pathway where the receptor dose might occur from incidental ingestion of contamination, for example, when a person comes in contact with contaminated soil, and subsequently proceeds to eat without washing his hands.

RESRAD was used to calculate the area factor for various hot-spot areas (10, 3, 1, 0.5, 0.1, and 0.01 m²) for the soil ingestion pathway. In each case it was apparent that the hot-spot dose (and therefore area factor) scaled directly with the hot-spot size. That is, if the hot-spot size is reduced by a factor of 1000, then the hot-spot dose is reduced by a similar factor, and therefore the area factor is 1000. Considering that the future occupant is likely to randomly occupy different locations within a survey unit, it seems plausible that the receptor dose would scale directly with the fraction of the survey unit actually contaminated. So, the hot-spot dose is essentially based on the total amount or inventory of radioactivity being in contact with a future receptor, and ultimately ingested by the future receptor. Therefore, this pathway is *not* considered to be hot-spot sensitive.

11.2.3.2 Ingestion of Drinking Water

The ingestion of drinking water is a water-dependent pathway, and as such, receptor dose will be delayed until radionuclides in soil can migrate to the groundwater and then reach a point of water withdrawal (e.g., well or pond). The water-dependent pathways are described by two segments— a water pathway segment and a food chain pathway segment. The water pathway segment connects the soil contamination zone with the point of water withdrawal (e.g., irrigation, drinking, or aquatic foods); the food chain segment connects the radionuclide concentration in water to the food chain and ultimately human exposure.

Time is an important consideration in the calculation of dose via water-dependent pathways. The time it takes for each radionuclide to reach the groundwater and produce dose via the plant irrigation pathway will vary. Therefore, the approach used for hot-spot dose assessment was to run RESRAD for time periods ranging from 0 to 5000 years for each radionuclide assumed to have contaminated area of 1000 m². RESRAD graphical output of total dose as a function of time (in years) was then reviewed to determine the time when the dose reached a peak due to the groundwater pathway. The following results per radionuclide were observed: uranium (about 700 years), Tc-99 (3 years), Ra-226 (about 700 years), I-129 (3 years), C-14 (2 years), and Am-241 (about 150 years). Five radionuclides did not exhibit a peak dose due to groundwater breakthrough: Co-60, Sr-90, Cs-137, Th-232, and Pu-239. That is, these radionuclides do not contribute to receptor dose via the ingestion of contaminated drinking water. Therefore, area factors for the drinking water pathway were limited to the six radionuclides that do have a water-dependent pathway dose component. It was observed that the time for maximum dose to occur for a particular radionuclide was dependent on the size of the contaminated area.

RESRAD was then used to calculate the area factors for the drinking water pathway for various hot-spot areas (10, 3, 1, 0.5, 0.1, and 0.01 m²). The area factor continued to increase with successively smaller hot spots, but there was no immediately obvious relationship with area. It turns out that the area factor is impacted by one parameter—WSR. This parameter is sensitive to area—although in a somewhat complicated fashion (Abelquist 2008).

An interesting outcome for the drinking water pathway is that the area factors for the six radionuclides that deliver receptor dose via this pathway are more restrictive than those that scale directly with the size of the contaminated area. That is, the area factors for the soil ingestion pathway for a 1 m² area were 1000 for all radionuclides. The drinking water pathway area factors for 1 m² area ranged from 90 to 119 for C-14, Tc-99, and I-129, and ranged from 20.3 to 21.9 for Ra-226, U-238, and Am-241. It is important to remember that area factors for this pathway were calculated based on the individual time that a maximum occurs for the water-dependent pathway for a particular contaminated area size. Given that the maximum for the external radiation pathway occurs at time $t = 0$, the area factor for a particular radionuclide will have a local maximum at $t = 0$, and another at the water-dependent time for maximum dose.

It is interesting to note that while the area factors for the drinking water are more restrictive than those that scale directly with contaminated area size, they are less restrictive than the external radiation pathway (for receptor located directly on the hot spot) area factor. The drinking water pathway is therefore regarded as mildly "hot-spot sensitive."

11.2.3.3 Ingestion of Plant Products Grown in Contaminated Soil

The ingestion of plant products grown in soil accounts for four food pathways: plant foods, meat, milk, and aquatic foods. The plant food pathway category can be divided into the following four subcategories: (1) root uptake from crops grown in the contaminated area, (2) foliar deposition uptake from the settling of contaminated dust on the plants, (3) root uptake from contaminated irrigation water, and (4) foliar uptake from overhead irrigation with contaminated water. In this section the focus is on the water-independent plant food ingestion pathways—that is, those pathways that deliver receptor dose via two common phenomena: (1) root uptake in contaminated soil, and (2) foliar deposition of contaminated dust. The first phenomenon considers the plant update of contamination via its root system, and the second considers the resuspension of contaminated material and its settling on plants.

RESRAD was used to calculate the area factor for various hot-spot areas for the plant food ingestion pathway. In each case it was apparent that the hot-spot dose (and area factor) scaled directly with the hot-spot size. That is, if the hot-spot size is reduced by a factor of 1000, then the hot-spot dose is reduced by a similar factor, and therefore the area factor is 1000. The RESRAD FA_3 parameter is significant in assessing how hot spots impact receptor dose. Both the RESRAD and validating hand calculations indicate that area factors scale directly with the size of the contaminated area. As such, this pathway is also not considered to be hot-spot sensitive.

Conceptually, considering that the crops are grown fairly uniform across a future survey unit, it seems reasonable that the receptor dose would scale directly with the fraction of the survey unit actually contaminated. Hence, the hot-spot dose is essentially based on the total amount or inventory of radioactivity getting into the plant food chain, and ultimately ingested by a receptor.

11.2.3.4 Ingestion-Based Pathway Conclusions

The receptor dose impact from hot spots via these environmental pathways is largely related to total source term. For example, the radioactivity present in the drinking water originates from the activity in the survey unit that is transported to the groundwater, and eventually to the drinking water. Also, there is a time lag for some pathways like the plant products irrigated with contaminated water because it might take hundreds of years, for example, for the contamination to travel to the groundwater. The results of this section point to another somewhat obvious conclusion—hot-spot dose assessment is more of a near term concern than for some future time (after breakthrough when they have reached the groundwater). That is, area factors derived from the external radiation pathway are typically more limiting.

The soil ingestion, water-independent animal product, and ingestion of plant products grown in contaminated soil all have area factors that scale directly with size of the contaminated area. As such, these pathways are *not* considered to be "hot-spot sensitive." The drinking water pathway is "mildly hot spot sensitive," having area factors somewhat smaller than those that scale directly with the size of the contaminated area.

Finally, what can we conclude about the cattle grazing at a rate of 50 m² per day—how do hot spots contribute to receptor dose? Well, as noted above, the ingestion dose scales directly with the size of the contaminated area. So if the contaminated area is only 1 m², the milk ingestion dose scales proportionately. Therefore, for a survey unit size of 1000 m², the hot-spot ingestion dose would be 1/1000 of that derived for the case when the entire survey unit is contaminated. The idea is that the cattle graze essentially randomly throughout the survey unit, so on average, the hot spot represents just 1/1000 of the total grazing area. It seems that this is a reasonable way to model hot spots for this pathway. Besides, the external radiation and drinking water pathways are more limiting, so the cattle grazing on small hot-spots argument turns out not to be that important in terms of deriving area factors.

11.2.4 Conclusion: Hot-Spot Limits in Soil

The overall conclusion is that hot-spot doses are much smaller under likely field conditions than assessed under the current practice outlined in MARSSIM. The external radiation pathway produced area factors that do *not* scale directly with size of the contaminated area. And when the predominant pathway is one based on source term inventory, regardless of whether the total activity is spread over 100 m² or concentrated in 0.1 m², the same amount of activity delivers the same dose—in this situation, hot spots are only important in the sense that they contribute to the total source term. The external radiation pathway is hot-spot sensitive, and often results in the smallest (most conservative) area factors.

The consistency of area factors across the radionuclides evaluated was noteworthy. For the external radiation pathway, the area factor ranged from roughly 7 to 11 for a 1 m² hot spot, from 12 to 21 for a 0.5 m² hot spot, and 60 to 100 for a 0.1 m² hot spot. From an application perspective, it might be beneficial to consider establishing the area factors for the external radiation pathway based on the most limiting radionuclide—which was Am-241 or I-129, depending on the model used to generate the area factor.

Finally, application of these area factors at cleanup sites would potentially result in substantial reductions in cleanup and survey costs. D&D practitioners and regulators are encouraged to establish hot-spot criteria on a stronger technical basis—for example, using area factors for hot-spot sizes less than 1 m² when the hot-spot size warrants, and considering that the receptor may be some distance from the hot spot.

11.3 HOT-SPOT LIMITS ON BUILDING SURFACES

The building occupancy scenario accounts for receptor exposure to fixed and removable surface contamination sources within a structure. This residual radioactivity is assumed to remain after decontamination and decommissioning activities have been completed, including the FSS. This section focuses on the hot-spot dose from the following building occupancy pathways—external radiation, inhalation of resuspended surface contamination, and inadvertent ingestion of surface contamination.

Both RESRAD-BUILD and MicroShield codes were used to assess hot-spot doses from building contamination, with MicroShield being used only for the external radiation pathway. Building occupancy is the primary scenario in RESRAD-BUILD—for example, an office worker spends roughly 2000 hours/year working in a building that may have residual radioactivity present. The hot-spot sizes considered are 3, 1, 0.5, 0.1, and 0.01 m². The default survey unit considered in this assessment is a floor area of 100 m². The smallest hot-spot size (0.01 m²) may be effectively considered to represent a discrete particle or contaminated area (10 cm × 10 cm, or 100 cm²) present on a building surface. For each of the three pathways, RESRAD-BUILD (and MicroShield for the external radiation pathway) was studied to determine how the code handled the hot-spot dose calculation.

11.3.1 External Radiation Pathway

The receptor dose from the external radiation pathway primarily depends on the radionuclide and the characteristics of its emitted radiation, quantity of radioactivity on the building surface (a time-dependent term due to physical removal and radioactive decay), geometry of the source term, source-to-receptor distance, and exposure duration. The approach used to assess this pathway's dependence on hot spots involved both RESRAD-BUILD and MicroShield codes to calculate dose when the receptor was located directly over the hot spot. Additionally, MicroShield was used for the case when the receptor was positioned 1 m from the hot spot.

RESRAD-BUILD was used to calculate hot-spot doses for areas as small as 0.01 m². This smallest hot-spot size (= 100 cm²) is also the conventional averaging area for a single direct measurement of surface activity, as well as the nominal size of many radiation detectors used to measure surface activity.

11.3.1.1 RESRAD-BUILD Area Factor Approach for External Radiation Pathway

RESRAD-BUILD calculates the hot-spot dose from the external radiation exposure pathway by treating an area source as a volume source of small thickness (0.01 cm) with unit density. The external radiation dose is estimated by assuming that the floor is an area source with the receptor located 1 m above the floor. The external dose at time t, $D_i(t)$, is calculated as follows:

$$D_i(t) = (ED/365) \times F_{in} \times F_i \times \bar{C}_s(t) \times DCF \times F_G$$

where
  ED is the exposure duration in days
  F_{in} is the fraction of time spent indoors
  F_i is the fraction of time spent in compartment i
  $C_s(t)$ is the average volume source concentration in pCi/g over the exposure duration
  DCF is the dose conversion factor from FGR-12 for an infinite volume source

F_G or the geometrical factor, is the ratio of the dose for the actual source geometry to the dose for the standard source—contaminated soil of infinite depth and lateral extent with no clean cover. This geometrical factor is effectively the product of the depth-and-cover factor (F_{CD}), an area and material factor (F_{AM}), and the off-set factor ($F_{OFF-SET}$).

RESRAD-BUILD was run for each radionuclide assumed to be uniformly present on the floor over a 100 m² survey unit. Unit concentration (1 pCi/m²) was input. The default indoor time fraction is 0.5 and exposure duration was 365 days. For this analysis, the source was positioned at the center of the room by specifying source coordinates at 5 m, 5 m, 0 (these are x, y, and z coordinates). The receptor was positioned at the same x and y coordinates as the source (e.g., 5 and 5 m), with z coordinate equal to 1 m—so the receptor dose location was 1 m above the center of the source. Next, RESRAD-BUILD was run to calculate area factors for various contamination areas.

The key parameter responsible for the difference in hot-spot doses is the geometrical factor. Given that RESRAD-BUILD uses a point-kernel approach, it is not surprising that MicroShield, which also uses a point-kernel calculation, produces similar results. This is discussed in the following section.

11.3.1.2 MicroShield Area Factor Calculation

MicroShield was used to calculate the exposure rate, with buildup, for each radionuclide assumed to be uniformly contaminated on the surface of a 100 m² survey unit. The disk geometry in the MicroShield model was used. Again, unit concentration in pCi/m² was converted into 1E–10 µCi/cm² and was input. The exposure rate result was calculated assuming the same exposure duration and indoor fraction as used by RESRAD (8760 h/y times 0.5 indoor fraction). The area factor was calculated for each radionuclide. The MicroShield area factor results were similar to the RESRAD results for each contaminated area evaluated. Again, it is important to remember that these results are for the case of the receptor located directly on the hot spot.

11.3.1.3 Receptor Location 1 m Distance from the Hot Spot

Probabilistic risk assessments were used to assess the likelihood of encountering a hot spot in a given area, given that all areas of a survey unit are equally likely to be occupied by a future receptor. Similar to the approach presented for the external radiation pathway for soil, a distribution of receptor-to-hot-spot distances was generated using Crystal Ball. Based on the output, a reasonably conservative distance was selected (i.e., 1 m).

The next step was to evaluate the receptor dose from a hot spot when the receptor is located some distance from the hot spot. An arbitrary distance of 1 m was selected to evaluate MicroShield calculations of the annual receptor dose from a hot spot. The MicroShield exposure rate was based on an indoor fraction of 0.5, and the receptor was assumed to be located 1 m from the hot spot for 0.5 × 8760 h/y. Tables 11.22 through 11.29 show the external radiation doses and area factors as a function of radionuclide, hot-spot size, and receptor distance from the hot spot for both RESRAD-BUILD and MicroShield.

TABLE 11.22 External Radiation Area Factors for Co-60 Hot Spots

	Hot-Spot Size (m²)					
	100	**3**	**1**	**0.5**	**0.1**	**0.01**
	Receptor on Hot Spot					
RESRAD-BUILD AF	1	5.19	12.6	23.6	111	1090
MicroShield AF	1	5.19	12.6	23.5	111	1100
	Receptor 1 m from Hot Spot					
MicroShield AF	1	7.49	21.8	43.4	217	2170

TABLE 11.23   External Radiation Area Factors for I-129 Hot Spots

	Hot-Spot Size (m²)					
	100	3	0.5	0.5	0.1	0.01
Receptor on Hot Spot						
RESRAD-BUILD AF	1	5.33	24.3	23.6	115	1130
MicroShield AF	1	5.26	23.9	23.5	113	1110
Receptor 1 m from Hot Spot						
MicroShield AF	1	22.0	43.9	43.4	219	2190

TABLE 11.24   External Radiation Area Factors for Cs-137 Hot Spots

	Hot-Spot Size (m²)					
	100	3	1	0.5	0.1	0.01
Receptor on Hot Spot						
RESRAD-BUILD AF	1	5.20	12.6	23.6	111	1100
MicroShield AF	1	5.20	12.6	23.6	111	1100
Receptor 1 m from Hot Spot						
MicroShield AF	1	7.50	21.8	43.5	217	2170

TABLE 11.25   External Radiation Area Factors for Ra-226 Hot Spots in Soil

	Hot-Spot Size (m²)					
	100	3	1	0.5	0.1	0.01
Receptor on Hot Spot						
RESRAD-BUILD AF	1	5.21	12.6	23.6	111	1100
MicroShield AF	1	5.20	12.6	23.6	111	1100
Receptor 1 m from Hot Spot						
MicroShield AF	1	7.50	21.8	43.5	217	2170

TABLE 11.26   External Radiation Area Factors for Th-232 Hot Spots

	Hot-Spot Size (m²)					
	100	3	1	0.5	0.1	0.01
Receptor on Hot Spot						
RESRAD-BUILD AF	1	5.21	12.6	23.7	111	1100
MicroShield AF	1	5.20	12.6	23.6	111	1100
Receptor 1 m from Hot Spot						
MicroShield AF	1	7.50	21.8	43.5	217	2170

TABLE 11.27   External Radiation Area Factors for U-238 Hot Spots

	Hot-Spot Size (m²)					
	100	**3**	**1**	**0.5**	**0.1**	**0.01**
	Receptor on Hot Spot					
RESRAD-BUILD AF	1	5.07	12.2	22.9	108	1070
MicroShield AF	1	5.25	12.7	23.8	112	1110
	Receptor 1 m from Hot Spot					
MicroShield AF	1	7.56	22.0	43.8	219	2190

TABLE 11.28   External Radiation Area Factors for Pu-239 Hot Spots

	Hot-Spot Size (m²)					
	100	**3**	**1**	**0.5**	**0.1**	**0.01**
	Receptor on Hot Spot					
RESRAD-BUILD AF	1	5.19	12.6	23.6	111	1090
MicroShield AF	1	5.30	12.9	24.1	114	1120
	Receptor 1 m from Hot Spot					
MicroShield AF	1	7.63	22.2	44.2	221	2210

TABLE 11.29   External Radiation Area Factors for Am-241 Hot Spots

	Hot-Spot Size (m²)					
	100	**3**	**1**	**0.5**	**0.1**	**0.01**
	Receptor on Hot Spot					
RESRAD-BUILD AF	1	5.03	12.2	22.8	107	1060
MicroShield AF	1	5.33	13.0	24.2	114	1130
	Receptor 1 m from Hot Spot					
MicroShield AF	1	7.67	22.3	44.4	222	2220

11.3.1.4 External Radiation Pathway Conclusions

The area factors calculated for the external radiation pathway are quite consistent for each of the radionuclides. For example, for the receptor located directly over a 0.1 m² hot spot, the area factors ranged from 107 to 115 for RESRAD-BUILD and ranged 111–114 for MicroShield. In this case, it may be appropriate to consider an area factor of 100 for all radionuclides for a 0.1 m² hot spot. For the case of the receptor located 1 m from the 0.1 m² hot spot, the area factors ranged from 217 to 222. Therefore, to conclude for the external radiation pathway, the area factor is largely independent of the radionuclide (i.e., area factor only depends on the size of the hot spot).

For the smallest hot spot studied (0.01 m² or 100 cm²), the area factors were approximately 1100. This compares to an area factor of 3 (3) cited in both Regulatory Guide 1.86 and DOE O 458.1. Thus, the area factors calculated based on dose modeling are much larger than the conservative, historical factor of three area factors used for decades. Therefore, conclude that this pathway is indeed "hot-spot sensitive."

11.3.2 Inhalation Pathway

Receptor dose from the inhalation exposure pathway is determined by performing three separate calculations: (1) the mechanical removal of material from the source and the rate of release of radionuclides into the indoor air; (2) the indoor airborne concentration of the radionuclides released into the air; and (3) the inhalation of airborne radioactive dust and the associated effective dose equivalent.

RESRAD-BUILD (Version 3.22) was used to calculate the receptor dose from the inhalation pathway for a uniformly contaminated area (e.g., 100 m² survey unit). Specifically, RESRAD-BUILD was used to determine the receptor dose from an area source of 1 pCi/m² (2.22E-2 dpm/100 cm²). The default room size in this model is 36 m², which was increased to 100 m². The source contamination area was also assumed to be 100 m². RESRAD-BUILD allows the user to specify both the receptor and contamination source locations in the building. For this analysis, the source was positioned at the center of the room by specifying source coordinates at 5 m, 5 m, 0 (x, y, and z coordinates). The receptor was positioned at the same x and y coordinates as the source (e.g., 5 and 5 m), with z coordinate equal to 1 m—effectively 1 m above the center of the source. Abelquist (2008) provides details of the RESRAD-BUILD results.

RESRAD-BUILD was then used to calculate the area factor for the inhalation pathway for various hot-spot sizes—the area factor for hot-spot areas (3, 1, 0.5, 0.1, and 0.01 m²) was calculated for the inhalation pathway. In each case it was apparent that the hot-spot dose (and therefore area factor) scaled directly with the hot-spot size. That is, if the hot-spot size is reduced by a factor of 1000, then the hot-spot dose is reduced by a similar factor, and therefore the area factor is 1000.

Hand calculations performed to validate the RESRAD-BUILD determined area factors were generally consistent. This is due to the fact that both RESRAD-BUILD and the hand calculation approach divide the hot-spot area by the survey unit area—which reduces the radionuclide source term available to deliver inhalation dose to the receptor. That is, for both approaches, the airborne contamination inhaled by the receptor is directly proportional to the hot-spot size.

It is interesting to compare the area factors obtained from the external radiation pathway (previous section), with those calculated for the inhalation pathway. Recall that for the receptor located directly over a 0.1 m^2 hot spot, the external radiation pathway area factors ranged from 107 to 115 for RESRAD-BUILD and ranged from 111 to 114 for MicroShield. For the same 0.1 m^2 hot spot, the inhalation pathway area factor is 1000—a consequence of the fact that as the size of the contaminated area is reduced from 100 to 0.1 m^2 (reduced by a factor of 1000), the hot-spot dose is similarly reduced by a factor of 1000, and therefore, the area factor is 1000. Therefore, conclude that the inhalation pathway is *not* "hot-spot sensitive."

11.3.3 Ingestion Pathway

The ingestion pathway of the building occupancy scenario considers two components of the receptor dose from the inadvertent ingestion pathway: (1) the inadvertent ingestion of radioactive material contained in removable material directly from the source (sometimes referred to as direct ingestion), and (2) the inadvertent ingestion of airborne radioactive particulates deposited on building surfaces (also called secondary ingestion).

RESRAD-BUILD (Version 3.22) was used to calculate the receptor dose from the inadvertent ingestion pathway for a uniformly contaminated area (e.g., 100 m^2 survey unit) using the same source and receptor parameters as for the inhalation pathway. Area factors for the ingestion pathway were calculated for various hot-spot sizes—3, 1, 0.5, 0.1, and 0.01 m^2. Again, it was apparent that the hot-spot dose (and therefore area factor) scaled directly with the hot-spot size. That is, if the hot-spot size is reduced by a factor of 1000, then the hot-spot dose is reduced by a similar factor, and therefore the area factor is 1000.

Hand calculations varied somewhat from the RESRAD-BUILD approach. The effective transfer rate for ingestion (GO), obtained from NUREG/CR-5512, vol. 3 (NRC 1999a), defines the parameter GO as the effective transfer rate of contamination from building surfaces via hands, food, and other items to the mouth—a process called secondary ingestion. The default value for GO is 1E-4 m^2/h. Note that GO is essentially the same parameter as the surface ingestion rate of dust particulates deposited on building surfaces (SER) used by RESRAD-BUILD. The ingestion dose is calculated using the following pathway equation:

$$D_{ing} = A_s \times \frac{A}{SU} \times GO \times t \times DCF_{ing}$$

where

A_s is the surface activity level in pCi/m^2

A is the contaminated area size in m^2

SU is the survey unit size in m^2—therefore A/SU represents the fraction of the survey unit area represented by the hot spot

t is the exposure time (97.5 days—based 45 h/week, 52 weeks per year)

DCF_{ing} is the dose conversion factor for ingestion

Federal Guidance Report No. 11 (EPA 1988) provides the ingestion DCF for Am-241 as 3.64E–3 mrem/pCi. The ingestion dose from Am-241 on 100 m^2 building floor surface is calculated as follows:

$$D_{ing} = \frac{1\,pCi}{m^2} \times \frac{100\,m^2}{100\,m^2} \times \frac{1E{-}4\,m^2}{h} \times \frac{97.5\,d}{y} \times \frac{24\,h}{d} \times \frac{3.64E{-}3\,mrem}{pCi}$$

$$= 8.52E{-}4\,mrem/y$$

This compares to the RESRAD-BUILD ingestion pathway dose of 9.06E–5 mrem/y. Thus, the hand calculation is nearly 10 times greater than the RESRAD-BUILD calculation. The two primary reasons for the difference between the RESRAD-BUILD and hand calculation results are the exposure time and surface contamination available for secondary ingestion. First, the RESRAD-BUILD model assumes that the receptor has an exposure time of 4380 h/y (assuming that 100% of indoor time is spent in the compartment of concern), while the hand calculation uses 2340 h. Second, the RESRAD-BUILD model uses an air quality model to determine the surface contamination that settles on horizontal surfaces. In this calculation the surface contamination turns out to be 5.7E–2 pCi/m^2. The hand calculation conservatively takes the surface contamination available for secondary ingestion as the initial source term on the surface (1 pCi/m^2). Overall, the hand calculation produces a receptor ingestion dose that is nearly a factor of 10 greater that that calculated with RESRAD-BUILD. However, even though the hot-spot *doses* are different, the *area factors* are very consistent between the RESRAD-BUILD code and hand calculations.

As with the inhalation pathway area factors, the ingestion pathway area factors are directly proportional to the hot-spot size. For a 0.1 m² hot spot, the ingestion pathway area factor is 1000—a consequence of the fact that as the size of the contaminated area is reduced from 100 m² to 0.1 m² (reduced by factor of 1000), the hot-spot dose is similarly reduced by a factor of 1000, and therefore, the area factor is 1000. Therefore, conclude that the ingestion pathway is *not* hot-spot sensitive.

11.3.4 Conclusion: Hot-Spot Limits on Building Surfaces

The receptor dose impact from hot spots via the three building occupancy pathways is either directly related to total source term (e.g., inhalation and ingestion pathways), or a more complex relationship holds (external radiation pathway). For example, the hot-spot dose via the inhalation and ingestion pathways scales directly with the size of the contaminated area, which means that the greater the hot-spot source term, the greater the receptor dose.

It is worthwhile to compare the area factors obtained from the external radiation pathway, with those calculated for the inhalation and ingestion pathways. Recall that for the receptor located directly over a 0.1 m² hot spot, the external radiation pathway area factors ranged from 107 to 115 for RESRAD-BUILD and ranged 111 to 114 for MicroShield. The area factors for the other two pathways were 1000. The external radiation pathway is the most limiting of the pathways, and it is certainly hot-spot sensitive. Therefore, a conservative application of these results would be to use the building occupancy area factors obtained from the external radiation pathway.

11.4 BAYESIAN STATISTICAL APPROACH TO ASSESS HOT SPOTS

This section presents a statistical approach to address compliance for hot spots present in a survey unit, for hot spots detected *and not detected*. This methodology may be thought of as a more comprehensive evaluation of potential hot spots, recognizing that both the contaminant mean and overall distribution are important parameters for demonstrating that the cleanup has satisfied the release criteria. To implement this hot-spot evaluation, dose modeling must be performed to generate both the derived guideline for the average residual radioactivity level ($DCGL_W$), and also the derived guideline level for residual radioactivity that equates to the upper percentiles of receptor dose. Once the dose modeling effort provides the DCGLs for the upper percentiles, the respective percentile of the contaminant distribution can be compared to the upper concentration limits to assess compliance. For instance, just like the 50th percentile (median) should be less than the $DCGL_{50th}$ (i.e., $DCGL_W$), the 98th percentile of the distribution should be less than the $DCGL_{98th}$.

The benefit of the proposed statistical compliance approach is that hot spots are considered in the overall context of the contaminant distribution in a survey unit. While the MARSSIM describes an approach for separately demonstrating compliance with both the mean contaminant concentration and elevated areas (hot spots), the approach described herein provides an integrated contaminant distribution concept where compliance is demonstrated for the contaminant distribution as a whole. The value of this approach is twofold. First, it addresses the issue of how to handle hot spots that may exist in the survey unit, but have not been found (which can be significant in terms of number and dose impact). Second, it conveniently evaluates multiple hot spots because they are accounted for in the overall contaminant distribution that is being assessed for compliance with the release criteria.

Plainly, hot spots are among the highest radiological concentrations from the distribution of survey unit concentrations. They are part of the true contamination distribution, and as such, occupy the right-hand tail of the distribution. Imagine an unrealistically high sampling density— for example, collecting a soil sample on a 0.3-m grid in a 1000 m^2 survey unit. This would produce more than 11,000 systematic soil samples and would reveal virtually all of the hot spots present in the survey unit. Rank ordering the concentration data from this mammoth data set and producing a histogram would quickly reveal the 90th, 95th, and 99th percentiles of the contaminant distribution. These upper percentiles could then be compared to regulatory limits for these upper percentiles (e.g., $DCGL_{95th}$, $DCGL_{98th}$, and $DCGL_{99th}$) to assess whether the upper tail of the distribution satisfied release criteria.

The development of an upper percentile DCGL (e.g., the $DCGL_{99th}$) should consider the dose modeling approach used to establish the hot-spot area factors described in Sections 11.2 and 11.3. The key is recognizing that at the time of the FSS, the upper percentile concentrations are, by definition, hot spots— assuming that hot spots are indeed present in the survey unit. Further, for the concentration to be considered an extreme value concentration (e.g., 99th percentile), it necessarily has to be associated with a *relatively* small area (e.g., 0.1 or 0.5 m²) as compared to the survey unit area. Consequently, for a survey unit area of 1000 m² a reasonably small hot-spot size might be 0.5 m² or less. The $DCGL_{99th}$ might be defined as the concentration equal to the $DCGL_W$ times the area factor for an area between 0.1 and 0.5 m².

For example, if the $DCGL_W$ for Cs-137 is 10 pCi/g, and the area factor for a 0.1 m² hot spot is 93.1 (Table 11.4), then the $DCGL_{99th}$ for Cs-137 is 930 pCi/g. So hot-spot compliance would be achieved by demonstrating that the 99th percentile of the distribution is less than the $DCGL_{99th}$ (930 pCi/g). Similarly, the $DCGL_{99th}$ might be defined based on a somewhat larger hot-spot area of 0.5 m². In this case, Table 11.4 indicates that the Cs-137 area factor associated with a 0.5 m² hot spot is 19.8, yielding a $DCGL_{99th}$ of 198 pCi/g. Thus, the determination of the $DCGL_{99th}$ depends on the assumed size of the hot-spot area for the 99th percentile concentration.

Challenges with this approach include the uncertainty in the determination of the upper percentile DCGL (as shown above) and the accuracy of estimates of percentiles from the upper tail of the contaminant distribution. Concerning the latter, a Bayesian statistical approach can be used to construct a posterior distribution of the contaminant concentration in a survey unit. The posterior distribution uses the sampling data generated during the FSS to produce an estimate of the desired upper percentile along with a confidence interval that reflects the uncertainty in the estimate.

11.4.1 Bayesian Statistical Approach

A Bayesian statistical approach is considered for describing the contaminant distribution in a survey unit, including hot spots. This statistical analysis combines current experimental information (data) in the form of a likelihood function with a distribution which summarizes the existing knowledge on the subject known as the prior distribution (refer to Section 12.4 for background on Bayesian statistics). To be precise, the likelihood function is a probability distribution which describes the probability of obtaining the data observed in the experiment as a function of the several unknown model parameters, for example, a mean and standard deviation. The prior distribution combines the existing subject matter knowledge into a distribution which describes the uncertainty in these model parameters. Bayes' Theorem explains how the current data, vis-à-vis the likelihood, updates the prior distribution to form the posterior distribution of the parameters.

This posterior distribution of the parameters summarizes our current state of knowledge about the probability distribution of the data.

A posterior distribution can be derived from survey data and used to make inferences on arbitrary percentiles of the contaminant distribution—for example, such as the 99th percentile. This provides a framework for the assessment of whether the hot spots comply with the release criteria. The posterior distribution is obtained from a prior distribution and likelihood function based on sampling data. The prior distribution in some cases might be considered an informative prior, which usually means the prior distribution is normal or lognormal based on the expected data. Conversely, a noninformative prior is often called a "flat prior." Flat priors have relatively minimal influence on the posterior distribution—that is, flat prior distributions add little to the contributions made by the survey data. The most popular noninformative prior is the uniform distribution.

The likelihood function based on sampling data is usually assumed to be normal (or lognormal) with some frequency of hot spots that results in a right-skewed distribution. Thus, after performing a final survey sampling campaign that includes both random sampling and judgmental sampling for hot spots, the expected result is a normal or lognormal underlying distribution, with a number of hot spots characterizing the upper tail of the distribution. This concept of viewing hot spots as part of the overall contaminant distribution provides a more comprehensive assessment of future receptor dose because the upper percentiles of the contaminant distribution include the contribution from hot spots.

The Bayesian approach to this problem allows both theoretical and practical modifications which provide an attractive method to evaluate upper percentiles of the distribution. Theoretically, treating an upper percentile (e.g., 99th) as a random quantity, we may derive its posterior distribution which provides both an estimate of this quantity and its uncertainty. In particular, we may understand the sensitivity of this quantity to the existence of hot spots in the data. From a practical point of view, the Bayesian methodology allows us to propose a different standard deviation for each observation in the data set, which allows us to handle the additional hot spots as discussed below.

In most problems, the posterior distribution is not available in a closed form, that is, it cannot be expressed analytically in terms of functions and mathematical operations. Rather, the resulting integrals are usually impossible to solve analytically, or difficult using standard techniques for numerical integration. Markov chain Monte Carlo (MCMC) algorithms are attractive solutions for the calculation of the posterior density. The use of MCMC techniques emerged in the late 1980s as the core of Bayesian computing, and it has since then revolutionized the field (Marin and Robert 2007). One common MCMC algorithm is the Gibbs sampler. This sampler can be used to estimate the posterior contaminant distribution—in particular, the upper percentiles of the distribution.

11.4.2 Bayesian Hot-Spot Assessment Using Robust *t* Distribution

A posterior distribution that incorporates hot spots must necessarily have thicker (wider) tails than the normal distribution. The *t* distribution was used to estimate parameters of the posterior distribution, such as the mean and upper percentiles. Albert (2007, p. 212) notes that "when there is a possibility of outliers, a good strategy assumes the observations are distributed from a population with tails that are heavier than the normal form." The possibility of outliers certainly holds for the situation of potential hot spots in a class 1 survey unit. The *t* distribution with small degrees of freedom is the most widely used heavy-tailed distribution and is easily adapted to Bayesian inference as described below (Albert 2007). In addition, as the degrees of freedom grow, the distribution converges to the normal distribution. Therefore, if outliers do not occur, the approach described below will mimic a conventional analysis based on the normal distribution. Assuming that the data are sampled from a *t* distribution with location μ and scale parameter σ, and known degrees of freedom v,

$$f(y|\mu,\sigma) = \prod_{i=1}^{n} \frac{1}{\sigma}\left(1 + \frac{(y_i - \mu)^2}{\sigma^2}\right)^{-(v+1)/2}$$

and a noninformative prior distribution (e.g., uniform distribution),

$$\pi(\mu,\sigma) \propto \left(\frac{1}{\sigma}\right)$$

then the posterior distribution is simply proportional to the product of these (Albert 2007):

$$\pi(\mu,\sigma|y) \propto \left(\frac{1}{\sigma}\right)\prod_{i=1}^{n} \frac{1}{\sigma}\left(1 + \frac{(y_i - \mu)^2}{\sigma^2}\right)^{-(v+1)/2}$$

Albert observes that the posterior can be derived as a mixture of normal distributions:

$$y|\lambda,\mu,\sigma \sim N\left(\mu,\frac{\sigma}{\lambda}\right), \quad \lambda|\mu,\sigma \sim \text{gamma}(v/2, v/2), \quad (\mu,\sigma) \propto (1/\sigma)$$

That is, by mixing together data from various normal distributions with differing scales, which are a function of a gamma distribution, the observed data follow Student's t. This can be verified by composing the product of the above distributions, the joint distribution, and integrating with respect to λ. The prior distribution of the mean and variance are represented by a noninformative flat prior. As mentioned, the noninformative flat prior is frequently used in situations where one has no prior information about the parameters at hand, or does not want to introduce any additional subjective information when consensus on any prior information does not exist.

In the case under consideration, it is possible that valuable prior information exists (e.g., from HSA, characterization, and remedial action support data). How such information could be conveniently integrated into the analysis is highly dependent upon the nature of the information and it is likely that the model introduced above would not be an appropriate choice to analyze such data. There are combinations of informative priors for μ and σ that could be integrated into the above analysis—for example, given σ, μ could be assumed to be normally distributed with some prior mean and scale. The difficulty would then become how to map the prior information which is usually understood intuitively into a detailed parametric form.

The joint distribution defined above can be factorized into the conditional distribution of each unknown parameter given the others by considering only terms that depend upon the random term. This results in the following set of distributions,

$$\lambda_i|y,\mu,\sigma \sim \text{gamma}\left(\frac{v+1}{2}, \frac{(y_i-\mu)^2}{2\sigma^2} + \frac{v}{2}\right)$$

$$\mu|y,\{\lambda_{i=1}^n\},\sigma \sim N\left(\frac{\sum_{i=1}^n \lambda_i y_i}{\sum_{i=1}^n \lambda_i}, \frac{\sigma}{\sqrt{\sum_{i=1}^n \lambda_i}}\right)$$

$$\sigma \sim \text{inv}-\text{gamma}\left(\frac{n}{2}, \frac{\sum_{i=1}^n \lambda_i(y_i-\mu)^2}{2}\right)$$

Gibbs sampling is simply the process of sequentially generating random variables from each conditional distribution in order. At each step in the sequence, the simulation generated is used as the conditional value of the fixed parameter in the next conditional distribution. For example, once the set of λ values is simulated from the gamma distribution, they along with a previous value of σ are plugged into the normal distribution and used to generate μ. The values of λ and μ are then used as inputs into the inverse gamma function and a new value of σ is generated. By considering only the joint samples of (μ, σ) and ignoring the set of λ at each iteration, the resultant sample will follow the robust t-distribution (Albert 2007). A Gibbs sampler coded in R programming language was used to assess the posterior distribution using the *robustt()* function of the LearnBayes package of R. (Refer to Abelquist 2008 for additional details on the Bayesian analysis.)

The following example illustrates a realistic FSS data set from a class 1 soil survey unit. The data set represents 16 soil sample results analyzed for Th-232. The FSS data shown in Table 11.30 ranged from 0.41 to 5.8 pCi/g, with two concentrations greater than the $DCGL_W$ of 1.5 pCi/g.

A log transform was applied to the data prior to performing the robust t analysis—a standard technique performed prior to conducting linear regression analysis. The Bayesian posterior analysis results are shown in Table 11.31.

The robust t posterior distribution exhibits the same mean and median value (0.96 pCi/g) because it is fit to a t distribution (mean and median are equal). Of course, we are most interested in the upper

TABLE 11.30   Th-232 Soil Concentrations from Class 1 FSS Survey

Th-232 Concentrations (pCi/g)							
0.68	0.98	0.76	1.32	0.94	0.58	0.73	1.30
0.86	0.52	0.41	0.76	0.60	4.95	5.8	0.85

TABLE 11.31   Robust t Posterior Distribution for Th-232 FSS Data

Statistic	FSS Data	Posterior Distribution
Mean	0.98	0.96
Median	0.80	0.96
90th percentile	2.56	2.53
95th percentile	5.15	3.35
99th percentile	5.66	5.79

percentiles of the distribution. The FSS data had a 99th percentile value of 5.66 pCi/g, while the 99th percentile of the posterior distribution was 5.79 pCi/g. In this example, the posterior distribution was only slightly greater than the FSS data at the 99th percentile. According to the proposed methodology, hot-spot compliance would be achieved by demonstrating that the 99th percentile of the distribution (5.79 pCi/g) is less than the $DCGL_{99th}$.

It is important to highlight that the FSS sample data have a substantial impact on the development of the posterior distribution. In particular, the method assumes that the sample data are representatively proportional to the true contaminant concentration distribution at the site so that if hot spots exist, they are proportionally reflected in the sample. If this is the case, the posterior distribution will accurately reflect the estimate and its uncertainty. In this process, it is not necessary to find 100% of the hot spots present; however, the larger an unbiased sample of data is collected the more accurate the estimates become. This point illustrates the relationship between the proposed statistical approach and the scan performance to identify hot spots. Ultimately, if "too many" of the hot spots are not detected—and obviously not included in the sample—then the estimate of the upper percentiles of posterior distribution will be inaccurate. Alternatively, if a disproportionate number of hot spots are collected through the systematic scan, the resulting posterior estimates may be biased high leading to conservative conclusions.

CASE STUDY   ASSESSMENT OF SITE WITH MULTIPLE HOT SPOTS

To wrap-up this section, the Shelwell Services site provides a case study concerning a situation with multiple identified and unidentified hot spots. The Nuclear Regulatory Commission and Oak Ridge Institute for Science and Education (ORISE) inspected the site in 1997 and identified over 200 separate locations of elevated radiation caused by very small discrete particles containing Cs-137.* On the basis of the survey results, the average Cs-137 concentration in soil was estimated as 2.5 pCi/g, which compared favorably to the guideline of 15 pCi/g. However, the results also indicated that additional, unidentified hot-spot particles remained in the soil.

NRC staff performed a risk assessment to determine if the site could be released for unrestricted use with hot spots remaining in the soil. To assess the risk to the public at the Shelwell site, the individual Cs-137 particles

---

* Shelwell Services, Inc. Risk Assessment, L. Joseph Callan, executive director for operations, SECY-98-117, May 27, 1998.

were evaluated. Some soil samples exhibited concentrations exceeding 1000 pCi/g, with individual particles having activity levels as high as 6 μCi. The risk was estimated by determining the annual expectation dose, calculated by multiplying the probability of inhaling or ingesting a particle in a given year by the dose from the particle. The maximum total risk was estimated to be 1E–2 mrem/y. The licensee was required to remediate identified hot spots that exceeded guideline levels. This case study illustrates the importance of assessing the entire contaminant distribution to effectively evaluate the risk involved in releasing a site.

11.4.3 Demonstrating Compliance with the Hot-Spot Limits

While the MARSSIM describes an approach for separately demonstrating compliance with both the mean contaminant concentration and elevated areas (hot spots), the method proposed here evaluates the contaminant distribution in an integrated fashion. That is, both the mean and upper percentiles of the contaminant distribution are compared to the $DCGL_W$ and upper percentile DCGLs to demonstrate compliance. Besides being able to statistically assess the effect of hot spots that may exist in the survey unit, this approach can conveniently evaluate multiple hot spots (since they are accounted for in the overall contaminant distribution).

This approach recognizes the connection between the mean and upper tail of the contaminant distribution in the survey unit, and uses a compliance test that compares the upper tail (e.g., 99th percentile) to the $DCGL_{99th}$. That is, hot spots are considered a further continuum of the contaminant distribution—that is, the upper tail of the distribution. Under this approach, the hot spots identified, as well as those not identified, are considered in the compliance demonstration.

This approach also addresses the issue of multiple hot spots. That is, the contaminant distribution that results from the Bayesian analysis accounts for multiple hot spots in the survey unit. For example, the range of contaminant concentrations that exist between the 98th percentile and the 100th percentile are likely to be defined as hot spots in a class 1 survey unit, depending on the value of the $DCGL_W$. Therefore, these "multiple" hot spots (whether they are identified or not) are handled during the assessment of the overall contaminant distribution. Again, it is important to recognize that the degree to which the posterior distribution represents the true (but unknown) contaminant distribution depends on the scan performance and inclusion of hot-spot samples (if present) into the data sample. Therefore, to the degree that the sample is representative, the posterior t distribution accounts for hot spots that may exist in the survey unit, but have not been identified. This is appealing from the perspective of regulatory compliance. Assuming that all the hot spots have been identified, the upper percentiles of the actual data would not be questioned. However, finding all of the hot spots is seldom the case, even with 100% scanning. Rather, the more appropriate question is how many hot spots have been missed and remain in the survey unit.

If the Bayesian analysis indicates that the 99th percentile exceeds the $DCGL_{99th}$, then additional investigation (sampling) is warranted to determine if noncompliant hot spots are truly present. The DQO process certainly helps to provide a framework for developing a sampling campaign that appropriately balances the risk of not finding potential hot spots with survey costs.

Performing another 100% scan of the survey unit should be considered; the initial scan may not have been conducted with sufficient quality. A graded approach may be to perform higher density systematic sampling in the vicinity of the higher radionuclide concentrations resulting from the FSS. A last resort might be to increase sampling densities across the entire survey unit, an expensive option for sure. Again, these additional investigations are only necessary when the Bayesian analysis predicts that significant hot spots exist in the survey unit. Finally, hot spots that are identified should be considered for remediation as part of an ALARA assessment.

11.4.4 Conclusions for Bayesian Statistical Approach

The ultimate goal of the FSS is to demonstrate that the contaminant concentration in the survey unit meets the release criterion (e.g., 25 mrem/y). The MARSSIM FSS design specifies that two aspects of the contaminant distribution must be assessed to demonstrate compliance with the release criteria—that is, the mean and the upper tail of the distribution (hot spots). Indeed, the FSS design specifies random or random-systematic sampling in the survey unit to determine the mean concentration, and radiation scanning and judgmental sampling to identify and assess any hot spots present in the survey uniform.

The proposed approach for hot-spot assessment is to recognize the connection between the average and upper tail of the contaminant distribution in the survey unit, and to use a compliance test that compares the upper percentiles of the contaminant distribution to the DCGLs for the upper percentiles. Specifically, a Bayesian statistical approach is employed to determine the posterior distribution, particularly the 99th percentile of the data distribution, and then compare the 99th percentile with the $DCGL_{99th}$. In fact, the Bayesian approach can be used for (1) checking that the posterior mean is less than the $DCGL_W$, and (2) that the posterior 99th percentile is less than the $DCGL_{99th}$. The proposed approach would improve the MARSSIM hot-spot assessment approach by providing a comprehensive compliance methodology that considers hot spots that may be present, but not found.

One final thought. The development of the DCGLs for upper percentiles might be based on dose modeling, rather than using an area factor based on an assumed hot-spot area ($0.1–0.5$ m^2) at the 99th percentile. Dose modeling codes usually treat the source term as a uniform distribution. That is, a unit radionuclide concentration is input in the modeling code, and the concentration that yields the release criterion is calculated—called the $DCGL_W$. Replacing the uniform distribution with a lognormal distribution for the source term (with unit mean and standard deviation estimated from the posterior distribution), a probabilistic dose modeling approach can be used to determine the radionuclide concentrations that equate to the release criterion for all percentiles of interest, for example, including the 50th, 90th, 95th, and 99th. These upper percentiles for the lognormal distribution that satisfy the release criterion can be used to test the upper percentiles obtained from the posterior distribution.

Admittedly, the proposed approach exhibits some weaknesses, albeit different than the shortcomings of the MARSSIM approach. It is worthwhile to point out that the scan data collected in the survey unit represents a rich source of information on the contaminant distribution and has yet to be incorporated directly into statistical assessments. A major goal of future work in this area would include integrating this information with the random soil sample data. This would allow the development of a much more accurate predictive distribution for the contaminant source term that might be probabilistically sampled to generate predicted receptor doses.

QUESTIONS AND PROBLEMS

1. Name two issues related to the current approach for deriving hot-spot limits.

2. Briefly summarize what is meant by pathways and parameters being identified "hot-spot sensitive."

3. One proposal to generate more realistic hot-spot limits is to assume that the receptor is some distance from the hot-spot location. How can this issue be resolved by generating a distribution of distances (l), and using this variable to calculate a distribution of doses that result when considering that the receptor will usually be located at varying distances from the hot spot?

4. The Bayesian approach was used for assessing hot-spot compliance. Consider a FSS data set that represents 45 soil samples analyzed for Cs-137. The FSS data are skewed to the right. The mean and standard deviation of the data are 9.0 and 17.2 pCi/g, respectively. A comparison of the robust t posterior distribution and FSS data are also shown below. Based on the posterior distribution, provide an assessment of the hot-spot potential in this survey unit. Is additional sampling for hot spots warranted?

Cs-137 Soil Concentrations in pCi/g									
14.7	6.88	17.0	47.8	5.65	0.008	0.008	0.051	0.056	0.044
0.189	0.065	0.99	18.5	3.72	1.8	0.18	6.1	72.7	0.20
2.98	20.0	10.3	3.22	0.41	12.9	0.77	5.52	1.57	3.65
1.85	4.66	2.71	80.9	3.49	3.20	1.82	22.8	11.1	6.57
7.17	0.48	0.42	0.071	0.058					

Statistic	FSS Data	Posterior Distribution
Mean	9.00	2.35
95th	42.80	50.74
99th	77.29	149.7

CHAPTER 12

Statistics and Hypothesis Testing

S TATISTICS AND SPECIFICALLY HYPOTHESIS testing are necessary tools for any decommissioning professional responsible for planning surveys and making decisions based on survey results. Statistics can be generally divided into (1) descriptive statistics and (2) inferential statistics. Descriptive statistics involve description of data sets by their central tendency (location) and variability (dispersion)—for example, mean, median, range, percentile, variance, standard deviation, and skewness. The MARSSIM DQA emphasizes the use of diagnostic tools such as histograms, posting plots, and quantile–quantile plots to better interpret the data. Inferential statistics consists of making informed decisions about various aspects of a set of values. Common inferential statistics include hypothesis testing and confidence interval testing (or estimation). Indeed, NUREG/CR-5849 recommends confidence interval testing for demonstrating compliance with release criteria, while the MARSSIM is based on hypothesis testing.

Very simply, statistics is the theory of information. It deals with the collection, analysis, interpretation, and reporting of numerical data. Two important terms that always come up in any discussion of statistics are "population" and "sample." A population contains all possible observations of a particular type; it is the largest collection of values (of a variable) for which there is an interest. A sample is any subset of measurements selected from the population—it is a limited number of observations collected randomly from a population.

12.1 BASIC POPULATION STATISTICS AND CONFIDENCE INTERVAL TESTING

This section reviews some of the basic statistical techniques that every decommissioning health physicist should be well versed—from the ability to determine the sample mean and standard deviation from a list of smear results, to being able to calculate the 95% upper confidence level from a set of Sr-90 soil concentrations.

12.1.1 Basic Statistics

We typically use sample statistics to infer population parameters. For example, the true population mean (μ) can be estimated by the sample mean ($\bar{x}$), and the population variance (σ^2) by the sample variance (s^2). The sample mean ($\bar{x}$) of n measurements is calculated as

$$\bar{x} = \frac{1}{n} \sum_{i=1}^{n} x_i$$

and the sample standard deviation (s) is calculated as

$$s = \sqrt{\frac{\sum_{i=1}^{n} (x_i - \bar{x})^2}{n-1}}$$

The divisor ($n - 1$) is used so that there is an unbiased estimator of the population variance. (One degree of freedom is used to determine sample mean.)

The contaminant standard deviation is an important parameter in the determination of the MARSSIM final status survey sample size. One MARSSIM strategy is to take actions to reduce the standard deviation because it reduces sample size. To do this, one must recognize that there are two components of the standard deviation—spatial variability and measurement uncertainty.

The spatial variability can be thought of as the underlying variability of a contaminant over a survey area. By selecting survey units that are uniform in their contamination levels, the spatial variability can be reduced. To emphasize this point, taking more samples does not necessarily reduce the spatial variability; rather, it reduces the uncertainty in the estimate of the true variability. The measurement uncertainty is of course related to the measurement technique employed. The measurement uncertainty component of the standard deviation can be reduced by using more precise measurement methods, or to use a common health physics practice—simply count the sample longer.

Other descriptive statistics include the median and the mode. The median is the middle value of a set of measurements (ordered lowest to highest) when the number of data points is odd, and is the average of the two middle values when the number of data points is even. The median, unlike the mean, is not affected by a few extremely large or small values. That is, while the mean is affected by the magnitude of the largest value, for the median, the largest value can only be the largest value, regardless of its value. Note that the median is the population parameter actually being tested in nonparametric hypothesis tests used in MARSSIM. The median equals the mean for a normal distribution; when there is a significant number of higher concentration values in the survey unit, the mean may be skewed to the right as compared to the median. Finally, the mode of a set of measurements is the measurement that occurs most frequently—somewhat ironic in that the mode is perhaps the least frequently used descriptive statistic. And perhaps the most elegant definition of a "statistic"—a statistic is simply a function of the data (overheard from Dr. Carl Gogolak teaching statistics during a MARSSIM course).

The field of statistical inference deals with generalizations and predictions. One of the fundamental theorems that is frequently applied to statistical inference is the central limit theorem, and its relationship to the sampling distribution of the mean. Consider the distribution of a "statistic" (e.g., mean, variance, and range) as a random variable that may be described by a probability distribution. The probability distribution of a "statistic" is called a sampling distribution (Walpole and Myers 1985). Consider the following example of a sampling distribution of means. The sample is defined as the mean of five rolls of a single die. If this experiment (rolling the die) is conducted repeatedly, the resulting frequency distribution of these sample means will be approximately normally distributed. This fortunate circumstance is a result of the central limit theorem:

The sampling distribution of the mean ($\bar{x}$) is approximately normal with mean μ and variance given by σ^2/n, provided that $n > 30$ for non-normal populations (the size of n necessary for this distribution to be normal depends on the shape of the distribution).

The standard error of the mean ($\sigma_{\bar{x}}$) is given by

$$\sigma_{\bar{x}} = \frac{\sigma}{\sqrt{n}}$$

Then, by the central limit theorem

- $\mu_{\bar{x}} = \mu$; the mean of the $\bar{x}$'s equals the mean of the individual observations from the population

- Sample means are approximately normally distributed (regardless of the shape of the parent population)

- Importantly, the central limit theorem allows us to use confidence interval estimates for populations that are not normal

Consider a normal population with $\mu = 100$ and $\sigma = 20$. For a sample size of $n = 16$, what is the probability that this sample will have a mean between 90 and 110? Thus, we need to find $P(90 < \bar{x} < 110) = ?$

z-Scores correspond to the measured value, x, and represents the number of standard deviations that x is from the population mean. We use z-scores to transform the given data, which in this example we are concerned with the distribution of sample means

$$z = \frac{x - \mu}{\sigma} = \frac{\bar{x} - \mu_{\bar{x}}}{\sigma_{\bar{x}}}$$

and because $\mu_{\bar{x}} = \mu$ and $\sigma_{\bar{x}} = \sigma/\sqrt{n}$

$$z = \frac{\bar{x} - \mu}{\sigma/\sqrt{n}}$$

Substituting the given data to calculate the z-scores

$$z_1 = \frac{90 - 100}{20/\sqrt{16}} = -2; \quad z_2 = \frac{110 - 100}{20/\sqrt{16}} = 2$$

and the probability corresponding to the area under the normal curve for these z-scores is 0.954.

So, $P(90 < \bar{x} < 110) = 95.4\%$.

Thus, z-scores provide a way of comparing normal curves by converting both curves to standardized normal curves, and allowing one to use z-score tables and probabilities.

12.1.2 Confidence Interval Testing

The confidence interval for the mean is commonly used in decommissioning applications. For example, to demonstrate compliance using NUREG/CR-5849, the upper 95% confidence interval for a particular radiological parameter must be less than the appropriate guideline. That is, not only must the mean radionuclide concentration in the survey unit be less than the guideline, but so must the 95% upper confidence bound of the mean. For those cases where the mean was less than the guideline, but the 95% upper bound was not, NUREG/CR-5849 recommended that one consider collecting additional samples to reduce the standard error of the mean ($\sigma_{\bar{x}}$). The reduction of the standard error of the mean may be sufficient to push the survey unit into compliance.

The confidence interval for the mean can be determined, using either the standard normal distribution or the t distribution (distributions are discussed in the next section).

Case 1. The $100(1 - \alpha)$% confidence interval for the mean (μ), when σ is known, or if σ unknown, and $n > 30$, uses s in place of σ

$$\bar{x} \pm z_{\alpha/2} \frac{\sigma}{\sqrt{n}}$$

where $z_{\alpha/2}$ is the z value leaving an area of $\alpha/2$ in each tail (Walpole and Myers 1985).

Case 2. The $100(1 - \alpha)$% confidence interval for the mean (μ), when σ is unknown and sample size $n < 30$, and assuming an approximate normal distribution

$$\bar{x} \pm t_{\alpha/2,\text{df}} \frac{s}{\sqrt{n}}$$

where $t_{\alpha/2}$ is the t value with $n - 1$ degrees of freedom (df), leaving an area of $\alpha/2$ in each tail (Walpole and Myers 1985).

The $100(1 - \alpha)$% is the confidence level that the true, but unknown, mean exists between the lower and upper bounds. In most decommissioning applications, the standard deviation is unknown, so, depending on the number of samples obtained, the confidence level is determined using either the standard normal or t distribution.

12.2 DATA DISTRIBUTIONS

The following sections briefly review some of the common distributions that are important to health physicists. These distributions include the binomial, Poisson, normal, and Student's t test; the first two distributions are often used to describe the fundamental process of radioactive decay, while the latter two are often used in evaluations of data sets via hypothesis testing.

12.2.1 Binomial Distribution

The binomial is an important discrete distribution that describes experiments that have only two possible outcomes—labeled as "success" or "failure." The probability of success is given by p. A common binomial experiment is the tossing of a coin, where the only two possible outcomes are heads or tails, and the probability of success (let us say getting "heads") is 0.5. When this experiment is repeated n times, and each outcome is independent of the next, the random variable x is defined as the number of successes observed in n independent trials. The binomial distribution of x is written as

$$P(x; n, p) = \binom{n}{p} p^x (1-p)^{n-x} = \frac{n!}{x!(n-x)!} p^x (1-p)^{n-x}$$

Remember, the binomial distribution has the following requirements:

- The experiment consists of n repeated trials
- Each trial results in success or failure
- The probability of success remains constant from trial to trial (items are replaced, if not from an infinite population)
- Repeated trials are independent

Radioactive decay is fundamentally a binomial process—a radioactive atom either decays (success) or does not decay (failure) in a specified time interval. The probability of success can be written as

$$p = 1 - e^{-\lambda t}$$

Consider the following example: given three radioactive atoms, determine the probability that an atom produces a count in $T = 10$ min, and then the probability of getting two counts. Thus, the probability of success is now defined as getting a count—so it must first decay and then the emitted radiation must be counted. Given that $T_{1/2} = 5$ min ($\lambda = 0.139$ min^{-1}) and the total efficiency is 0.5 c/dis.

$$p = (1 - e^{-0.139\,\text{min}^{-1}\times10\,\text{min}}) \times 0.5\,\text{c/dis} = 0.3755$$

The probability of getting two counts is determined from the binomial distribution of x; specifically, we calculate $P(x; n, p)$ for $x = 2$, $n = 3$, and $p = 0.3755$:

$$P(2;3,0.3755) = \frac{3!}{2!(3-2)!}0.3755^2(1-0.3755)^{3-2} = 0.264$$

This process can be performed for each possible outcome of x ($= 0, 1, 2,$ and 3 counts) to determine the respective probability of each:

Outcome (# Counts)	Probability
0	0.244
1	0.439
2	0.264
3	0.053
Total	1.000

Finally, the mean and variance of the binomial distribution are provided:

$$\text{Mean } (\mu)\text{: } \mu = np \quad \text{Variance}(\sigma^2)\text{: } \sigma^2 = np(1-p)$$

So, the mean number of counts expected above is $n \times p$ or $(3) \times (0.3755) = 1.13$ counts, and the standard deviation is 0.84 counts. This is consistent with the probability distribution shown above, which is clearly centered about $x = 1$. Note that the binomial distribution is symmetric when $p = 0.50$.

In radioactive decay, where the number of radioactive atoms, n, is frequently very large, the binomial is computationally cumbersome. The Poisson distribution is used as a mathematical simplification of the binomial, provided that the probability of success, p, is small. Note: For counting situations characterized by a very short-lived radionuclide with high counting efficiency, the probability of success is not small and the binomial must be used (Knoll 2010).

Another use of the binomial distribution is in the area of acceptance sampling. Acceptance sampling is traditionally used to decide whether or not manufactured parts are acceptable based on measurements of some attribute in the lot. Could there be an application of acceptance sampling to the clearance of materials (Chapter 15)? The objective of acceptance sampling is to estimate the characteristics of a lot; perhaps this technique can be used to estimate the surface area that exhibits contamination levels in excess of the release criterion.

Acceptance sampling is used to "accept" or "reject" materials in terms of lots, each consisting of N parts. In the clearance survey context, a "part" may be a certain 100 cm^2 area of the material surface in the lot that is subjected to a direct measurement of surface activity. Using this definition of "part," the number of parts in a lot (N) can be quite large. For example, if the estimated total material surface area in a lot is 100 m^2, then we would have 10,000 "parts" or 100 cm^2 areas. Some of these parts may be defective, which perhaps can be defined as the level of residual activity in excess of release criteria, or possibly some fraction of the criteria.

Assume that p represents the fraction defective in the lot—therefore, if p is small, the materials are released, while a large p indicates that the lot is disposed of as radioactive waste or decontaminated. Since we know neither p nor the number of defective parts in a lot (i.e., given by $N \times p$), an estimate of p is needed for the decision to accept or to reject the lot, and ensure a predetermined quality level—this estimate would likely come from process history and previous clearance survey results. The regulator would also provide input on an acceptable value of p that would allow the lot to be released.

The probability of accepting the lot is called the operating characteristic (OC) curve, and is a function of p. The true probability of accepting the lot is given by the hypergeometric probability; however, as long as the lot (N) is large compared to the sample size (n), for example $n < N/10$, the binomial probability provides an excellent approximation to the hypergeometric. The sample size (n) is the number of surface activity measurements performed in the lot.

Applying the binomial distribution, x is the number of defectives in the sample (n), and p is the probability of "success," which in this context is the probability that a randomly selected item is defective—exceeds the release criteria. Thus

$$OC(p) = \sum_{x=0}^{A_c} \frac{n!}{x!(n-x)!} p^x (1-p)^{n-x}$$

where A_c is the acceptance number. If the sample yields more than A_c defective items, the entire lot is rejected.

12.2.2 Poisson Distribution

The Poisson is another important discrete distribution that is applicable for computing probabilities of events in time and space. The probability distribution of the Poisson random variable, x, is

$$P(x;\mu) = \frac{e^{-\mu} \mu^x}{x!}$$

where μ is the average number of outcomes occurring in the given time interval or specified region (Walpole and Myers 1985). Note that μ is the constant value of $n \times p$ when used as the approximation to the binomial.

The Poisson distribution has the following requirements applicable to both time intervals and regions of space (Walpole and Myers 1985):

- The number of outcomes (x) in one time interval is independent of the number that occurs in any other equivalent time interval.

- The probability (p) that a single outcome will occur during a very short time interval is proportional to the length of the time interval, and does not depend on the number of outcomes occurring outside this time interval; that is, p is constant from one small interval to another.

- The probability (p) that more than one outcome will occur in such a short time interval is negligible; that is, p is small.

As mentioned previously, the Poisson distribution can be used as an approximation to the binomial (which is the proper distribution describing radioactive decay), provided that $p \ll 1$ and np is a constant, equal to μ. Many texts state that if $n \geq 20$ and $p < 0.05$, the Poisson approximation is acceptable, and that it is very good if $n > 100$ and $np < 10$ (NRC 1988).

Thus, we only need to know the mean of a distribution (μ) to determine $P(x; \mu)$ at all other values of x. Further, for the Poisson distribution, the mean and variance are equal:

$$\mu = \sigma^2 = np$$

Note that the variance obtained earlier for the binomial distribution [$\sigma^2 = np(1 - p)$] reduces to the above result in the limit of $p \ll 1$. As most health physicists will attest, this result makes the determination of the standard deviation for a number of radiation counts rather convenient—the standard deviation is simply the square root of the number of counts ($\sigma = \sqrt{c}$). This identity is useful when propagating errors, which is necessary when calculating the standard deviation in any derived quantity related to counts, such as a count rate or an average value (Knoll 2010).

> On the basis of the Poisson identity, the standard deviation of a count is the square root of the count—*but*, this technique only works for the directly measured number of counts, the association does not apply to counting rates, averages of independent counts, or any other derived quantity.

12.2.3 Normal Distribution

The normal distribution (also called a Gaussian distribution) is unquestionably the most commonly used distribution in statistical applications, especially inferential statistics. The normal is a continuous distribution, with the probability density function of random variable x given by

$$n(x; \mu, \sigma) = \frac{1}{\sigma\sqrt{2\pi}} e^{-\frac{1}{2}\left(\frac{x-\mu}{\sigma}\right)^2} \quad \text{for } -\infty < x < \infty$$

Thus, once the mean and variance are specified, the normal curve is completely described.

The normal distribution has the following properties (Daniel 1977; Walpole and Myers 1985):

- The total area under the curve and above the horizontal axis is equal to 1.

- The distribution is symmetric about its mean (μ).

- The mean, median, and mode are all equal.

- The normal distribution is a "family" of distributions, since there is a unique distribution for each value of μ and σ.

A standard normal distribution—which has $\mu = 0$ and $\sigma^2 = 1$—is used to facilitate the determination of area under the normal curve. That is, given a normal curve with specified μ and σ, we transform the normal curve to a standard normal distribution by

$$z = \frac{x - \mu}{\sigma}$$

This "z-score" can now be looked up in standard statistical tables of the standard normal distribution. For example, many tables provide the area under the standard normal curve corresponding to $P(Z < z)$ for values of z ranging from −3.0 to 3.0. However, some tables only provide areas under the curve for positive values of z. Thus, determining the area under the curve—which represents the probability—depends on how the z-scores are presented in the statistical tables.

Finally, we will consider the normal approximation to the binomial. The normal distribution provides an accurate approximation to the binomial distribution when n is large and p is not extremely close to 0 or 1. A rule of thumb: The normal approximation to the binomial is appropriate when np and $n(1 - p)$ are both greater than five (Daniel 1977).

To use the normal approximation, we determine the z-scores for $\mu = np$ and $\sigma^2 = np(1 - p)$ and use the values of z to find the probabilities of interest. That is, the z-score is calculated by

$$z = \frac{x - np}{\sqrt{np(1 - p)}}$$

Earlier, we discussed that the Poisson distribution can be used as an approximation to the binomial (e.g., $p \ll 1$). Note that the normal approximation to the binomial does not require that the Poisson approximation also apply.

12.2.4 Student's t Distribution

The Student's t distribution (or just t distribution) is also a continuous distribution and it plays an important role in hypothesis testing. As discussed in the previous section, once the mean and variance are specified, the normal curve is completely described. However, the variance is usually unknown. For samples of sufficient size (e.g., $n > 30$), the sample variance (s^2) often provides a reasonable estimate of the population variance (σ^2), and it is appropriate to assume that the statistic z is distributed as a standard normal variable, from the central limit theorem:

$$z = \frac{\bar{x} - \mu}{s/\sqrt{n}}$$

If the sample size is small ($n < 30$), the variance can fluctuate considerably from sample to sample, and the random variable z is no longer normally distributed (Walpole and Myers 1985). Now, if the population is approximately normal and $n < 30$, we deal with the t distribution, which is defined as

$$t = \frac{\bar{x} - \mu}{s/\sqrt{n}}$$

with $v = n - 1$ degrees of freedom. The number of degrees of freedom is used for purposes of identifying the correct distribution of t to use, that is, this distribution is a function of sample size.

The Student's t distribution has the following properties (Johnson 1980):

- The distribution is symmetric about its mean (μ) of zero.

- The variance is greater than 1 and depends on the sample size n; as the sample size n increases, the variance approaches 1.

- The t distribution is less peaked at the mean compared to the normal distribution, and like the normal, it is a "family" of distributions—a separate distribution for each sample size.

- As n approaches infinity, the t distribution becomes the same as the normal distribution.

The probability that a random sample results in a value of the t statistic falling between any two specified values can be looked up in standard statistical tables of t distribution critical values. Usually, t_α represents the t value above which we find an area equal to α. For example, suppose that our sample size is 10, and we wish to find the critical values of t that correspond to $\alpha = 0.05$ (0.025 in each tail of the distribution). We look up the critical value of t for 9 degrees of freedom for $\alpha/2 = 0.025$—we find that $t_{0.975}$ equals 2.262. Since the t distribution is symmetric about the mean of zero, $t_{0.025}$ equals $-t_{0.975}$, or -2.262.

12.3 HYPOTHESIS TESTING

Hypothesis testing is an extremely valuable aspect of statistical inference. As stated in NUREG/CR-4604, a statistical hypothesis is a formal statement that specifies the assumed parameter values. Furthermore, a statistical hypothesis is an assertion or conjecture concerning one or more populations (Walpole and Myers 1985). Many types of decommissioning decisions can be formulated as hypotheses about populations or process parameters—for example, the determination of a survey unit's compliance with release criteria, evaluating the normality of a population, and even the framework for determining an instrument's MDC is based on hypothesis testing. Chapter 18 provides a number of decommissioning applications of hypothesis testing.

Hypothesis testing can be performed on both parametric and nonparametric distributions. The four distributions in the previous section are examples of parametric statistics, that is, specific parameter values define these distributions (e.g., mean and variance), and they are usually normally distributed. Nonparametric distributions have no particular distinguishing distribution; in fact, they are often called distribution-free. Their utility in hypothesis testing stems from the fact that if a set of survey data are not normally distributed, a condition for parametric statistical tests, the nonparametric statistics can be used to evaluate these data. Nonparametric statistics as described in MARSSIM provides a way of comparing a survey unit population to either a $DCGL_W$ (using Sign test) or an appropriate background reference area (using the WRS test). Both these nonparametric statistics are covered in detail in Chapter 13. For those desiring to read more about nonparametric statistics outside the specific MARSSIM application, Conover (1980) is a good start.

12.3.1 Hypothesis Testing Fundamentals and Examples

It is helpful to have an understanding of the big picture when it comes to setting up a statistical hypothesis test. Daniel (1977) provides a useful overview of statistical hypothesis testing as follows:

1. Statement of hypothesis

2. Selection of the significance level (type I error)

3. Description of the population of interest and statement of necessary assumptions

4. Selection of the relevant statistic

5. Specification of the test statistic and consideration of its distribution

6. Specification of rejection and acceptance regions

7. Collection of data (from sample size determination) and calculation of necessary statistics

8. Statistical decision

9. Conclusion

It would seem that even though "data quality objectives" may not have been a coined term a couple of decades ago, the concept of planning for the necessary data to test a null hypothesis was certainly in use.

Let us now consider an example to illustrate the concepts involved in hypothesis testing. Suppose that it is of interest to calculate the total uranium concentration in soil samples, for a site contaminated with 3% enriched uranium. (Note that the total uranium concentration refers to the sum of U-238, U-235, and U-234 concentrations in the sample.) Gamma spectroscopy will be used to quantify the U-238 and U-235—the U-234 concentration is calculated based on an assumed U-234 to U-235 ratio (R). The standard ratio for U-234 to U-235 is 22, based on natural isotopic uranium abundances. Let us assume that the accuracy of this ratio is being questioned by the regulator, who believes that the total uranium concentration calculation is in error because of this ratio. Your job is to use a hypothesis test to demonstrate to the regulator that an appropriate U-234 to U-235 ratio is being applied. The null hypothesis (H_0) concerning the population of U-234 to U-235 ratios (R) is stated as

$$H_0: R = 22$$

against the alternative hypothesis (H$_a$):

$$H_a: R \neq 22$$

In this example, the alternative hypothesis includes values of R that are either greater than or less than 22 (the alternative hypothesis is two-sided).

Note: If the regulator is only concerned when the U-234 concentration is being underestimated by U-235, then the alternative hypothesis might be written as

$$H_a: R > 22$$

indicating that if the alternative hypothesis is selected, the U-234 concentration is higher than that calculated by the licensee using a ratio of R equal to 22.

In the hypothesis testing framework, a decision is made based upon an evaluation of the data obtained from a random sample from the population of interest. If the data indicates that it is reasonable that the random sample could have come from a distribution described by the parameter values specified in H$_0$, the decision is made to continue operating as though the null hypothesis were true. Otherwise, the decision is to reject H$_0$ in favor of the alternative hypothesis.

The hypothesis testing approach requires that a test statistic be calculated from the sample data. This test statistic is then compared to a value from a statistical table (critical value), and used to make a decision about H$_0$. Returning to our enriched uranium example, suppose that 15 samples were selected for alpha spectroscopy analyses to determine the U-234 to U-235 ratio. The ratios generated from these analyses will be averaged to determine the mean ratio—our test statistic, R_{ave}. Let us assume that if our test statistic $R_{ave} \leq 20.5$ or $R_{ave} \geq 23.5$, we conclude that the U-234 to U-235 ratio is not 22, and we reject H$_0$. This decision rule was based on earlier studies on the variance in R and the type I error rate (it will be shown later how the type I error rate is related to the size of the acceptance region). By rejecting the null hypothesis, we conclude that it is false, and imply that the alternative hypothesis is true. The set of possible values of $R_{ave} \leq 20.5$ or $R_{ave} \geq 23.5$ is the critical region (or rejection region) of the test. Conversely, if $20.5 < R_{ave} < 23.5$ (the acceptance region), then we fail to reject (hence we accept) the null hypothesis.

Acceptance of a hypothesis merely implies that we have insufficient evidence to believe otherwise; it does not necessarily imply that it is true. It is more appropriate to report that we "fail to reject" the null hypothesis, rather than to report we have "accepted" the H_0. Further, a conclusion based on a rejected null hypothesis is more decisive than one based on an "accepted" null hypothesis. Thus, an investigator will often choose to state the hypothesis in a form that will hopefully be rejected (Walpole and Myers 1985).

The truth or falseness of a statistical hypothesis is never known with 100% certainty, unless we examine the entire population—which is usually impossible or at least impractical. The random sample that is used to base this decision is subject to error. There are two types of errors that can be made when hypothesis testing. Type I errors (also referred to as α) result from incorrect rejections of the H_0—the null hypothesis is rejected when it is in fact true. Type II errors (referred to as β) are incorrect nonrejections of H_0—the null hypothesis is not rejected when it is in fact false. Table 12.1 summarizes the possible situations in hypothesis testing.

Depending on how the null hypothesis is stated, the type I and type II errors usually have "stakeholders" valuing one error over the other. For example, the MARSSIM null hypothesis is stated "residual radioactivity in the survey unit exceeds the release criterion." The type I error is an incorrect rejection—meaning that the conclusion is that the survey unit passes, but in error. The type II error is an incorrect nonrejection—meaning that the conclusion is that the survey unit does not pass, when it really does. So what error is the regulator more concerned with? Of course it is the type I error, while the MARSSIM user is more concerned with the type II error.

Note that in hypothesis testing, the probability of committing a type I error, denoted by α, is commonly called the "significance level." The type I error reflects the amount of evidence the investigator would like to see before abandoning the null hypothesis—α is also called the "size"

TABLE 12.1 Hypothesis Test Outcomes

Decision	True Condition	
	H_0 is True	H_0 is False
Accept H_0	No error	Type II error
Reject H_0	Type I error	No error

of the test. The smaller the type I error, the greater the acceptance region. The probability of correctly rejecting a false H_0 is called the "power of the test" and is given by $1 - \beta$. The probability of committing a type II error is impossible to compute unless we have a specific alternative hypothesis (Walpole and Myers 1985).

Let us return to our example to further illustrate these concepts. We have previously defined our acceptance and rejection regions—$R_{ave} \leq 20.5$ or $R_{ave} \geq 23.5$ is the critical region, and $20.5 < R_{ave} < 23.5$ is the acceptance region (Figure 12.1). Considering these decision criteria, we can calculate the type I and type II errors when performing the hypothesis test for the U-234/U-235 ratio. (Note that we could have selected the type I and type II errors, and calculated the acceptance region instead of calculating the decision errors—the process works both ways.)

We assume that we know the population distribution well and that the standard deviation (σ) of the population of ratios (R) is 3.2. From our sample size of 15, and assuming that the sampling distribution of R_{ave} is approximately normally distributed, we can determine the standard error in the mean ($\sigma_{R_{ave}}$):

$$\sigma_{R_{ave}} = \frac{\sigma}{\sqrt{n}} = \frac{3.2}{\sqrt{15}} = 0.83$$

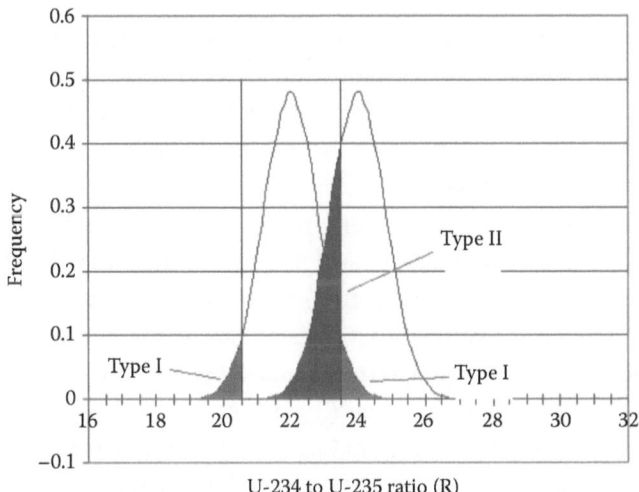

FIGURE 12.1   Distributions of R_{ave} for both the null and alternative hypotheses.

The probability of committing a type I error is equal to the sum of the areas in each tail of the distribution of R_{ave}, or $\alpha = P(R_{ave} \leq 20.5$, when $R = 22) + P(R_{ave} \geq 23.5$, when $R = 22)$. Converting to z-scores facilitates the calculation of this probability:

$$z_1 = \frac{20.5 - 22}{0.83} = -1.8; \quad z_2 = \frac{23.5 - 22}{0.83} = 1.8$$

Therefore, $\alpha = P(Z \leq -1.8) + P(Z \geq 1.8) = 2P(Z \leq -1.8) = 2(0.036) = 0.072$. This is the sum of the areas under each tail on the $R = 22$ curve on Figure 12.1. The licensee would like the type I error as small as possible since it means rejecting the null hypothesis when it is true.

This means that the proposed sampling design ($n = 15$ samples) will, on average, incorrectly reject the null hypothesis 7.2% of the time. To reduce the type I error, we can either increase the sample size or increase (widen) the acceptance region—in fact, that is why α is also called the "size" of the test. For example, if we increased our sample size to 25 samples from 15, the standard error in the mean becomes 0.64, the z-score becomes -2.3, and the resulting type I error is 0.02. Similarly, if we increased the acceptance region to $20 < R_{ave} < 24$, the resulting z-score is -2.4, and the type I error is 0.016.

We must now focus our attention on the type II error. This is the regulator's error of concern—accepting the null hypothesis that $R = 22$ when it is false. Simply adjusting the type I error does not ensure that we have an appropriate testing procedure—we must evaluate the type II error. As mentioned previously, a specific alternative hypothesis (H_a) must be stated before we can evaluate the type II error. Let us assume that it is important to reject H_0 when the true mean ratio is some value $R \geq 24$ or $R \leq 20$. We are then interested in the probability of committing a type II error for alternatives $R = 24$ and $R = 20$ (because of symmetry it is necessary to only evaluate one of these alternatives, they will be equal). Thus, we will calculate the probability of not rejecting (accepting) the H_0 ($R = 22$) when the alternative ($R = 24$) is true. It is assumed that the standard error in the mean for this distribution ($R = 24$) is equal to the standard error in the mean for $R = 22$.

A type II error will occur when R_{ave} falls between 20.5 and 23.5 (acceptance region), when H_a is true. That is, we have already made the decision to accept H_0 if $20.5 < R_{ave} < 23.5$. Therefore, the area under the H_a distribution ($R = 24$), bounded by the acceptance region, is the type II probability. Figure 12.1 shows this type II probability as the area under the alternative hypothesis curve ($R = 24$) bounded by the vertical line at R_{ave} equal 23.5 on the right and R_{ave} equal 20.5 on the left. The type II or β error can be written as

$$\beta = P(20.5 < R_{ave} < 23.5 \text{ when } R = 24)$$

Calculating the z-scores that correspond to $R_{ave} = 20.5$ and $R_{ave} = 23.5$ when $R = 24$:

$$z_1 = \frac{20.5 - 24}{0.83} = -4.2; \quad z_2 = \frac{23.5 - 24}{0.83} = -0.6$$

Therefore, $\beta = P(-4.2 < Z < -0.6) = P(Z < -0.6) - P(Z < -4.2) = 0.27$.

Consequently, the proposed sampling plan has a 27% probability of incorrectly accepting the H_0 ($R = 22$) when the true mean ratio (R) is 24. The power of the test ($1 - \beta$) is 73%, which is the probability of rejecting a false null hypothesis (this should usually be at least 90% at an alternative hypothesis value of concern). Indeed, it may take quite a sales pitch for the regulator to approve this sampling plan. The type II error will be the same for the alternative hypothesis $R = 20$ due to symmetry. The type II error may be reduced by reducing the size of the acceptance region (which increases the type I error), or by increasing the sample size.

We can now summarize some points related to hypothesis testing (Walpole and Myers 1985):

- The type I and type II errors are related—a decrease in the probability of one generally results in an increase in the probability of the other.

- The type II error cannot be calculated unless there is a specific alternative hypothesis.

- An increase in sample size will reduce the type I and type II errors simultaneously.

12.3.2 Chi-Square Test: A Hypothesis Test for Evaluating Instrument Performance

One of the more popular statistical tests that is employed to evaluate the performance of survey instruments is the chi-square test. The chi-square test is usually performed to ensure that the survey instrument's response, specifically the spread of the measurement data (in cpm), is consistent with the predicted spread of the data based on statistical theory (namely, the Poisson distribution). This test is not performed on a daily basis, but rather at a frequency that may be similar to the calibration frequency.

The chi-square distribution is the underlying distribution of the sample variance determined from the observations in a random sample from a normal distribution. Therefore, tests about the variance are quite sensitive to departures from the normality assumption. If there is doubt as to the normality of the distribution, it is recommended that a normality test be performed and/or nonparametric methods used to test hypotheses about the variance.

When performing a chi-square test, the null hypothesis is usually stated in a manner such that the variance of a normal distribution is equal to a specified value, such as the variance predicted from some theory that is to be tested. As mentioned above, the chi-square test of a counter is a method to compare the measured variance to the predicted variance from a Poisson distribution. Thus, for the chi-square test of a counter, the null hypothesis is written as

$$H_0: \sigma^2 = \sigma_0^2$$

against the alternative $H_a: \sigma^2 \neq \sigma_0^2$ at the α level of significance.

The test statistic, χ^2, is computed

$$\chi^2 = \frac{(n-1)s^2}{\sigma_0^2}$$

and when H_0 is true, χ^2 has a chi-square distribution with $n-1$ degrees of freedom.

Let us consider an example of the chi-square test of a counter to ensure that the measurement variability is consistent with the random nature of radioactive decay—which is assumed to exhibit a Poisson distribution. Thus, we compare the measured variance to the predicted variance from the Poisson distribution. Assume that 10 measurements were obtained with a radiation counter and that the sample mean was 144 counts per second, with a sample standard deviation of 9.9 c/s.

As before, the null hypothesis is written as

$$H_0: \sigma^2 = \sigma_0^2$$

where σ_0^2 is that variance predicted by the Poisson distribution—
$\sigma_0^2 = \mu \approx \bar{x}$, against the alternative $H_a: \sigma^2 \neq \sigma_0^2$ at the α level of significance of 0.10.

The test statistic, χ^2, is computed as

$$\chi^2 = \frac{(10 - 1)(9.9)^2}{144} = 6.12$$

At the 0.10 level of significance for a two-tailed test, we need the χ^2 quantile values at $\chi_{0.05}^2$ and $\chi_{0.95}^2$—which can be found as 3.33 and 16.9. Because the χ^2 test statistic falls between these bounds, we conclude that H_0 is true, and that the counter is functioning properly (i.e., it does not give rise to abnormal fluctuations).

Note that $s^2/\bar{x}$ (equals 0.68 in our example) provides a measure of the extent to which the observed sample variance differs from the Poisson predicted variance. That is, a value of unity for $s^2/\bar{x}$ provides a perfect fit to the Poisson, while very low values may indicate abnormally small fluctuations, and very high values may indicate abnormally large fluctuations in the data. For instance, if the sample standard deviation in the above example was 4.5 c/s, $s^2/\bar{x}$ is 0.14, and the χ^2 value would be calculated as 1.3—the counter fails because it is less than the χ^2 quantile value at $\chi_{0.05}^2$ of 3.33.

When reporting these results, it is conventional to report the test statistic (e.g., χ^2), along with a statement as to whether or not it was significant at the chosen level of significance. In our example, we would report χ^2 equals 6.12, not significant at the 0.10 level (because we did not reject the H_0).

One may wish to know whether the counter would have failed at a different level of significance. To determine this, one can report the "p value." A p value is the smallest value of α (type I error) for which the null hypothesis may be rejected (Daniel 1977). Similarly, Walpole and Myers state that the p value represents the lowest level of significance at which the observed value of the test statistic is significant—that is, causes H_0 to be rejected.

Returning to our example, the test statistic was 6.12, so the question is "what is the smallest value of α that will cause the null hypothesis to be rejected?" To answer this question, we need to look up the critical values for a number of χ^2 quantiles. For example, the χ^2 quantile values at $\chi_{0.2}^2$ and $\chi_{0.8}^2$ are 5.38 and 12.2, respectively; thus, our χ^2 (6.12) is still not significant at the 0.40 level of significance. However, the χ^2 quantile values at $\chi_{0.3}^2$ and $\chi_{0.7}^2$ are 6.39 and 10.7, respectively; thus, our χ^2 (6.12) is significant at the 0.60 level of significance. We conclude that the p value is between 0.4 and 0.6, and would report: $p > 0.4$. The smaller the p value, the less likely it is to observe such an extreme value and the more significant is the result. Conversely, the larger the p value, the less likely it is that the null hypothesis will be rejected.

12.4 BAYESIAN STATISTICS

The statistics discussed to this point may be referred to as classical statistics or frequentist statistics. They can be viewed as an evaluation of population parameters (e.g., mean and variance) through repeated sampling from the population, and drawing conclusions from statistical samples. Uncertainty in these parameter estimates is calculated by examining how these estimates vary from one sample to the next. Classical statistics treats the parameter value as fixed, but unknown. That is, one might calculate the confidence interval of the mean—for example, the true but unknown mean has a 95% probability of existing within a specified confidence interval. Bayesian statistics, on the other hand, treats the parameter as having a probability distribution (i.e., it is not constant).

NRC's NUREG-1475, rev. 1 (Lurie et al. 2011) provides a convenient way to think about Bayesian inference, recognizing that it starts with three elements—a hypothesis (H), data (D), and relevant information (X). The hypothesis is a statement about a parameter, the data come from a random sample from a population determined by H, and X is prior information about H before D is observed. Using Bayes' theorem, the conditional probability of H based on both D and X can be written as

$$P(H \mid D, X) = P(H \mid X) \frac{P(D \mid H, X)}{P(D \mid X)}$$

Bayesian analysis relies on the investigator's past knowledge about a parameter, and allows probabilities to be inferred from that information. Note that in the above equation all probabilities are conditioned on X, the prior information. The investigator begins with a probability distribution reflecting the current state of knowledge about a parameter—called the prior distribution. Priors may be classified as either informative or noninformative; the noninformative priors are also called flat priors. As new data become available from sampling campaigns, a likelihood expression is determined that represents the likelihood of the parameter given the observed data. The resulting probability distribution is called the *posterior distribution*—it is calculated based on two sources, the prior distribution and the observed data (likelihood).

In Bayesian estimation, the prior, the likelihood, and the posterior have the following interpretations. The prior reflects the state of our knowledge about the parameter(s) before we have seen the data. The likelihood is the probability of observing the data if the parameter of interest would have the current value. And the posterior considers the state of our knowledge about the parameter(s) after we have observed (and treated) the data. So, we can write Bayes' theorem as an expression for the posterior density of the distribution of the parameter of interest (θ), $f(\theta \mid x)$ (Sorensen and Gianola 2002):

$$f(\theta \mid x) = \frac{f(\theta)f(x \mid \theta)}{\int_{-\infty}^{\infty} f(\theta)f(x \mid \theta)d\theta}$$

where $f(\theta)$ is the density of the prior probability distribution for parameter(s) of interest, and the likelihood $f(x \mid \theta)$ is the density of x as a function of θ (i.e., the likelihood is the density of values that x takes at a given, albeit unknown, value of θ). The denominator is the normalization that ensures that the integral of the posterior is equal to 1.

The hot spot assessment covered in Chapter 11 used Bayesian statistics to effectively express the contaminant distribution, recognizing that existing hot spots comprise the upper tail of the distribution. Using the 99th percentile as the parameter of interest, a posterior distribution can be constructed to assess the probability that hot spots above specified concentrations exist in the survey unit.

The attractiveness of Bayes' theorem is that it reflects the dynamics of learning and accumulation of the knowledge. The prior distribution captures the state of current knowledge; the posterior distribution reflects the latest knowledge as new data is observed. And the posterior distribution becomes the prior for future experiments.

A common approach in Bayesian analysis is to select the form of a prior such that when it is used in conjunction with the likelihood they provide simple, elegant forms for posterior distributions. These types of priors are called conjugate priors, and are essentially models with convenient analytic properties. A prior is a conjugate to the likelihood if the posterior is in the same class of distributions as the prior. They depend on the form of the likelihood. Here is a list of some common conjugate priors used for one-dimensional cases:

Likelihood	Parameter	Prior/Posterior
Normal	Mean (μ)	Normal
Normal	Variance (σ^2)	Inverse gamma
Poisson	λ	Beta
Binomial	π	Gamma

One of the challenging aspects of Bayesian statistics is identifying an appropriate prior distribution that adequately captures the current state of knowledge. Oftentimes, convenient priors can be incorporated into calculations, but if they do not represent the relevant information, then the outcome may be incorrect results and interpretation.

CASE STUDY   USING BAYESIAN ANALYSIS IN
             SAFETY AND RELIABILITY DECISIONS
             (KAPLAN AND GARRICK 1979)

Bayes' theorem is a useful tool when safety decisions have to be made on incomplete, outdated, or only partially relevant data. The probability of A given information B is equal to P(A), the probability of A prior to having information B, times the correction factor in brackets:

$$P(A \mid B) = P(A) \times [P(B \mid A)/P(B)]$$

This shows how our state of confidence with respect to A changes upon getting new information.

The railroad transport of spent nuclear fuel (SNF) has been a contentious issue for decades. One aspect of the debate centered on whether SNF must be transported in "special trains" that are much more expensive than an ordinary shipment. The railroads (RR) sought to show that a shipment of SNF is highly dangerous, and argued that a strong safety record established to date (in 1979) should have no effect on the confidence of future shipments of SNF.

The RR argument against ordinary SNF shipments was that only 4000 or so SNF shipments had been made to date (1979)—all without incident (i.e., no rad release). In fact, releases are expected to occur once in 10^8 to 10^9 shipments, and since 4000 is infinitesimal relative to these numbers, we cannot expect to improve that statistic in any reasonable foreseeable future. However, from a Bayesian standpoint, 4000 release-free shipments should *not* be dismissed.

The Bayesian argument is that while it is true that we cannot expect to amass enough statistics to distinguish between radioactive release frequencies of 10^{-8} or 10^{-10}, the SNF shipping data do have value. What do we really want to know? Not whether the release frequency is 10^{-8} or 10^{-10} (both considered by most to be safe shipments), but whether it is on the order 10^{-3} to 10^{-4}—and 4000 release-free shipments very important in this context. We begin by assuming we did not have the evidence of 4000 release-free SNF shipments.

The objective is to predict the frequency of rad releases during SNF shipments. A probability distribution is constructed that expresses our state of knowledge, prior to having data.

We now assume that we have 4000 release-free SNF shipment data—B equals 4000 shipments with no releases. We also define the following rad release frequency rates:

A_1 = Release frequency rate is 10^{-3}
A_2 = Release frequency rate is 10^{-4}
A_3 = Release frequency rate is 10^{-5}
A_4 = Release frequency rate is 10^{-6}
A_5 = Release frequency rate is 10^{-7}
A_6 = Release frequency rate is 10^{-8}

Thus, the probability distribution prior to having evidence B is simply given by probabilities in Figure 12.2 for $P(A_i)$, $i = 1,2,3,\ldots,6$. For example, the probability is 40% for release rate of 10^{-5}.

After having evidence B, the distribution becomes

$$P(A_i|B), \quad i = 1,2,3,\ldots,6 \text{ (probability of } A_i \text{ given } B)$$

Bayes' theorem allows us to correct $P(A_i)$ now that we have evidence B (4000 release-free shipments):

$$P(A_i|B) = P(A_i) \cdot [P(B \mid A_i)/P(B)]$$

Probability calculations are shown for the event defined by A_1, a rad release frequency of 10^{-3}. Recall that the prior probability for A_i is 0.01 (Figure 12.2).

If the frequency of rad release was 10^{-3}, the probability of B (4000 release-free trips) would be

$$P(B|A_1) = (1 - 10^{-3})^{4000} = (0.999)^{4000} = 0.0183$$

The calculation is based on the binomial, where each event is assumed to be independent. For A_1, the probability of a single release-free shipment is 0.999, and the probability of getting 4000 release-free shipments is given by $(0.999)^{4000}$. This result indicates that we have <2% probability

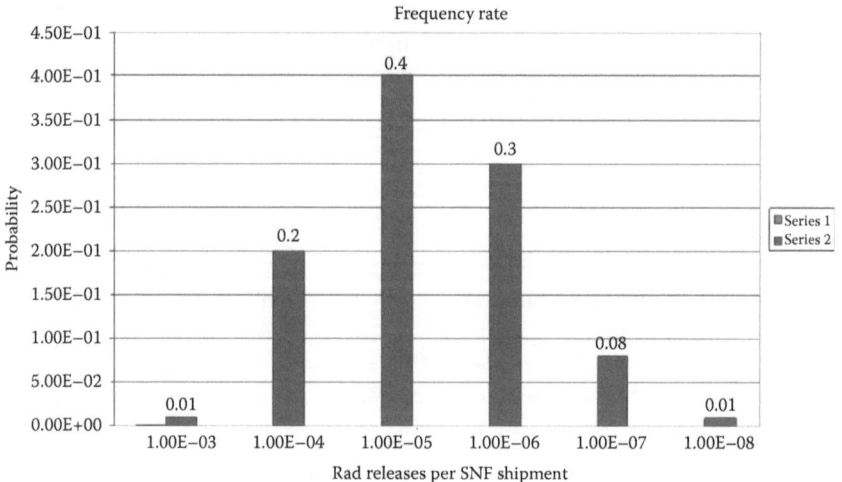

FIGURE 12.2   Probability distribution $P(A_i)$ prior to actual SNF shipment data.

of obtaining 4000 release-free trips if rad release frequency really is 10^{-3} (not likely).

The same probability calculations were performed for events A_2 to A_6:

$$P(B|A_2) = (1 - 10^{-4})^{4000} = 0.670$$

$$P(B|A_3) = (1 - 10^{-5})^{4000} = 0.961$$

$$P(B|A_4) = (1 - 10^{-6})^{4000} = 0.996$$

$$P(B|A_5) = (1 - 10^{-7})^{4000} = 0.9996$$

$$P(B|A_6) = (1 - 10^{-8})^{4000} = 0.99996$$

Now, we can calculate $P(B)$, the probability of getting 4000 release-free shipments using the data in Table 12.2:

$$P(B) = \sum_{i=1}^{6} P(A_i)P(B/A_i) = 0.907$$

So, what can we conclude from this analysis? First, introducing information B certainly has an impact on the probability of release, particularly for A_1 and A_2. The prior probabilities were 0.01 and 0.2, respectively. The updated probabilities are 0.0002 and 0.148. Information B reduced probability A_2 by ~26% and substantially reduces probability A_1. However, information B had

TABLE 12.2   Summary of Bayesian Calculations

i	1	2	3	4	5	6
A_i	10^{-3}	10^{-4}	10^{-5}	10^{-6}	10^{-7}	10^{-8}
$P(A_i)$	0.01	0.2	0.4	0.3	0.08	0.01
$P(B\|A_i)$	0.0183	0.670	0.961	0.996	0.9996	0.99996
$P(B\|A_i)/P(B)$	0.0202	0.739	1.06	1.098	1.102	1.102
$P(A_i\|B)$	0.0002	0.148	0.424	0.329	0.0882	0.01102

a relatively small effect on our degree of confidence in the propositions A_3, A_4, A_5, and A_6. Overall, specifically having 4000 release-free shipments of SNF shows it is highly unlikely that the release frequency will turn out to be 10^{-3}, and further, it reduces our degree of belief that the frequency is as high as 10^{-4}; it is more likely that the frequency is 10^{-5} or less.

In conclusion, Bayes' theorem is a useful approach for quantifying the effect of new evidence on a pre-existing state of knowledge. While the priors used to obtain probability estimates are subjective, it is worthwhile to note that many model input parameters are equally subjective.

QUESTIONS AND PROBLEMS

1. Name the two components of standard deviation of the contamination level in a survey unit. What measures can be taken to reduce each component?

2. Briefly summarize the main attributes of the binomial, Poisson, and normal distributions, and their applications to health physics.

3. What are some considerations in stating the null hypothesis in a certain manner?

4. State the meaning of the type I and type II errors. Why do you suppose it is not a good idea to view the type I error as a "false positive" as is so common in health physics?

5. Given the following information, determine if the null hypothesis can be rejected using the Sign test. The sample size (n) was 18.

   H_0: Residual radioactivity in the survey unit exceeds the DCGL$_W$

   Type I error = 0.05; Type II error = 0.10

   Test statistic ($S+$) = 14 (equals the number of measurements less than DCGL$_W$)

What if the type I error is changed to 0.025? (Hint: It is necessary to look up critical values for the Sign test in MARSSIM Appendix I.)

6. Discuss the use of the p value.

7. Explain the relationship of the posterior distribution to the prior and the likelihood in Bayesian statistics. What is the value of a flat prior?

SUGGESTED READING

Elementary statistics textbooks
Nonparametric Statistics by Conover
Statistics chapter in Knoll's *Radiation Detection* textbook
Any good textbook on Bayesian statistics

CHAPTER 13

MARSSIM Final Survey Design and Strategies

T HE ULTIMATE GOAL OF decommissioning projects is the assurance that future site or building uses will not result in unreasonable risk to human health and the environment due to residual contamination from past operations. This goal is achieved through the remediation, if necessary, of the facility/site, and then demonstrated by performing an FSS which shows that the appropriate radiological release criteria have been achieved. The MARSSIM offers Federal agency consensus guidance on how to design and perform FSSs for radiological contamination. The MARSSIM approach is fundamentally based on the DQO process.

In this chapter, the focus is on the design of MARSSIM FSSs and to discuss a number of example applications of MARSSIM. Depending upon the specifics of the particular decommissioning project, various MARSSIM strategies can be employed to design the most cost-effective survey possible. For instance, it may be more appropriate to use scenario B, than the standard scenario A approach—How can one decide which approach is more economical? What about the balance between the number of survey units and the spatial standard deviation used to determine sample size? Or the larger sample size that results from the use of a default DCGL as compared to a site-specific DCGL—when should resources be expended into site-specific modeling parameters to derive a more realistic DCGL? Should surface-activity assessment be performed using the WRS test or the sign test? These and other issues will be discussed in this chapter, but first it is worthwhile to revisit the survey guidance that preceded the MARSSIM, as this guidance was the basis for the latest survey concepts now found in the MARSSIM.

13.1 FSS PROTOCOLS PRIOR TO MARSSIM

There are several sources of guidance that describe how one should plan and implement an FSS. One example, "Monitoring Programmes for Unrestricted Release Related to Decommissioning of Nuclear Facilities" (IAEA 1992) discusses the use of stratified random sampling and systematic sampling, and states that the division of a site into survey grids should consider the operating history of the plant, as well as the nature and quantity of known contamination. This IAEA report references NUREG/CR-2082, "Monitoring for Compliance with Decommissioning Termination Survey Criteria" (1981a) for further detail in performing FSSs. Apparently, NUREG/CR-2082 never gained the widespread popularity achieved by its successor guidance document draft NUREG/CR-5849 in 1992. We will now spend a little time getting acquainted with the MARSSIM predecessor guidance documents—NUREG/CR-2082 and NUREG/CR-5849.

13.1.1 NUREG/CR-2082 Guidance

It is instructional to revisit the FSS procedures outlined in NUREG/CR-2082. This survey design guidance served as the precursor to both NUREG/CR-5849 and MARSSIM, and introduced several key concepts. The first concept that has direct application to the MARSSIM approach is that of the preliminary survey. According to NUREG/CR-2082, the preliminary survey is conducted "for the purpose of designing a final survey plan to establish whether or not a site is decontaminated sufficiently to warrant unrestricted release." The preliminary survey is used to help divide the site into logical divisions called survey units. This use of the preliminary survey is similar to the use of scoping and characterization surveys in MARSSIM.

As with NUREG/CR-5849 and MARSSIM, NUREG/CR-2082 states that survey units may consist of a parcel of land, one story of a building, a roof, a loading dock, or any area naturally distinguishable from the remainder of the site. This NUREG further states that since some minimum number of measurements should be made in each survey unit, the site should not be divided into a prohibitively large number of survey units. Identical guidance is offered in the MARSSIM. Finally, the preliminary survey is supposed to aid the survey designer in how to sample the site, using random, stratified-random, or systematic sampling approaches.

The NUREG/CR-2082 survey design for indoor and outdoor areas is summarized next. Each indoor survey unit is divided into subunits—(1) lower surfaces that include floors and wall surfaces up to 2 m, and (2) upper surfaces that include ceilings and wall surfaces greater than 2 m. Further, the floor and lower wall surfaces are divided into survey blocks on a rectangular grid system, and the following guidance is provided on the size and number of grid blocks in a survey unit:

- No survey block should be less than 1 m on a side

- No survey block should be greater than 3 m on a side

- There should be at least n survey blocks in the population (assume population is the survey unit), unless it violates the first point

The last point bears further discussion. For a fixed survey unit size, the guidance in the last bullet may conflict with the first bullet, so preference is given to the first point, namely, the survey blocks will not be smaller than 1 m on a side. For example, assume the survey unit is 40 m² and that n equals 54. In this case dividing the survey unit into 54 survey blocks results in survey blocks that are less than 1 m on a side; therefore, only 40 survey blocks would be necessary in this example, regardless of the determination of n.

How does NUREG/CR-2082 specify the calculation of n? The sample size, n, is given by

$$n = 45 M^2$$

where M is the maximum of the set $(s/x)_\alpha$, $(s/\bar{x})_\beta$, $(s/\bar{x})_\gamma$, and 0.82. Some explanation of the coefficient of variation terms (and 0.82) is necessary. First, the NUREG/CR-2082 user must understand the radiological conditions that are to be measured in these interior survey blocks. They include the mean ($\bar{x}$) and standard deviation (s) of alpha surface activity in dpm/100 cm², beta–gamma dose rates at 1 cm in mrad/h, and gamma exposure rates at 1 m. Therefore, to calculate $(s/\bar{x})_\alpha$, $(s/\bar{x})_\beta$, $(s/\bar{x})_\gamma$ it is necessary to perform a preliminary survey in each survey unit to estimate these values. Note that if each coefficient of variation term is less than 0.82, then M is 0.82 (which is the smallest value that M can take on).

This translates to a minimum sample size of $n = 30$ in each survey unit (i.e., 45×0.82^2). Of course, the sample size can be greater depending on the largest value of the coefficient of variation for a particular measurement type.

This statistical approach is based on a t test that requires that the true population mean of the survey unit to be known with less than a 25% error at the 90% confidence level. This can be written as an inequality:

$$t_{90\%,\mathrm{df}} \frac{s_x}{\sqrt{n}} < 0.25\overline{x}$$

where the value of the t distribution for n greater than 30 is about 1.7. This inequality can be solved for n as follows:

$$n > \left(\frac{1.7s}{0.25\overline{x}} \right)^2 > \frac{45s^2}{\overline{x}^2}$$

So there we have the rather simplistic statistical basis for the sample size calculations in NUREG/CR-2082.

Some final notes on the FSS guidance in NUREG/CR-2082 are warranted. While we focused on the interior survey design, the outdoor survey design for land areas is very similar. The survey blocks are larger outdoors—that is, no smaller than 5 m on a side, but no more than 15 m on a side. The sample size equation is the same and the radiological conditions to measure include beta–gamma levels at 1 cm, gamma exposure rates at 1 m and soil samples. Surface scanning is discussed for both interior and exterior survey units, with the preference given to GM scans indoors and NaI scans outdoors. NUREG/CR-2082 also points out the difference between unbiased measurements that result from the systematic measurements in each survey block, as contrasted to the biased measurements that result from surface scans.

One of the difficulties with NUREG/CR-2082 was the format of the survey guidance—that is, several sections of the document have to be referenced before the overall survey design can be understood. In this regard, it is not unlike the current DOT guidance for radiological shipping requirements. Perhaps this is too strong a criticism of NUREG/CR-2082, as MARSSIM too has suffered the same criticisms. It may be that NUREG/CR-5849's popularity was due in no small measure to its clear format and presentation of survey requirements. We will now consider how the NRC expanded upon the survey unit and *t* test statistical basis in NUREG/CR-5849.

13.1.2 NUREG/CR-5849 Guidance

The U.S. Nuclear Regulatory Commission published NUREG/CR-5849, "Manual for Conducting Radiological Surveys in Support of License Termination (draft)" in June 1992 which effectively replaced NUREG/CR-2082 and provided licensees with an acceptable process for conducting FSSs during decommissioning. This manual has received widespread application at many NRC, U.S. DOE and DOD sites over the past dozen years, but it is increasingly unlikely to be approved as the final survey guidance at ongoing and future D&D project sites. The NRC has announced that the MARSSIM has superseded NUREG/CR-5849. An overview of the FSS design and data reduction in NUREG/CR-5849 will be introduced, followed by a discussion on the MARSSIM approach.

The NUREG/CR-5849 guidance recommends that the site be divided into affected and unaffected areas for purposes of designing the FSS. These areas are then divided into a number of indoor and outdoor survey units. Areas are classified as affected or unaffected based on contamination potential in the particular survey unit. Affected areas have potential radioactive contamination (based on operating history) or known contamination (based on past surveys)—and include areas where radioactive materials were stored, where records indicate spills occurred, and any areas adjacent to these areas. Unaffected areas simply include all areas that are not classified as affected.

Affected area survey units are further divided into grid blocks (1 m^2 for structure surfaces and 100 m^2 for land areas). The guidance then recommends that scanning, direct measurements and sampling be performed at specific frequencies depending on the classification. For example, 4 soil samples per 100 m^2 in the affected land areas and one direct measurement per 1 m^2 for affected structure surfaces. A total of 30 direct measurements and soil samples are required in unaffected indoor survey units and unaffected outdoor survey units, respectively. Thus, NUREG/CR-5849 provides a prescriptive survey approach to determining the sample size for both direct measurements and soil samples—the survey design is rigid and does not provide a statistical basis for sample collection. The number of samples required in a particular survey unit is based on its classification—affected or unaffected; there is little flexibility based on how difficult it may be to detect the contaminant in the field (e.g., through scanning), or based on the guideline relative to the background concentration (for those radionuclides present in the background).

Once the survey data are collected per the NUREG/CR-5849 approach, it is necessary to demonstrate compliance with the release criteria. First of all, it is important to note that decommissioned sites are released survey unit by survey unit; survey units are the fundamental compliance units. That being said, additional conditions must be met before a statistical determination can be made to demonstrate that a survey unit meets the release criteria. Specifically, these conditions include:

1. Individual measurement limit—each soil sample and direct measurement must not exceed the average guideline by more than a factor of 3.

2. Elevated areas in soil limit—each elevated area between 1 and 3 times the guideline value in soil must not exceed the average guideline by more than $(100/A)^{1/2}$.

3. Area-weighted average—the area-weighted average over each 1 m² of structure surface and 100 m² of land area must satisfy the average guideline.

4. Statistical evaluation (*t* test for survey units)—finally, all of the data within a survey unit is further tested to determine compliance with the release criteria at the 95% confidence level.

Decommissioned sites are released survey unit by survey unit; survey units are the fundamental compliance units.

An example is provided to demonstrate the data reduction required for NUREG/CR-5849. Assume that four systematic soil samples were collected from a 100 m² grid block in an affected survey unit. The radionuclide of concern is natural thorium, with a guideline of 10 pCi/g for total thorium (e.g., sum of Th-232 and Th-228). The individual soil samples, after correcting for the background concentration, measured 2.7, 5.0, 1.6, and 3.8 pCi/g for total thorium. In addition, gamma scanning has identified an area of elevated direct radiation measuring approximately 15 m². Three additional samples were collected from this area and measured 28, 26, and 21 pCi/g total thorium—with an average concentration in the elevated area of 25 pCi/g, total thorium. Does this grid block satisfy the first three conditions stated above? (Note: The compliance status of the survey unit is determined only after all grid blocks in the survey unit have been evaluated.)

1. Since each sample was less than 3 times the limit (highest was 28 pCi/g), the individual measurement limit is satisfied.

2. The elevated area factor for the 15 m² area is calculated $(100/15)^{1/2}$ equals 2.6. Thus, the average concentration in the elevated area must not exceed 26 pCi/g. Since this concentration is 25 pCi/g, this condition is satisfied.

3. Now, we must calculate the area-weighted grid block average. The average of the four systematic samples is 3.3 pCi/g and represents 85 m², and the elevated area represents 15 m², thus:

$$\bar{x}_w = 3.3 \times \frac{85 \, m^2}{100 \, m^2} + 25 \times \frac{15 \, m^2}{100 \, m^2} = 6.6 \, pCi/g$$

Therefore, the grid block does indeed satisfy the first three conditions.

Now, assume that there are 26 grid blocks in this particular affected survey unit, and that each grid block is determined to satisfy the first three conditions. One can now evaluate the final condition—performing the t test to determine that the data from the survey unit provides a 95% confidence level that the true mean activity concentration satisfies the guidelines. To perform the t test, the thorium concentrations are used from all samples collected in the 26 grid blocks. The number of samples is calculated assuming that four systematic soil samples were collected from each grid block, unless locations of elevated direct radiation are identified, then only the area-weighted average is used for those grid blocks. Assuming that 20 grid blocks had no elevated areas identified (thus six had hot spots), the total number of data points evaluated in the survey unit is 4 × 20 plus 6 (a total of 86). From these data, the mean and standard deviation of the total thorium concentration in the survey unit are calculated—for example, assume a mean of 4.7 pCi/g and a standard deviation of 2.5 pCi/g. The following equation is used in NUREG/CR-5849 to test the data, relative to the guideline value, at the 95% confidence level:

$$\mu_\alpha = \bar{x} + t_{95\%,df} \frac{s_x}{\sqrt{n}}$$

where $t_{95\%,df}$ is the critical value of the t distribution at $1 - \alpha$ (=95%) for the $n - 1$ degrees of freedom (df). Specifically, the above formula is for an upper one-sided $100(1 - \alpha)$ percent confidence limit around the survey unit mean. This one-sided confidence interval is used to test whether the survey unit has attained the appropriate guideline.

We can now determine μ_α given the soil sample results from our survey unit. For, $n - 1$ equal 85, the critical value of the t distribution at the 95% confidence level is about 1.66. Substituting our survey data into the equation, we obtain:

$$\mu_\alpha = 4.7 + 1.66\frac{2.5}{\sqrt{86}} = 5.1\,\text{pCi/g}$$

This result indicates that the survey unit is in compliance with the total thorium guideline of 10 pCi/g at the 95% confidence level.

The final area of guidance in NUREG/CR-5849 that will be considered pertains to the number of background data points. Section 2.3.1 of NUREG/CR-5849 states that "experience has indicated the variance in the average background value from a set of 6 to 10 measurements that will usually not exceed ±40% to 60% of the average at the 95% confidence level." Further, NUREG/CR-5849 offers that as long as the average background level is insignificant as compared to the guideline level (defined as <10% of the guideline), 6–10 background samples are sufficient. For instances where the background levels are significant (>10% of guideline), the stated objective is that the average background level "should accurately represent the true background average to within ±20% at the 95% confidence level." The number of background samples needed to satisfy this objective can be computed:

$$n_B = \left(\frac{(t_{97.5\%,df})(s_x)}{(0.2)(\bar{x})}\right)^2$$

where the mean and standard deviation of the background distribution are from the initial 6–10 background measurements. The $t_{97.5\%,df}$ value corresponds to an overall α of 0.05 considering 2.5% for each tail of the distribution.

As discussed previously for the need for additional samples in the survey unit, the number of background samples already collected (6–10 in this guidance) are subtracted from n_B to determine the number of additional background samples to collect to satisfy this objective.

Consider an example for Th-232, assuming that our initial six background samples resulted in a mean and standard deviation of 1.0 and 0.7 pCi/g Th-232, respectively. Assuming that the background is significant relative to the guideline, we can calculate the number of samples necessary to determine the background value to within ±20% at the 95% confidence level:

$$n_B = \left(\frac{(2.571)(0.7)}{(0.2)(1.0)} \right)^2 = 81 \text{ samples}$$

Therefore, an additional 75 samples would be required to meet the objective for the background determination.

Admittedly, NUREG/CR-5849 states that these "criteria for defining an acceptable accuracy for background determinations is arbitrary, based on the natural variations ... and (site planners) need to keep the effort and cost devoted to background determination reasonable." In fact, the overriding consideration in determining an acceptable accuracy for background determinations is the guideline level. That is, we used to think in terms of "what fraction of the guideline is background," but with the decommissioning criteria in the United States now established on a dose basis (NRC's decommissioning rulemaking establishes the release criterion at 25 mrem/y), the guideline may be on the order of the background level (e.g., 1.1 pCi/g for Th-232 from the NRC's DandD, ver. 1.0 screening model). The importance of accurately determining the background level has increased significantly. Instead of a background determination accuracy objective at ±20% at the 95% confidence level, potentially far greater accuracy was necessary. Consider the number of samples for reducing the background uncertainty to ±10% at the 95% confidence level for our Th-232 example:

$$n_B = \left(\frac{(2.571)(0.7)}{(0.1)(1.0)} \right)^2 = 324 \text{ samples}$$

Not only was there concern in the tremendous increase in the number of background samples, but also because of the reliance of the *t* test on the assumption of normality—background levels of radionuclides frequently exhibit a nonnormal distribution (which means that the underlying normality assumption for the t test may be absent in many cases). With that, the serious discussions on nonparametric statistics began, along with a multiagency effort to produce consistent survey guidance (MARSSIM).

13.2 OVERVIEW OF MARSSIM SURVEY DESIGN

MARSSIM stresses the use of DQOs and recommends a hypothesis testing approach using nonparametric statistics. Unlike the parametric statistics that we have discussed so far, nonparametric statistics do not require the assumption that the data are normally or log-normally distributed (frequently a problem for the t test, which requires normality). The default null hypothesis in MARSSIM is that the residual radioactivity in the survey unit exceeds the release criterion. Therefore, the null hypothesis must be rejected for the survey unit to pass.

While the MARSSIM approach to FSS design recommends the use of nonparametric statistics, it is similar to the guidance offered in NUREG/CR-5849 in many regards. The MARSSIM survey design begins with the identification of the contaminants and the determination of whether the radionuclides of concern exist in the background. The site is divided into class 1, class 2, and class 3 areas, based on contamination potential, and each area is further divided into survey units. Class 1 areas are the most likely to be contaminated, such that the contamination levels are likely to exceed the $DCGL_W$ in some areas. Class 2 areas are expected to have residual radioactivity levels greater than the background, but not at levels that exceed the $DCGL_W$. Finally, class 3 areas are expected to be quite similar to the background areas, but if contamination is identified, it should be at a small fraction of the $DCGL_W$. As with NUREG/CR-5849, the decommissioned site is released survey unit by a survey unit. Appropriate reference areas for indoor and outdoor background measurements are selected.

In general, for contaminants that are present in background (or measurements that are not radionuclide-specific), the WRS test is used; for contaminants that are not present in the background, the sign test is used. The number of data points needed to satisfy these nonparametric tests is based on the contaminant DCGL, the expected standard deviation of the contaminant in the background and in the survey unit, and the acceptable probability of making type I and type II decision errors. The MARSSIM also requires a reasonable level of assurance that any hot spots that could be significant relative to regulatory limits are not missed during the FSS.

For situations where the contaminant is not present in the background, the one-sample sign test replaces the two-sample WRS test. The sign test may also be used in circumstances where the contaminant is present in the background at such a small fraction of the $DCGL_W$ as to be considered insignificant; and therefore, a background reference area is not necessary.

13.2.1 Sign Test Example: Co-60 in Soil

A rather simple example will be covered in sufficient detail to illustrate the MARSSIM process and define all necessary terms. The reader is encouraged at this point to study the following example with the MARSSIM manual opened to the corresponding sections (MARSSIM Chapter 5). This example assumes that Co-60 is the contaminant of concern in soil and that the sign test will be used to demonstrate compliance with the release criteria. The objective of the FSS is to demonstrate that residual radioactivity levels meet the release criterion. In demonstrating that this objective is met, the null hypothesis recommended in MARSSIM is that residual contamination exceeds the release criterion; the alternative hypothesis is that residual contamination meets the release criterion.

Null Hypothesis (H_0): Residual radioactivity in the survey unit exceeds the release criterion.

Therefore, the MARSSIM user must reject the null hypothesis to pass the survey unit.

To determine data needs for the sign test, the acceptable probability of making type I and type II decision errors are established. The type I decision error (also termed the alpha error) occurs when the H_0 is rejected when it is true—results in concluding that survey units incorrectly satisfy release criterion (regulator's risk). The type II decision error occurs when the H_0 is accepted when it is false—results in unnecessary investigations and possibly remediation (licensee's risk). The acceptable type II decision error rate is determined during the DQO process to reflect the anticipated level of Co-60 contamination in the survey unit. For this example, the type I error (α) is specified as 0.05 and type II decision error (β) is set at 0.10.

13.2.1.1 Derived Concentration Guideline Levels

The exposure pathway modeling is used to translate the release criterion (25 mrem/y) to measurable quantities called DCGLs. There are two types of DCGLs used in MARSSIM: $DCGL_W$ is derived based on the average concentration over a large area, while $DCGL_{EMC}$ is derived separately for small areas of elevated activity (hot spots). (Note: Whenever the term DCGL is used without a subscript it is understood to mean $DCGL_W$.)

The results from previous scoping and characterization surveys are used to estimate the Co-60 contamination levels in the areas of interest, specifically the standard deviation of Co-60 in the survey unit being considered. The applicable DCGL for residual Co-60 concentrations in soil can be obtained from dose pathway models, such as RESRAD and DandD. For example, the RESRAD modeling (Version 5.95) using default parameters can be used to determine the $DCGL_W$ for Co-60 in soil. The result is

Co-60: 3.3 pCi/g

13.2.1.2 Sign Test: Determining Number of Data Points

The following steps detail the procedure for determining the number of data points for the sign test. The key parameters affecting sample size include the magnitude of the type I and type II decision errors, and the relative shift (Δ/σ). The mechanics in determining sample size are discussed, followed by the strategy for taking advantage of the MARSSIM flexibility.

13.2.1.2.1 Calculate the Relative Shift   The contaminant $DCGL_W$, LBGR, and the Co-60 standard deviation in the survey unit are used to calculate the relative shift, Δ/σ. (Note that $\Delta = DCGL_W - LBGR$.)

The following information was used in the determination of relative shift, $(DCGL_W - LBGR)/\sigma$:

- *DCGL.* The $DCGL_W$ for Co-60—3.3 pCi/g in soil

- *Standard deviation.* The standard deviation of the contaminant can be obtained from (1) previous surveys—scoping or characterization, or remedial action support surveys in the case where remediation was performed, (2) limited preliminary measurements (5–20) to estimate the distributions, or (3) reasonable estimate based on site knowledge (assume 30% coefficient of variation). Note that the estimate of the standard deviation includes both spatial variability of the Co-60 and the precision of the measurement system, which in this case would likely be gamma spectroscopy.

    On the basis of the remedial action support survey, the standard deviation of Co-60 in the survey unit is about 1 pCi/g. (Note:

Characterization data for an area are rendered obsolete once the site area is remediated; that is, suppose that the soil that was characterized is now in a B-25 waste container. New radiological conditions must be determined with the remedial action support survey.)

- The LBGR is the concentration at which the type II error rate is set. The MARSSIM recommends that the LBGR be initially set at 50% of the $DCGL_W$. This makes sense only in the absence of information about the concentration level in the survey unit. If the median concentration in the survey unit can be determined from a preliminary survey, use this concentration for setting the LBGR. In fact, the same data set used to determine the standard deviation can be used to estimate the median concentration.

On the basis of the remedial action support survey, let us assume that the median Co-60 concentration in the survey unit is 1.8 pCi/g (following remediation in this unit). Thus, $\Delta = DCGL_W - LBGR = 3.3 - 1.8 = 1.5$. The relative shift was then calculated directly—3.3 – 1.8/1 equals 1.5. (Note that the relative shift is a unitless value.)

The LBGR is the concentration at which the type II error rate is set. It is advantageous to set the LBGR at or above the expected median contaminant concentration in the survey unit.

13.2.1.2.2 Determine Sign p Table 5.4 in MARSSIM (NRC 2000a) contains a listing of relative shift values and values for sign p. Sign p is the probability that a measurement at a random location from the survey unit will be less than the $DCGL_W$ when the survey unit median concentration is actually at the LBGR (NRC 1998a). In other words, as the LBGR is lowered then the probability (given by sign p) that a random measurement from the survey unit will be less than the $DCGL_W$ increases. Sign p is a function of relative shift (($DCGL_W - LBGR$)/σ)—sign p increases as Δ/σ increases.

Using the relative shift value calculated previously, the value of sign p was obtained from the tabulated values. Therefore, for a relative shift value of 1.5, the value of sign p is 0.933193.

(Note: The MARSSIM user will typically skip this step because the sign test sample size can be determined directly from tabulated values of relative shift and decision errors in MARSSIM Table 5.5. This example provides the calculational steps to illustrate how Table 5.5 is derived.)

13.2.1.2.3 Determine Decision Error Percentiles   The next step in this process was to determine the percentiles, $Z_{1-\alpha}$ and $Z_{1-\beta}$, represented by the selected decision error levels, α and β, respectively. As stated earlier, α was selected at 0.05 and β was selected at 0.10. From Table 5.2 (NRC 2000a), the percentile $Z_{1-\alpha}$ equals 1.645, and $Z_{1-\beta}$, equals 1.282.

13.2.1.2.4 Calculate Number of Data Points for Sign Test   The number of data points, N, to be obtained for the survey unit is calculated below:

$$N = \frac{(z_{1-\alpha} + z_{1-\beta})^2}{4(\text{sign } p - 0.5)^2}$$

Substituting in the values determined above, N was calculated as

$$N = \frac{(1.645 + 1.282)^2}{4(0.933193 - 0.5)^2} = 11.4$$

To assure sufficient data points to attain the desired power level with the statistical tests and allow for possible lost or unusable data, it is recommended that the number of calculated data be increased by 20%, and rounded up, for further assurance of sufficient data points. This yielded 14 to be collected in the survey unit.

Note: As mentioned previously, the above equation is shown for completeness. The reader is encouraged to use Table 5.5 in MARSSIM (NRC 2000a) to obtain the sign test sample size. In this case, the sample size that corresponds to the values of α, β, and Δ/σ is 15 (the author is not certain as to why there is an apparent discrepancy in sample sizes between the above calculation and Table 5.5). Hence, the table has already increased the sample size by the MARSSIM-recommended 20%.

Perhaps, the most important element of a particular FSS design is to assess if the design meets the DQOs for this site area. This can be accomplished by constructing a prospective power curve. The power curve provides a graph of the survey unit's probability of passing versus the median contamination level in the survey unit. It is called "prospective" because it is based on the predicted standard deviation and the resulting planned sample size. The COMPASS software developed by ORAU with funding provided by the U.S. Nuclear Regulatory Commission will be used to illustrate the power curve for this example design.

The power curve plots the probability of passing the survey unit as a function of median concentration in the survey unit. The power curve is anchored by two points, the LBGR and the DCGL$_W$—the curve crosses the DCGL$_W$ at the type I error probability and should go through the LBGR concentration at type II error. The shape of the prospective power curve is affected by the planned sample size and the estimated contaminant standard deviation. The reason why the power curve has a higher probability (power) at the LBGR concentration than the designed 10% type II error is because the MARSSIM recommends a 20% sample size increase above that which the statistical test requires. Therefore, this survey design aspect provides a built-in hedge against the pitfalls of underestimating the standard deviation or failing to collect the planned sample size.

Note that the type II error is determined at a specific concentration measured from the top of the curve (100% power) to the actual power intersected by the curve. For example, the type II error (shown in Figure 13.1) at a concentration of 2.5 pCi/g is 25% and the type II error at 3.0 pCi/g is approximately 77%. Of course, the type II error is the complement of the power; in other words, the power equals 1 minus the

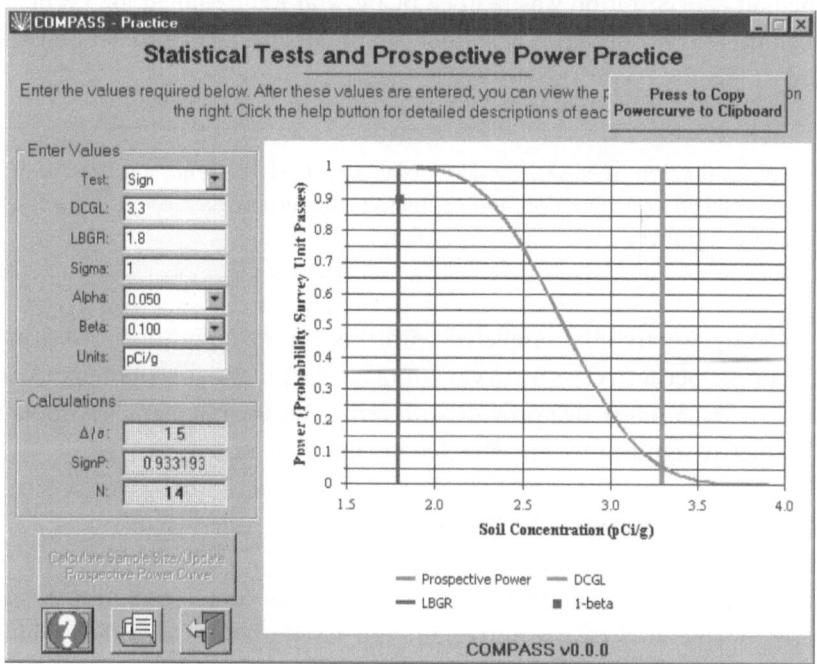

FIGURE 13.1   Prospective power curve for the sign test example.

type II error. The type I error (5% in this example) is set at the $DCGL_W$, and it always goes through this point regardless of actual sample size or the actual standard deviation. Also notice how quickly the type I error decreases with increasing concentrations above the $DCGL_W$; at 3.5 pCi/g the type I error is only about 2%, and it is asymptotically zero at 4 pCi/g.

Returning to our example, the remedial action support survey results indicate that the median concentration of Co-60 in the survey unit is 1.8 pCi/g. The prospective power at this concentration shows a nearly 100% probability of passing the survey unit. From the perspective of the MARSSIM designer, this is a superb FSS design—only 14 samples to have nearly a 100% chance of passing. The only potential problem at this point is like the old saying goes: "garbage in equals garbage out." First point—this example prospective power curve is only as accurate as the estimated standard deviation used to generate it, and it is always better to overestimate the standard deviation than to underestimate it. Second point—assuming the power curve is reasonably accurate, the practical use of it for evaluating the merits of the particular survey design demands that the user knows the median concentration in the survey unit. For instance, if the median concentration was really 3 pCi/g, and not designed for 1.8 pCi/g, then this survey design is absolutely abysmal with an estimated power of only 23%.

> The evaluation of MARSSIM FSS designs critically depends on how well the median concentration and estimated standard deviation in the survey unit are known.

Finally, we are not completed with this survey design for class 1 survey units. Because class 1 survey units have a potential for hot spots, the survey design must also provide sufficient assurance that any hot spots of dosimetric significance are detected during the FSS.

13.2.1.3 Determining Data Points for Areas of Elevated Activity

For class 1 areas, the number of data points required by the sign test for uniform levels of contamination may need to be supplemented to ensure a reasonable level of assurance that any small areas of elevated residual radioactivity are not missed during the FSS. Soil sampling on a specified grid size, in conjunction with surface scanning, are used to obtain an adequate assurance level that hot spots will still satisfy DCGLs—applicable to small areas ($DCGL_{EMC}$). Maybe a review of this somewhat confusing process is in order.*

First, and foremost, this additional consideration for hot spots only needs to be performed in class 1 survey units. Why? By definition, only class 1 areas can have a potential for hot spots. If one identifies hot spots in a class 2 survey unit, it is probably a good decision to reclassify the survey unit to class 1.

Let us introduce this subject by saying that it is necessary to evaluate the scan MDC of the selected scanning instrument to determine if additional direct measurements may be required. These measurements would be in addition to the sample size determined by the sign or WRS test. Note: These additional samples, if warranted, are not to be collected after the statistical sample size has been collected, but rather are integrated with the statistically based sample size to provide one sample size that is collected on a systematic-random sampling grid.

The first step in this process is to assess the scan MDC for the scanning instrument—this is usually called the "actual scan MDC." The next step is to determine the "required scan MDC." The required scan MDC is the product of the $DCGL_W$ and the area factor. The area factor is obtained from dose modeling, and is determined based on the size of the area (a') bounded by the sample size in the survey unit. This bounded area (a') is simply the survey unit area divided by the number of samples from the statistical test.

The actual scan MDC is compared to the required scan MDC. If the actual scan MDC is less than the required scan MDC, then no additional samples are necessary. If the actual scan MDC is greater than the required scan MDC, the user must either select an instrument with a lower scan MDC or increase the sample size in the survey unit. To determine the increased sample size, the actual scan MDC is divided by the $DCGL_W$, to yield the area factor that is necessary for the chosen scan instrument. This area factor is used to obtain the area bounded by samples (a'), via the area factors table generated in a previous step. The new sample size is given by the survey unit area divided by a'. If this is still confusing, the MARSSIM itself provides a detailed discussion, as well as including a couple of examples.

---

* Chapter 11 presented an alternative approach for handling hot spots. The reader should not confuse the proposed statistical assessment approach for hot spots in Chapter 11 with that discussed in MARSSIM. The hot-spot approach discussed herein is consistent with MARSSIM.

For the example at hand, the number of survey data points needed for the sign test (14 in the survey unit) is positioned, on a scale map of each survey unit, using a random-start triangular pattern. The number of calculated survey locations, 14, was used to determine the grid spacing, L, of the triangular pattern. Specifically, the spacing, L, of the triangular pattern was given by

$$L = \sqrt{\frac{A}{0.866\,n}}$$

where A is the area of the class 1 survey unit (let us assume 1800 m²) and n is the number of data points in the survey unit. The spacing equals 12.2 m. The grid area (a') bounded by these survey locations was calculated by dividing the survey unit area A by the number of samples. In this case a' equals 129 m². This area represented the largest elevated area that could exist and not be sampled by the random-start triangular grid pattern established for the sign test. Specifically, the area factor by which the concentration in this potential elevated area (129 m²) can exceed the $DCGL_W$ value while maintaining compliance with the release criterion was determined using the RESRAD model.

RESRAD, version 5.95 was run to generate the area factors shown in Table 13.1. The input parameters included 1 pCi/g of Co-60 to a depth of 15 cm. The initial contaminated area used was 10,000 m and the length parallel to the aquifer was the square root of the contaminated area. So for each contaminated area, the length parallel to the aquifer was changed to maintain a square-shaped contaminated area in the model.

TABLE 13.1 Co-60 Area Factors Table Using RESRAD 5.95

Contaminated Area (m²)	Length (m)	Dose (mrem/y)	Area Factor
10,000	100	7.525	1
3000	54.77	7.348	1.02
1000	31.62	7.188	1.05
300	17.32	6.696	1.12
100	10	6.039	1.25
30	5.477	4.640	1.62
10	3.162	3.147	2.39
3	1.732	1.485	5.07
1	1	0.6341	11.87

Once the area factors table is complete, the area factor for the specific area of concern (129 m²) can be obtained by interpolation. Better yet, while the RESRAD code is still warm, input the contaminated area as 129 m² and determine the resulting dose. In this manner, the area factor can be obtained directly—the area factor is 1.21.

Just for practice, we can again illustrate the ease of calculating the $DCGL_W$. Basically, it simply requires the dose rate at the initial contaminated area (10,000 m²) to be divided into the release criterion of 25 mrem/y. Calculationally, we have 25 mrem/y divided by 7.525 mrem/y per 1 pCi/g, which results in 3.3 pCi/g.

The MDC of the scan procedure that is required to detect a hot-spot area at the $DCGL_{EMC}$ for the bounded areas (129 m²) was determined. Note that the $DCGL_{EMC}$ for the bounded area and the required scan MDC denote the same quantity. The required scan MDC for Co-60 was calculated by

$$\text{Scan MDC (required)} = (DCGL_W) \times (\text{area factor}) = 3.3 \times 1.21 = 4.0\,pCi/g$$

The actual MDCs can be determined for NaI scintillation detectors. NUREG-1507 provides scan MDCs for both 1.25″ × 1.5″ NaI detectors and 2″ × 2″ NaI detectors. Let us assume that the survey technicians are more comfortable scanning with the 1.25″ × 1.5″ NaI detectors. The scan MDC for this instrument can be looked up in NUREG-1507; for the conditions cited, the scan MDC is 5.8 pCi/g. This scan MDC is not sensitive enough to achieve the required scan MDC of 4.0 pCi/g. Therefore, additional soil samples are warranted if this scanning instrument is used.

The process of determining the number of additional samples (i.e., above the 14 required for the sign test) begins by calculating the area factor that corresponds to the actual scan MDC:

$$\text{Area factor} = \frac{\text{Actual scan MDC}}{DCGL_W} = \frac{5.8\,pCi/g}{3.3\,pCi/g} = 1.76$$

The area factor table (Table 13.1) must be logarithmically interpolated to determine the area that has an area factor of 1.76. The interpolation can be set up as follows:

$$\frac{\ln x - \ln 10}{\ln 1.76 - \ln 2.39} = \frac{\ln 30 - \ln 10}{\ln 1.62 - \ln 2.39}$$

Solving for *x*, the area is calculated to be 23.7 m².

Therefore, if this scanning instrument is used the total number of samples required is 1800 m² divided by 23.7 m², or 76 samples. The MARSSIM user has a decision to make: accept the fact that the sample size has increased from 14 to 76 by using the selected scan instrument, or try to identify an instrument that has greater sensitivity, but possibly at a greater cost.

Fortunately, the 2″ × 2″ NaI detector is not significantly more expensive, and indeed is more sensitive—its scan MDC is reported in NUREG-1507 as 3.4 pCi/g for Co-60. Because 3.4 pCi/g is less than the required scan MDC of 4.0 pCi/g, using this instrument results in no additional soil samples. That is, the NaI scintillation gamma scan survey using the 2″ × 2″ NaI detector has adequate sensitivity to detect any elevated areas of concern. The decision in this case is straightforward, survey technicians will become acquainted with the 2″ × 2″ NaI detector.

The MARSSIM design for Co-60, thanks in large measure to the DCGLs based on 25 mrem/y, is certainly straightforward. The scan MDC is also pretty good for Co-60, which further helps to keep the sample size to a minimum. While this initial MARSSIM example is somewhat simplistic in that it assumed only a single contaminant, it does provide an indication of the complexity involved in designing the MARSSIM surveys, especially for class 1 survey units. The next section provides an example of the WRS test, which is used when the contaminants are present in the background.

13.2.2 WRS Test Example: Uranium and Thorium in Soil

The WRS test is generally performed whenever the contaminant(s) are present in the background. Other than the fact that the WRS test requires the use of a background reference area to perform the statistical evaluation of the survey unit, this test is very similar to the sign test. The very same null hypothesis is used to demonstrate that the release criterion has been met:

Null hypothesis (H_0): Residual contamination in the survey unit exceeds the release criterion

The scenario for this example is the resultant uranium and thorium contamination in soil at a rare earths processing facility. During the recovery of zirconium and other rare earths, the land areas surrounding the plant have become contaminated with thorium; specifically, Th-232 in secular equilibrium with its progeny. The uranium contamination at the facility dates back to the 1950s when the company processed uranium for the Atomic Energy Commission for a period of 30 months. The uranium contamination consists of U-238, U-234, and U-235 at natural isotopic ratios. After 2 y of remedial actions, FSS activities to demonstrate compliance with the 25 mrem/y release criterion are now underway at this site. The outdoor soil areas have been classified and survey units have been identified.

13.2.2.1 Derived Concentration Guideline Levels

One particular exterior soil survey unit, contaminated with both thorium and uranium, has been classified as class 1. The survey unit measures 3000 m². It is anticipated that soil samples will be collected and analyzed by gamma spectroscopy for Th-232 and U-238. RESRAD modeling was used to determine DCGLs for these contaminants (results are provided in Chapter 5, Section 5.4); the results were 2.8 pCi/g for Th-232 and 178 pCi/g for U-238. The Th-232 $DCGL_W$ is based on the Th-232 decay series in secular equilibrium, while the U-238 $DCGL_W$ is based on U-238 serving as a surrogate for processed natural uranium. Again, refer to Section 5.4 for a more complete discussion on the nature of these DCGLs.

13.2.2.2 WRS Test: Determining Numbers of Data Points

The following steps detail the procedure for determining the number of data points for the WRS test. The number of data points necessary in this survey unit to satisfy the WRS test is based on the contaminant DCGLs, the expected standard deviation of the contaminants in the survey unit and background reference area, and the acceptable probability of making type I and type II decision errors. The mechanics in determining sample size are discussed, followed by the strategy for taking advantage of the MARSSIM flexibility. Because there are two DCGLs—that is, for Th-232 and U-238—it is necessary to use the unity rule for the determination of WRS test sample size. Note that when using the unity rule, no assumptions about the relationship between the two contaminants (Th-232 and U-238) are necessary—just the opposite for the case when surrogates are used to modify the DCGL. The following survey information and DQOs were used for this FSS design:

- There are two DCGLs used in this survey unit: the Th-232 $DCGL_W$ is 2.8 pCi/g and the U-238 $DCGL_W$ is 178 pCi/g.

- Results from characterization and remedial action support surveys were used to estimate the mean and standard deviation of Th-232 and U-238 in the survey unit. The background reference area samples were also collected and analyzed. The results were as follows:

	Survey Unit pCi/g (1σ)	Background Reference Area pCi/g (1σ)
Th-232	1.9 ± 0.5	1.1 ± 0.3
U-238	37.4 ± 8.8	1.4 ± 0.7

- Type I and type II decision errors were set at 0.05.

The mechanics of performing the WRS test are discussed below.

13.2.2.2.1 Calculate the Relative Shift The contaminant DCGL value, LBGR, and the standard deviation in the background level of the contaminant were used to calculate the relative shift, Δ/σ. When the estimated standard deviation in the reference area and survey units are different, the larger of these values should be used to calculate the relative shift. As shown above, both the Th-232 and U-238 standard deviations are greater in the survey unit.

The following information was used in the determination of relative shift, $(DCGL_W - LBGR)/\sigma$:

- *DCGL*. The $DCGL_W$ for Th-232 is 2.8 pCi/g and for U-238 is 178 pCi/g. Yet, whenever multiple radionuclides are measured in the same sample, the unity rule must be used. In this case, the $DCGL_W$ is one.

- *Standard deviation*. The standard deviation for each of the contaminants was determined from soil samples collected from the characterization and remedial action support surveys. The standard deviations are shown above—0.5 pCi/g for Th-232 and 8.8 pCi/g for U-238. The standard deviation when using the unity rule involves combining the individual standard deviations using the following equation:

$$\sigma = \sqrt{\left(\frac{\sigma_{Th\text{-}232}}{DCGL_{Th\text{-}232}}\right)^2 + \left(\frac{\sigma_{U\text{-}238}}{DCGL_{U\text{-}238}}\right)^2} = \sqrt{\left(\frac{0.5}{2.8}\right)^2 + \left(\frac{8.8}{178}\right)^2} = 0.185$$

- *LBGR*. The LBGR is set based on the expected net median concentration (by subtracting the background), put in terms of the unity rule, that is, sum of the fractions: $(1.9 - 1.1)/2.8 + (37.4 - 1.4)/178 = 0.488$; therefore, the LBGR will be set slightly above the expected net median concentration, at 50% of the $DCGL_W$. The type I and type II decision errors will be set at 0.05.

  Therefore, the relative shift, Δ/σ, is equal to $(1 - 0.5)/0.185$, or 2.7.

13.2.2.2.2 Determine P_r Table 5.1 in MARSSIM (NRC 2000a) contains a listing of values for P_r as a function of the relative shift. P_r is the probability that a random measurement from the survey unit is greater than a random measurement from the background reference area, by less than the $DCGL_W$, when the survey unit median concentration is actually at the LBGR above background (NRC 1998a). P_r is also a function of the relative shift $[(DCGL_W - LBGR)/\sigma]$, and increases as Δ/σ increases.

Using the relative shift value calculated above, the value of P_r can be obtained from tabulated values. Therefore, for a relative shift value of 2.7, which must be rounded off to the next lowest table entry of 2.5, the value of P_r is 0.961428.

(Note: As with the sign test, the MARSSIM user will typically skip this step because the WRS test sample size can be determined directly from tabulated values of relative shift and decision errors in MARSSIM Table 5.3.)

13.2.2.2.3 Calculate Number of Data Points for WRS Test The number of data points, N, to be obtained from each reference area/survey unit pair for the WRS test was calculated using

$$N = \frac{\left(z_{1-\alpha} + z_{1-\beta}\right)^2}{3\left(P_r - 0.5\right)^2}$$

Note that the N data points are divided between the survey unit (n) and the reference area (m), and that they are split equally ($n = m = N/2$). Substituting in the values determined above, N was calculated as

$$N = \frac{\left(1.645 + 1.645\right)^2}{3\left(0.961428 - 0.5\right)^2} = 16.9$$

The decision error percentiles used in the numerator for selected type I and type II error rates were obtained from MARSSIM Table 5.2.

Again, to assure sufficient sample size to attain the desired power level with the statistical tests and allow for possible lost or unusable data, or underestimated standard deviation, it is recommended that the number of calculated data be increased by 20%, and rounded up. This yielded N equal to 21, and splitting between the survey unit and background reference area evenly, $N/2$ equals 10.5 or 11 samples each. Remember that each sample will be analyzed for both Th-232 and U-238.

The prospective power curve can now be reviewed for this FSS design (Figure 13.2). Note that the survey unit concentration is in terms of the unity rule; this is necessary to interpret the survey design. On the basis of our characterization and remedial action support data, and correcting for the anticipated background levels, the estimated median concentration was 0.488. The power at this point is nearly 100%, in fact, even if the median concentration were actually closer to 0.7, the power is still greater than 95%. As long as our survey unit data are sufficiently estimated, this WRS test design seems more than reasonable.

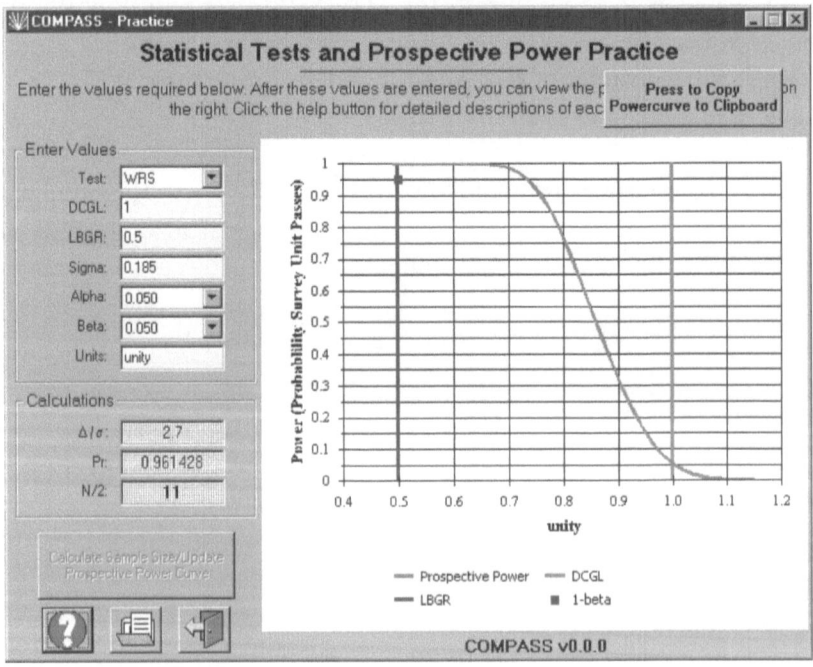

FIGURE 13.2   Prospective power curve for the WRS test example.

13.2.2.3 Determining Data Points for Areas of Elevated Activity

While the WRS test for this FSS design only requires 11 samples in the survey unit, because the survey unit is class 1, it is necessary to ensure that any potential hot spots of dosimetric concern are not missed during the FSS. As discussed in the previous example, this assessment requires the comparison of the actual scan MDC with the required scan MDC, which depends on area factors for Th-232 and U-238.

One of the first things we can do is to determine the average area in the survey unit that is bounded by soil samples. This can be quickly calculated by dividing the survey unit area (A) by the number of soil samples in the survey unit (n). In our example, the survey unit area is 3000 m² and n is 11 samples; therefore, the average area bounded by samples (a') is equal to 3000/11, or 273 m². Since the 11 samples will likely be laid out on a triangular pattern, this area (273 m²) represents the largest elevated area that could exist and not be sampled by the random-start triangular grid pattern established for the WRS test.

It is then necessary to determine the magnitude by which the concentration in this potential elevated area (273 m²) can exceed the $DCGL_W$ while maintaining compliance with the release criterion—this value is the area factor, and was calculated in Section 5.4 for Th-232 and U-238. Table 13.2 provides the soil concentration area factors for Th-232 and U-238. Note that these area factors are specific to the conditions specified in the problem statement, and in Section 5.4.

The scan MDC that is required to detect an elevated area at the limit determined by the area factor, which is the definition of the $DCGL_{EMC}$, is then determined. The required scan MDC for Th-232 and U-238 is given by the following equation:

$$\text{Scan MDC (required)} = (DCGL_W) \times (\text{area factor})$$

TABLE 13.2   Area Factors for Th-232 and U-238 for 3000 m² Example Survey Unit

Source Area (m²)	U-238 Area Factors	Th-232 Area Factors
3000	1	1
300	1.36	1.19
100	1.61	1.36
30	2.07	1.78
10	2.89	2.63
3	5.36	5.49
1	9.66	12.4

The area factors for both Th-232 and U-238 associated with a' of 273 m² could be interpolated from Table 13.2. However, since the area is close enough to 300 m², it is much easier to use the area factors for this value that is in the table, especially since it is not overly conservative to do so. (The area factors determined from interpolation might be slightly higher than this simple approximation). The area factors for Th-232 and U-238 are 1.19 and 1.36, respectively, for Th-232 and U-238. Thus, the required scan MDCs are calculated as

$$\text{Th-232: } (2.8 \text{ pCi/g}) \times (1.19) = 3.33 \text{ pCi/g}$$
$$\text{U-238: } (178 \text{ pCi/g}) \times (1.36) = 242 \text{ pCi/g}$$

The question now becomes: Can the survey instrument used to perform a 100% scan in this survey unit identify potential hot spots at the required scan MDCs above? The actual MDCs of scanning techniques can be determined from Chapter 9. The scan MDCs for 1.25″ × 1.5″ NaI scintillation detectors are 2.8 pCi/g for Th-232 and 115 pCi/g for processed natural uranium, which when based on the U-238 component is $(0.485) \times (115 \text{ pCi/g})$, or 56 pCi/g for U-238.

Because the actual scan MDC is less than the required scan MDCs for both Th-232 and U-238, no additional measurements will be required for this survey unit. In other words, the NaI scintillation detector used to perform the gamma scan exhibits adequate sensitivity to detect any elevated areas of concern. The FSS design will consist of 11 soil samples each in the survey unit and background reference area, and the survey unit will be 100% scanned.

Now the question may have come up; what if one or both of the actual scan MDCs were greater than the required scan MDC. In this case, the contaminant that requires the higher number of samples based on its scan MDC and area factor becomes the driver for the number of additional samples needed in the survey unit. (You may want to review the process that was used in the first example with Co-60.) But in this present example, since the scan MDCs for both Th-232 and U-238 were sufficiently sensitive, no matter what the mixture of Th-232 and U-238 is in the survey unit, the selected instrument will be sensitive enough to detect any hot spots of dosimetric concern.

One final technical point may come up in the course of class 1 FSS designs. Let us assume that unlike in this present example the two contaminants did indeed exhibit some fashion of a relationship between their relative concentrations. Perhaps the Th-232-to-U-238 ratio was between 0.3% and 0.85, 95% of the time. The manner that one might choose to assess whether or not the actual scan MDC was less than the required scan MDC is as follows. The lower bound ratio (0.3) is selected first, and this specific mixture is modeled using RESRAD to determine the $DCGL_W$ and area factors. The actual scan MDC can then be determined for this specific mixture using the scan MDC equation for multiple radionuclides covered in Chapter 9. Therefore, the comparison of actual scan MDC to the required scan MDC is specific for the ratio of Th-232 to U-238 selected. The same process can be repeated at two or three more Th-232-to-U-238 ratios to empirically prove that regardless of the specific ratio, the selected scanning instrument has sufficient sensitivity.

13.3 SURFACE-ACTIVITY MEASUREMENTS: WILCOXON RANK SUM TEST OR SIGN TEST?

In this section, we will explore the options related to the statistical design for surface-activity measurements—using either the WRS test or the sign test. While a case can be made for using either nonparametric test as the design basis for the number of surface-activity measurements, certain situations will usually dictate which test is more advantageous. For example, if the survey unit can be characterized as consisting of predominantly one surface material (e.g., concrete), then the WRS test may make more sense. Conversely, if the building area being considered as a survey unit consists of a number of different surface materials, then the sign test will likely provide a more optimal FSS design. The technical bases and considerations for making this decision are presented in this section.

The MARSSIM guidance for FSSs recommends that the WRS test be used when the contaminant is present in the background or whenever gross (as contrasted with radionuclide-specific) measurements are performed. The WRS test evaluates whether the difference between the median of the survey unit data and the median of the background reference data is greater or lesser than the release criterion. As discussed in Chapter 10, survey techniques for measuring surface activity during decommissioning surveys involves the gross measurement of alpha and/or beta activity on surfaces, routinely using GM or gas proportional detectors. Consequently, strictly following the MARSSIM guidance for performing the WRS test requires that a background reference area be selected for each sufficiently distinct surface type, such as concrete, drywall, metal, and wood. The disadvantage of this approach is that each building surface area must be divided into survey units not only based on contamination potential—which of course is the intent of survey unit identification—but also on the basis of different surface materials present within the area. The potential impact of using the WRS test for surface-activity assessments in this circumstance is an overall increase in the number of measurements due to more survey units. This predicament becomes particularly acute for larger decommissioning sites with many survey units composed of different surface materials, such as power reactors and fuel cycle facilities.

Problem Statement: The WRS test may result in building surface areas that logically comprise a single survey unit—because of their similar contamination potential throughout—and may in fact require further division into multiple survey units due to the need for different background reference areas for each of the various material surface types present in the area.

The decision to use the WRS test or sign test also impacts the manner in which the surface-activity measurement data are collected and how the results are reported. Basically, when using the WRS test, the gross measurement results are not corrected for the background; rather they are simply reported in units of cpm. Conversely, when the sign test is employed, the gross measurements are corrected for the background, and the results are reported in dpm/100 cm². The next section recaps the surface-activity equation that was covered in Chapter 10.

13.3.1 Surface-Activity Measurements

The sign test approach is consistent with the conventional approach for assessing surface-activity levels. This approach results in the calculation of the surface activity following an assessment of both the gross radiation levels on surfaces in the survey unit and the background radiation levels from appropriate surface materials. A well-known background for the detector and surface type is subtracted from the gross surface-activity measurement. The background level accounts for both the ambient exposure rate background (from natural gamma radiation) and the surface material background. The DQO process should be used to determine the number of background measurements necessary for each material type.

Therefore, the expression for surface activity per unit area as given in Chapter 10 is ideally suited for the desired sign test results in dpm/100 cm²:

$$A_s = \frac{R_{S+B} - R_B}{\varepsilon_i \varepsilon_s W}$$

where R_{S+B} is the gross count rate of the measurement in cpm, R_B is the background count rate in cpm, ε_i is the instrument or detector efficiency (unitless), ε_s is the efficiency of the contamination source (unitless), and W is the area of the detector window (cm^2).

As discussed in Chapter 7, the background response of the detector is the sum of the ambient exposure rate background (commonly referred to as instrument background) and the contribution from naturally occurring radioactivity in the surface material being assessed. In most cases, it is the instrument background that produces the majority of the overall detector background response. Nevertheless, background levels for surface-activity measurements do vary because of the presence of naturally occurring radioactive materials in building surfaces, and the possible shielding effect that these construction materials can provide. The surface material background considerations discussed in Chapters 7 and 10 should guide the determination of the proper background count rate, R_B.

Surface-activity assessments can be further complicated due to the sometimes significant variation of ambient background levels, in addition to variable surface material backgrounds. In these circumstances, shielded measurements can be used to distinguish the ambient background from the surface material background. This approach requires the subtraction of the two background components—ambient background and surface material background—from each gross surface-activity measurement in the survey unit. This situation probably dictates the use of the sign test, since it is very unlikely that an appropriate background reference area could be found that has the same ambient background and surface material background.

The next two sections describe the specific applications of the WRS test and the sign test to surface-activity assessments.

13.3.2 WRS Test for Surface-Activity Assessment

The WRS test approach for performing surface-activity assessments will be illustrated using the MARSSIM example in Appendix A (of MARSSIM). This example shows how the WRS test sample size is determined for surface-activity measurements. The example assumes that gross beta activity on structure surfaces is being measured, and that a building of similar construction was identified on the property that can serve as a reference area. Two reference areas—one for concrete surfaces and one for drywall surfaces—were required based on the surface materials present in the building. Proceeding with the example, direct measurements of gross beta activity were made using 1-min counts with a gas flow proportional counter with an MDC of 425 dpm/100 cm². This MDC satisfies the DQOs for instrument selection, as it is actually less than 10% of the $DCGL_W$ for Co-60—which is 5000 dpm/100 cm² for this example.

A gas flow proportional counter with 20 cm² probe area (note: this is likely a MARSSIM typo since GM pancake detectors have 20 cm² probe areas, while gas proportional detectors are typically much larger) and 16%, 4π response (efficiency) was placed on the surface at each direct measurement location, and a 1-min count was taken. The $DCGL_W$, adjusted for the detector size and efficiency, is

$$DCGL_W = (5000 \text{ dpm}/100 \text{ cm}^2)\ (20 \text{ cm}^2/100 \text{ cm}^2)\ (0.16) = 160 \text{ cpm}$$

The decision to use gross activity measurements for building surfaces means that the survey of all the interior survey units was designed for use with the two-sample WRS test for comparison with an appropriate reference area.

In the example in Appendix A, the site has 12 concrete survey units in the building that are compared with one reference area. Drywall surfaces were not further considered in this example. The same type of instrument (e.g., gas proportional detector) and survey method were used to perform measurements in each area. The DQO process was used to select parameters necessary to design the survey. The standard MARSSIM null hypothesis was used for the WRS test:

H_0: residual radioactivity in the survey unit exceeds the release criterion (e.g., $DCGL_W$)

The LBGR was selected to be one-half of the $DCGL_W$, and type I and type II decision errors associated with the null hypothesis were selected at 0.05. The relative shift, Δ/σ, was calculated by first determining the width of the gray region, Δ, which is simply the $DCGL_W$ minus the LBGR, or 80 cpm.

Data from previous scoping and characterization surveys indicated that the background level is 45 ± 7 (1σ) cpm. The standard deviation of the contaminant in the survey unit (σ_s) is estimated at ± 20 cpm. When the estimated standard deviation in the reference area and the survey units are different, the larger value should be used to calculate the relative shift. Thus, the value of the relative shift, Δ/σ, is (160 − 80)/20 or 4. From Table 5.3, for the sample size corresponding to the type I and type II decision errors selected, the calculated relative shift, was 18 measurements. This means 18 data points total for each reference area and survey unit pair. Of this total number, nine were planned from the background reference area and nine from each survey unit.

A similar calculation would be required for each of the other surface materials in a building survey unit—for example, drywall, wood, and ceramic. That is, if this example included a survey unit that comprised more than one surface material, say three different material surface types, each one would constitute a new survey unit for the WRS test. In this case, all factors being equal, it may be necessary to have nine measurements in each surface material survey unit within a room that would logically be one survey unit due to the contamination potential. It is precisely this real-life scenario of having many different surface materials in a room/building that may make the sign test an attractive alternative.

13.3.3 Sign Test for Surface-Activity Assessment

When using the sign test for surface-activity assessments, each gross measurement made in the survey unit is corrected by an appropriate background value. The resulting net surface-activity levels represent the data that are evaluated by the sign test. The benefit of this approach is obvious—survey units based on the contamination potential do not require further division based on surface material composition. The expressed concern with this approach is that the true background variability may not be accounted for when the mean value of the background is subtracted from each gross measurement; that is, the WRS test, by comparing the survey unit and background distributions, provides more statistical power than a simple background subtraction. Section 13.3.4 discusses the results of a simulation study that was performed to address this concern.

Because the MARSSIM framework is based on the data quality objectives approach, it seems reasonable that alternative approaches, provided that they can be justified, may be used for survey design. Therefore, when using the sign test to determine the sample size, one must consider the selected type I and type II errors, and standard deviation of the contaminant in the survey unit. The standard deviation of the contaminant in the survey unit must be estimated to calculate sample size. It is the standard deviation of the contaminant as well as the background standard deviation that must be carefully considered when using the sign test for this application.

For survey units that truly contain residual radioactivity (e.g., class 1 and 2 survey units), the variability due to the surface activity will likely surpass the background variability, and using the sign test with well-known background subtraction should not be much of a concern. The concern that arises is that the background variability may not be properly accounted for when taking the average of some number of background measurements and subtracting this point estimate of background from each gross measurement. For survey units that do not have residual radioactivity (likely class 3 survey units), the standard deviation must include the background measurement and spatial variability. For example, if a survey unit contains three distinct surface types, then the overall variability in background presented by these surfaces must be determined and used in the sample size determinations. The mechanics of this calculation is discussed next.

The DQO process should be used for determining the number of background measurements needed. The overall variability should be determined when using the sign test for surface-activity measurement assessment. Because the average background is subtracted from each gross surface-activity measurement, the overall uncertainty has two components—the standard deviation in the survey unit and the background standard deviation. Since the sign test can be used for multiple surface types in a survey unit, it is prudent to use the background surface type present in the survey unit that exhibits the greatest standard deviation. The overall variability can be determined from the propagated errors:

$$\sigma_{total} = \sqrt{\sigma_s^2 + \sigma_r^2}$$

Now, let us look at the sign test applied to the MARSSIM Appendix A for example. As compared to the WRS test—which compares instrument response in cpm in both the survey unit and the reference area—the sign test is performed by comparing the net dpm/100 cm^2 value directly to the DCGL$_W$. As provided in the Appendix A example, the standard deviation of the contaminant for the concrete surfaces was 20 cpm. This value can be translated to surface-activity units: 20 cpm/(0.16 c/dis)/0.20 = 625 dpm/100 cm^2. However, we will assume that this survey unit includes both concrete and drywall surfaces, so it is necessary to consider the standard deviation over the entire survey unit. Let us assume that the standard deviation that covers both surfaces is increased to 800 dpm/100 cm². It should be recognized that for nearly all surface-activity assessments, the standard deviation resulting from the presence of contamination will greatly exceed the variation from background surface-activity levels.

Returning to our MARSSIM example, assume that a number of measurements were performed on each background surface material. Type I and II errors are still 0.05, and the relative shift is calculated, again setting LBGR to one-half of the DCGL. The relative shift, Δ/σ, is given by (5000 − 2500)/800, or 3.1. Table 5.5 in MARSSIM requires 14 measurements for the sign test.

Therefore, using the sign test to design the survey requires 14 measurements in the survey unit, as compared to 9 measurements using the WRS test. Assuming that each room had three different surface materials, each room using the WRS test results in 27 measurements (9 measurements for each material, which is a separate survey unit), while only 14 are needed for the sign test. For this one room, the WRS test would require a total of 54 measurements—27 in the three survey units comprising the room, and 27 for the required three background reference areas. The sign test would require the 14 direct measurements in the survey unit, and then about 10 measurements for each of the 3 background reference surface areas, for a total of 44 measurements. Therefore, this example shows a reduction of 10 direct measurements for a typical room using the sign test in place of the WRS test. More importantly, recognize that facilities usually have more than one room. The reduction in the total number of direct measurements due to the sign test grows as the size of the facility increases. Moreover, the difference in sample size becomes even greater as the number of different surface materials is increased.

CASE STUDY   SIGN TEST EXAMPLE FOR MULTIPLE
BACKGROUNDS IN SURVEY UNIT

Let us consider the MARSSIM survey design for contaminated building surfaces using the sign test. The class 2 survey unit (360 m²) is located on the second floor of the machine shop, and consists of both concrete and wood surfaces. The contamination has been identified as processed natural uranium from miscellaneous machining operations. The project engineer has determined that a gas proportional detector operated in the alpha plus beta mode will be used to conduct the survey.

The regulator has provided the following uranium surface-activity DCGLs:

U-238	1700 dpm/100 cm²
U-235	1600 dpm/100 cm²
U-234	1600 dpm/100 cm²

The project engineer used uranium ratios of 0.488 for both U-238 and U-234, and 0.023 for U-235, along with the gross activity DCGL equation in Chapter 6 to calculate the gross activity DCGL for processed uranium:

$$\text{Gross activity DCGL} = \frac{1}{(0.488/1700) + (0.023/1600) + (0.488/1600)}$$
$$= 1650 \, \text{dpm}/100^2$$

The gas proportional total efficiency for processed natural uranium (Table 13.3) was determined using the approach outlined in Chapter 8.

The gross activity $DCGL_W$ can be converted to the net count rate by multiplying by the total efficiency and the physical probe area, resulting in a $DCGL_W$ of 581 cpm.

Background surface-activity measurements were performed on concrete and wood. Each measurement was 1 min long.

Concrete (cpm)	Wood (cpm)
426	308
392	317
402	320
415	322
385	315
400	299
392	312
410	308
418	312
407	319

From these data, the background means and standard deviations (1σ) were 405 ± 13 cpm for concrete and 313 ± 6.9 cpm for wood.

During characterization survey activities, a number of measurements were performed in the machine shop. The mean and standard deviation of these gross measurements was 612 ± 120 cpm.

The overall standard deviation can be determined—propagating the error in the survey unit gross mean and the background surface material exhibiting the greatest standard deviation (concrete in this example):

TABLE 13.3   Weighted Total Efficiency for Processed Natural Uranium Using
Gas Proportional Detector

Radionuclide	Uranium Fraction	Instrument Efficiency	Surface Efficiency	Total Efficiency
U-238	0.488	0.38	0.25	4.64E–02
Th-234	0.488	0.18	0.25	2.20E–02
Pa-234 m	0.488	0.66	0.50	1.61E–01
U-234	0.488	0.38	0.25	4.64E–02
U-235	0.023	0.38	0.25	2.2E–03
Th-231	0.023	0.22	0.25	1.3E–03
			Weighted total efficiency	0.28

$$\sigma_{total} = \sqrt{\sigma_s^2 + \sigma_r^2} = \sqrt{(120)^2 + (13)^2} = 121\,cpm$$

Returning to the other survey unit DQOs, the type I and type II error rates were each set at 0.05. Setting the LBGR in this case requires a bit of thought. Since the DCGL$_W$ is in terms of the net count rate, the LBGR must also be in net cpm. We also desire to set the LBGR somewhat close to the expected median concentration in the survey unit. The survey unit information we have is the gross count rate of 612 ± 120 cpm, and to convert to the net count rate we need to subtract the background. Alas, we have two different backgrounds with concrete and wood. If we subtract the higher-background material (concrete), then we might underestimate the net count rate in the survey unit, and if we subtract the lower-background material (wood), then we will have a conservative estimate of the net cpm (median concentration) in the survey unit. Of course, we could determine how much of the total surface area is represented by concrete and wood, and determine a weighted background to subtract from the estimated gross mean in the survey unit. This being said, the LBGR will be selected at 300 cpm as a result of subtracting the lower background from the survey unit gross count rate (612 − 313 cpm), thereby assuming that the lower background value prevails in the survey unit. As seen in the COMPASS results in Figure 13.3, the sign test sample size (N) is 15, and the prospective power curve is shown.

Let us assume that we implemented this FSS design and collected 15 samples. Each of the random gross measurements in this survey unit were converted to surface activity by subtracting the average background for the particular surface material, and correcting for total efficiency and physical probe area. The survey unit results are shown in Table 13.4.

The actual gross mean and standard deviation of these FSS measurements were 622 and 205 cpm, respectively. Thus, we underestimated the actual variability in the survey unit (remember that the standard deviation from characterization survey was 120 cpm). This means that the actual power of the survey

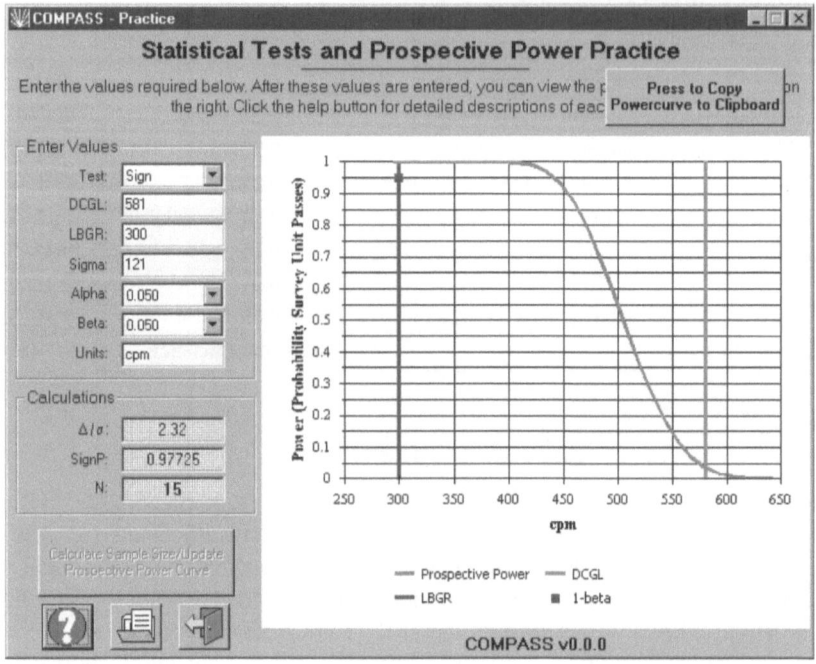

FIGURE 13.3 Prospective power curve for sign test used for surface-activity assessment.

TABLE 13.4 Surface-Activity Measurement Results from FSS

Sample #	Gross Measurement (cpm)	Surface Material	Net Surface Activity (dpm/100 cm²)
1	667	Concrete	740
2	452	Wood	390
3	1089	Concrete	1940
4	658	Concrete	720
5	392	Concrete	−40
6	377	Wood	180
7	444	Wood	370
8	561	Concrete	440
9	1012	Concrete	1720
10	702	Wood	1100
11	633	Concrete	650
12	555	Concrete	430
13	589	Wood	780
14	489	Wood	500
15	712	Concrete	870

design will be somewhat less than that predicted by the prospective power curve. Only when the sign test is performed on these data will we know for sure whether the survey unit passes or not. Chapter 14 describes how the sign test is performed. Do you think this survey unit will pass?

13.3.4 Simulation Study Conceptual Design

To conduct the simulation studies, a typical room within a facility was modeled. It was assumed that the room had been contaminated by a mixture of Co-60 and Cs-137, and was classified as class 2 for the FSS. From the class 2 definition in MARSSIM, these areas have, or had, a potential for radioactive contamination or known contamination, but are not expected to exceed the $DCGL_W$ (NRC 2000a). The gross activity $DCGL_W$, based on the radionuclide mixture of Co-60 and Cs-137, is 6500 dpm/100 cm^2.

This room was composed of four walls constructed of drywall (each was 3 m high) and the floor was a poured concrete slab, measuring 20 m × 25 m. The northern portion of the room (8 m × 20 m) was covered with linoleum flooring. It was assumed that the class 2 survey unit includes the floor and lower walls up to 2 m (the upper walls and ceiling would likely be a class 3 survey unit). Thus, the concrete floor area in this survey unit is 340 m^2, the total drywall area is (2) × (2 m)(20 m) + (2) × (2 m) (25 m) = 180 m^2, and the linoleum floor covering is 160 m^2.

The visitor's center has been identified as an appropriate background reference area. Twenty background measurements have been performed, uniformly spaced, for each surface material in this building, using a gas proportional detector (total efficiency equals 0.25 c/dis; physical probe area is 126 cm^2). The mean and standard deviation for each background material were reported as

Concrete floor: 344 cpm ± 38 cpm (1σ)

Drywall: 312 cpm ± 46 cpm (1σ)

Linoleum: 366 cpm ± 35 cpm (1σ)

Using the WRS test, each of the three surface materials in this room would constitute one survey unit, and therefore require a specific sample size determination. The sample size calculation provides $N/2$ samples for the concrete, drywall, and linoleum flooring. The sign test considers the entire class 2 area as a single survey unit, composed of three separate surface materials. The modeling assumed that there were varying levels of contamination within the room (floor and lower walls). For instance, median contamination levels at 0.7 $DCGL_W$, 0.9 $DCGL_W$, and 1.15 $DCGL_W$ were modeled. The simulation study was performed by the Statistical Sciences Group at Los Alamos National Laboratory (LANL) to objectively evaluate the statistical power of each test.

Martz and Keller-McNulty (1999) describe the simulation study conceptual design and comparison results of the sign and WRS tests. The report concludes that for the example problem considered in the simulation, *the power of the sign test is essentially the same as that of the WRS test, and that the subtraction of the mean background when using the sign test appears to have no appreciable effect on either the power or the probability of a type I error.* Therefore, it was concluded that the power of the sign test is sufficiently large relative to that of the WRS test and that the sign test is an acceptable alternative to the WRS test.

It is therefore suggested that the sign test be viewed as a legitimate alternative for surface-activity assessments. The sign test typically results in significant savings in the number of surface-activity measurements required for the MARSSIM FSS due to an overall reduction in the number of survey units needed, and appears to have no appreciable effect on the power or probability of a type I error. It is recognized that some circumstances may warrant the use of the WRS test, such as in the event that the survey unit is composed of only one surface material type.

13.4 COMPARISON OF MARSSIM AND NUREG/CR-5849 FSSS FOR NUCLEAR POWER PLANT DECOMMISSIONING PROJECTS

When the MARSSIM became publicly available in December 1997, one of the first questions raised concerned its impact on the rigor of FSSs. Specifically, D&D contractors were interested in the comparison of FSS designs using the new MARSSIM strategy and the NRC's NUREG/CR-5849 guidance at nuclear power reactor facilities. To satisfy this inquiry, an FSS design comparison was performed on two structure surface survey units at the Fort St. Vrain (FSV) nuclear power plant. The completed survey units were selected from the turbine building and the reactor building, where both survey units had been classified as "affected" and surveyed using the NUREG/CR-5849 designation. These same survey units were then evaluated using the MARSSIM FSS guidance to provide a basis for comparison.

A quick overview of each FSS design may prove useful for comparing the NUREG/CR-5849 and MARSSIM FSS designs. Following the NUREG/CR-5849 design, the site is divided into affected and unaffected survey units, and the sample size for each survey unit simply depends on the survey unit classification; one measurement per 1 m² for affected survey units and 30 measurements for unaffected survey units. Therefore, the number of direct measurements of surface activity in each of these power reactor survey units was set at the NUREG/CR-5849 prescribed frequency of one measurement per 1 m².

The MARSSIM design starts with the site being divided into class 1, 2, and 3 survey units depending on the contamination potential. The MARSSIM FSS sample size depends on several variables, including the $DCGL_W$, LBGR, type I and II decision errors, and the variability of contamination in the survey unit. Furthermore, MARSSIM offers a choice of statistical tests to base the survey design, either the WRS test or the sign test. The WRS test will be the basis for the MARSSIM FSS used in this comparison. The instrument selection and survey procedure for making surface-activity measurements will be considered the same for both survey designs.

The FSS design for the turbine building will be studied first. This survey unit included floor, lower walls, and equipment surfaces, and was classified as "affected" using NUREG/CR-5849. The total surface area in this survey unit was about 573 m². According to the survey design guidance, direct measurements of surface activity were collected on a 1 × 1 m grid, for a total of 573 measurements using a gas proportional detector. To determine the MARSSIM sample size for this survey unit, a number of inputs are necessary, including the survey unit's classification and the $DCGL_W$.

The FSV site-specific guideline considering the radionuclide mix was 4000 dpm/100 cm²—this value will be taken as the $DCGL_W$. While the DCGL has no bearing on the NUREG/CR-5849 sample size, it is the starting point for determining the MARSSIM sample size. Concerning the MARSSIM classification, it is beneficial to review the results from the FSV FSS for this survey unit:

- Mean = 82 dpm/100 cm²

- Standard deviation = 238 dpm/100 cm²

- Maximum value = 676 dpm/100 cm²

- Upper 95% confidence level = 98 dpm/100 cm²

It is interesting to note that NUREG/CR-5849 required more than 570 measurements; simply based on intuition, a sample size much smaller should suffice, considering the mean and maximum value in the survey unit as compared to the guideline.

On the basis of these survey data, it is likely that this survey unit would be classified as class 2. Also, this area constitutes only one survey unit because the approximate floor area (400 m²) is less than 1000 m². DQO inputs for the MARSSIM sample size include an initial LBGR of 50% of the DCGL, and type I and II errors of 0.05. The standard deviation (238 dpm/100 cm²) is used from the NUREG/CR-5849 survey results. The relative shift equals (4000 − 2000)/238, or 8.4. Since this is rather high, we can take advantage of high relative shift by moving LBGR closer to DCGL$_W$ by setting the LBGR equal to 3600 dpm/100 cm². (This is making use of step 7 of the DQO process—Optimizing the Survey Design.) The relative shift is now (4000 − 3600)/238, or 1.68. Table 5.3 in MARSSIM provides the WRS test sample size for these DQOs: only 16 direct measurements required in this class 2 survey unit. To summarize, the FSS comparison for the turbine building survey unit:

- NUREG/CR-5849 573 direct measurements

- MARSSIM   16 direct measurements

What about comparing a MARSSIM class 1 survey unit to the NUREG/CR-5849 survey design?

To facilitate this comparison, a survey unit within the reactor building was selected. This survey unit included the floor, the lower walls, and the equipment surfaces, for a total surface area of 390 m². This survey unit was classified as "affected." The NUREG/CR-5849 survey design in this survey unit resulted in 474 direct measurements of surface activity. Again, the site-specific guideline for the radionuclide mix was 4000 dpm/100 cm². Survey unit summary results were as follows:

- Mean = 105 dpm/100 cm²

- Standard deviation = 416 dpm/100 cm²

- Maximum value = 2422 dpm/100 cm²

- Upper 95% confidence level = 136 dpm/100 cm²

Once again, the survey unit easily satisfies the release criteria, hardly justifying the exorbitant number of samples required by NUREG/CR-5849.

While it may be possible to justify classifying this survey unit as class 2, we will designate this survey unit as class 1. This means that in addition to satisfying the statistical test data needs, the overall sample size must also consider the sample size for the EMC (which involves an assessment of the scan MDC and dose-based area factors). For this approach to work, the MARSSIM design requires a $DCGL_W$ and area factors based on dose modeling, and an estimate of the actual scan MDC. To perform the MARSSIM FSS design for these units, it was necessary to calculate the DCGLs for building surface contamination. RESRAD-BUILD was used to calculate the surface contamination levels based on the expected radionuclide mixture in the survey units and default modeling parameters that corresponded to an annual dose of 25 mrem. The source term identified at FSV was as follows:

Fe-55	74.2%
H-3	10.9%
Co-60	8.6%
C-14	1.0%

This source term was input into RESRAD-BUILD at the fractional amounts given for each radionuclide, and the DCGL based on a release criterion of 25 mrem/y was calculated. The $DCGL_W$ for the FSV mixture resulted in 60,370 dpm/100 cm². Area factors were also determined for this radionuclide mixture using RESRAD-BUILD.

The survey instrumentation used for direct measurements was the same as that used for the NUREG/CR-5849 survey design, and the scan sensitivity was evaluated using the approach outlined in Chapter 9. The total efficiency for the gas proportional detector was weighted for the radionuclide mix in Table 13.5.

TABLE 13.5 Total Efficiency Weighted for the Radionuclide Source Term

Nuclide	Fraction	Total Efficiency	Weighted Total Efficiency
Fe-55	0.742	0	0
H-3	0.109	0	0
Co-60	0.086	0.21	0.018
C-14	0.01	0.05	5E–4
		Weighted total efficiency = 0.02	

The MARSSIM survey design is based on the development of data quality objectives—including decision errors for testing the null hypothesis and the expected standard deviation of the surface contamination in the reactor building survey unit (416 dpm/100 cm^2). This standard deviation was based on the gas proportional detector used at FSV, which had a total efficiency of 21%. The standard deviation used for the MARSSIM design must correct this value for the total efficiency of 2%—this yields a standard deviation of approximately 4500 dpm/100 cm^2. Setting the LBGR to 52,000 dpm/100 cm^2, the relative shift is given by (60,370 − 52,000)/4500, or 1.9.

The MARSSIM WRS test sample size for this relative shift and type I and II decision errors is 13 direct measurements. However, it is necessary to evaluate the need for additional samples due to hot spots in this class 1 survey unit.

The gas proportional detector scan MDC was determined for a scan rate of 10 cm/s (observation interval was 1 s), a background of 500 cpm, d' of 2.12, and a total efficiency of 0.02. The scan MDC for the gas proportional detector was 25,900 dpm/100 cm^2. Because the scan MDC is less than the DCGL$_W$, no additional samples above that required by WRS test are needed. So, again the comparison is completely one-sided: 474 direct measurements using NUREG/CR-5849 versus 13 direct measurements using MARSSIM.

Now, one may be thinking at this point that the MARSSIM comparison to NUREG/CR-5849 for this survey unit is unfair, given the large DCGL of 60,370 dpm/100 cm^2 based on 25 mrem/y. In one sense, this is a reasonable concern, because the low sample size may be more a result of the relatively high DCGLs based on 25 mrem/y rather than on the MARSSIM survey design. But at the same time, the increase in the release criterion may not be as egregious as it appears at first glance. That is, while the DCGL$_W$ is higher (60,370 dpm/100 cm^2 vs. 4000 dpm/100 cm^2), the total efficiency is only 2% as compared to 21% used by FSV. Now, if the comparison is made accounting for the efficiencies, to put the release criteria in terms of net counts (an equal footing for the comparison), we find that the DCGL is only 44% higher than the NUREG/CR-5849 guideline (i.e., 60,370 × 0.02 vs. 4000 × 0.21, or 1200 cpm vs. 840 cpm).

In conclusion, the MARSSIM approach resulted in a significant reduction in the required number of direct measurements of surface activity. The increased DCGL value as a result of going to a dose-based release criterion was a significant factor as well. Remember that the principal contaminants at power reactor D&D projects are beta emitters, and that the DCGLs for beta emitters generally increased when compared to

Regulatory Guide 1.86 guidelines. The same is not true for alpha emitters; so, a comparison using MARSSIM for a nuclear power plant with significant alpha contamination may have a very different conclusion.

The MARSSIM survey design implemented at reactor D&D sites may greatly reduce sample sizes, but these potential savings come at the expense of significantly more up-front planning and design time. The prescriptive approach presented by NUREG/CR-5849 resulted in the same number of samples, regardless of the release criterion, a situation entirely inconsistent with the DQO process.

13.5 ANNOTATED MARSSIM EXAMPLES

This section provides examples of MARSSIM survey design. These examples cover both interior and exterior scenarios, and involve both class 1 and 2 survey units. The survey designs in these examples consider the use of surrogates and the unity rule.

13.5.1 Example 1: Class 1 Interior Survey Unit

An interior survey unit is potentially contaminated with Co-60. The survey unit has been classified as class 1, and encompasses a floor and lower wall surface area of 320 m^2 (the floor area alone is about 100 m^2). The surfaces are composed of concrete floor and concrete block surfaces—it is assumed that the surface material backgrounds are reasonably similar for the purposes of background reference area selection. It is anticipated that GM detectors will be used to perform surface-activity measurements for gross beta radiation, while a gas proportional floor monitor will be used for scanning building surfaces. This example will demonstrate the WRS test approach to MARSSIM FSS design. Prospective power curves will also be generated using COMPASS to illustrate the survey design.

RESRAD-BUILD was used to obtain the DCGL$_W$ and area factors for Co-60. The data below are reproduced from Table 5.2.

Source Area (m^2)	Area Factor	DCGL$_W$
100	1	11,400 dpm/100 cm^2
36	1.45	
25	1.68	
16	2.06	
9	2.77	
4	4.58	
1	13.7	

13.5.1.1 Survey Instrumentation

Before the number of direct measurements can be determined, it is necessary to calculate the static MDC for the GM detector to demonstrate that it has sensitivity less than 50% of the $DCGL_W$. The expression of the MDC equation used assumes equivalent (paired) observations of the sample and blank (i.e., equal counting intervals for the sample and background). The expression for MDC may be written as

$$MDC = \frac{3 + 4.65\sqrt{C_B}}{KT}$$

where C_B is the background count in time, T, of a paired blank. The quantities encompassed by the proportionality constant, K, include both the instrument efficiency, ε_i, and the surface efficiency, ε_s, and physical probe area. Assume that the GM detector was calibrated to Tc-99 (similar beta energy as Co-60). Suppose that 20, 1-min background measurements were performed, and that the average background, C_B, was equal to 72 counts on concrete. The quantities encompassed by the proportionality constant, K, include $\varepsilon_i = 0.21$, $\varepsilon_s = 0.25$ (from ISO-7503) and the physical probe area is 20 cm². The MDC is calculated as

$$MDC = \frac{3 + 4.65\sqrt{72}}{(0.21)(0.25)20/100} = 4040 \text{ dpm}/100 \text{ cm}^2$$

Therefore, the GM detector, calibrated to Tc-99 and used to make 1-min measurements of surface activity has sufficient sensitivity, that is, it is <35% of the $DCGL_W$.

As mentioned in the problem statement, a gas proportional floor monitor will be used for scanning building surfaces. Assume that the scan MDC is determined for a background level of 1200 cpm and a 1 s observation interval. For a specified level of performance of 90% true-positive rate and 20% false-positive rate, d' equals 2.12 and it can be shown that the MDCR is 570 cpm. Using a surveyor efficiency of 0.5, and assuming instrument and surface efficiencies of 0.24 (from Table 9.2) and 0.25, respectively, the actual scan MDC is calculated as

$$\text{Scan MDC} = \frac{570}{\sqrt{0.5}(0.24)(0.25)} = 13,400 \text{ dpm}/100 \text{ cm}^2$$

Since the scan MDC is greater than the $DCGL_W$, it is not immediately known whether additional samples above that are needed by the statistical tests will be necessary.

13.5.1.2 WRS Test Sample Size Determination

The number of data points necessary in this survey unit to satisfy the WRS test is based on the contaminant $DCGL_W$, the expected standard deviation of the contaminant in the background and in the survey unit, and the acceptable probability of making type I and type II decision errors. (Note that the results of the characterization survey affect each of these parameters needed for the design of the FSS.)

- The $DCGL_W$ for Co-60 for this example is 11,400 dpm/100 cm².

- Results from characterization surveys were used to estimate the standard deviation of gross beta measurements in both the survey unit and an appropriate background reference area. The survey unit results (which just includes the concrete floor) were 126 ± 58 cpm (1σ), while the background reference area indicated 72 ± 19 cpm (1σ). The higher standard deviation is used for survey design purposes.

- Other DQO inputs include the LBGR set initially at 50% of the $DCGL_W$, type I error of 0.05, and a type II error of 0.05.

The following steps detail the procedure for determining the number of data points for the WRS test. The key parameters affecting sample size include the magnitude of the decision errors (types I and II) and the relative shift (Δ/σ). The contaminant $DCGL_W$ and the standard deviation of the contaminant in the survey unit are used to calculate the relative shift, Δ/σ.

It is necessary to convert the gross beta $DCGL_W$ into the same units as the standard deviation:

$$\text{Gross } \beta \text{ DCGL} = (11{,}400 \text{ dpm}/100 \text{ cm}^2)(0.21)(0.25)\, 20/100 = 120 \text{ cpm}$$

Note that these are net counts above the background.

The estimated median concentration is 126 cpm − 72 cpm, or 54 cpm net count rate. This is 54/120 or 45% of the $DCGL_W$. Therefore, the LBGR being set at 50% of the $DCGL_W$, or 60 cpm is favorable; it ensures that the power at the estimated concentration will be quite high. The estimated concentration in the survey unit is used to evaluate the survey design—essentially, one desires a high probability of passing the survey unit at the median concentration in the survey unit. The relative shift equals (120 cpm − 60 cpm)/58 cpm = 1.0.

Table 5.3 in MARSSIM (NRC 2000a) provides a list of the number of data points to demonstrate compliance using the WRS test for various values of α, β, and Δ/σ. These values were determined using the WRS test sample size equation and have already been increased by 20%. These numbers represent $N/2$, to be conducted in each survey unit and corresponding reference area—and results in 32 direct measurements using the GM detector. The prospective power curve for this design is shown in Figure 13.4. The statistical power at the estimated median of 45 cpm is nearly 100%.

However, before we proceed to implement this survey design, we must assess the data requirements for the EMC. This involves a comparison of the actual scan MDC to the required scan MDC. The required scan MDC is calculated as the product of the area factor for the area bounded by samples, and the $DCGL_W$. The area bounded by samples is simply 320 m²/32 samples, or 10 m². The area factor associated with this area

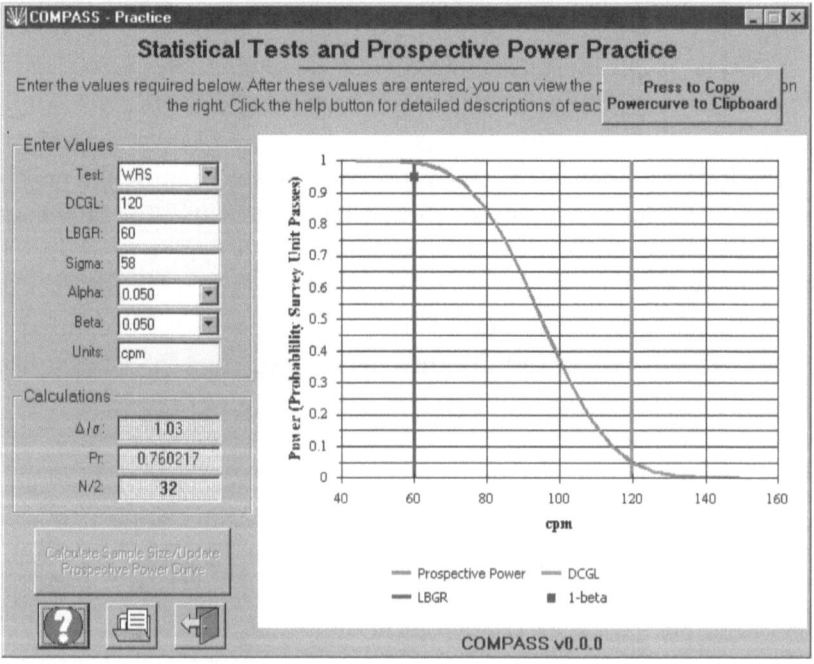

FIGURE 13.4 Prospective power curve for WRS test for surface-activity assessment.

can be determined by interpolating the area factors table shown earlier. Studying this table, we can see that the area factor for 10 m² lies between 2.06 and 2.77, based on area factors for 16 and 9 m², respectively. Using the more conservative smaller value of 2.06, the required scan MDC is 11,400 dpm/100 cm² × 2.06 or 23,500 dpm/100 cm². Because the actual scan MDC (13,400 dpm/100 cm²) is less than this value, no additional direct measurements are required.

13.5.2 Example 2: Class 2 Interior Survey Unit

An interior survey unit is potentially contaminated with Am-241, Cs-137, and SrY-90. The survey unit has been classified as class 2, and encompasses a floor area of 820 m². The floor surface is composed of poured concrete. It is anticipated that gas proportional detectors will be used to perform surface-activity measurements for both gross alpha (in alpha-only mode) and gross beta radiation (in beta-only mode). The sign test will be used for this example. The DCGLs for these contaminants are as follows: 30 dpm/100 cm² for Am-241; 4000 dpm/100 cm² for SrY-90; and 18,000 dpm/100 cm² for Cs-137.

Note that if this were a class 1 survey unit, then it would be necessary to obtain area factors—and since the survey design would likely comprise the use of a gross beta DCGL for the SrY-90 and Cs-137, the area factors table would necessarily be specific for the stated mixture of these two beta emitters. An illustration of this concept is offered in Example 4.

13.5.2.1 Gross DCGLs

FSS design for multiple contaminants often involves the unity rule, surrogate approach, gross activity DCGL, or any combination of the former. Because gross measurements of both alpha and beta are planned, it is desired to establish a gross DCGL for both alpha and beta activity. For the gross alpha DCGL it is simply the Am-241 DCGL since it is the only alpha emitter. The gross beta DCGL is more involved because we have two potential beta contaminants, SrY-90 and Cs-137. The results of our characterization survey in this survey unit have identified the relative ratios of these contaminants as 0.35 SrY-90 and 0.65 Cs-137. The gross beta DCGL is calculated as

$$\text{Gross } \beta \text{ DCGL} = \frac{1}{(0.35/4000) + (0.65/18,000)} = 8090 \text{ dpm/100 cm}^2$$

13.5.2.2 Survey Instrumentation

Prior to determining the number of direct measurements, it is necessary to calculate the static MDC for both the gross alpha and gross beta detectors to demonstrate that they have sensitivities less than 50% of the DCGL$_W$. The expression for MDC may be written as

$$MDC = \frac{3 + 4.65\sqrt{C_B}}{KT}$$

where C_B is the background count in time, T, of a paired blank. The quantities encompassed by the proportionality constant, K, include both the instrument efficiency, ε_i, and the surface efficiency, ε_s, and the physical probe area.

We assume that the gas proportional detector used for alpha measurements was calibrated to the Am-241 alpha emission. A 5-min background count was performed, and C_B was equal to 3 cpm on concrete. The quantities encompassed by the proportionality constant, K, are as follows: $\varepsilon_i = 0.45$, $\varepsilon_s = 0.25$ (from ISO-7503), and the probe area is 126 cm². The MDC is calculated as

$$MDC = \frac{3 + 4.65\sqrt{(3\,\text{cpm})*(5\,\text{min})}}{(5\,\text{min})(0.45)(0.25)(126/100)} = 30\,\text{dpm/100 cm}^2$$

Therefore, making gross alpha measurements for 5 min provides detection sensitivity equal to about the DCGL$_W$, not 50% of the DCGL$_W$. However, because we are comfortable that we have not overestimated the MDC, we will proceed with this measurement approach for gross alpha.

We assume that the gas proportional detector used for beta measurements was calibrated to the weighted beta energy from the two contaminants to determine the instrument efficiency. Suppose that a 1-min background count was performed, and C_B was equal to 325 counts on concrete. The quantities encompassed by the proportionality constant, K, are as follows: $\varepsilon_i = 0.52$ (from 35% × 0.56 for SrY-90 added to 65% × 0.5 for Cs-137), $\varepsilon_s = 0.50$ (from ISO-7503, both these beta emitters have endpoint beta energies greater than 400 keV) and the probe area is 126 cm². The MDC is calculated as

$$MDC = \frac{3 + 4.65\sqrt{325}}{(0.52)(0.5)(126/100)} = 265\,\text{dpm/100 cm}^2$$

Therefore, the gas proportional detector used for gross beta measurements and used to make 1-min measurements of surface activity has sufficient sensitivity—it is <5% of the gross beta $DCGL_W$.

13.5.2.3 Sign Test Sample Size Determination

Because there are two modified DCGLs—that is, one for gross alpha measurements of Am-241 and one for gross beta measurements that account for SrY-90 and Cs-137—it is necessary to use the unity rule for the determination of sample size. The number of data points necessary in this survey unit to satisfy the sign test is based on the contaminant DCGLs, the expected standard deviation of the contaminants in the survey unit, and the acceptable probability of making types I and II decision errors.

- There are two DCGLs used in this survey unit: the gross alpha $DCGL_W$ is 30 dpm/100 cm², and the gross beta $DCGL_W$ is 8090 dpm/100 cm².

- Results from characterization surveys were used to estimate the standard deviation of gross alpha and beta measurements in both the survey unit and an appropriate background reference area. For both the gross alpha and gross beta measurements, the standard deviation was higher in the survey unit as compared to background reference area: 310 cpm for the gross beta and 1.7 cpm for the gross alpha.

- Other DQO inputs include the LBGR set initially at 50% of the $DCGL_W$, type I error of 0.05 and a type II error of 0.05.

The following steps detail the procedure for determining the number of data points for the sign test using the unity rule. The key parameters affecting sample size include the magnitude of the decision errors (types I and II) and the relative shift (Δ/σ). The contaminant $DCGL_W$ and the overall standard deviation of the contaminants (normalized for the unity rule) in the survey unit are used to calculate the relative shift, Δ/σ.

It is necessary to convert the gross alpha and beta DCGLs into the same units as the standard deviation:

Gross α DCGL = (30 dpm/100 cm²)(0.45)(0.25)126/100 = 4.25 cpm

Gross β DCGL = (8090 dpm/100 cm²)(0.52)(0.5)126/100 = 2650 cpm

Because we will be applying the unity rule, it is also necessary to calculate the normalized standard deviation for gross alpha and gross beta measurements:

$$\sigma = \sqrt{\left(\frac{1.7}{4.25}\right)^2 + \left(\frac{310}{2650}\right)^2} = 0.417$$

The relative shift can now be calculated as $(1 - 0.5)/0.417 = 1.2$.

Table 5.5 in MARSSIM (NRC 2000a) provides a list of the number of data points to demonstrate compliance using the sign test for various values of α, β, and Δ/σ. This sample size ($N = 23$ direct measurements) is obtained from the survey unit.

Is this a reasonably good survey design? Believe it or not, you cannot answer this question without some additional information about the survey unit. Namely, what is the anticipated median concentration of gross alpha and gross beta in the survey unit? Suppose that the same characterization data used to obtain the standard deviations was used to provide estimates of the mean net count rates of gross alpha and gross beta, as follows:

Gross alpha mean net count rate: 2.3 cpm

Gross beta mean net count rate: 825 cpm

The LBGR was set at 50% of the DCGL, without any idea of the median concentration in the survey unit. This median value can be estimated by dividing each value above by the respective gross alpha and gross beta DCGL:

2.3 cpm/4.25 cpm + 825 cpm/2650 cpm = 0.85 (in terms of unity)

Reviewing the prospective power curve in Figure 13.5 illustrates an important lesson in MARSSIM survey design—the survey design is as good as the degree to which the median concentration and standard deviation in the survey unit is known. Reviewing the prospective power curve, one can determine the power at the median concentration of 0.85. Specifically, the probability of passing the survey unit using this survey design is only 35%. To answer the original question, no, this certainly is not a good survey design given the survey unit characteristics.

But at the same time, what if the median concentration in the survey unit was 0.65? In this case, the probability of passing the survey unit is 90%, and one would conclude that this survey design is fine. Here is the point: there is more to designing a MARSSIM FSS than simply crunching through the sample size calculations. To fully employ the DQO process to

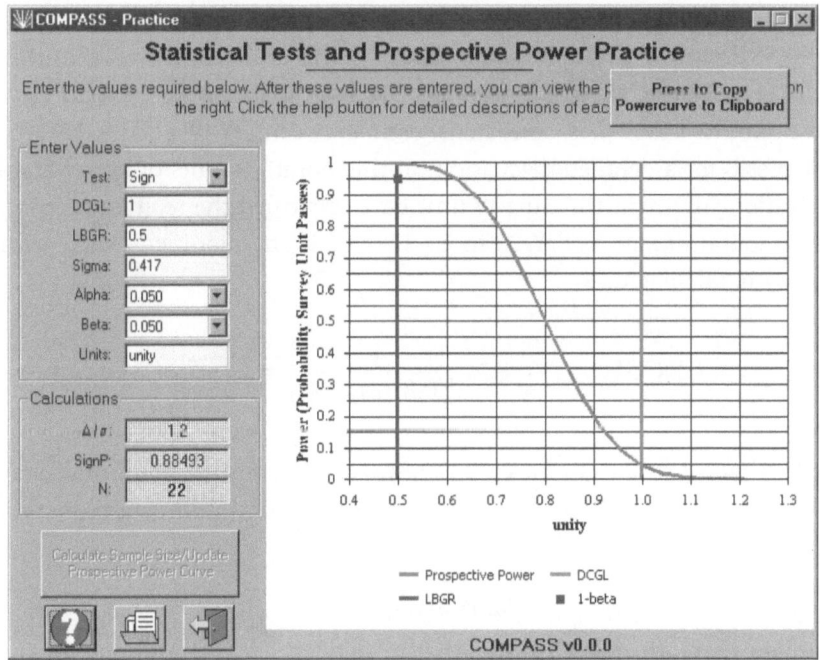

FIGURE 13.5   Prospective power curve for sign test using unity rule for surface-activity assessment in class 2 survey unit.

optimize the survey design, one must have a good understanding of the median concentration and standard deviation of the contaminants in the survey unit; do not underestimate the importance of HSA and characterization data.

13.5.3 Example 3: Class 1 Exterior Survey Unit

An exterior soil survey unit is potentially contaminated with Am-241, Cs-137, and SrY 90. The survey unit has been classified as class 1, and encompasses a soil area of 1600 m². It is anticipated that soil samples will be collected and analyzed by gamma spectroscopy for Am-241 and Cs-137, and that the Cs-137 will also serve as a surrogate for the SrY-90. The DCGLs for these contaminants are as follows: 1.5 pCi/g for Am-241; 4 pCi/g for SrY-90; and 8 pCi/g for Cs-137. Cs-137 has been identified in background at levels ranging from <0.1 to 0.6 pCi/g. We will use the sign test, and in effect, decide to "eat" the Cs-137 background. That is, we make the assumption that Cs-137 is entirely from the D&D site, and therefore we do not have to make background measurements of Cs-137.

13.5.3.1 Modified DCGL

The FSS design for multiple contaminants in this exercise involves the unity rule, as well as the surrogate approach. To use the surrogate approach, it is necessary to establish a "consistent" ratio between SrY-90 and Cs-137 (see Chapter 6 for a detailed discussion on this point). Results of our characterization survey in this survey unit have identified the relative ratios of these contaminants as SrY-90 to Cs-137 = 0.8. The modified $DCGL_W$ for Cs-137 is calculated as

$$DCGL_{Cs,\,mod} = 8 * \left(\frac{4}{[(0.8)(8)] + 4} \right) = 3.1 \, pCi/g$$

This modified DCGL for Cs-137 now accounts for the amount of SrY-90 in the survey unit, and is used for survey design and data reduction purposes.

13.5.3.2 Sign Test Sample Size Determination

Because there are two DCGLs—that is, one for Am-241 and one for the modified Cs-137—it is necessary to use the unity rule for the determination of sample size. The number of data points necessary in this survey unit to satisfy the sign test is based on the contaminant DCGLs, the expected standard deviation of the contaminants in the survey unit, and the acceptable probability of making types I and II decision errors.

- There are two DCGLs used in this survey unit: Am-241 $DCGL_W$ is 1.5 pCi/g and the modified Cs-137 $DCGL_W$ is 3.1 pCi/g.

- Results from characterization surveys were used to estimate the mean and standard deviation of Am-241 and Cs-137 in the survey unit. The results were as follows:

$$Am\text{-}241 \quad 0.4 \pm 0.3 \ pCi/g \ (1\sigma)$$

$$Cs\text{-}137 \quad 1.6 \pm 1.1 \ pCi/g \ (1\sigma)$$

- The LBGR is set based on the expected median concentration, put in terms of the unity rule: $0.4/1.5 + 1.6/3.1 = 0.78$, or 78% of the $DCGL_W$. Types I and II decision errors are set at 0.1.

The overall standard deviation of the contaminants (normalized for the unity rule) in the survey unit must be calculated to determine the relative shift, Δ/σ:

$$\sigma = \sqrt{\left(\frac{1.1}{3.1}\right)^2 + \left(\frac{0.3}{1.5}\right)^2} = 0.41$$

The relative shift for the unity rule equals $(1 - 0.78)/0.41 = 0.54$ (rounded off to 0.5).

Table 5.5 in MARSSIM (NRC 2000a) provides the sign test sample size for these DQOs. The sample size results in 54 soil samples. This sample size is somewhat large because the anticipated median concentration is rather close to the DCGL.

As before, the class 1 survey unit must be assessed to determine if the data requirements for the FMC are satisfied. This involves a comparison of the actual scan MDCs for both Am-241 and Cs-137 to the required scan MDCs. For this survey unit, we will scan using a $2'' \times 2''$ NaI detector. Scan MDCs are reported in NUREG-1507 as 31.5 pCi/g for Am-241 and 6.4 pCi/g for Cs-137.

The required scan MDC is calculated as the product of the area factor and the $DCGL_W$. The area bounded by samples is simply 1600 m²/54 samples = 30 m². The area factor associated with this area can be determined directly from MARSSIM Table 5.6, which results in an area factor of 1.7 for Cs-137 and 44.2 for Am-241. Thus, the required scan MDCs are calculated as

$$\text{Cs-137:} \quad (3.1 \text{ pCi/g}) \times (1.7) = 5.3 \text{ pCi/g}$$

$$\text{Am-241:} \quad (1.5 \text{ pCi/g}) \times (44.2) = 66.3 \text{ pCi/g}$$

Because the actual scan MDC is greater than the required scan MDC for Cs-137 (6.4 pCi/g vs. 5.3 pCi/g), additional measurements will be required for this survey unit. The actual scan MDC for Am-241 is capable of detecting the required scan MDC.

To proceed, Cs-137 becomes the "driver" for the number of additional measurements. The next step is to calculate the necessary area factor: $(6.4 \text{ pCi/g})/(3.1 \text{ pCi/g}) = 2.06$. We now interpolate Table 5.6 to determine the area that corresponds to this area factor; it is 19.1 m². So, the new sample size is determined by dividing this area into the survey unit area: 1600 m²/19.1 m² = 84 soil samples. This represents an increase of 30 samples above the 54 samples required by the sign test. Thus, the overall survey design requires 84 soil samples to be laid out on a systematic grid, and each sample analyzed for Cs-137 and Am-241. Additionally, some number of samples should be analyzed for Sr-90 as well, to confirm the surrogate ratio used to modify the Cs-137 DCGL.

Another possible solution to this EMC design problem would be to determine a reasonable relationship between the Cs-137 and Am-241. Using dose modeling, DCGLs and area factors appropriate for the assumed mixture of these radionuclides are determined. The scan MDC that represents the specific combination of these two radionuclides is calculated (see Chapter 9 for scan MDCs for multiple radionuclides). Finally, this actual scan MDC that accounts for both radionuclides is compared to the required scan MDC that is based on the area factors table specific to this mix of Cs-137 and Am-241. The key, of course, is being able to justify a reasonable relationship between the two radionuclides—the relationship would be known after the fact because each sample will be analyzed for both Cs-137 and Am-241.

13.5.4 Example 4: Class 1 Interior Survey Unit with Multiple Contaminants

A building surface survey unit at a sealed source facility is potentially contaminated with Am-241, Co-60, Cs-137, and SrY-90. The survey unit has been classified as class 1, and encompasses a total floor and lower wall surface area of 320 m². The floor surface is composed of poured concrete and measures 115 m². Gas proportional detectors will be used to perform surface-activity measurements for both gross alpha (in alpha-only mode) and gross beta radiation (in beta-only mode). The sign test will be used for this example. The DCGLs for these contaminants were determined using RESRAD-BUILD and are as follows: 130 dpm/100 cm² for Am-241; 34,400 dpm/100 cm² for SrY-90; 11,400 dpm/100 cm² for Co-60; and 44,000 dpm/100 cm² for Cs-137.

13.5.4.1 Gross Activity DCGLs and Area Factors

The FSS design for multiple contaminants on building surfaces often involves the unity rule and the gross activity DCGLs. Because gross measurements of both alpha and beta are planned, it is desired to establish a gross DCGL for both alpha and beta activities. For the gross alpha DCGL, it is simply the Am-241 DCGL since it is the only alpha emitter (same situation as Example 2). The gross beta DCGL is more involved because we have three potential beta contaminants, Co-60, SrY-90, and Cs-137. The results of HSA and characterization survey activities in this survey unit have identified the relative ratios of these contaminants as

0.30 Co-60, 0.20 SrY-90, and 0.50 Cs-137. The gross beta DCGL is calculated as

$$\text{Gross } \beta \, \text{DCGL} = \frac{1}{(0.3/11,400) + (0.2/34,400) + (0.5/44,000)}$$
$$= 23,000 \, \text{dpm}/100 \, \text{cm}^2$$

It is necessary to determine area factors for this class 1 MARSSIM survey design using RESRAD-BUILD. Table 5.4 in Chapter 5 provides the area factors for this particular mix of beta contaminants, which are the area factors specific to this precise mix of beta emitters:

Source Area (m²)	Area Factor	DCGL$_W$
100	1	23,100 dpm/100 cm²
36	1.60	
25	1.85	
16	2.14	
9	3.12	
4	5.22	
1	16.0	

Note that the DCGL$_W$ for this radionuclide mixture was determined directly from RESRAD-BUILD to be 23,100 dpm/100 cm², the small difference from the DCGL calculated in the above equation is likely from the impact of modeling multiple radionuclides at the same time (and the resulting differences in the time for peak dose to occur).

RESRAD-BUILD was also used to generate area factors for Am-241:

Source Area (m²)	Area Factor	DCGL$_W$
100	1	130 dpm/100 cm²
36	2.78	
25	4.01	
16	6.25	
9	11.1	
4	25.0	
1	100	

Note the simplistic interpolation of the Am-241 area factors table—each product of the source area and the area factor is 100, a result of the area factors being linear with source area. This phenomenon occurs when the inhalation pathway is dominant, because the inhalation dose is directly related to the contaminated source area that could be removed and rendered airborne. The area factors table for the beta emitters has a sizable external radiation component, so interpolation is not as direct.

13.5.4.2 Instrumentation, Static MDC, and Scan MDC

Before the number of direct measurements can be determined, it is necessary to calculate the static MDC for both the gross alpha and gross beta detectors to demonstrate that they have sensitivities less than 50% of the $DCGL_W$, using the expression for MDC:

$$MDC = \frac{3 + 4.65\sqrt{C_B}}{KT}$$

where C_B is the background count in time, T, of a paired blank. The quantities encompassed by the proportionality constant, K, includes both the instrument efficiency, ε_i, and the surface efficiency, ε_s, and physical detector area.

Assume that the gas proportional detector used for alpha measurements was calibrated to a Th-230 alpha calibration source. Assume that a 1-min background count was performed, and C_B was equal to 2 cpm on concrete. The quantities encompassed by the proportionality constant, K, are as follows: $\varepsilon_i = 0.44$, $\varepsilon_s = 0.25$ (from ISO-7503), and the physical probe area is 126 cm². The MDC is calculated as

$$MDC = \frac{3 + 4.65\sqrt{2}}{(0.44)(0.25)126/100} = 69\,dpm/100\,cm^2$$

Hence, performing gross alpha measurements for 1 min provides an adequate detection sensitivity, just a little more than 50% of the DCGL.

Now, we must focus on using the gas proportional detector used for beta measurements. The first step is to determine the weighted instrument and surface efficiencies for these beta contaminants. Assume that instrument efficiencies were determined for each of the beta contaminants and resulted in 0.41, 0.59, and 0.46, respectively, for Co-60, Sr-90, and Cs-137. The weighted instrument efficiency is calculated as follows:

$$\varepsilon_i = (0.30)(0.41) + (0.20)(0.59) + (0.50)(0.46) = 0.47$$

The surface efficiency is determined in a similar fashion. On the basis of guidance in ISO-7503, the surface efficiency for Sr-90 and Cs-137 is 0.50 based on their beta energies, and is 0.25 for the lower-energy Co-60. The weighted surface efficiency is calculated as follows:

$$\varepsilon_s = (0.30)(0.25) + (0.20)(0.50) + (0.50)(0.50) = 0.425$$

Now, suppose that a 1-min background count was performed, and C_B was equal to 360 counts on concrete. The MDC is calculated as

$$MDC = \frac{3 + 4.65\sqrt{360}}{(0.47)(0.425)126/100} = 360\,dpm/100\,cm^2$$

Therefore, the gas proportional detector has a sufficiently low MDC when used for gross beta measurements (<2% of the gross beta $DCGL_W$).

Because this example comprises a class 1 survey unit, it is necessary to evaluate the scan MDCs for both the alpha and beta detectors. We will assume that the same instruments used for static measurements will also be used for scanning. Let us first focus on the alpha scan MDC. Assume that the scan speed (2 cm/s) is such that a residence time of about 5 s is maintained over the contamination (for a nominal hot-spot size of 10×10 cm). The instrument efficiency is 0.44, and the surface efficiency is 0.25. Therefore, the scan MDC is based on a 95% probability of detecting one count:

$$\alpha\ scan\ MDC = \frac{[-\ln(1 - 0.95)]60}{(0.44)(0.25)(5)} = 330\,dpm/100\,cm^2$$

The beta scan MDC is determined next. The scan MDC will be determined for the same background level of 360 cpm and a 1-s interval using the gas proportional detector. The scanning level of performance is given by 95% true-positive rate and 20% false-positive rate, so d' equals 2.12 (from MARSSIM Table 6.5) and the MDCR is calculated as follows:

$$b_i = (360\ cpm)(1\ s)(1\ min/60\ s) = 6\ counts$$

$$s_i = (2.12)(6)^{\frac{1}{2}} = 5.19\ counts$$

$$MDCR = (5.19\ counts)[(60\ s/min)/(1\ s)] = 312\ cpm$$

Using a surveyor efficiency of 0.5, and instrument and surface efficiencies of 0.39 (scanning instrument efficiencies are assumed to be less than static instrument efficiencies; see Table 9.3) and 0.425 (from static calculation of surface efficiency), respectively, the scan MDC is calculated as

$$\text{Scan MDC} = \frac{312}{\sqrt{0.5}\,(0.39)(0.425)} = 2660\,\frac{\text{dpm}}{100\,\text{cm}^2}$$

13.5.4.3 Sample Size Determination

Because there are two DCGLs—that is, one for gross alpha measurements based on Am-241 and one modified for the gross beta measurement of three beta emitters—it is necessary to use the unity rule for the determination of sample size. The number of data points necessary in this survey unit to satisfy the sign test is based on the contaminant DCGLs, the expected standard deviation of the contaminants in the survey unit, and the acceptable probability of making types I and II decision errors.

- The two DCGLs used in this survey unit: gross alpha $DCGL_W$ is 130 dpm/100 cm², and the gross beta $DCGL_W$ is 23,000 dpm/100 cm². To facilitate the sample size determination, it is necessary to convert the gross alpha and beta DCGLs into the same units as the standard deviation (cpm):

  Gross α DCGL = (130 dpm/100 cm²)(0.44)(0.25)126/100 = 18 cpm

  Gross β DCGL = (23,000 dpm/100 cm²)(0.47)(0.425)126/100 = 5790 cpm

- The results from recent characterization surveys, and supplemented by remedial action support survey measurements in those areas requiring decontamination, were used to estimate the mean and standard deviation of gross alpha and beta measurements in the survey unit. The standard deviations for gross alpha and gross beta measurements in the survey unit were as follows:

  Gross alpha standard deviation: 5 cpm

  Gross beta standard deviation: 562 cpm

- The mean gross alpha and beta measurements in the survey unit were 8 cpm and 1544 cpm, respectively.

- Since the unity rule will be used, the $DCGL_W$ is one. The LBGR will be set slightly above the expected median contaminant concentration in the survey unit. This value can be estimated by using the characterization/remedial action support data, and dividing each value by the gross alpha and gross beta DCGL:

$$8 \text{ cpm}/18 \text{ cpm} + 1544 \text{ cpm}/5790 \text{ cpm} = 0.71$$

Therefore, an LBGR value of 0.72 will be selected. Additional DQOs include a type I error of 0.05 and a type II error of 0.10.

- It is also necessary to calculate the normalized standard deviation for gross alpha and gross beta measurements to facilitate the determination of the relative shift.

$$\sigma_{overall} = \sqrt{\left(\frac{5}{18}\right)^2 + \left(\frac{562}{5790}\right)^2} = 0.294$$

The relative shift for the unity rule equals $(1 - 0.71)/0.294 = 0.985$.

Table 5.5 in MARSSIM (NRC 2000a) provides the sign test sample size for various values of α, β, and Δ/σ. The sign test requires 26 direct measurements for the desired DQOs.

Check out the prospective power curve for this survey design (Figure 13.6). So as long as the estimated mean and standard deviations for the gross alpha and beta measurements are reasonably accurate, this survey design seems appropriate.

But what about the potential for elevated measurements—are additional direct measurements needed for this? It is necessary to evaluate the scan MDC of the selected alpha and beta scanning instruments to determine if additional direct measurements may be required. The first step in this process is to assess the scan MDC for the scanning instrument—called the "actual scan MDC." The actual scan MDC for the gross beta scan (2660 dpm/100 cm²) is less than the gross beta DCGL of 23,000 dpm/100 cm², so no additional measurements are necessary on account of the gross beta scanning instrument. However, this is not the case for the gross alpha scanning instrument. Here, the actual scan MDC is 330 dpm/100 cm², while the gross alpha DCGL is only 130 dpm/100 cm².

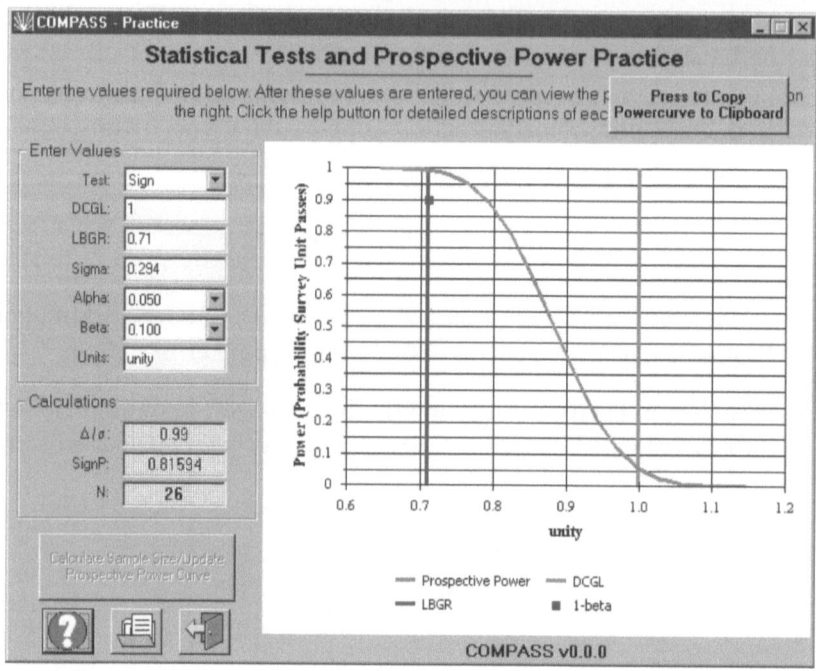

FIGURE 13.6 Prospective power curve for sign test using unity rule for surface-activity assessment in class 1 survey unit.

The next step is to determine the "required scan MDC" for the gross alpha scan instrument. The required scan MDC is the product of the $DCGL_W$ and the area factor. The area factor is obtained from dose modeling, and is determined based on the size of the area (a') bounded by the sample size in the survey unit. This bounded area (a') is simply the survey unit area (320 m²) divided by the number of samples from the statistical test 26, which equals 12.3 m². This area represented the largest elevated area that could exist and not be sampled by the random-start triangular grid pattern established for the sign test.

The area factor that corresponds to this 12.3 m² area is obtained from the Am-241 area factors table. Because of the linear relationship between the area factors, where the product of the two is equal to 100, the area factor is simply 100 divided by 12.3, or 8.1. Thus, the Am-241 concentration in this potential elevated area (12.3 m²) can exceed the $DCGL_W$ value by a factor of 8.1 and still maintain compliance with the release criterion.

The MDC of the scan procedure that is required to detect a hot-spot area at the $DCGL_{EMC}$ for the bounded areas (12.3 m²) is determined. The required scan MDC for the gross alpha scanning instrument is given by

$$\text{Scan MDC (required)} = \text{DCGL}_w \times \text{area factor} = 130 \times 8.1 = 1051 \frac{\text{dpm}}{100\,\text{cm}^2}$$

If the actual scan MDCs is less than the required scan MDC, no additional measurements are necessary. Well, since 330 dpm/100 cm² is certainly less than 1050 dpm/100 cm², we conclude that no additional samples are necessary, and that the sign test sample size will suffice.

The survey design in this example will be implemented by identifying a random-start location, and then laying out 26 measurement locations on a triangular pattern. At each location, separate measurements of gross alpha and gross beta radiation will be performed. The survey data will be assessed using the unity rule (refer to Chapter 14). If either the gross beta scan or the gross alpha scan was insufficiently sensitive, to the degree that additional measurements for one of the radiation types (gross alpha or gross beta) were necessary, it would likely be appropriate to also perform the additional measurements for the radiation type not responsible for increasing the sample size.

For example, let us assume that the scan MDC for gross alpha was higher than it was for this example, and that the sample size increased from 26 to 42 because of the poor alpha scan MDC. Because the grid pattern will be established by this larger required sample size, it is recommended that the sample size for gross beta measurements also be increased to 42, even though the gross beta scan MDC is sufficiently sensitive. Not only would this increase the overall statistical power of the survey design, but from a practical standpoint, it alleviates the concern of tracking which alpha measurement locations did not have a beta measurement.

13.6 MARSSIM FSS DESIGN STRATEGIES: UNDERSTANDING THE POWER CURVE

One of the more valuable improvements in the MARSSIM survey design is the flexibility that it offers. Unlike previous FSS guidance documents, the MARSSIM does not recommend a prescriptive sample size for a specified area. Recall that the sampling frequency for "affected" survey units in NUREG/CR-5849 was one direct measurement per 1 m² for structure surfaces and four soil samples per 100 m² of land area. MARSSIM, on the other hand, bases the sample size on the release criterion (DCGL), contaminant variability in the survey unit and the background reference area, and the magnitude of type I and type II decision errors. Additional parameters that impact the MARSSIM sample size include the LBGR and the sensitivity of measurement methods, particularly the scan MDC in class 1 survey units. The increased flexibility in FSS design is most directly the result of MARSSIM being undergirded by the DQO process. In essence, MARSSIM provides sufficient guidance for the D&D contractor to plan an effective FSS, while at the same time fully expressing the fact that not all sites and radiological conditions warrant the same approach. This is the spirit of the DQO process in MARSSIM—design the most cost-effective, technically defensible FSS possible for your D&D site. Hopefully, the MARSSIM strategies discussed in this textbook, and particularly, in this section, will heighten your awareness of the design options available.

In one sense, the flexibility in MARSSIM can be put in the context of a trade-off between the sample size and the type II decision error. Obviously, the D&D contractor desires a survey plan that calls for as few samples as possible, while at the same time minimizes the type II error, that is, incorrectly failing to release a survey unit that satisfies the release criteria. Unfortunately, these two parameters are inversely related. Reducing the sample size tends to increase the probability of incorrectly failing a survey unit, while reducing the type II error generally increases the sample size. What is one to do? In this section, we will discuss each of the parameters that affect this balance. Note that Appendix D in MARSSIM provides an excellent discussion of the flexibility offered by MARSSIM.

The null hypothesis in MARSSIM, sometimes called scenario A, is stated as

H_0: Residual radioactivity in the survey unit exceeds the release criterion (i.e., DCGL)

The D&D contractor strives to reject the null hypothesis, and thereby pass the survey unit. In this process of determining the fate of the survey unit, two decision errors may be made. Decision errors can occur when H_0 is rejected when it is true (type I error), or when H_0 is not rejected (accepted) when it is false (type II error). Note that committing either one of these errors is not desirable, but one type of error may be less desirable than the other—and it depends entirely on one's perspective as to which error is more troublesome. Clearly, the regulator is more concerned with type I errors, while the D&D contractor pays more attention to minimizing type II errors. To make matters more interesting, all things being equal, a decrease in the type I error increases the type II error, and vice versa.

So, how is this dilemma in setting decision errors solved? Well, let me put it this way: while the regulator may be willing to listen to any and all D&D contractor arguments on this issue, the regulator will ultimately ensure that the selected type I error is acceptable. In fact, it is likely that the type I errors will be established, or at least approved, by the responsible regulator. For example, the NRC has stated in Appendix A of NUREG-1757, volume 2 that type I errors of 0.05 are acceptable, and further, that larger type I errors may be approved under certain conditions (NRC 2006a). That is, it may seem reasonable to the D&D contractor to propose a type I error rate of 0.20 to help reduce the sample size in a difficult survey design situation (perhaps the survey design is exacerbated by a very low DCGL). Although the regulator may not buy-in to this increased type I decision error, it is nonetheless worthwhile to communicate these concerns and ideas to the regulator. One effective argument is to point out that the type I error set during the survey design process only occurs at the $DCGL_W$ value, and that the type I errors at concentrations greater than the $DCGL_W$ are even lower.

In reality, the D&D contractor only has unilateral control over the type II error rate because the regulator will have a major say in the type I error (α). So, let us explore some issues in setting the type II error (β) rate. First, selecting a relatively low value of β minimizes the risk of unnecessary investigations as a result of incorrectly failing a survey unit—thus, the allure of going with a low type II error rate. Be that as it may, the primary trade-off for selecting a low type II decision error is increased sample size. One needs to only review MARSSIM Tables 5.3 and 5.5 reproduced in Appendix B to see the impact. Conversely, if the D&D contractor strives to minimize the sample size, then a relatively high value of β should be selected. How high? Well, that depends on one's aversion to the higher risk of failing survey units that actually meet the release criterion. To put these error values into context, the "standard" value of the type II error is 0.05, so higher values may be 0.10 or 0.20, usually no greater than 0.25, while low values may be 0.025 or 0.01.

Perhaps it is beneficial to consider some situations that impact setting the type II error rate:

- Is the sample size already high due to a low relative shift (Δ/σ)?

  If this is the case, it is unlikely that a low type II decision error will be selected. A low type II error will only serve to exacerbate an already large sample. The D&D contractor may consider somewhat higher values of the type II error.

- How expensive are sample analyses?

  If sample analyses are relatively inexpensive, the arguments presented above begin to lose some of their importance. For example, if the statistical sample is a direct measurement of surface activity, then the type II error rate will likely be set low, because more value would be placed on not incorrectly failing the survey unit, rather than minimizing the sample. Of course, if the statistical sample is an expensive alpha spectrometry measurement, then the converse may be true—the type II error would be increased to minimize the sample size, accepting higher error of failing clean survey unit.

- How will rejected survey units be handled?

  Remember, all survey units must ultimately satisfy the release criteria. Survey units that "fail" must be further investigated as to the possible reasons for failure—Chapter 14 addresses this topic. Therefore, if it is a major headache trying to pass survey units that initially failed, maybe the regulator demands that the entire survey unit be resurveyed no matter the circumstances, then there would be a strong bias toward keeping the type II error rate as low as possible. Conversely, if the regulator shows willingness toward "working" with the D&D contractor, then the type II error may not be set as low.

Another parameter that deserves attention is LBGR. The LBGR is best described as the concentration in the survey at which the type II error is set. In other words, the D&D contractor should have a good idea what the median concentration is from HSA and characterization, and therefore, recognizes the consequences of setting a particular type II error at this concentration. For example, assume that the median concentration for Am-241 in the survey unit is 2.3 pCi/g and that the DCGL is 4 pCi/g. If the MARSSIM planner sets the LBGR at the median concentration, and sets the type II error at 0.05, then the planner knows that if the median concentration truly is 2.3 pCi/g, then there is a 5% chance of "incorrectly" failing the survey unit. We say "incorrectly" because if the median Am-241 concentration in the survey unit was truly 2.3 pCi/g, it very well should pass. This points out the fact that type II errors can only occur at concentrations less than the $DCGL_W$, while type I errors can only occur at concentrations greater than the $DCGL_W$.

In setting the LBGR, the MARSSIM planner must also consider the balance between type II errors and the relative shift, which of course directly affects sample size. First, it is critical to understand the mechanics of the relative shift, Δ/σ. By setting the LBGR at a lower concentration, the Δ is increased, as is the Δ/σ, and the sample size is minimized (review MARSSIM Tables 5.3 and 5.5 as necessary). However, this comes at the expense of setting the type II error at a lower concentration—which translates into an increased probability of committing a type II error at all concentrations less than the DCGL$_W$. It is important to remember that the design type II error is specified at the LBGR, so by having a small LBGR, we are setting the type II error at a lower radionuclide concentration—or conversely, increasing the LBGR means we set the type II error at a higher radionuclide concentration. It is beneficial to study an example prospective power curve (Figure 13.7) at this point.

The power curve plots the probability of passing the survey unit as a function of median concentration in the survey unit. It is important to recognize that it is the probability being plotted, so even if the probability is only 30% at a particular concentration (3.6 pCi/g in this example), the D&D contractor may pass the survey unit exhibiting a median concentration of 3.6 pCi/g, although it is not very likely.

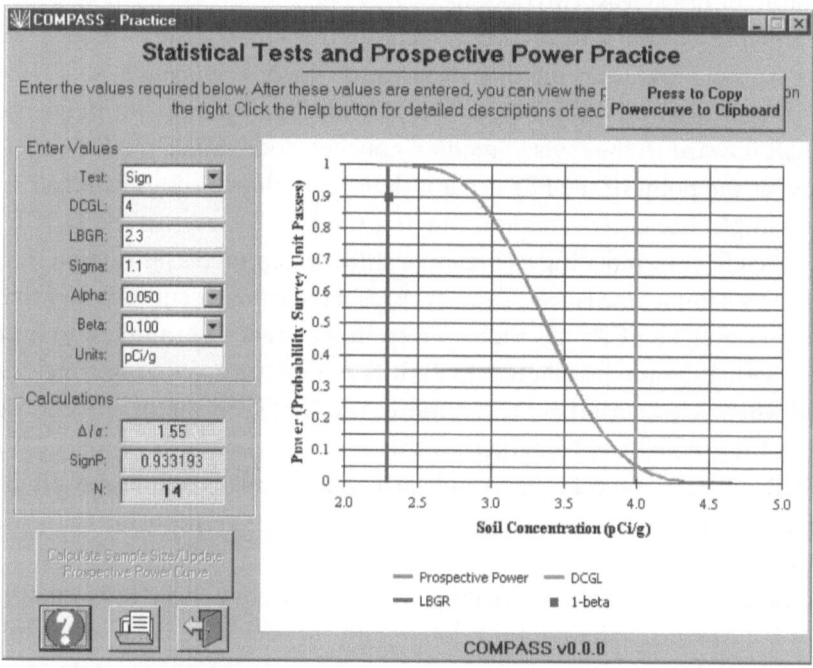

FIGURE 13.7   Prospective power curve for sign test for surface soil assessment.

OK, so the probability of passing is described by the power curve, and it is anchored by two points, the LBGR and the $DCGL_W$. That is, the power curve passes through the $DCGL_W$ at the type I error probability and should pass through the LBGR concentration at the type II error. The reason the power curve has a higher probability at the LBGR than the survey design would suggest is due to the 20% sample size increase above that which the statistical test requires. The shape of the prospective power curve is influenced by the sample size and the contaminant standard deviation. Notice how rapidly the type I error decreases with increasing concentrations above the $DCGL_W$; this is a valuable consideration in petitioning the regulator to allow larger type I errors.

Let us continue with our Am-241 example. If we were to implement this survey design, we would collect 14 soil samples and have roughly 100% chance of passing the survey unit if the median concentration is truly 2.3 pCi/g. By all accounts, this seems to be a very reasonable FSS design. Now, suppose that new data comes to light, suggesting that the median concentration is actually a lot closer to 3.6 pCi/g. Would the D&D contractor still proceed with the survey design that is on the table? Not advisable, given that the probability for success is about 30%. The D&D contractor has at least two options.

The first option is to further remediate the survey unit to reduce the median concentration. In this instance, a reasonable sample size (like 14 in this case) may still be possible. The second option is to move the LBGR upward to the revised median concentration. Since the $DCGL_W$ is fixed at the point of the FSS design, the LBGR plays a significant role in the sample size determination. Moving the LBGR closer to the $DCGL_W$ has the effect of reducing the relative shift, which increases sample size. The new sample size is 185! The new prospective power curve shows that the power at 3.6 pCi/g has increased from 30% (with 14 samples) to over 95% with 185 samples (Figure 13.8). It is probably a safe bet to conclude that when faced with these two options, the D&D contractor will choose to further remediate. Nonetheless, this example shows the power of the LBGR—it both impacts the sample size via the relative shift and establishes the shape of the power curve by virtue of its relationship with the type II error.

The MARSSIM states a design goal for the relative shift: Try to maintain $1 < \Delta/\sigma < 3$. Values of the relative shift that are less than 1 will result in a relatively large number of measurements needed to demonstrate compliance. If Δ/σ is greater than 3 or 4, then take advantage of this flexibility

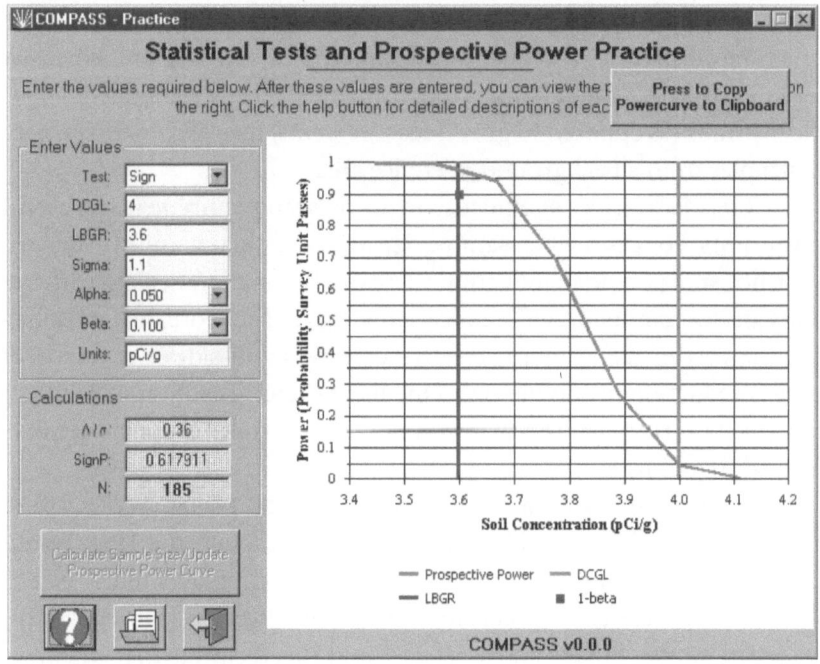

FIGURE 13.8 Prospective power curve for sign test for surface soil assessment when median concentration close to $DCGL_W$.

by increasing the LBGR, which reduces the type II error rate. This latter circumstance usually arises as a result of having a large DCGL.

Another parameter that affects the sample size is the number (or nature) of survey units. We have seen that large values of the relative shift reduce the sample size—and one way to increase the relative shift (Δ/σ) is to reduce the contaminant variability, σ. Survey units that are uniform in their levels of residual radioactivity help to reduce the spatial variability; however, this may increase the number of survey units (and therefore number of samples). Most importantly, regardless of the total number of survey units identified, it is of paramount importance not to underestimate the standard deviation in the survey unit. This is best achieved by performing effective characterization surveys and not simply guessing on the contaminant's variability. While the additional 20% on sample size provides some cushion for inaccuracies in estimating the standard deviation, this is one area that should not be short-changed.

The contaminant variability may also be reduced by using more precise measurement methods. This may be achieved by simply increasing the count time or by using a more sensitive laboratory analysis or instrument. The trade-off involves a reduction in sample size versus more expensive measurements. A related point is that using more costly radionuclide-specific methods (as opposed to gross measurements) may eliminate the need for samples from a background reference area.

MARSSIM strategies should also focus on scan requirements in class 1 survey units. For cases when sample size in class 1 survey units is "driven" by the hot-spot potential, reductions in scan MDC—more sensitive instruments and/or methods—will reduce sample size. Therefore, one must balance the cost of using more sensitive detectors and slower scan speeds versus a reduction in sample size. One must exercise caution in reducing scan speed for the purposes of lowering the scan MDC; there is a point of diminishing returns.

The classification of site areas is another important decision that affects the FSS design. It is probably fair to say that although the HSA results and data from scoping and characterization surveys play a major role in classification, there exists a certain amount of subjectivity in classifying areas. One strategy that may prove useful in this regard is that whenever the classification of a particular area is being debated, and both sides of the debate truly have merit, select the higher classification and move on. For example, if the D&D project manager and the project engineer are discussing how best to classify the turbine deck, class 1 versus class 2, and good points are being made by both parties—then go with class 1. A similar debate may come up between classifying a land area as nonimpacted versus class 3. A good rule of thumb is that performing the minimal class 3 FSS is well worth heading off potentially endless debates over why some land areas received no FSS effort whatsoever.

Let us entertain one final MARSSIM strategy that involves the use of new survey technology. We might begin this discussion by asking the question: How can *in situ* gamma spectrometer measurements be worked into a MARSSIM survey design? While this question has been posed a number of times over the past few years, the best answer seems to come back to the DQO process. That is, it is the responsibility of the survey designer to demonstrate that this technology can indeed satisfy the survey DQOs. A survey design using *in situ* gamma spectrometer measurements, albeit general, is proposed as follows.

First, the specific design of each *in situ* gamma spectrometer measurement would require research and experimentation—for example, the issues of collimated versus uncollimated measurements, distance of the detector above the surface, and counting time must all be optimized. It is also necessary to consider the correspondence between the measurement and the dose modeling. Once the individual *in situ* measurement design has been decided, the measurement data in pCi/g would be evaluated via statistical tests. For class 1 survey units, it is particularly important to consider the detector's field of view, which is a function of both the collimation and detector height above the surface, to assess the relationship to the area factor.

Given the concern that small areas of elevated radiation may be masked when averaged over the large soil volumes associated with uncollimated *in situ* measurements, one strategy that might gain acceptance is outlined as follows. The integrated FSS design consists of *in situ* gamma spectrometer measurements as a replacement for soil samples, complemented with gamma scans using a NaI scintillation detector. This improves the chances for regulatory concurrence for two reasons. First, collimated measurements are performed on a random-systematic grid pattern and the data are evaluated via the statistical tests. The use of collimation reduces the legitimate concern of missing elevated areas due to "over-averaging," since the detector's field of view is kept relatively small (perhaps maintained at 0.5 m² or less by design). Second, although the *in situ* gamma spectrometer can be used as a scanning device, it cannot outperform the NaI scintillation detector's ability to identify isolated hot spots. Therefore, in class 1 survey units, a 100% conventional scan would be performed using NaI scintillation detectors. The required scan MDC would be determined in the same manner as always—that is, the product of the DCGL$_W$ and area factor, where the area factor would be based on the area bounded by the center of the *in situ* gamma spectrometer measurements. Again, this adds a bit of conservatism because it does not take credit for the detector's actual field of field (which of course is much more than a point). The only soil samples necessary using this approach would be those collected initially to help define the depth distribution of the contamination, and those collected as a result of scan hits once the FSS is implemented.

Finally, Section 13.5 provides a number of MARSSIM examples that serve to illustrate the various strategies that can be employed. One theme illustrated was the value of being able to establish a surrogate radionuclide (Cs-137 for Sr-90) and the impact that strategy had on the surrogate's DCGL (Cs-137 DCGL was reduced from 8 to 3.1 pCi/g). These examples also demonstrated that many MARSSIM survey designs rely on the unity rule. In fact, using the unity rule might be described as the default position, since most D&D projects do involve multiple radionuclides and it is not often possible to justify that relative ratios exist between radionuclides. Perhaps the most important lesson in these examples is that the MARSSIM survey design is only as good as the degree to which the median concentration and standard deviation are known. Without this information, it is impossible to intelligently review the prospective power curve. The bottom line can be stated as follows: there is more to designing a MARSSIM FSS than simply crunching through the sample size calculations.

> MARSSIM uses the DQO process to optimize the FSS design—this can only be accomplished by having a thorough understanding of the survey unit's radiological characteristics, that is, the contaminant median concentration and standard deviation.

13.7 RANKED SET SAMPLING

The final section in this chapter focuses on a statistical approach that has applications to FSSs, such as designing surveys that involve hard-to-detect radionuclides in soil, and independent verification surveys (refer to Section 2.4). An important independent verification (IV) activity involves verifying the licensee's reported mean concentration in selected survey units. Ranked set sampling provides an unbiased estimator of the true mean, and it is a recent addition to the statistical tool box for conducting independent verification surveys. ORAU has used ranked set sampling for several years to reduce the number of samples required for statistical analysis. Ranked set sampling, or simply RSS, combines simple random sampling with HSA and process knowledge, and scanning results, to select judgment locations.

Ranked set sampling offers an efficient way to obtain a representative sample prior to performing expensive soil sampling and analysis. For survey unit populations with heterogeneous distributions (class 1) that are expensive to sample, the ranked set sampling approach can be an effective tool for reducing sampling costs. EPA (2002) provides an example use of ranked set sampling to obtain samples for estimating the mean plutonium concentration in surface soil. The idea is to rely on Am-241 for field screening (a field surrogate) for plutonium. The soil areas can be measured using a FIDLER (field instrument for the detection of low-energy radiation) to quantify and rank the concentrations of Am-241 in surface soil, which is correlated to plutonium. Thus, many more inexpensive field measurements of Am-241 can be conducted to reduce the number of soil samples needed for the laboratory analysis of plutonium.

The basic RSS method consists of a number of cycles to ultimately identify the number of required soil samples for laboratory analysis. A cycle involves the random selection of m^2 units from the survey unit population, where m is the set size. These units are randomly partitioned into m subsets, each containing m sampling units. The members of every subset are ranked according to a field measurement characteristic—for example, 1-min gamma measurement in contact with soil using a NaI detector. The lowest ranked member is quantified (i.e., soil sample collected and analyzed) from the first set, the second lowest ranked member is quantified from the second set, and so on until the highest ranked member of the last set is quantified. This yields m quantification from among the m^2 selected units. Since m is usually small (2–5) to facilitate the ranking (e.g., low, medium, and high count rate), there may not be enough measurements in one cycle to satisfy the statistical sample size (n), so the basic cycle is repeated r times to give $n = m \times r$ quantifications out of r selected units. In other words, the number of ranking cycles is defined by the statistical sample size (n) divided by the set size (m). Confusing? An example will help.

We assume that the statistical sample size for the independent verification of a selected survey unit is $n = 6$. Ranked set sampling will be used to identify the soil sample locations required for laboratory gamma spectrometry analysis (Vitkus 2011). The set size will consist of three locations—the gamma count data that are ultimately collected from the three locations associated with each set will be ranked as low, medium, or high. So, $n = m \times r$, where the set size (m) is 3 and the number of cycles, r, is 2. Thus, n samples will be identified from 2 cycles (r) of 3 sets (m) each. OK so far?

The number of field assessment locations per cycle is a function of the set size and is simply m^2; therefore, the total number of field assessment locations where gamma measurements will be performed is given by $m^2 \times r$, or 18. These locations are randomly grouped into cycle/sets and distributed in the survey unit. For independent verification surveys, ORAU often uses a quasi-random location distribution to prevent spatial clustering of the data. The nomenclature for identifying a specific assessment location is cycle#-set#-arbitrary sequence # (1, 2, or 3). The first gamma measurement location in cycle 1 of set 1 would be designated as 1-1-1. Figure 13.9 illustrates the use of color-coded mapping (based on cycle ID) and geometric shapes (based on set ID) to identify field assessment locations (Vitkus

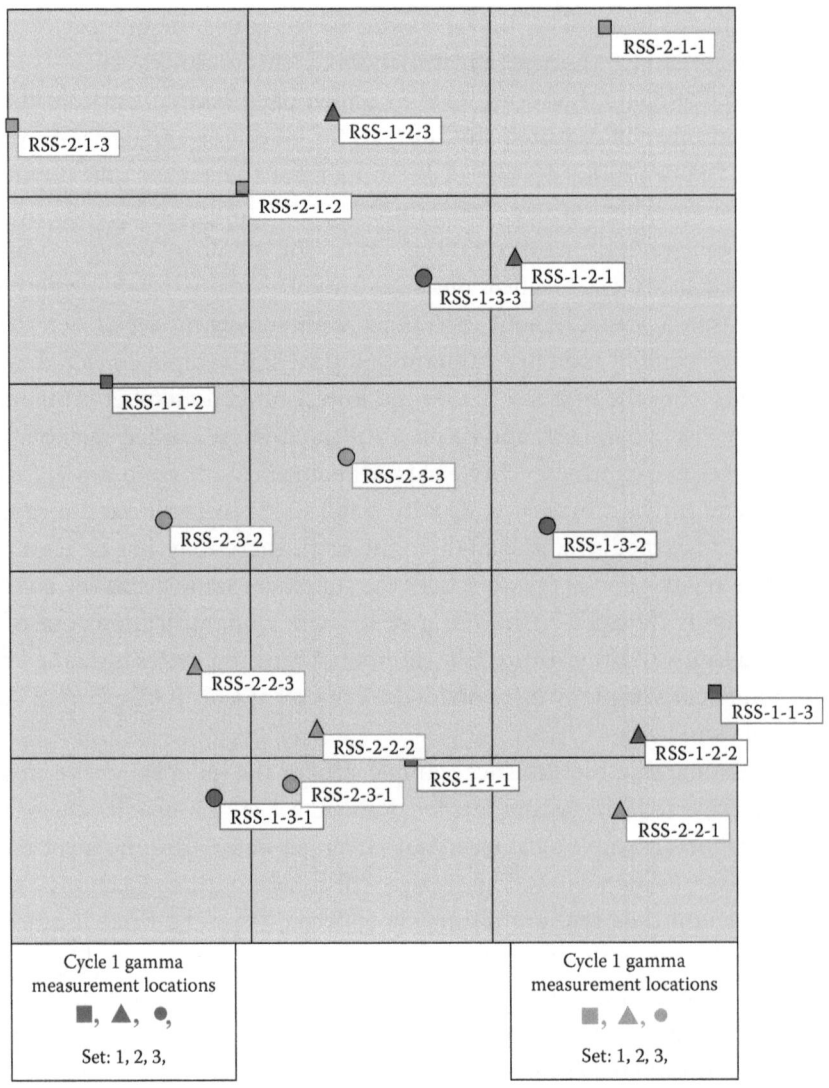

FIGURE 13.9 Ranked set sampling locations across the survey unit.

2011). The use of survey unit mapping software such as visual sampling plan (VSP) to randomly identify sampling locations is recommended.

Finally, the gamma measurements are collected at each RSS location, the data are ranked, and samples collected based on the ranking (Table 13.6). For example, soil sample S007 is selected from the lowest gamma measurement of the first set, sample S008 is selected from the middle gamma measurement of the second set, and so on. Three soil samples will

TABLE 13.6   Ranked set sampling gamma measurements and soil sample locations

Ranked Set Sampling (Cycle-Set-Loc)	Gamma (cpm)	Sample Select/ID (L = Low, M = Medium, H = High)	Soil Sample
1-1-1■	12,147	L	
1-1-2■	9916	L	S007
1-1-3■	11,142	L	
1-2-1▲	11,638	M	
1-2-2▲	15,535	M	S008
1-2-3▲	23,545	M	
1-3-1●	10,366	H	
1-3-2●	14,587	II	
1-3-3●	24,369	H	S010
2-1-1■	14,543	L	
2-1-2■	10,598	L	S012
2-1-3■	12,897	L	
2-2-1▲	12,823	M	S009
2-2-2▲	10,866	M	
2-2-3▲	23,545	M	
2-3-1●	8494	H	
2-3-2●	11,850	H	S011
2-3-3●	11,826	H	

be collected during the first cycle. This process is repeated for the second cycle to yield six soil samples. Hence, ranked set sampling combines random sampling with the use of professional judgment to select sampling locations. This statistical approach provides a more precise estimate of the mean concentration and requires fewer soil samples.

QUESTIONS AND PROBLEMS

1. An engineer is preparing a final survey plan consistent with NUREG/CR-2082 guidance at a thorium processing facility. The engineer has identified six survey units in the high bay, each measuring approximately 100 m² and consisting of floor and lower wall surfaces. A preliminary survey has been conducted over this 600 m² area and the following results were obtained:

Alpha surface-activity measurements $\bar{x}_\alpha = 430$ dpm/100 cm²

$s_\alpha = 620$ dpm/100 cm²

Beta–gamma dose rates	$\bar{x}_{\beta\gamma} = 0.06$ mrad/h
	$s_{\beta\gamma} = 0.05$ mrad/h
Gamma exposure rate	$\bar{x}_\gamma = 12$ μR/h
	$s_\gamma = 4$ μR/h

On the basis of these preliminary data, what is the sample size for each of the 100 m² survey units? Is this sample size consistent with the NUREG/CR-2082 guidance for survey blocks that states they should not be less than 1 m on a side or greater than 3 m on a side?

2. Given the following FSS data that were obtained using the NUREG/CR-5849 survey design, does the survey unit satisfy the release criterion? Assume that this particular building survey unit consisted of 60 1-m² grid blocks. The surface-activity guideline for this site is 4000 dpm/100 cm². The survey unit mean was 3750 dpm/100 cm², with a standard deviation of 540 dpm/100 cm². The critical value of the t distribution at the 95% confidence level ($t_{95\%,df}$) for the $n - 1$ degrees of freedom (df) is approximately 1.671.

3. State the conditions for using the sign test and the WRS test—for example, conditions such as contaminant being present in the background, gross or radionuclide-specific measurements, expected radionuclide concentration level in background versus DCGL$_W$, and so on.

4. Design a MARSSIM FSS given the following information. A building surface survey unit at a fuel fabrication facility is potentially contaminated with 4% enriched uranium. The survey unit has been classified as class 2, and encompasses a total floor area of 820 m² (concrete). Gas proportional detectors will be used to perform surface-activity measurements in alpha plus beta mode. Assume that the average background for the concrete is 378 cpm, and that the instrument and surface efficiencies determined for 4% enriched uranium are 0.41 and 0.27, respectively. Results from recent characterization surveys were used to estimate the mean and standard deviation of gross alpha plus beta measurements in the survey unit: 782 ± 467 cpm. Assume that the DCGLs for the uranium isotopes are as follows: 2000 dpm/100 cm² each for U-234 and U-235, and 2200 for U-238. Set the LBGR value at the expected median concentration in the survey unit, and select a type I error of 0.05 and a type

II error of 0.10. What is the necessary sample size using the sign test for this example survey unit?

5. Design a MARSSIM FSS given the following information. An exterior soil survey unit is potentially contaminated with Ra-226. The survey unit has been classified as class 1, and encompasses a soil area of 2200 m². The $DCGL_W$ for Ra-226 is 1.2 pCi/g. Results from characterization surveys were used to estimate the mean and standard deviation of Ra-226 in both the survey unit and background reference area:

Ra-226 in survey unit 1.9 ± 0.6 pCi/g (1σ)
Ra 226 in background reference area 1.1 ± 0.4 pCi/g (1σ)

Set the LBGR value at the expected median concentration in the survey unit, and select a type I error of 0.05 and a type II error of 0.05. What is the necessary sample size using the WRS test for this example survey unit? Since the survey unit is class 1, determine if any additional samples are necessary using a scan MDC of 4.5 pCi/g and the default area factors provided in MARSSIM.

6. What field conditions warrant the use of scenario B in place of scenario A for the FSS design approach?

7. State four MARSSIM design strategies involved in selecting values for the type II error and the LBGR. How would you describe what happens to the type I error as the concentration exceeds the $DCGL_W$?

8. Generally describe the proposed survey design approach that makes use of *in situ* gamma spectrometer measurements in a MARSSIM FSS.

CHAPTER 14

MARSSIM Data Reduction

IN THIS CHAPTER, WE WILL ILLUSTRATE the five-step DQA process for reducing the FSS data. Basically, the initial steps following the implementation of the final status survey are to reduce the collected survey data and determine if each survey unit satisfies the release criterion. The major change using the MARSSIM DQA approach is the extensive use of exploratory data analysis. This may be a departure from our desire to simply rush to the bottom line—does the survey unit satisfy the release criteria? However, before we answer the bottom-line question, there are several reasons why taking the time to deliberately review and assess the data makes a lot of good practical sense. What was the nature of the sampled distribution—normal, bimodal, and so on? Does the background appear representative to the survey unit data? Are assumptions of statistical test validated? Only after we have thought about these and other questions should we perform the statistical tests to answer the bottom-line compliance question.

The MARSSIM recommends that DQA be performed to determine if the data are of the right type, quality, and quantity to support their use (NRC 2000a). The DQA process is the scientific and statistical evaluation of data and includes (1) review of the DQOs, (2) preliminary data review, (3) selection of statistical tests, (4) verification of assumptions necessary for using the statistical tests, and (5) performance of the statistical tests and drawing conclusions from the data. An extremely useful guidance document to supplement the MARSSIM for performing DQA is the

EPA's Guidance for Data Quality Assessment, Practical Methods for Data Analysis (EPA 2000).

The MARSSIM approach to data reduction will be illustrated using the final status survey examples in Chapter 13. We are going to assume that a couple of the MARSSIM final status survey designs in Chapter 13 were actually implemented and that final status survey data are available. Hypothetical data obtained for a number of these survey designs will then be evaluated using the DQA process. Our first example is the sign test example from Section 13.2.1 for Co-60 in soil.

14.1 DATA QUALITY ASSESSMENT FOR THE SIGN TEST FOR Co-60 IN SOIL

We will briefly summarize the final status survey design for Co-60 in this class 1 soil survey unit. First, the D&D contractor selected the sign test since Co-60 is not present in the background. The $DCGL_W$ for Co-60 was 3.3 pCi/g. Based on the results of remedial action support surveys, the mean and standard deviation of Co-60 in the survey unit were 1.8 and 1 pCi/g, respectively. The LBGR was set at the estimated median Co-60 concentration in the survey unit (1.8 pCi/g), and the type I and type II errors were 0.05 and 0.10, respectively. The sign test required a sample size of 15 soil samples, and the actual scan MDC satisfied the required scan MDC for the $2'' \times 2''$ NaI scintillation detector, so no additional samples were necessary.

This final status survey design was implemented and 15 soil samples were collected from triangular grid spacing in this class 1 survey unit. The $2'' \times 2''$ NaI scintillation detector was used to perform 100% scans over the survey unit soil surface. We will now reduce the final status survey data using the DQA approach described in MARSSIM Chapter 8.

14.1.1 Review of the DQOs

The sampling design and data collection documentation were reviewed for consistency with the DQOs. During the allocation of samples on a triangular grid pattern, it was determined that the survey unit size and shape supported three rows with five samples in each row. Therefore, a total of 15 soil samples were collected and analyzed. The sign test DQOs only required 14 samples. These soil sample results were reviewed to determine if the survey unit was properly classified. The survey results indicated that each survey unit was properly classified as class 1, since three of the 15 systematic samples exceeded the $DCGL_W$, while the 100% scan of the 1800 m² survey unit identified two areas of elevated radiation that were further investigated. A total of six additional judgmental soil samples were collected from these two hot spot areas.

The accuracy of the prospective power curve depends on the number of samples collected and estimates of the standard deviation for each survey unit. Strictly speaking, for scenario A the assessment of the retrospective power curve is only necessary when the null hypothesis is not rejected (i.e., survey unit does not pass release criteria). For scenario B, it is just the opposite when the null hypothesis is rejected, then the survey unit fails. Therefore, when the null hypothesis is not rejected (and the survey unit passes), then the retrospective power analysis is necessary. As discussed in Chapter 18, the general rule of thumb is that a retrospective power analysis is necessary whenever the null hypothesis is not rejected (i.e., accepted), regardless of how the hypothesis may be stated.

The estimated standard deviation of Co-60 in the survey unit was 1.0 pCi/g. The standard deviation obtained during the final status survey was 1.67 pCi/g; Table 14.1 provides the results of basic statistical quantities. Somewhat disturbing is the actual mean of 2.33 pCi/g versus that estimated (1.8 pCi/g), though the actual median was right at 1.8 pCi/g. Therefore, the class 1 survey unit mean and standard deviation assumed during the survey design were underestimated. This may indicate that an insufficient number of samples were collected to achieve the desired statistical test power $(1 - \beta)$ at the expected concentration in the survey unit, and result in a survey unit that fails the statistical test. That is, if the class 1 survey unit fails the sign test, it may be due to an insufficient sample size that lowers the probability of passing at the actual median concentration. It is important to recognize that since we are discussing the *probability* of passing the survey unit or not, we will not know if the survey unit *actually* passes until the sign test is performed. At this point, we will use COMPASS to construct the retrospective power curve based on the actual survey unit results (Figure 14.1); the fact that we collected an extra sample during the grid layout may prove helpful.

The retrospective power curve is shown in Figure 14.1, and it is evident that there is less power than that planned using the prospective power curve.

TABLE 14.1 Basic Statistical Quantities for Co-60 in the Class 1 Survey Unit

	Mean (pCi/g)	Median (pCi/g)	Standard Deviation (pCi/g)
Survey unit (actual)	2.33	1.8	1.67
Survey unit (estimated)	1.8	1.8	1.0

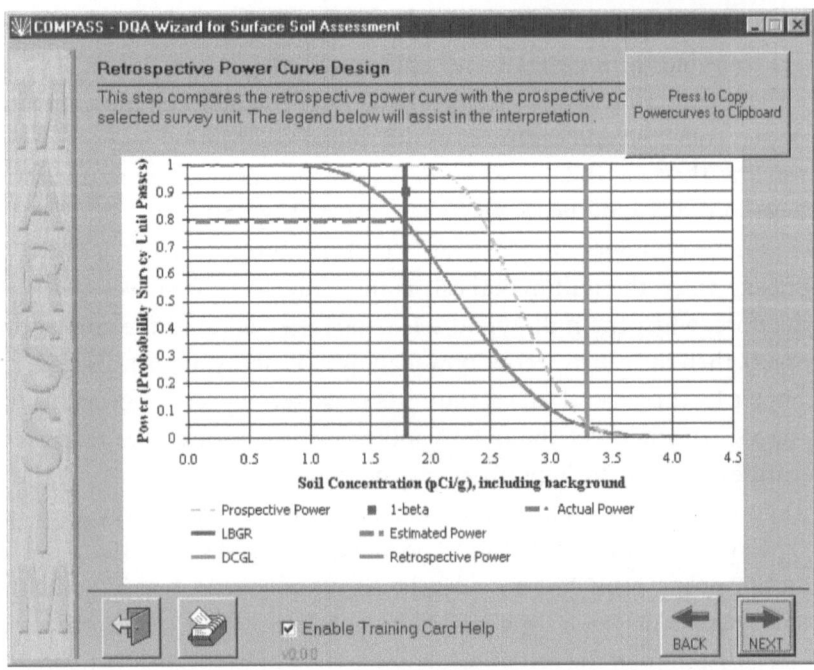

FIGURE 14.1 Retrospective power curve for sign test for Co-60 in soil.

The major factor responsible for the retrospective power curve drooping below the prospective power curve is the increased actual standard deviation versus the planned standard deviation. The additional sample offset the power loss to some degree, but not enough to make up for the underestimated Co-60 standard deviation. Perhaps of greater concern is that while the retrospective power at the estimated concentration (1.8 pCi/g) is about 80%, the actual mean was 2.33 pCi/g, and the retrospective power at that concentration is only 45%. At this point in our DQA, we realize that our probability of passing the survey unit has taken hits from both an underestimated mean and standard deviation. Our probability of passing this survey unit, based on the retrospective power curve, is actually closer to 80% than 45%, because the nonparametric statistical tests work on the median. Usually, the median and the mean are close enough for us to conclude that the statistical tests are working on the mean as well.

14.1.2 Preliminary Data Review

The first statistical calculation should be that of the mean contaminant concentration in the survey unit. If the mean exceeds the $DCGL_W$, then the survey unit fails, regardless of the statistical test outcome. Obviously, if the contaminant is present in the background, then the net mean (not the gross mean) would be compared to the $DCGL_W$. Remediation of the survey unit is required when the mean exceeds the $DCGL_W$. In addition to the mean, other basic statistical quantities are calculated such as the median and the standard deviation. Basic statistical quantities for Co-60 in the survey unit are provided in Table 14.1.

To evaluate the structure of the data, such as identifying patterns and rela- tionships, the survey data should be graphed. These graphical displays of the data will typically include histograms and posting plots. The posting plot is useful for identifying spatial relationships among the data. Figure 14.2 shows a posting plot of the 15 soil samples collected from the class 1 survey unit. The posting plot simply shows the radionuclide concentration at each sample location. Inspection of the class 1 data posting plot does not clearly reveal any systematic spatial trends. Perhaps the two Co-60 concentrations that exceed the $DCGL_W$ on the western edge of the survey unit (north is pointing up) pro- vide some minimal evidence of a spatial trend. However, the fact that both the elevated areas identified were on the eastern side of the survey unit does not lend further support to the notion of a spatial trend.

If the survey unit mean exceeds the $DCGL_W$, then the survey unit fails, regardless of the statistical test outcome.

Figure 14.3 provides a histogram of the Co-60 soil sample results in the survey unit. Histograms are used to provide a picture of the data structure. Are the data normally distributed (i.e., bell shaped)? Does there appear to be two or more distributions in the data? In our small data set, the Co-60 data appears to be somewhat normally distributed. As evidenced in Figure 14.3, the difference between the mean and the median provides an indi- cation of skewness in the data. This skewness is to the right because the mean is greater than the median. With only 15 data points used to con- struct the histogram, one should not read too much into it.

0.7		1.4		6.6		2.2		0.4	
	3.9		2.6		2.9		0.9		1.5
4.8		1.8		2.3		1.7		1.2	

FIGURE 14.2 Posting plot for Co-60 concentrations in class 1 survey unit.

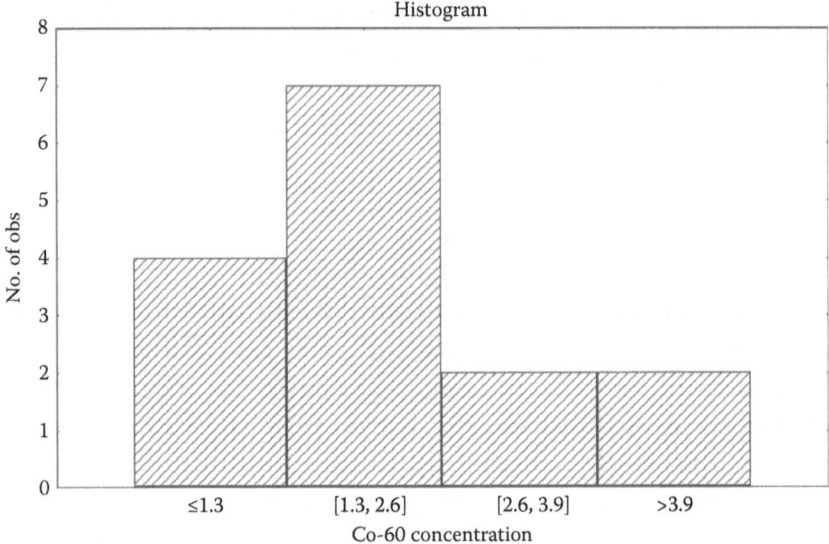

FIGURE 14.3   Histogram for Co-60 concentrations in class 1 survey unit.

14.1.3 Selection of Statistical Test

The selection of the statistical test is the third step in the DQA process. However, this step is probably not applicable for the MARSSIM survey design approach because the statistical test has already been selected during the DQO process—and there is not likely to be a compelling reason for changing statistical tests at this point.

The sign test is used to evaluate the Co-60 concentrations in the class 1 survey unit. The WRS test is used only when the contaminant is going to be compared to a background reference area. The null hypothesis tested by the sign test is that "the median concentration of residual radioactivity in the survey unit is greater than the $DCGL_W$." Therefore, rejection of this null hypothesis results in a decision that the survey unit passes—that is, satisfies the release criterion. Specifically, the result of the sign test determines whether or not the survey unit as a whole is deemed to meet the release criterion.

14.1.4 Verification of Statistical Test Assumptions

The underlying assumption for the nonparametric statistical tests is that each measurement is independent of every other measurement—regardless of the set of samples from which it came. Each of the samples from the class 1 survey unit was collected on a random-start triangular grid pattern—note that biased samples are not included in statistical tests—thus, the assumption of independent random samples is valid. Further, the posting plot does not suggest that spatial dependencies exist in the survey unit.

EPA's Guidance for Data Quality Assessment (EPA 2000) provides an excellent discussion on how the various statistical test assumptions can be validated. These assumptions include those concerning distributional form (e.g., normality), independence of data, and dispersion. One of the major advantages of nonparametric statistics over the more common parametric statistics is that they are less dependent on distributional assumptions.

14.1.5 Perform Statistical Test and Draw Conclusions from the Data

The sign test (or WRS test, as the case may be) evaluates the survey unit concentrations taken as a whole—that is, the random/systematic samples collected during the final status survey provide an unbiased sampling of the survey unit population. Therefore, once the statistical test is performed, the results indicate the overall status of residual radioactivity in the survey unit (above or below the $DCGL_W$). In other words, the statistical tests are performed to demonstrate compliance with release criteria by considering the general residual radioactivity level that exists in the survey unit, as indicated by the survey unit median or mean. Conversely, the statistical tests are not good tools for assessing the acceptability of small areas of elevated contamination (hot spots) that may be present in the survey unit. Instead, the EMC test is performed for these outliers.

We will now perform the sign test on our Co-60 concentration data. It should be reemphasized that the sign test (or any statistical test for that matter) is only performed on the random/systematic survey unit data collected. In other words, only the 15 systematic soil sample results are analyzed using the sign test; the additional six samples collected as a result of investigating the two hot spot areas are analyzed using the EMC test (which is not a statistical test).

The sign test is performed by listing the 15 systematic soil sample results and subtracting each result from the $DCGL_W$ for Co-60 (3.3 pCi/g). If the difference is positive, then the sign is " + ," and if the difference is negative, then the sign is " − ." The test statistic, S+, is simply the number of positive differences. Table 14.2 shows the sign test for our Co-60 example.

The test statistic, S+ , as shown in Table 14.2 is 12. The value of S+ is compared to the appropriate critical value from MARSSIM Appendix I,

TABLE 14.2   Sign Test for Co-60 Soil Sample Results

Co-60 Data (pCi/g)	DCGL$_W$-Data	Sign (pCi/g)
0.7	2.6	+
1.4	1.9	+
6.6	−3.3	−
2.2	1.1	+
0.4	2.9	+
3.9	−0.6	−
2.6	0.7	+
2.9	0.4	+
0.9	2.4	+
4.8	−1.5	−
1.8	1.5	+
1.5	1.8	+
2.3	1.0	+
1.7	1.6	+
1.2	2.1	+
		S+ = 12

Table I.3. In this case, for $N = 15$ and $\alpha = 0.05$, the critical value is 11. Since S+ = 12 is greater than the critical value, the null hypothesis that the survey unit exceeds the release criterion is rejected. Thus, even though our retrospective power curve (Figure 14.1) indicated a reduced probability of passing the survey unit than we had planned, nonetheless, the survey unit passed. Note that if the test statistic (S+) equals the critical value, the null hypothesis is not rejected, and the survey unit does not pass.

OK, so the survey unit passed the statistical test, but what about the hot spots that were found? Specifically, we must perform the EMC test for all of the hot spots that we identified and confirmed via sample results. We should first note that in addition to the two hot spots identified during surfaces scans, three systematic soil results exceed the DCGL$_W$ value of 3.3 pCi/g: 6.6, 3.9, and 4.8 pCi/g. The first way that we might address these concentrations is to evaluate whether they are part of the survey unit distribution defined by the mean and the standard deviation. That is, if these measurement results do not exceed the mean of the data plus three standard deviations (used as an investigation level in MARSSIM), then we can conclude that the sign test has adequately accounted for these outliers. The mean (2.33 pCi/g) plus three standard deviations (1.67 pCi/g) is 7.3 pCi/g. Thus, these Co-60 values appear to reflect the overall variability of the concentration measurements rather than to indicate hot spot locations. As a check, each location should be quickly investigated (via scanning, since scan MDC is sufficiently low) to ensure that these measurements are not indicative of hot spots. As it turned out, the highest systematic measurement was actually part of one of the elevated areas determined by scanning. We are now ready to perform the EMC test on these two hot spots.

The sign test is performed in the following manner:

1. List the survey unit measurements, X_i, $i = 1, 2, 3, \ldots, N$.
2. Subtract each measurement, X_i, from the $DCGL_W$ to obtain the differences:

$$D_i = DCGL_W - X_i, \quad i = 1, 2, 3, \ldots, N.$$

3. Discard each difference that is exactly zero and reduce the sample size, N, by the number of such zero measurements.
4. Count the number of positive differences—the result is the test statistic S+. Note that a positive difference corresponds to a measurement below the $DCGL_W$ and contributes evidence that the survey unit meets the release criterion.
5. Large values of S+ indicate that the null hypothesis (that the survey unit exceeds the release criterion) is false. The value of S+ is compared to the critical values in the MARSSIM. If S+ is greater than the critical value, k, in that table, the null hypothesis is rejected.

The EMC was performed for soil sampling data from the two hot spot locations. Hot spot #1 was located in the north center of the survey unit, and included the initial soil sample result at the highest NaI reading of 13.2 pCi/g. Additional soil samples and scanning determined the hot spot area to be approximately 3 m². A total of four soil samples were collected from this area: 13.2 pCi/g, 6.6 pCi/g (systematic sample), 3.2 pCi/g, and 2.2 pCi/g. The average Co-60 concentration in this hot spot is 6.3 pCi/g. The derived concentration guideline level for the EMC—$DCGL_{EMC}$—is obtained by multiplying the DCGL (3.3 pCi/g) by the area factor that corresponds to the actual area (3 m²) of the elevated concentration. The area factor from Table 13.1 is 5.07 for a 3 m² area. Therefore, the $DCGL_{EMC}$ is calculated: (3.3 pCi/g) × (5.07), or 16.7 pCi/g. Therefore, hot spot #1 is deemed acceptable since it does not exceed the appropriate $DCGL_{EMC}$ (i.e., 6.3 pCi/g is less than 16.7 pCi/g).

The second hot spot was somewhat larger and much more difficult to bound with samples. The initial sample collected at the highest scan reading in this area was 6.3 pCi/g. Two additional samples were collected in an attempt to bound the hot spot—sample results were 5.4 and 3.2 pCi/g. However, the process of scanning and collecting samples to adequately bound the hot spot determined that the hot spot area was at least 10 m², and possibly even larger. The D&D contractor made the decision at that point to simply remediate the hot spot rather than to continue the process of bounding the hot spot area. Once the hot spot was remediated, an additional three samples were collected and analyzed to confirm that the remediation was successful and did not recontaminate the survey unit.

Oftentimes, the best solution for dealing with hot spots identified during the final status survey is to simply remediate them. Trying to satisfy the EMC evaluation can be extremely difficult and time consuming; especially considering that any combination of hot spot area and radionuclide concentration that exceeds the appropriate $DCGL_{EMC}$ must be identified, and then remediated. The point is that simply remediating the hot spot may save time and resources trying to demonstrate that the hot spot does not exceed the EMC.

So far the survey unit satisfied the nonparametric statistical test and the EMC for the one remaining hot spot. We are almost done demonstrating that this survey unit satisfies the release criterion, but not until we evaluate the impact from multiple hot spots (using Equation 8-2 in MARSSIM). In essence, this final condition provides a safety net to ensure that the presence of multiple hot spots in a survey unit still satisfies the dose criterion. Two conditions must be met prior to evaluating the condition of multiple hot spots, namely, the null hypothesis must be rejected for the survey unit and each individual hot spot area must satisfy its specific $DCGL_{EMC}$. Once these two conditions have been met, the unity rule hot spots is performed using the following equation:

$$\frac{\delta}{DCGL_W} + \frac{(\text{average concentration in hot spot}) - \delta}{DCGL_{EMC}} < 1$$

where δ is the average concentration in the survey unit.

Applying the above equation to our Co-60 data provides the following results:

$$\frac{2.33}{3.3} + \frac{6.3 - 2.33}{16.7} = 0.94 < 1$$

Therefore, this final test for assessing the acceptability of this hot spot in consideration of the general survey unit Co-60 concentration also indicates that the survey unit should pass, barely. To summarize, this survey unit has to pass the sign test, EMC, and the unity rule for hot spots to demonstrate that the release criterion had been met.

14.2 DATA QUALITY ASSESSMENT FOR THE WRS TEST FOR URANIUM AND THORIUM IN SOIL

We will briefly summarize the final status survey design for uranium and thorium contamination in soil at a rare earths processing facility. DCGLs for these contaminants were 2.8 pCi/g for Th-232 and 178 pCi/g for U-238 based on dose modeling. The class 1 survey unit measured 3000 m², and the following survey information and DQOs were used to design the final status survey:

Results from characterization and remedial action support surveys were used to estimate the mean and standard deviation of Th-232 and U-238 in the survey unit. Background reference area samples were also collected and analyzed. The results were as follows:

	Survey Unit	Background Reference Area
Th-232	1.9 ± 0.5 pCi/g (1σ)	1.1 ± 0.3 pCi/g (1σ)
U-238	37.4 ± 8.8 pCi/g (1σ)	1.4 ± 0.7 pCi/g (1σ)

On the basis of these data, the median concentration in terms of the unity rule was 0.488 and the overall standard deviation was 0.18. The LBGR was set at 0.5 based on these data. Type I and type II decision errors were set at 0.05.

The WRS test sample size ($N/2$) was 11, and it was determined that the scan MDC was sufficiently sensitive, so no additional samples were required.

This final status survey design was implemented and 11 soil samples were collected in both the survey unit and the background reference area. Surface scans were performed over 100% of the survey unit soil area and no locations of elevated direct radiation were detected. Table 14.3 provides the Th-232 and U-238 concentration results of the 22 soil samples collected.

The DQA for the WRS test consists of the same five steps as described for the sign test above. The MARSSIM COMPASS software will be used to perform the DQA for this survey unit. The basic statistical quantities are

TABLE 14.3   Th-232 and U-238 Soil Sample Results in the Survey Unit (S) and Reference Area (R)

Location	U-238 (pCi/g)	Th-232 (pCi/g)	Sum-of-the-Ratios
S	54.7	2.8	1.31
S	3.4	3.1	1.13
S	13.6	0.5	0.25
S	34.2	1.1	0.58
S	23.2	3.1	1.24
S	12.9	1.4	0.57
S	10.2	1	0.41
S	21.9	2.2	0.91
S	8.7	1.7	0.66
S	44	1	0.60
S	23.9	2.1	0.88
R	1.4	1.1	0.40
R	1.9	0.7	0.26
R	0.6	1.3	0.47
R	0.8	0.7	0.25
R	1.1	0.9	0.33
R	1.6	1.6	0.58
R	2.3	1.3	0.48
R	1.8	0.8	0.30
R	1.2	0.9	0.33
R	0.9	1.5	0.54
R	1.6	1.2	0.44

shown in Table 14.4. Notice how well the Th-232 concentrations were estimated versus the U-238 concentrations in the survey unit. The estimated mean U-238 was nearly twice as high as the actual U-238 concentration, but the standard deviation was underestimated by a factor of two. This means that the actual power will be less, but at the same time we have less contamination than anticipated. Figure 14.4 shows the retrospective power curve for these data.

TABLE 14.4   Basic Statistical Quantities for Th-232 and U-238

	Th-232	U-238
Survey unit (actual)	1.8 ± 0.9 pCi/g (1σ)	22.8 ± 15.9 pCi/g (1σ)
Reference area (actual)	1.1 ± 0.3 pCi/g (1σ)	1.4 ± 0.5 pCi/g (1σ)
Survey unit (estimated)	1.9 ± 0.5 pCi/g (1σ)	37.4 ± 8.8 pCi/g (1σ)
Reference area (estimated)	1.1 ± 0.3 pCi/g (1σ)	1.4 ± 0.7 pCi/g (1σ)

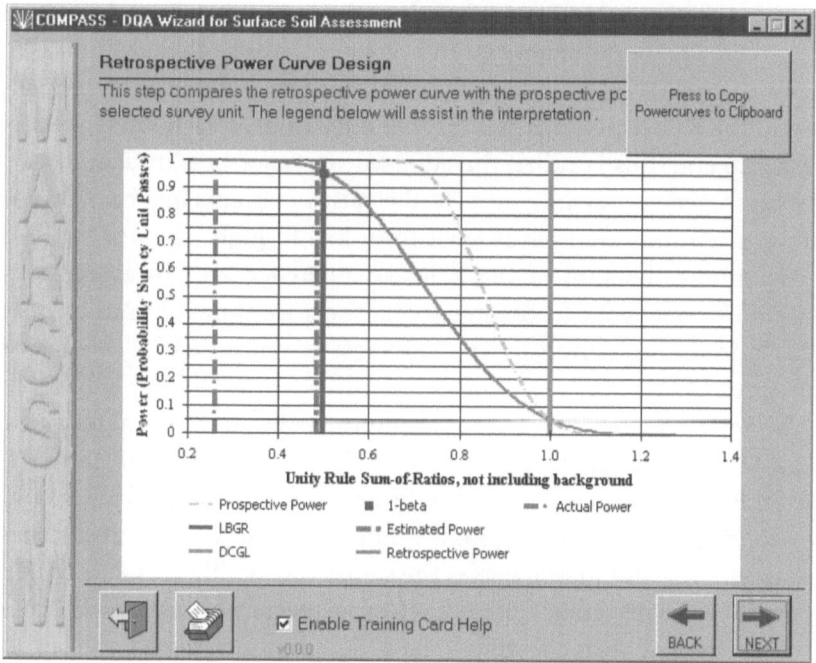

FIGURE 14.4   Retrospective power curve for Th-232 and U-238 in the survey unit.

As expected, the retrospective power curve is certainly less powerful than the prospective power curve, but the redeeming factor is that the actual median concentration in the survey unit was less than 0.4, while that planned was 0.5. The important point is that the power is most important at the concentration that actually exists in the survey unit. The fact that the retrospective power curve shows a reduced probability of passing the survey unit at a median concentration of 0.8 is immaterial if the actual concentration in the survey unit is less than 0.4. Based on the retrospective power curve, it appears that the WRS test will pass the survey unit.

Before the WRS test can be performed using the unity rule, each of the sample data must be transformed. This transformation involves calculating the sum of the ratios for Th-232 and U-238 in each sample (shown in Table 14.3); that is, the Th-232 concentration is divided by 2.8 pCi/g and added to the U-238 concentration divided by 178 pCi/g. Once the sum of the ratios has been determined for all 22 soil samples, the WRS test is performed to compare the two populations. To demonstrate compliance with release criteria, we are not interested in simply whether the survey unit is greater than the background reference area, but rather, whether the survey unit is greater than the background reference area plus $DCGL_W$. To accomplish the proper statistical comparison, we add the $DCGL_W$ (which is 1 for the unity rule) to each reference area measurement. This is called the adjusted reference area measurements. The combined survey unit sum-of-the-ratios and adjusted background sum-of-the-ratios are then ranked from low to high. The test statistic, W_r, is simply the sum of the adjusted reference area measurements. MARSSIM Appendix I provides spreadsheet formulas for ranking the data. Indeed, the analysis for the WRS test is well suited for calculation on a spreadsheet.

The test statistic (W_r) for our example is 184; the critical value for $N/2 = 11$ and type I error of 0.05 is 152 (MARSSIM Appendix I, Table I.4). The decision rule is that if the W_r exceeds the critical value, then the null hypothesis is rejected, and the survey unit passes. Since 184 is greater than 152, the survey unit indeed passes the statistical test. This result is consistent with our expectation from the retrospective power curve. Since there were no elevated areas identified, the EMC is not performed; the survey unit satisfies the release criterion.

The specific details for conducting the WRS test are as follows:

1. Obtain adjusted reference area measurements by adding the $DCGL_W$ to each background reference area measurement.
2. Rank the pooled adjusted reference area measurements (m) and survey unit measurements (n) from 1 to N ($N = m + n$).
3. If several measurements are tied (i.e., have the same value), they are assigned the average rank for that group of tied measurements.
4. Add the ranks of the adjusted measurements from the reference area, W_r.
5. Compare W_r with tabulated critical value (MARSSIM Appendix I; based on n, m, and α) (Note: For m or n greater than 20, the critical value is calculated by

$$\frac{m(n + m + 1)}{2} + z\sqrt{nm(n + m + 1)/12}$$

where z is the $(1 - \alpha)$ percentile of a standard normal distribution.)
6. Reject H_0 if W_r > critical value.

14.3 WHAT IF THE SURVEY UNIT FAILS?

The first, and perhaps obvious, point to make is that a failed survey unit is a temporary condition. That is, in the final analysis, all survey units comprising the D&D site must pass (satisfy release criteria). The real question is: What needs to be done to move the failed survey unit into the ledger of passed survey units? Appropriate actions in this regard can include remediation, complete redo of the final status survey, collecting additional final status survey samples, or possibly just a review of the data. The issue of survey unit failures is addressed in MARSSIM Chapter 8. In fact, the reader is encouraged to review the several paragraphs that MARSSIM covers on this issue of survey unit failures and possible remedies. In this section, will review the MARSSIM guidance, and expand on it in the area of statistical failures of the survey unit.

14.3.1 Why Survey Units Fail

First of all, a survey unit may fail to satisfy the release criteria for a few reasons: (1) the survey unit mean exceeds the $DCGL_W$, (2) it fails the nonparametric test (WRS or sign test), or (3) it fails the EMC test or hot spot unity rule. Additionally, investigation levels may be triggered which result in concluding that a particular survey unit has been misclassified, but this is somewhat different than failing one of the above criteria.

When a survey unit fails to demonstrate compliance with the release criterion, the first step is to review and confirm the data that led to that decision. Once the data review has been completed, MARSSIM recommends using the DQO process to identify and evaluate potential solutions to the problem. It may be that additional data are needed to document that the survey unit demonstrates compliance with the release criterion. Unfortunately, MARSSIM does not offer specific guidance to a situation that is likely to arise by the very nature of the survey design itself—the type II error rate, by design, will result in a definite probability (e.g., type II error of 0.10) of some survey units failing in error. How will the licensee ultimately demonstrate compliance for the 10% of survey units that incorrectly failed due to the survey design?

First, consider an example for a class 1 survey unit that passes the nonparametric statistical tests, but contains two areas that were flagged for investigation during scanning. This exact situation was described in our first example for Co-60. Further investigation, sampling, and analysis indicated that one hot spot area satisfied the release criteria—that is, the hot spot concentration was less than the $DCGL_{EMC}$. The second hot spot area was difficult to bound using scanning and sampling, so the D&D contractor decided to remediate the hot spot. Remedial action control surveys were performed to demonstrate that the residual radioactivity was removed, and that no other areas were contaminated with the excavated material. In this case, the survey unit passes and one would likely document the original final status survey, the fact that remediation was performed, the results of the remedial action support survey, and the additional remediation data.

Now consider a survey unit that fails the nonparametric statistical tests. The final status survey data were reviewed and indicated that the residual radioactivity in the survey unit does exceed the $DCGL_W$ over a majority of its area. In this situation, the survey unit probably deserves to fail—that is, the statistical test produced the proper result. (Remember, that whenever the survey unit mean exceeds the $DCGL_W$, the survey unit automatically fails.) This indicates remediation of the entire survey unit is necessary, followed by another final status survey. Perhaps an investigation into why this survey unit was initially considered ready for final status survey should be performed.

On the other hand, what if the review of the data in the failed survey unit did not indicate that the residual radioactivity in the survey unit exceeded the $DCGL_W$ over a majority of its area? Perhaps an inspection of the retrospective power curve indicates that the estimated standard deviation was underestimated, and the actual probability of passing was only 60% at the median concentration in the survey unit, rather than that designed for 90% power. Let us suppose that if the survey designer had planned appropriately for the actual standard deviation in the survey unit, then six more samples would have been collected. Should we let the surveyor simply collect six more random samples and add these to the overall sample size? Should they redo the final status survey and do a better job of estimating parameters next time around? The MARSSIM only offers the DQO process—stating that the DQOs should be revisited to plan how to attain the original objective—that is, to safely release the survey unit by showing that it meets the release criterion.

14.3.2 Double Sampling

Obviously, an attractive solution to the above problem is to resample. This allows the surveyor another chance to pass the survey unit, while still making use of the initial sample size. Should this practice be allowed? The answer is clearly "yes" if this double sampling strategy is planned ahead of time, that is, prior to implementing the final status survey. NRC provides guidance for double sampling in Appendix C of NUREG-1757, vol. 2.

Double sampling usually results in less samples being collected overall, while still allowing for the MARSSIM user to resample the survey unit if it fails to reject the null hypothesis the first time. Conversely, double or even multiple sampling schemes can be developed, which allow a questionable survey unit another chance. The attractiveness of double sampling is that it can be designed to have a smaller average sample number (i.e., the average total sample size from the initial and second sample sizes)—after assessing the need for the second sample (i.e., survey unit may pass following the initial sampling during the final status survey, thus negating the need for resampling effort)—when compared to an equivalent single sampling plan. As would be expected, the double sampling plan is more difficult to administer than a single sampling plan, and it is recommended that a statistician be involved with the design and implementation.

Studies have shown that the type I error increases marginally—so it is possible that the regulator will allow resampling in limited cases provided that the alpha error does not exceed some agreed-upon value. Perhaps the regulator sets a type I error of 0.05, but will allow the overall type I error (after resampling) to go as high as 0.10 in a limited number of cases. According to *Use of Two-Stage or Double Sampling in Final Status Decommissioning Surveys* (Gogolak 2000), "increasing the probability that clean survey unit passes (power) by the use of double sampling will also tend to increase the probability that a survey unit that is not clean will pass (type I error)" (http://cvg.homestead.com/files/cvg_s_notes__2_-_double_sampling.pdf). He continues, "the increase in the type I error rate is probably less than a factor of two." In other words, the option to permit double sampling should be planned ahead of time as part of the DQO process; a high degree of communication must exist between the regulator and the MARSSIM user if double sampling is allowed.

Let us consider a double sampling example. Suppose the MARSSIM survey has been planned for a soil area potentially contaminated with Th-232, and the $DCGL_W$ is 7 pCi/g. Limited characterization results for this survey unit indicates a mean and standard deviation of 4.1 and 1.7 pCi/g, respectively. The Th-232 concentration in the background reference area is of 1.1 pCi/g, resulting in an average net of 3 pCi/g in the survey unit. The LBGR is set at 3 pCi/g. The WRS test sample size is determined as follows:

- Relative shift: $\Delta/\sigma = (7 - 3)/1.7 = 2.35$

- Type I error = 0.025; Type II error = 0.1

- $N/2 = 11$ samples

This survey design was implemented and 11 samples were collected in both the survey unit and the reference area. Upon inspection of the survey data, the actual mean and standard deviation were 5.3 and 3.1 pCi/g, respectively. Both the mean and the standard deviation are larger than planned. It appears that the survey unit may be headed for failure. Statistical failures of clean survey units will often occur due to poor characterization efforts in estimating the radiological status of the survey unit. Therefore, while the mean may be less than $DCGL_W$, the sample size may be too small to reject the null hypothesis for the actual standard deviation in the survey unit—the result of inadequate statistical power.

A retrospective power curve can be generated to illustrate the impact of underestimating the standard deviation. For example, the planned probability of passing the survey unit (or power) with a net mean of 3 pCi/g is greater than 90%. However, the actual probability of passing the survey unit with an actual net mean concentration of 4.2 pCi/g (not the planned 3 pCi/g), and standard deviation of 3.1 pCi/g is about 47%. As expected, the survey unit fails the statistical test (even though the net mean is less than the $DCGL_W$). The MARSSIM user in this case realizes that more samples (given the actual mean and standard deviation in the survey unit) would likely have resulted in a more pleasing outcome. Here is how the double sampling comes into play.

The WRS sample size based on actual standard deviation and mean (e.g., set the LBGR at the actual net mean and use the actual standard deviation) is calculated. The probability for passing might be set at 70% or so to maintain a reasonable number of additional samples. In this case, $N/2$ is 17 for the revised DQOs and actual data. Since 11 samples have already been collected in each area, an additional six samples are collected in both the survey unit and the reference area. These samples should be collected randomly, and are added to the initial sample data set to yield a total of 17 samples.

The WRS test is performed a second time on all 17 samples. The survey unit passes this time. MARSSIM is once again your friend. What about your regulator? Well, the overall type I error has just increased—but no more than double. In this example, the initial type I error was 0.025; therefore, the final type I error is still less than 0.05. The regulator is a friend too.

Double sampling—it is worth another look.

Now for the fine print. Double sampling allows for one more set of additional samples—for example, triple sampling is not allowed. Double sampling should never be needed for class 2 or 3 survey units (i.e., data should all be less than $DCGL_W$ in these survey units). Finally, the option to perform double sampling should be planned during the DQO process and discussed with the regulator.

QUESTIONS AND PROBLEMS

1. Provided below are the soil sample results from the final status survey performed in the soil area surrounding the power reactor waste disposal facility. Use the DCGLs provided in Table 4.3. The unity rule using the sign test was used to design the survey. The selected type I and type II decision errors were both 0.10. Please perform the sign test on these data to demonstrate compliance with the release criteria. Does the survey unit satisfy the release criteria?

Sample Number	Co-60 (pCi/g)	Cs-134 (pCi/g)	Cs-137 (pCi/g)	Eu-152 (pCi/g)
1	0.7	0.3	1.5	1.7
2	1.4	0.4	0.4	1.6
3	0.3	1.3	1	0.8
4	0.2	0.3	2.7	0.4
5	0.6	0.4	2.2	1
6	0.2	0.4	1.9	0.7
7	2.6	0.1	2	1.5
8	2	0.1	2	0.7
9	0.9	0.1	2	2.2
10	1.1	0.9	1.6	0.8
11	1.8	0.7	1.4	1.6
12	0.3	1.1	0.7	1.8
13	3.8	0.7	0.3	1.2
14	0.7	0.4	1.7	1.1
15	0.7	0.2	1.8	0.6
16	0.1	0.2	3	2.2
17	0.5	0.6	2.5	2.1

18	2.8	0.5	0.9	1.3
19	2.2	1.4	1.3	2.3
20	0.7	0.6	1.7	0.6
21	0.2	0.5	1.9	1
22	0.9	0.3	1.5	1.4
23	1.1	0.5	1.8	0.6
24	2	0.8	1.9	1.7
25	0.1	1	2	0.6
26	0.4	1	1.4	1.6

2. Perform the sign test for the surface activity data provided in Table 13.4. Does the survey unit demonstrate compliance with the release criteria? How does the retrospective power curve compare to the prospective power curve?

3. Using the Shapiro–Wilk test described in Section 18.1, determine whether the Co-60 data in Section 14.1 are normally distributed at $\alpha = 0.05$ level of confidence. The coefficients needed to solve the problem are 0.5150, 0.3306, 0.2495, 0.1878, 0.1353, 0.0880, and 0.0433.

CHAPTER 15

Clearance of Materials

T HE RELEASE OF SOLID MATERIALS for unrestricted use—or clearance as it is referred to in the D&D vernacular—has been a hot-button issue for years. The debate primarily centers on whether there is an acceptable (i.e., safe) level of residual radioactivity that can remain on (or in) scrap metal, concrete construction materials, soil, and so on so that they can be released for continued use in commerce, or whether they must be disposed of as radioactive waste. Over the years, many organizations have studied and provided guidance on the clearance of materials and equipment, including the DOE, EPA, NRC, States, HPS/ANSI, as well as international bodies, such as the IAEA, European Commission (EC), and Nuclear Energy Agency (NEA).

Many expected the NRC clearance rulemaking to finally settle the issue in the United States as we entered the new millennium. However, in August 2000, the NRC decided to defer a final decision on whether to proceed with the rulemaking on controlling the release of solid materials. Instead, the NRC requested a study by the National Academy of Sciences (NAS) on possible alternatives for controlling the release of slightly contaminated materials. The NAS report "The Disposition Dilemma" (2002) resulted in a series of findings and recommendations. The report concluded that NRC's current approach is workable and protective, but is inconsistent and lacks a risk basis. The report highlighted the need for broad stakeholder involvement, stating that participation in the decision-making process is critical. The NRC concurred with 1 mrem/y as a "starting point" in assessing alternatives, and acknowledged the importance of significant stakeholder involvement in the clearance process.

At the same time, a change in the nuclear industry was about to significantly affect the need for a clearance rulemaking. The decommissioning of commercial nuclear power plants and the expected tons of scrap metal and concrete rubble potentially being released was put on hold in favor of 20-y nuclear plant license extensions. The impact of these nuclear power plant license renewals has significantly diminished the projected volumes of scrap materials expected from decommissioning activities. By 2012, more than 50% of the current nuclear power plant fleet had been granted 20-y license extensions; the anticipated glut of scrap metals and other solid materials from the decommissioning of nuclear power plants will be a couple of decades late.

15.1 CLEARANCE: A CONTROVERSIAL HISTORY

The NAS report accurately noted that we have been releasing materials using the current approach to make clearance decisions. For instance, release surveys have been performed to demonstrate compliance with Regulatory Guide 1.86 and DOE O 5400.5 (now DOE O 458.1) limits for surface-contaminated materials, and on a case-by-case basis for volumetric releases. These guidelines are not dose-based, but rather, founded largely on survey instrument capabilities. So why upset the applecart, or perhaps more aptly, stir up a hornet's nest? Releases of solid materials have been going on for decades, largely unnoticed. The game changer was decommissioning—D&D activities at commercial nuclear power and DOE weapon complex sites were expected to increase the volume of scrap steel and concrete to be released.

The NRC successfully promulgated the license termination rule in 1997 that included a 25 mrem/y cleanup standard. This beneficial outcome contrasted with their earlier efforts to declare materials with low concentrations of radioactive contamination to be below regulatory concern (BRC). Specifically, in 1990, the NRC attempted to categorize extremely low-level radioactive waste as "below regulatory concern"; solid materials would be eligible for release without further regulation if the projected dose was less than some *de minimis* level (e.g., 1 mrem/y). Many comments were received from the general public, the states, and the Congress concerning the BRC expressing disfavor. The United States Congress officially revoked the NRC's BRC policy in the Energy Policy Act of 1992 due to public opposition (NAS 2002).

For years, citizen groups and individuals have harbored strong opinions on both sides of the clearance debate. Generally in favor of clearance were licensees (i.e., universities, fuel cycle facilities, nuclear power plants), many states, and the Health Physics Society, among others. They have argued for the need to have a national standard for recycle, reuse, or disposal. They support a clearance standard based on negligible doses that are a small fraction of the 10 CFR Part 20 public dose limit. These proponents claim that prohibiting the release of materials would disrupt operations and waste resources without accompanying health benefit. Their mantra might be "Why fill up limited LLW disposal sites with mildly contaminated materials?" Indeed, the DOE EM cleanup chief stated that "saving a lot of material, not throwing it away, not filling up disposal sites, whether it be Nevada or on-site disposal cells, I'm a big supporter of it" (Huizenga, 2012).

Those who oppose the notion of clearance include citizens groups and individuals, the Steel Manufacturers Association (SMA), among others. The SMA opposes clearance for both the potential for negative economic impacts and environmental perception. The economic impact stems from inadvertently melting a shielded radioactive source. This can result in multimillion dollar losses, so mills have "zero tolerance" policy toward incoming shipments of scrap that set off alarms (disrupts operations). The environmental perception is more subtle; recycling radioactive materials tarnishes the image of recycling, and that perception hurts the scrap business. Meanwhile, tons of relatively clean steel are being buried in landfills and radioactive waste disposal sites across the United States. Additionally, DOE has vast amounts of valuable copper, aluminum, and other materials at its gaseous diffusion plants that are minimally contaminated, if at all.

Those opposing clearance call for the isolation of radioactive wastes and the strict prohibition of releasing radioactive materials that may wind up in consumer products. They argue that the risks are too high, completely avoidable, and involuntary. They cite issues of trust (lack of) with licensees and government, and advocate for the recapture material previously released. Comment letters to the NRC following their 1999 Issues Paper and public meetings were captured in SECY-00-0070, attachment 2, "Summary of Written and Public Meeting Comments" (NRC 2000c). More than 800 public comment letters were received from stakeholders; one example: "Surely this is not true! Is it? I hope it is just a hoax, but if it is indeed the case that this is a real plan, I oppose it vehemently." Articles on clearance were commonplace—"NUCLEAR SPOONS: Hot metal may find its way to your dinner table" (Cusac 1998), and "Radioactive Fork, Dear?" (Auster 1999).

Around the same time NCRP was preparing Report No. 141 Managing Potentially Radioactive Scrap Metal (2002). Their recommendations included, in part, the following: (1) comprehensive and consistent national and international risk-based policies for managing potentially radioactive scrap metal need to be developed, (2) regulatory control over orphan sources must be improved, and (3) steps should be taken to enhance public understanding of clearance process. Section 5.4 of NCRP Report No. 141 provides a concise background on the development of clearance standards, and Appendix A provides a comprehensive overview of the scrap metal making process—an interesting summary of the technologies for processing scrap iron, steel, aluminum, and others.

More recently, the Department of Energy issued a draft Programmatic Environmental Assessment (PEA) for the Recycle of Scrap Metals Originating from Radiological Areas (DOE/EA-1919) for public comment. This PEA (2012) addresses the Secretary of Energy suspension issued on July 13, 2000 for the unrestricted release for the recycling of scrap metals from radiological areas within DOE facilities. Recall that the suspension was imposed in response to public concerns about the potential for radioactivity to end up in metal recycled from DOE facilities, and was planned to be in effect until December 31, 2000. As this process moves forward, DOE will consider various alternatives for the disposition of uncontaminated scrap metal. Note that metals with volumetric contamination and scrap metals produced from RCRA and CERCLA clean-up activities are not included in the scope of the PEA.

To conclude, clearance of solid materials has been a polarizing topic. The NRC has weighed in more than a few times. The NAS report points to the need for strong stakeholder involvement. No doubt that it will take that and strong leadership for a clearance rulemaking to become a reality. Interestingly, the principal regulatory agencies continued to discuss the clearance process, and published the MARSAME in 2009. Simply stated, the MARSAME supplement expands the scope of MARSSIM to include methods and processes to support the disposition of materials and equipment (which are specifically stated to be outside MARSSIM's scope).

15.2 MARSAME AND DQOs FOR THE RELEASE OF MATERIALS

The MARSAME has been used for the release of materials and equipment for a few years now. While it has not taken off in popularity like MARSSIM (yet), it is *the* reference for designing and conducting material and equipment release surveys. It belongs to the bookshelf of everyone engaged in the planning and conduct of clearance surveys. Now, the MARSAME roadmap provides an overview of all that is involved in the release survey process. Yet, an ominous sign concerning MARSAME's complexity may be that two-thirds of the 20-page roadmap is composed of Rube Goldberg-like flow diagrams. These detailed descriptions are meant to summarize MARSAME. Thumb through these pages, even a few times. One diagram box directs you to a previous figure, while a second diagram box welcomes you back from visiting a future roadmap figure. The flow diagram logic is sound; yet something is missing in the attempt to assist MARSAME users in negotiating the information. Suffice to say once you understand the process detail inherent in the flow diagrams, you will understand MARSAME.

While the size and scope of the MARSAME guidance document may at first seem a bit surprising, the level of detail (particularly the statistical detail) and plethora of real-life clearance examples is rather pleasing (well, to me at least). That said, I do understand why some have asked if MARSAME is somewhat disproportional to its scope. Specifically, why was so much written about how to appropriately survey and release materials? After all, releasing material can be boiled down to several major activities and considerations: (1) make sure you have sufficient detector sensitivity relative to the release criteria, (2) consider where this material came from (process knowledge), (3) have trained survey technicians follow approved survey procedures, (4) evaluate survey data relative to the release criteria, and (5) document the survey release process and results, including where/how the material was finally dispositioned.

MARSAME may not be as popular as MARSSIM because there is no clearance rule—recall that the NRC's license termination rule was a driver for MARSSIM's use. In the United States, we continue to release materials largely on a case-by-case basis. In fact, draft NUREG-1761, Radiological Surveys for Controlling Release of Solid Materials (NRC 2002a) was written 10 y ago in anticipation of the NRC-proposed rulemaking on the release of solid materials. In 1999, the NRC published, for public comment, an Issues Paper indicating that they were examining their approach for the control of solid material. The Issues Paper presented alternative courses of action for controlling the release of solid materials that have very low amounts of, or no, radioactivity. MARSAME ensures that release surveys are rigorously designed, implemented, and documented precisely because of public concerns over release of materials to general commerce (consumer products).

The DQO process (introduced in Chapter 3) is the underpinning for the design and implementation of clearance surveys for solid materials. A number of clearance survey approaches are discussed using a variety of survey technologies and survey instrumentation. The DQO process provides the strategic basis for determining the most appropriate clearance survey protocol based on the solid material being released, available survey instrumentation, and applicable release criteria. Effective survey design considers the available process knowledge of the solid materials— for example, materials may have inaccessible surfaces or that they may not be in directly accessible areas. Pertinent characteristics of the solid material that impact its release include the material's physical description, potential for contamination, nature of the contamination, and degree of inaccessible areas. The overall objective is to provide guidance on the selection and proper application of clearance survey strategies.

MARSAME covers a number of different survey approaches, including conventional scanning, automated scanning using a CSM, and *in toto* techniques, such as *in situ* gamma spectrometry and tool monitors. *In toto* refers to the fact that solid material being cleared is surveyed totally, or in its entirety. *In situ* survey techniques (i.e., *in situ* gamma spectrometry) can be a subset of *in toto* techniques; a good example is that of a tool monitor where small tools are placed in a detection chamber and counted long enough to achieve measurement quality objectives.

The general release survey approaches identified in MARSAME include (1) scan-only survey designs (using conventional survey instrumentation or automated scanning/conveyorized surveys), (2) MARSSIM-type surveys using conventional instrumentation that incorporate both scanning and statistical designs for determining sample sizes, and (3) *in situ* survey designs such as performed with gamma spectrometers, bag monitors, tool monitors and portal monitors. MARSAME also provides guidance on survey data reduction, demonstrating compliance with clearance release criteria and documenting results.

15.3 RELEASE CRITERIA, PROCESS KNOWLEDGE, AND OTHER SURVEY DESIGN CONSIDERATIONS

The following sections address specific areas of consideration necessary to prepare for radiological surveys for controlling release of solid materials. Information was cited from MARSAME, ANSI/HPS N13.12, and draft NUREG-1761 and covers the following topics: release guidelines and their application, the nature of solid materials and concept of survey units, process knowledge, inaccessible areas, nature and distribution of contamination, and material classification. These topics should be addressed during the planning stages of radiological surveys for solid materials. In essence, Chapter 15 may be viewed as a companion document for use with MARSAME; it is characterized by a large sampling of MARSAME guidance complemented with information contained in draft NUREG-1761.

Release guidelines can either be in the form of activity concentrations or be based on a potential dose to an individual. Currently, an example of surface-based guidelines is contained in Regulatory Guide 1.86 (AEC 1974). The guidelines provided in Regulatory Guide 1.86 are generally based on the detection capabilities of commercially available survey instruments. Table 4.1 in Chapter 4 provides the Regulatory Guide 1.86 surface activity guidelines and conditions for implementation and it is reproduced here to provide historical perspective on clearance criteria. Removable surface activity guidelines are 20% of the average surface activity guidelines for each grouping. Hence, Regulatory Guide 1.86 not only provides the release criteria for surface activity but also averaging conditions for the application of these criteria. Also note that Regulatory Guide 1.86 does not include the volumetric release criteria.

As with MARSSIM, MARSAME does not specify the release criteria for solid material. Rather, MARSAME identifies potential sources of action levels for use in developing clearance surveys. These include dose-based (or risk-based) regulatory standard, waste acceptance criteria at a disposal site, indistinguishable from background, activity-based standard, administrative limits, or limits on technology. Whatever the source, the release criteria should not only include an action level upon which to base the $DCGL_C$ (clearance DCGL), but should also include any necessary conditions on the implementation of the $DCGL_C$. For example, any limits on the area or volume averaging of solid materials should be clearly expressed. Restrictions on the averaging area or volume of solid materials will necessarily impact the material survey unit or batch size. Area and volume factors, as determined from dose modeling, can be used to determine activity concentrations greater than the $DCGL_C$ that could exist in smaller surface areas (or volumes) than those modeled to derive the $DCGL_C$, and still demonstrate compliance with the dose criterion. These area and volume factors are used to determine maximum limits on contamination that may be present in smaller areas and volumes.

As an example of area and volume averaging conditions, ANSI N13.12 (1999), "Surface and Volume Radioactivity Standards for Unconditional

Release," provides a primary dose criterion (1 mrem/y) and derived screening levels for groups of similar radionuclides, for both surface and volume contamination. This guidance discusses both direct measurements of surface activity and scanning, as well as the use of *in situ* measurements. Averaging conditions are stated for the derived screening levels—multiple surface activity measurements can be averaged over 1 m^2 area (but not greater) and multiple volumetric measurements can be averaged over a 1 m^3 or mass of 1 metric ton. N13.12 states that no single measurement of surface or volume can exceed 10 times the screening level.

A lot of research and assessment work went into NRC's comprehensive NUREG-1640 report "Radiological Assessments for Clearance of Materials from Nuclear Facilities" (NRC, 2003a). This exhaustive four-volume report considers reuse, recycle, and disposal scenarios, and was written to provide a description of how potential annual radiation doses to individuals were estimated. The assessments address four types of material—scrap iron and steel, copper, aluminum, and concrete rubble, and evaluate exposures of individuals to such materials at key steps in recycling and disposal processes. Each exposure scenario considered was characterized by a set of parameters that provide a mathematical description of that scenario. Some example scenarios include the processing steel scrap at a scrap yard, truck driver hauling cleared aluminum scrap, and processing concrete rubble at satellite facility, and the use of recycled concrete in road construction.

Annual radiation doses were normalized to unit concentrations (to generate dose factors) for the clearance of these specific materials. Indeed, a major benefit of NUREG-1640 is that it provides dose factors for a number of different metals and concrete for many radionuclides—and these dose factors address contamination, both surficially on equipment and volumetrically (mass-based) in scrap materials. The dose factor calculations are based on data from the U.S. industrial and commercial practices that describe the anticipated recycling and disposal of cleared materials. The variability and uncertainty in parameter values that characterize these practices, such as the processes for melting and refining of scrap metals, are expressed as probability distributions. The dose factors in NUREG-1640 (Tables 2.1 and 2.2) present the mean values of normalized effective dose equivalents (EDEs) and effective doses, respectively, to the critical groups in units of annual dose per unit of radioactivity (e.g., in μSv/y per Bq/g and or μSv/y per Bq/cm^2).

MARSAME recognizes various disposition criteria (MARSAME Appendix E) and action levels (ALs) that might be used for the clearance of materials and equipment. The "preapproved" action levels were effectively summarized at the Waste Management Conference (Boerner 2012). DOE O 458.1 (2011) is used to provide average, maximum, and removable surface contamination levels for several different groups of radionuclides, and the NRC cites specific ALs in Regulatory Guide 1.86 (1974). For volumetric considerations, ANSI/HPS N13.12 (1999) provides screening levels. In addition to these preapproved ALs, derived limits may be developed through the use of exposure pathway models (e.g., the RESRAD family of codes), and NUREG-1640 dose factors.

15.3.1 Solid Material Description and Survey Units

This section discusses the physical nature of the solid materials being cleared. The physical nature of the material refers to attributes such as the size of the material and composition (or homogeneity) of the material—and it directly impacts the handling issues, as well as the selection of the clearance survey approach and determination of survey unit boundaries. For example, large pieces of metal can be surveyed using conventional hand-held survey instruments, while pea-sized pieces of copper chop are perhaps best surveyed using a CSM or via laboratory analyses. These smaller solid materials that consist of many small, regular pieces are best handled and released as a bulk material, perhaps using a CSM or an *in toto* clearance technique. Furthermore, a concrete slab may be released on the basis of a surface scan using a large-area gas proportional detector, as compared to rubblized concrete that is cleared on the basis of a number of representative samples analyzed in a laboratory. Therefore, it may be appropriate to consider solid materials as being composed of one of the following categories: (1) many small regular pieces, (2) individual, large pieces of equipment and metal, or (3) medium-sized items and materials that fit on a pallet (e.g., perhaps 10–100 pieces of cut pipe and fan blades).

It may be advantageous for the material to be processed prior to it being surveyed. Solid materials that can be made homogeneous via melting, chopping, cutting, and so on are more easily surveyed (see the section on inaccessible areas). For example, copper wire can be surveyed with hand-held survey instruments, but if the wire is chopped into small pieces, it can be more effectively surveyed using a CSM. Similarly, material processing might include decontamination techniques (e.g., grit blasting and melting) that can homogenize and reduce the material's contamination potential.

One of the technically challenging areas is in defining a "survey unit" for clearance surveys of materials. The material survey unit (or batch) concept is at the core of statistical designs for release surveys. In MARSSIM, the survey unit represents a specific land area or building surface area. For clearance of solid materials, the survey unit may consist of equipment surface area, volume of bulk material (soil or rubblized concrete), number of small items, lengths of pipe, and so on. Like the survey unit concept in MARSSIM, any relationship between the survey unit size (i.e., batch size) and the modeling input used to establish the $DCGL_C$ should be adhered to. Thus, the definition of a material survey unit (or batch) for solid materials released using a CSM may relate to the amount of material scanned as it passes under the detector(s) for a specified observation interval and given belt speed. On the basis of the material's classification, 10–100% of the material might be selected for analysis on the CSM. Another example might include a few large pieces of equipment of metal. In this case, the survey unit might consist of the entire piece itself, such as a large electrical panel. Therefore, material survey unit selection is ultimately based on the DQO process, in consideration of the nature of the material, clearance survey technique selected, and its potential for contamination.

The clearance survey design of solid materials must consider the nature of the material, which is more complex than MARSSIM's simple categorization of land (soil) or building surface. MARSAME suggests that four attributes should be addressed when describing the solid material—dimensions, complexity, accessibility, and inherent value. The minimum information suggested for each attribute is listed in MARSAME Table 2.1. For example, the complexity attribute considers the need for material segregation, disassembly, and whether the materials also contain hazardous chemicals (e.g., PCBs in transformer oil). MARSAME states that the primary factor in determining survey unit sizes/boundaries are the assumptions used to develop the action levels.

Draft NUREG-1761 considered various solid materials in terms of material composition and the attribute of survey unit dimensions (NRC 2002a). This description of common materials and equipment considered for release is a useful guide for designing and executing clearance surveys.

1. *Concrete rubble* consists of crushed concrete of a soil-like consistency from the demolition of buildings and structures. The reinforcing steel rebar has been removed from the concrete rubble.

2. *Concrete slab* consists of 30-cm-thick medium-density concrete (2.4 g/cm³), with surface dimensions of 1.2 m × 1.8 m. If volumetric

contamination is expected, then alternative clearance survey techniques are warranted, such as concrete core samples.

3. *Small-bore pipe* (<6 cm diameter pipe) from piping systems and electrical conduit are assumed to be sectioned into 1.2–1.8 m lengths. It is assumed that conventional survey instrumentation cannot access the pipe interiors. The surface area for pipe section exteriors per survey unit is 17 m² (based on a pipe diameter of 6 cm and lengths of 1.5 m).

4. *Large-bore pipe* (>6 cm diameter pipe) from piping systems assumed to be sectioned into 1.2–1.8 m lengths (see Figure 15.1). It is assumed that conventional survey instrumentation can access the pipe interiors. The surface area for pipe section interiors and exteriors per survey unit is 72 m² (based on a pipe diameter of 30 cm and lengths of 1.5 m).

5. *Structural steel* consists of light and heavy gauge steel that may require sizing to fit on a pallet (1.2–1.8 m lengths). The structural steel may consist of I-beams, structural members, decking, ductwork, tanks, and other containers.

6. *Copper wire* consists of insulated and noninsulated wire (0.6 cm or larger), copper windings, and bus-bars. It is assumed that this amount of copper weighs 0.75 tons. *In toto* clearance techniques may also be useful to assess copper wire.

FIGURE 15.1   Large-bore piping segmented prior to clearance survey.

7. *Copper ingots* consist of size-reduced pieces of copper and ingots. The total surface area of the bulk copper when spread out to a height of 5 cm is about 15 m².

8. *Soil* includes soil and soil-like materials that are a finely divided mesh size. The total surface area of the soil when spread out (to facilitate scanning) to a height of 15 cm is about 50 m².

9. *Large items for reuse* include transformers, specialty equipment (e.g., lathes), electrical panels, and other complete systems. These materials are assumed to require some amount of disassembly to provide access to interior surfaces, but consideration must be given to the fact that these items are valued for their function, so cutting is usually not an option. The nominal weight of the large item is 1.5 tons.

10. *Scrap metal pile* consists of miscellaneous mixed metals with no common configuration. The scrap may require sizing to fit on a pallet. The nominal weight of the material on a pallet is assumed to be 1 ton. The total surface area of the scrap metal pile is assumed to be about 10 m².

11. *Scrap equipment and small items for reuse* includes small pumps, motors, hand tools, power tools, scaffolding, and the like. These materials are often associated with operational releases and are assumed to require some amount of disassembly to provide access to interior surfaces. The nominal weight of the material on a pallet is assumed to be 1.5 tons.

Survey units for these materials should be selected in consideration of the nature of the material, clearance survey technique selected, and considering any relationship between the survey unit size (i.e., batch size) and the modeling input used to establish the $DCGL_C$. The size of the material survey unit may also be a function of the material's classification (i.e., potential for contamination)—that is, the amount of material comprising class 1 survey units being smaller than either class 2 or 3 survey units. (However, it is important to note that MARSAME suggests that the material survey unit does not change based on classification.) Table 4.2 in draft NUREG-1761 (NRC 2002a) provides typical survey unit sizes. For bulk materials (soil, concrete rubble, and copper ingots), the survey unit size ranges from 1 to 7.5 m³; for concrete slabs and other large items, the survey unit size is the item itself; and for palletized items (pipe sections, structural steel, scrap metal), the survey unit size ranges from 10 to 100 m².

15.3.2 Process Knowledge

The release of solid materials can occur both during normal operations of a facility and during decommissioning. Releases during operations typically involve smaller quantities of materials than during facility decommissioning (when tons of materials can be released), and the material's potential for having contamination is usually better known for operational releases than for decommissioning releases since the material's origin is more certain. Of course, our focus is on material releases during decommissioning, and process knowledge is important for properly assessing the material's likelihood of being contaminated.

MARSAME describes the term initial assessment (IA) as the investigation to collect *existing* information describing the solid materials. This of course is very similar to the historical site assessment process used in MARSSIM. The IA allows for the initial categorization of the material and equipment as impacted or nonimpacted (covered in Section 15.3.5). An initial step is to use process knowledge to determine whether the solid material was impacted during facility operations—or perhaps during decommissioning operations. Specifically, process knowledge is obtained through a review of the operations conducted in facilities where materials may have been located and the processes where the materials were involved. Were solid materials in direct contact with radioactive materials by design, such as ventilation ductwork and process piping? This question is answered by reviewing operational and survey records to evaluate whether spills, fires, and airborne contaminant releases had occurred that may indicate the presence of contamination in the vicinity of the solid materials being released. Process knowledge information is usually collected during the initial assessment or historical site assessment process described earlier in Chapter 3, Section 3.4.

When process knowledge is limited, or perhaps even nonexistent, sentinel measurements can be performed on the material to provide information for making a decision regarding the material's categorization. These sentinel measurements are biased measurements performed at locations that are most likely to exhibit contamination, if solid material contamination is present at all. MARSAME Section 2.2.4 provides examples on the best locations for sentinel measurements—for example, access points on difficult-to-measurement locations and prefilters used to capture airborne particulate radioactivity. Note that sentinel measurements are not the same as scoping or characterization surveys, the latter surveys being conducted to specifically establish the contamination potential and the radionuclide identity of the contamination on these solid materials. See Chapter 3 for the design of characterization surveys.

15.3.3 Inaccessible Areas

How to address inaccessible areas is another important issue that impacts the decision of whether or not to clear the material in the first place. If material preparation activities include dismantlement (i.e., cutting, disassembly), or the use of specialized survey instruments to gain access to inaccessible areas, then it may be deemed too expensive to survey and release the material. In these situations, disposal may be a more appropriate option. Obviously, if the material surfaces are inaccessible, then it is unlikely to demonstrate that the release criteria have been satisfied using routine clearance survey activities. For instance, owing to the small openings on some items, conventional survey activities to address the potential for internal contamination are nearly impossible. In this case, one option is to dispose of as radioactive waste—where the DQO process should be used for evaluating the cost of surveys versus disposal. For the disposal option, these materials would be assumed to have contamination at levels greater than the release criteria, and therefore would not be cleared. This is often a practical approach for dealing with materials that have inaccessible surfaces.

Materials and equipment that potentially have inaccessible areas include small-bore pipe (interior surfaces), internal surfaces of items (pumps, motors, machinery, electrical panels, etc.) being released for reuse, and scrap equipment. ANSI HPS N13.49 (2001) provides guidance on situations and conditions that may indicate the presence of inaccessible contamination:

> (a) reason to suspect inaccessible contamination based on operational history; (b) direct or smear surveys show that surface contamination is present on accessible surfaces; (c) the item was located in the vicinity of an airborne radioactivity release, or the item was located in an airborne radioactivity area; (d) the item was connected to a contaminated system; (e) the item is painted and paint may be covering or shielding radioactivity; (f) the item contains inaccessible dissimilar materials and was exposed to a neutron flux or particle activation.

For materials with inaccessible surfaces, the question becomes how to assess the inaccessible surfaces that may be contaminated in excess of the release criteria.

A common inaccessible material scenario is the interior surfaces of pipes that are difficult to access—such as buried or embedded pipes. In general, the small diameter of embedded piping makes it extremely difficult to access their interior surfaces. Buried and embedded pipes likely have contamination by virtue of their function of transporting radioactive process liquids. Buried pipes may be at some depth beneath the soil surface and cannot be accessed unless they are excavated, or via *in situ* characterization of pipe internals using specialized survey equipment. Small detectors such as miniature GM detectors and other "pipe-crawling" detector systems have been used to assess surface contamination in pipe systems. Process piping, such as that associated with nuclear power reactor systems, can be embedded in concrete. In fact, nuclear reactor plants have extensive lengths of embedded piping and conduit that can be very challenging to access and assess.

Draft NUREG-1761 (NRC 2002a) identifies a less common inaccessible material scenario—the material surfaces in a scrap metal (or other material) pile. In this scenario, the surfaces can be made accessible, but the effort required to separate the materials for survey might be considered too labor intensive to warrant conventional clearance surveys (again, think DQO process). Therefore, it might be worthwhile to consider releasing a pile of scrap metal by taking *in situ* gamma spectrometry measurements of the scrap metal pile. In this case, some of the scrap metal surfaces are considered inaccessible because they do not directly contribute to the detector's response. However, *in situ* gamma spectrometry might satisfy the clearance survey DQOs for scrap metal piles provided that a sufficient fraction of gamma radiation from the contamination is detected (refer to Section 15.5.5. on *in toto* clearance techniques for detail on this survey approach).

Finally, we can discuss making the surfaces accessible either by cutting or dismantling the material, or by using specialized survey equipment (e.g., small detectors) to make the surfaces accessible. This option usually requires the use of additional resources beyond that required for conventional clearance surveys. For example, this can be accomplished by dismantling scrap equipment or by excavating the buried or embedded pipes. Inaccessible areas that might require disassembly include small pumps, motors, power tools, and electrical control panels. These materials are assumed to require some amount of disassembly to provide access to their interior surfaces. If functional reuse is an option, the item should be carefully dismantled when attempting to gain access to internal surfaces. Conversely, cutting techniques to gain access can be employed to expedite the process if reuse is not an option.

15.3.4 Nature of Contamination

The nature of contamination on materials and equipment can be described in terms of the radionuclides comprising the contamination, and its distribution on the material. For example, the contamination distribution on most items and materials is generally spotty, though some materials (particularly those that were designed to have intimate contact with radioactivity like process piping and ducting) exhibit a more uniform contamination distribution. As mentioned earlier in Chapter 3, process knowledge supported by subsequent scoping and characterization surveys can be used to determine the identity of the radionuclides present, and the extent and location of contamination on the solid material.

Clearly, the type of facility from which the materials originated helps to determine the identity of the contaminants. For example, if the solid materials came from a nuclear power reactor, the likely radionuclides include fission and activation products (e.g., Co-60, Cs-137, Ni-63, and Fe-55); if the materials were from a gaseous diffusion plant, the radioactivity may include enriched uranium and Tc-99. The radionuclide mixtures for each facility type (or industry category) should be known to effectively design the clearance survey.

Scoping and characterization surveys would likely be performed, and may include field measurements and sample collection with laboratory analysis, to determine specific radionuclides present and their radiation characteristics. The identification of radionuclides is generally performed through laboratory analyses—such as alpha and gamma spectrometry, and other radionuclide-specific analyses. For instance, the radionuclide mixture of contamination on solid materials that originate from a power reactor facility may be assessed by collecting representative samples, and performing gamma spectrometry analyses to determine the relative fractions of activation and fission products present. As mentioned in Section 6.2, calculating the gross activity DCGLs for surface activity depends on radionuclide analyses to determine the relative ratios among the identified radionuclides, as well as providing information on the isotopic ratios and percent equilibrium status for uranium and thorium decay series. This information is necessary for establishing and applying the $DCGL_C$ for the materials being released.

15.3.5 Material Classification

All materials can be divided into two types—nonimpacted and impacted. Nonimpacted solid materials have no contamination potential based on process history, while impacted solid materials have some contamination potential based on operations and process knowledge. MARSAME defines the term *categorization* for determining whether the solid materials are impacted or nonimpacted—a nonimpacted decision means that no further action is necessary to release the material. Conversely, impacted materials are further subdivided into three classes based on the solid material's known contamination levels or contamination potential, and are outlined below.

Once the material's process knowledge has been reviewed, and the characterization is completed, an initial classification is performed. The material classification selected should be based on the known process knowledge, as well as previous operational records and survey data, to establish the potential for solid materials to have contamination. This may include the consideration of function and use of the material, location where the solid material was used, determinations as to whether previous surveys were performed to supplement the process knowledge, and if there is a potential for internal contamination and how it affects the classification. Additionally, the potential for the materials to have been exposed to a neutron fluence resulting in the formation of long-lived activation products should be evaluated.

Materials that have never been in a radiological area are typically classified as nonimpacted. For example, virgin steel I-beams that resulted from the demolition of an office building that was located outside control areas and had never housed radiological activities of any type would be classified as nonimpacted. Impacted solid materials are those items that were, at any period of time, stored or used within a radiological area. These items could have contamination and therefore require further evaluation prior to release. The contamination potential of the solid material is used to further classify the material as class 1, 2, or 3. The specific classification will assist in defining the survey approach prior to release. Those materials having the highest potential for contamination would receive the greatest clearance survey effort.

The classification of solid materials is based on the material's contamination potential and is used to determine the clearance survey coverage for that material. The basic idea is that the greater the potential for the material to have contamination, the greater the clearance survey effort. This is the same philosophy in the MARSSIM manual as well. The solid material classification will specify, for example, how much metal scrap on a pallet must be surveyed, or what fraction of soil must be processed through a CSM.

Improper classification of materials has serious implications, particularly when it leads to the release of materials with contamination in excess of clearance criteria. For example, if materials are mistakenly thought to have a very low potential for having contamination, then these materials will have a minimal survey rigor. This misclassification results in a higher potential for releasing materials in error. To minimize these potential errors, investigation levels should be established and implemented to indicate when additional investigations are necessary. For example, a measurement that exceeds an appropriately set investigation level may indicate that the material survey unit has been improperly classified.

15.3.5.1 Class 1 Solid Materials

Class 1 solid materials are those materials that have, or had a potential for contamination (based on process knowledge) or known contamination (based on previous surveys) above the release criterion (DCGL$_C$). These solid materials include materials that comprise processing equipment or components that may have been affected by a spill or an airborne release. Class 1 materials and equipment exhibit the highest potential for containing hot spots.

Basically, class 1 solid materials are those materials that were in direct contact with radioactive materials during the operations of the facility or may have become activated. Additionally, solid materials that have been cleaned to remove contamination are likely considered to be class 1. An exception may be considered if there are no inaccessible areas and any contamination is readily removable using cleaning techniques. Examples of such methods may include vacuuming, wipe downs, or chemical etching that confidently remove all contamination such that surface activity levels would be less than the release criteria.

15.3.5.2 Class 2 Solid Materials

Class 2 solid materials are those materials that have, or had a potential for known contamination, but are not expected to have concentrations above the release criteria. These materials include those items that are within radiologically posted areas but are not expected to have contamination. This class of materials might consist of electrical panels, water pipe, conduit, ventilation ductwork, structural steel, and other materials that might have come in contact with radioactive materials. These materials have little or no potential for hot spots.

Any class 2 solid materials that exceed the release criteria, from previous surveys, should be reclassified as class 1 for clearance surveys. For items of unknown or questionable origin, scoping surveys should be performed to determine if residual surface contamination is present. Provided that no activity is identified, the minimum classification for such materials should be class 2.

15.3.5.3 Class 3 Solid Materials

Class 3 solid materials are those materials that are not expected to contain any contamination, or expected to contain contamination less than some small specified fraction of the release criteria based on process knowledge or previous surveys. Any solid materials that exceed the specified fraction of the release criterion, from previous surveys, should be reclassified as class 2 for clearance surveys. Additionally, if the historical assessment data or process knowledge is insufficient to clearly document that an item or area is nonimpacted, then the minimum classification for such materials would be class 3.

15.3.6 Application of Release Guidelines

This section addresses how individual DCGLs for clearance can be applied when more than one radionuclide is potentially present. Options may include the use of gross activity DCGLs for surface activity compliance and use of surrogate measurements or the unity rule for volume activity compliance. Regardless of the option used to modify the DCGLs to account for multiple radionuclides, it is necessary to identify the potential radionuclides, as well as the relative ratios of these radionuclides. Section 15.3.4 presented the approach for determining the nature of the contamination, as well as calculating the relative ratios among the multiple radionuclides and state of equilibrium for decay series radionuclides.

Surface activity DCGLs for clearance apply to the total surface activity level. For cases where the surface contamination is entirely due to one radionuclide, the $DCGL_C$ for that radionuclide is used for comparison to clearance data. The clearance data may be obtained from direct measurements of surface activity, scanning with data logging, and CSM surveys. For situations where multiple radionuclides with their own DCGLs are present, a gross activity $DCGL_C$ can be developed using the approach described in Section 6.2.2. As a reminder, the gross activity DCGL for material and equipment surfaces with multiple radionuclides is calculated as follows:

1. Determine the relative fraction (f) of the total activity contributed by the radionuclide.

2. Obtain the $DCGL_C$ for each radionuclide present.

3. Substitute the values of f and DCGL_C in the following equation:

Gross activity DCGL_C

$$= \frac{1}{((f_1/\text{DCGL}_1) + (f_2/\text{DCGL}_2) + \cdots + (f_n/\text{DCGL}_n))}$$

where f_i is the radionuclide fraction for each radionuclide present.

Next, consider the assessment of volumetric activity when multiple radionuclides are present. DCGLs correspond to a release criterion, such as the regulatory limit in terms of dose or risk. When multiple radionuclides are present, the total of the DCGLs for all radionuclides could exceed the release criterion. In this case, the unity rule is satisfied when radionuclide mixtures yield a combined fractional concentration limit (called the weighted sum) that is less than or equal to one. This was covered in Section 6.2.3. Another method to consider for adjusting the DCGLs is to modify the assumptions made during exposure pathway modeling to account for multiple radionuclides.

The use of surrogate measurements is another approach for demonstrating compliance. In this situation, the individual DCGLs would need to be adjusted to account for the presence of multiple radionuclides contributing to the total dose. Surrogate measurements were discussed previously in Section 6.2.1 for adjusting the DCGL to account for multiple radionuclides when radionuclide-specific laboratory analyses of media samples or *in toto* measurements are performed. In this case, it may be possible to measure just one of the radionuclides and still demonstrate compliance for all of the other radionuclides present through the use of surrogate measurements.

Equation 6.2 is reproduced here to illustrate how the DCGL for the measured radionuclide is modified ($\text{DCGL}_{\text{meas,mod}}$) to account for the inferred radionuclide:

$$\text{DCGL}_{\text{meas,mod}} = (\text{DCGL}_{\text{meas}}) \left(\frac{\text{DCGL}_{\text{infer}}}{(C_{\text{infer}}/C_{\text{meas}})\text{DCGL}_{\text{meas}} + \text{DCGL}_{\text{infer}}} \right)$$

where $C_{\text{infer}}/C_{\text{meas}}$ is the surrogate ratio for the inferred to the measured radionuclide.

And when it is necessary for the measured radionuclide to be used as a surrogate for more than one radionuclide, equation I-14 on MARSSIM page I-32 can be used to calculate the modified DCGL for the measured radionuclide.

$$\text{DCGL}_{\text{meas,mod}} = \frac{1}{((1/D_1) + (R_2/D_2) + (R_3/D_3) + \cdots + (R_n/D_n))}$$

where D_1 is the DCGL$_C$ for the measured radionuclide by itself, D_2 is the DCGL$_C$ for the second radionuclide (or first radionuclide being inferred) that is being inferred by the measured radionuclide, and R_2 is the ratio of the concentration of the second radionuclide to that of the measured radionuclide. Similarly, D_3 is the DCGL$_C$ for the third radionuclide (or second radionuclide being inferred) that is being inferred by the measured radionuclide, and R_3 is the ratio of the concentration of the third radionuclide to that of the measured radionuclide.

15.4 DETECTION LIMITS FOR MATERIAL RELEASE SURVEYS

Chapter 9 provides a thorough discussion on the detection sensitivity concepts of static and scan MDC. These concepts are an important driver for the selection of suitable clearance survey instrumentation and measurement procedures. Fundamentally, the MDC is calculated for a particular survey instrument and commensurate standard operating procedures to assess whether that particular measurement methodology exhibits sufficient detection capability, often defined relative to the DCGL$_C$. Recall that the MARSSIM recommends that the MDC be sufficiently less than the DCGL—for example, no greater than 10–50% of the DCGL.

The MARSAME takes a somewhat different tack, focusing on the broader application of MQOs, of which the MDC is but one MQO parameter. (Note that MQOs are a subset of DQOs—while DQOs apply to both sampling and measurement activities, the MQOs address the measurement portion of the project DQOs.) The MARSAME warns that the MDC "should be used only as a MQO for the measurement method," reinforcing the fact that the MDC strictly addresses the capabilities of the measurement process, not the measurement results. Notably, the MARSAME (NRC 2009) spends considerable time distinguishing between *detectable* radioactivity and *measurable* radioactivity; the latter being defined as "radioactivity that can be quantified using known or predicted relationships developed from historical information, process knowledge, or preliminary measurements ... as specified in the DQOs and MQOs."

The MQOs identify the characteristics of a measurement method, and include measurement uncertainty, detection capability, quantification capability, range, specificity, and ruggedness. The measurement method uncertainty—an important MQO for survey planning—is part and parcel of both the MDC and the minimum quantifiable concentration (MQC). The MQO that expresses the survey method's quantification capability is the MQC—defined as the smallest concentration that can be measured with a specified relative standard deviation. That is, the MQC (y_Q) is "the concentration at which the measurement process gives results with a specified relative standard deviation, $1/k_Q$, where k_Q is usually chosen to be 10 for comparability" (NRC 2009). Applying this guidance, the MQC is simply the concentration at which the relative measurement uncertainty is 10%. The MARSAME specifies that the MQC should be less than the action level (i.e., $DCGL_C$).

In general, the MDC is approximately 3–5 times the standard deviation of the blank (σ_B), while the MQC is defined here as 10 times the standard deviation of the blank. Thus, the MQC is usually 2 or 3 times the MDC and is considered the lowest concentration that can be accurately measured, as opposed to the concentration that can be just detected. The relationship between MDC and MQC may sound familiar to many health physicists. Indeed, Currie (1968) defined the determination limit (L_Q) as the level at which the measurement precision will be satisfied for quantitative determination. The oft-cited Table 1 in Currie (1968) provides "working" expressions for the critical level (L_C), detection limit (L_D), and determination limit (L_Q) as 1.64 σ_B, 3.29 σ_B, and 10 σ_B, respectively, for a well-known blank.

Finally, and somewhat interesting as it pertains to application, the MARSAME illustrates the practical relationship between the MDC and MQC. Requiring the MQC to be less than the $DCGL_C$ (or other action level) is loosely equivalent to requiring the MDC to be less than 10–50% of the $DCGL_C$. To summarize, both the MDC and the MQC address the capabilities of the clearance measurement process. The MQC is a stricter criterion compared to the MDC since it (MQC) is the concentration where measurements are defined to have only a 10% relative standard deviation—while the MDC concentration will likely have a relative standard deviation two to three times greater. Given this admittedly brief "minimum understandable dissertation" on the relationship of MQC and MDC, and how MQC is applied in MARSAME, hopefully it is not as clear as MUD. Regardless, the reader is referred to MARSAME Sections 5.8, 7.6, and 7.10

for a comprehensive overview of MQCs, including illustrative examples of MQC calculations (MARSAME, Section 7.6.1 and Chapter 8).

The next two sections address the measurability, or more precisely, the detectability of contamination under two general survey approaches: (1) static measurements and (2) scanning. *Static MDCs* are calculated when the clearance survey approach includes conventional direct measurements of surface activity, *in toto* measurements, or laboratory analyses of media samples. *Scan MDCs* are calculated when the clearance survey approach includes scanning with conventional detectors or when using automated scanning equipment such as the CSM. For example, the formula for calculating the MDC for a technician scanning large-bore pipe sections for Pu-239 (for alpha contamination) would be different than for the MDC for Co-60 and other fission products in soil using an *in situ* gamma spectrometer.

15.4.1 Static MDCs

As defined in Chapter 9, the MDC corresponds to the smallest activity concentration measurement that is practically achievable with a given instrument and type of measurement procedure. So the MDC depends not only on the particular instrument characteristics (instrument efficiency, background, integration time, etc.), but also on the factors involved in the survey measurement process, which include surface material type, source-to-detector distance, source geometry, and surface efficiency (backscatter and self-absorption). Additional detail on detectability, detection limits, and equations to compute MDCs are available in Chapter 9, and other references, particularly the MARSAME.

For starters, the calculation of MDC for surface activity assessments in support of clearance using conventional survey instrumentation was given in Chapter 9 (Equation 9.10):

$$\text{MDC} = \frac{3 + 4.65\sqrt{C_B}}{KT}$$

where C_B is the background count in time, T, for paired observations of the sample and blank. Recall that the quantities encompassed by the proportionality constant, K, include the instrument efficiency, surface efficiency, and probe geometry.

In general, the methodology to determine an MDC for a given instrument, radionuclide, matrix or surface, and measurement protocol is based on the specific formulation of the MDC for the application in question. This is an important consideration when the particular clearance survey approach is more involved than conventional surface activity measurements. NCRP (1985) notes that MDC equations have the following structure:

$$\text{MDC} = k \frac{L_D}{(\text{efficiency})(\text{sample size})}$$

where k is a units conversion (from instrument response to activity and the desired units).

The reader is again referred to Chapter 9 for further details on the detection limit (L_D). Recall that the detection limit considers both the instrument background and backgrounds from other sources, such as interfering radiations from the environment, in determining the response of the instrument that is statistically different from background radiation levels. The efficiency term includes both the instrument and surface efficiency. The surface efficiency accounts for solid material conditions such as rusty metal, porous surfaces, and dusty/dirty surfaces (Chapter 10 covers surface efficiency in greater detail).

The sample size term takes on different values depending on the type of measurement. For field survey instruments, this usually means the physical probe area of the detector. For laboratory measurements, it is again a well-defined quantity—it is a measured amount of the sample (e.g., mass of soil and volume of water). It becomes somewhat more involved for *in situ* or *in toto* measurements—in this case, the sample size is a function of the detector's field of view, which can be difficult to define accurately.

Draft NUREG-1761 presents the results of research used to determine the detection sensitivity (MDC) of an *in situ* gamma spectrometer used to release materials. The NRC-funded research conducted by ORAU/Oak Ridge Institute for Science and Education (ORISE) empirically determined the ISGS detection capabilities for a release of potentially contaminated scrap metal. A pallet loaded with 1 metric ton of 5-in-diameter steel conduit was studied. A nonuniform contamination geometry was simulated via random placement of 20 sources of a particular radionuclide (Co-60 and Cs-137 sources were used in this research) within the conduit interiors. Each source had a radioactivity level of one-twentieth of a microcurie (0.05 μCi); therefore, a total of 1 μCi (37 kBq) was positioned within the steel conduit for each test.

The measurement protocol consisted of the ISGS positioned at the mid-point of each of the pallet for 10 minutes (i.e., total count time of 40 min). The ISGS efficiency was calculated for each region of interest by accounting for the Compton continuum counts in the region of interest (ROI) and dividing the resultant net count in the ROI by the total activity (1 μCi). The MDA was calculated as follows:

$$ \text{MDA} = \frac{3 + 4.65\sqrt{C_B}}{(T)(\text{efficiency})} $$

where the efficiency was determined from the net peak counts per minute per total activity.

The measurement process was repeated 10 times for each radionuclide; the sources were randomly placed within the steel conduit each time. The efficiency for Cs-137 ranged from 8.5 to 21.9 net counts per min per μCi, and averaged 15.2. The resulting MDA for Cs-137 ranged from 0.19 to 0.51 μCi, and averaged 0.30 μCi. The efficiency for Co-60 (1173 keV line) ranged from 7.0 to 17.4 net counts per min per μCi, and averaged 12.2. The resulting MDA for Co-60 ranged from 0.19 to 0.59 μCi, and also averaged 0.30 μCi (Table 5.3 in draft NUREG-1761 provides complete research results).

With the same Co-60 and Cs-137 sources, a second experimental configuration consisting of a pallet of insulated copper wires with a total weight of 490 kg was prepared. The insulated copper wires increased the attenuation of the gamma-rays, resulting in a lower efficiency and higher MDA. The MDA for Cs-137 was 0.89 μCi, while the MDA for Co-60 (1173 keV) was 1.0 μCi (Table 5.5 of draft NUREG-1761).

In summary, the approach to calculate the MDC for any clearance measurement method is to determine the detection limit, appropriate efficiencies, and sample size for the given instrument and measurement protocol. The use of empirical studies can be helpful for nonuniform geometries.

15.4.2 Scanning-Based MDCs

Scan MDCs must be calculated for those clearance survey approaches that are based on scanning—both conventional instrument hand-held scans and CSM scans. The initial step when calculating the scan MDC is to determine the MDCR—defined as the detector signal level, or count rate for most equipment, that is likely to be flagged by a surveyor as being "greater than background." The MDCR accounts for the background level, performance criterion (d') adopted by the surveyor, and observation interval (which in turn is based on scan speed). Section 9.3 details the derivation of scan MDC and related concepts.

15.4.2.1 Hand-Held Detector Scan MDCs
It is recognized that many solid materials will be cleared via hand-held scanning; therefore, it is necessary to calculate the scan MDC. The scan MDC for survey instruments scanning material surfaces may be calculated as

$$\text{Scan MDC} = \frac{\text{MDCR}}{\sqrt{p\varepsilon_i\varepsilon_s}}$$

where the MDCR, in counts per minute, can be written as

$$\text{MDCR} = d'\sqrt{b_i}\left(\frac{60}{i}\right)$$

d' is the detectability index (the value can be obtained from MARSSIM Table 6.5)
b_i is the background count in the observation interval
i is the observational interval (in seconds) based on the scan speed
ε_i is the instrument efficiency (unitless)
ε_s is the surface efficiency (unitless)
p is the surveyor efficiency (usually taken to be 0.5)

Common hand-held survey instruments used for performing release surveys include gas proportional and GM detectors. Draft NUREG-1761 provides a useful example for evaluating the scan MDC for a GM detector used to clear stainless-steel materials. In this example, the GM detector scan MDC was calculated for a background level of 70 cpm and a 1-s observation interval. For a specified level of performance at the first scanning stage of 95% true-positive rate and 25% false-positive rate, d' equals 2.32 (from MARSSIM Table 6.5); the MDCR is calculated from b_i as follows:

$$b_i = (70 \text{ cpm})(1 \text{ s})(1 \text{ min}/60 \text{ s}) = 1.2 \text{ counts}$$
$$\text{MDCR} = d'\sqrt{b_i}\left(\frac{60}{i}\right) = 2.32\sqrt{1.2}\left(\frac{60}{1}\right) = 150 \text{ cpm}$$

Using a surveyor efficiency of 0.5, and using the total weighted efficiency of 0.018 (based on GM efficiency for enriched uranium), the scan MDC is calculated as

$$\text{Scan MDC} = \frac{\text{MDCR}}{\sqrt{p \varepsilon_i \varepsilon_s}} = \frac{150}{\sqrt{0.5}\,(0.018)} = 12{,}000 \frac{\text{dpm}}{100\,\text{cm}^2}$$

For this particular example, a GM detector scan is expected to reliably (with 95% confidence) identify contamination from enriched uranium at 12,000 dpm/100 cm². Section 9.3.3.2 provides additional examples of scan MDC calculation.

15.4.2.2 Conveyor Survey Monitor Scan MDCs

The use of CSMs to release materials should only be used if the CSM exhibits scan sensitivity less than the DCGL_C (or other appropriate action level). The scan MDC for a CSM can be calculated similarly to conventional hand-held instruments. Generally speaking, the CSM scan MDC is based on the background counts obtained over the counting interval, the detectability index (d'), and the detection efficiency. The background level is dependent on the type of media, and the counting interval is a function of both the detector's field-of-view and the CSM's belt speed—which establishes the length of time that the detector(s) can respond to a certain amount of material on the belt. The detection sensitivity for potential hot spots located on the conveyor belt will vary with the square root of the observation interval for any segment of material being monitored—this is the same relationship as discussed above for conventional scans. So reducing the belt speed will increase the detection sensitivity for hot spots.

The scan MDC is obtained by dividing the detection efficiency into the MDCR, and the MDCR can be calculated for the CSM in as much as the same manner as shown above for conventional scans. (Note that the primary difference being that automated systems interpret the data using a computer-based analysis algorithm rather than by calculation). The detection efficiency for a CSM depends on the detector characteristics, nature of the contamination, the material being surveyed, and the source-to-detector geometry. The gamma and beta detection sensitivity for the CSM analysis of soil was determined via modeling in draft NUREG-1761. In this example, the CSM detection efficiency for Cs-137 was determined for NaI detectors (3″ × 3″ cylindrical crystals) operating in a gross count rate mode. The model assumed that three such NaI detectors are operated in tandem in a detector bank, and that the total detector volume per bank will be roughly about 1000 cm³. Two empirical parameters to characterize this detector bank are the background count rate and the detector efficiency—the total system background is roughly 2.7×10^4 cpm and the total detector efficiency for Cs-137 is about 1.2×10^4 cpm per µR/h (NRC 2002a). Similar to the approach introduced in Section 9.3.4 for scan MDCs for land areas, these parameters can be coupled to calculated exposure rates that result from materials passing along a conveyor system to determine the detection sensitivity as a function of the material geometry and radionuclide.

Draft NUREG-1761 provides an example application where a CSM is used to scan Cs-137 in the soil. The CSM's three NaI detectors were assumed to be placed approximately 15 cm above a 76-cm wide conveyor belt such that they are evenly spaced across the width of the belt. The soil was assumed to be 2.5 cm thick. The modeled soil geometry yielded an exposure rate of 0.120 µR/h per pCi/g at the two outside detectors, and approximately 0.140 µR/h per pCi/g for the center detector. Coupling these data with the expected detector efficiency given above (1.2×10^4 cpm per µR/h), the total efficiency for the CSM in this geometry is obtained by multiplying detector efficiency by the modeled exposure rate, producing roughly 1.5×10^3 cpm per pCi/g of Cs-137. Finally, the scan MDC can be estimated by assuming a false-positive detection rate of 5% and a true detection rate of 95% (results in d' of 3.28 from MARSSIM Table 6.5). For an observation interval of 6 s, the MDCR can be calculated as

$$b_i = (2.7 \times 10^4 \text{ cpm})(6\,\text{s})(1\,\text{min}/60\,\text{s}) = 2700 \text{ counts}$$

$$\text{MDCR} = d'\sqrt{b_i}\left(\frac{60}{i}\right) = 3.28\sqrt{2700}\left(\frac{60}{6}\right) = 1.70 \times 10^3 \text{ cpm}$$

Finally, the scan MDC for the CSM for the 2.5-cm-thick layer of soil containing Cs-137 is calculated as follows:

$$\text{Scan MDC} = \frac{\text{MDCR}}{\varepsilon_{\text{CSM}}} = \frac{1.70 \times 10^3 \text{ cpm}}{1.5 \times 10^3 (\text{cpm/pCi/g})} = 1.1 \,\text{pCi/g}$$

Accepting a lower false alarm rate (perhaps 1%) and increasing the belt speed (which lowers the observation interval) would increase the CSM scan MDC.

In a similar manner, draft NUREG-1761 provides the beta detection efficiency and scan MDC for a CSM equipped with thin-window proportional detectors. Specific geometries were modeled including flat surfaces contaminated with Tc-99, Sr-90, and Y-90, and soil contaminated with these radionuclides, in addition to Cs-137 (evaluating the beta emission). The CSM consisted of an 80-cm-wide conveyor with five proportional counters (500 cm^2 active area each, total sensitive area of 2500 cm^2) that are placed 5 cm above the belt surface. Table 5.1 in draft NUREG-1761 shows the detection efficiencies and scan MDCs (assuming 6 s observational interval, 1% false alarm rate, and 95% detection). The surface MDCs for Tc-99 (294 keV), Sr-90 (546 keV), and Y-90 (2280 keV) were 5, 2, and 1 dpm/cm^2, respectively. (Note: To obtain the results in more conventional units of dpm/100 cm^2, the results are multiplied by 100).

The differences in the scan MDCs are directly attributable to the differences in detection efficiency (which is a function of beta energy). The detection efficiencies were 300, 650, and 1300 cpm/(dpm/cm^2), respectively for Tc-99, Sr-90, and Y-90. Detection efficiencies for other beta energies can be determined by constructing an efficiency versus energy curve (refer to Section 8.1), and calculating the efficiency for the beta energy of interest. In summary, the CSM may be a viable clearance survey approach for beta emitters provided that the scan MDC is sufficiently less than the $DCGL_C$.

15.5 CLEARANCE SURVEY APPROACHES

Clearance survey approaches include (1) scan-only survey designs using conventional survey instrumentation or automated scanning/conveyor survey monitors, (2) MARSSIM-type surveys using conventional instrumentation that incorporate both scanning and statistical designs for determining sample sizes, and (3) *in toto* survey designs. The MARSAME briefly mentions a fourth type of clearance survey design—method-based survey designs that combine a required measurement method with a scan-only, *in situ*, or MARSSIM-type survey design. Draft NUREG-1761 notes that scan-only survey designs are the preferred clearance approach whenever the scan MDC is less than the $DCGL_C$.

A significant factor in determining how much survey effort should be expended to release the material is the potential for the material to have contamination in excess of the release criterion. As covered in Section 15.3.5, the solid material's classification (class 1, 2, or 3) impacts the degree of survey effort warranted to release the material. This graded survey approach based on the contamination potential is consistent with the MARSSIM approach and is used to design an appropriate level of survey rigor.

The decision to implement a particular clearance survey approach depends on the material characteristics, nature of the contamination, detectability of the radiation emitted, and availability of survey instrumentation. The reader is encouraged to revisit the DQOs discussion in Chapter 3 prior to selecting a particular clearance survey approach. Another useful reference is ANSI/HPS N13.49, Performance and Documentation of Radiological Surveys.

15.5.1 Background Radiation Levels for Clearance Measurements

The release criteria for the clearance of solid materials are usually expressed in radioactivity concentrations in excess of background levels. Additionally, background measurements are necessary to calculate the MDC of the instruments/detectors selected for clearance surveys. Therefore, an important aspect of clearance surveys is to assess the background levels associated with materials and equipment. Chapter 7 addresses background determination and background reference areas, and that information is applicable to this present section. The universal guidance for selecting background reference materials is to ensure that they are nonimpacted (i.e., materials that have no reasonable potential to be contaminated), and are *representative* of the solid materials being released. The number of background measurements should be based on the requirements of the statistical test (for MARSSIM-type releases), or on the DQO process. If background levels are a small fraction of the $DCGL_C$, then one might consider ignoring the background in demonstrating compliance. Recall that this conservative practice of not taking credit for the background was discussed in Section 7.1.

For hand-held survey instruments, the background data sets may be pooled or analyzed individually according to surface material types. Table 7.1 illustrates the background count rate for a number of common surface materials. When pooling background data, the mean and the variance of the background measurements should be calculated for the combined data set. At a minimum, materials with very dissimilar background levels (e.g., dry wall and ceramic tile) should *not* be grouped together. As a rule of thumb, background means should not differ by more than 25–30% to be considered for grouping.

According to draft NUREG-1761 (NRC 2002a), background measurements for the CSM should be determined for each type of nonimpacted solid material being released. For instance, soil from a background reference area, or nonimpacted copper chop, should be analyzed by the CSM repeatedly to develop a background database for these specific materials.

The radiation background at the clearance survey location is an important consideration for *in situ* gamma spectrometry measurements. Cosmic ray intensity varies with location primarily as a result from changes in altitude, while the gamma radiation levels inside a building result from the penetration of radiation from outside and the contribution from natural radioactivity from materials in the building itself. The most variable component of background over time is radon; over the course of a day, the outdoor radon concentrations can change by more than a factor of two (NRC 1994a). Considering these background variations, at least one ambient background measurement for the *in situ* gamma spectrometer should be performed in the area where clearance surveys will be conducted (using the same counting time as that determined to provide sufficient sensitivity for the radionuclide(s) and material being cleared).

15.5.2 Clearance Survey Activities: Measurement and Sampling Methods

Many clearance surveys will undoubtedly be performed using the same survey instruments used to conduct MARSSIM surveys. After all, the release of materials and equipment is part and parcel of the overall decommissioning process. Conversely, for large D&D projects, where tons of materials will be generated, more efficient, higher-throughput survey processes such as CSMs may be the best approach.

Similar to MARSSIM, clearance survey methods using conventional instrumentation can be classified into three survey categories: scanning, direct measurements of surface activity, and smear and miscellaneous sampling. Measurement techniques should be based on the radionuclides (radiations) of concern and use appropriately sensitive instrumentation for field use. Recall that the selection and calibration of field survey instrumentation is described in Chapter 8. These hand-held survey instruments should have an MDC that is less than 10–50% of the applicable release criterion ($DCGL_C$) as discussed in Section 15.4, and to maintain sufficient survey instrument detection capabilities, release surveys should be conducted in low background areas.

Common clearance surveys include both scans and surface activity measurements. Chapter 10 describes these survey procedures, along with the interpretation of measurement data. Materials and equipment being prepared for clearance should include a historical site assessment (considering process knowledge) to determine the nature of the potential radionuclides (Section 15.3.4). When scanning, any locations of elevated direct radiation should be marked for further investigation to include judgmental measurements of surface activity. Investigation levels should be established and implemented for evaluating elevated radiation levels from hot spots.

Direct measurements of surface activity should be performed for materials being released. The type of surface activity measurement (gross alpha or gross beta) should be selected in consideration of the potential radionuclides present. Surface activity measurements should be performed using calibrated survey instruments, including gas proportional, GM, and ZnS detectors coupled to survey meters. Material-specific background measurements should be performed for each surface material type as mentioned in the preceding section.

Miscellaneous sampling may be performed to support the clearance of materials and equipment. Media samples may include smear, residue, and/or swab samples, depending on the nature of the materials being surveyed (e.g., inaccessible surfaces on equipment). Residue and/or swab samples may also be collected at specific locations where the surface area is inaccessible for direct measurements. The number of samples will usually be based on the statistical tests used for the clearance survey design.

15.5.3 MARSSIM-Type Clearance Survey Design

MARSSIM-type clearance survey design generally implies the use of a statistical sampling paradigm in conjunction with scanning performed using conventional survey measurements. Similar to MARSSIM, the clearance survey design is based on a statistical sample size. Scans are performed to identify potential hot spot contamination. The scan coverage should be graded based on the classification, and according to MARSAME guidance—100% of surfaces should be scanned for class 1 materials, 10–100% for class 2, and up to 10% for class 3.

The size of the material survey unit (see Section 15.3.1) may also be a function of the material's classification—the amount of material comprising class 1 survey units being smaller than either class 2 or 3 survey units. The rationale being the same as articulated in MARSSIM—because the number of statistical samples is independent of the area (or mass or volume) that they are allocated over, varying the size of the material survey unit allows the user to adjust the sampling or measurement density.

This flexibility in sampling density as a function of classification may be warranted based on the DQO process. Again, the MARSAME guidance is that the survey unit size does *not* change based on classification; consequently, proposing variable survey unit sizes as a function of contamination potential (classification) would certainly need to be approved by the responsible regulatory authority.

MARSAME recommends the same nonparametric statistical tests as used in the MARSSIM, and for the same reasons. The criteria for choosing between the sign test and the WRS test are the same as discussed in Chapter 13. The general MARSSIM guidance is that when the radionuclide is *not* in the background (or its background concentration is negligible) and radionuclide-specific measurements are made, the sign test is used; otherwise, the WRS test is used. These nonparametric statistical tests can be used for both surface activity assessments and volumetric concentrations in materials. For clearance, it is expected that the sign test will be performed for surface activity measurements. However, given that many material survey units will be composed of the same material types, using the WRS test should be relatively straightforward. Therefore, the WRS test is likely much more practical for clearance surveys than it would be for indoor building surveys (recall the discussion in Section 13.3.3).

Two possible scenarios are considered in MARSAME—scenario A and scenario B. In scenario A, the survey data are tested against the $DCGL_C$ (or action level) to determine if the concentration in the material survey unit exceeds that value. In this scenario, the upper bound of the gray region (UBGR) is equal to the $DCGL_C$, and the LBGR is typically set at the expected radionuclide concentration in the materials and equipment survey unit. The test should have sufficient power ($1 - \beta$, based on the type II error) to detect radioactivity concentrations at the LBGR. If σ is the standard deviation of the measurements in the material survey unit, then Δ/σ expresses the size of the shift (i.e., $\Delta = DCGL_C - LBGR$) as the number of standard deviations that would be considered "large" for the distribution of measurements in the survey unit. Recall that MARSSIM Table 5.5 provides sample sizes for the sign test as a function of relative shift and type I and II decision errors. This of course is very similar to scenario A in the MARSSIM. The formal hypothesis tested by the sign test under scenario A is

H_0: The median concentration of contamination in the material survey unit is greater than the $DCGL_C$

versus

H_a: The median concentration of contamination in the survey unit is less than the $DCGL_C$

The two-sample statistical test (WRS test) under scenario A is also similar to the approach described in MARSSIM. The WRS test is most effective when the contamination is uniformly present throughout a material survey unit; it is designed to detect whether or not this contamination exceeds the $DCGL_C$. The formal hypothesis tested by the WRS test under scenario A is

H_0: The median concentration in the material survey unit exceeds that in the reference material by more than the $DCGL_C$

versus

H_a: The median concentration in the material survey unit exceeds that in the reference material by less than the $DCGL_C$

As described in MARSSIM, the WRS test assumes that any difference between the reference material and material survey unit concentration distributions is due to a shift in the survey unit concentrations to higher values (i.e., due to the presence of contamination in addition to background). The result of the hypothesis test determines whether or not the material survey unit as a whole is deemed to meet the release criterion. The test should have sufficient power $(1 - \beta)$ to detect radionuclide concentrations at the LBGR, which should be set at the expected mean residual contamination level in the material survey unit. MARSSIM Table 5.3 provides sample sizes for the WRS test as a function of relative shift and type I and II decision errors.

For both the Sign and WRS tests, individual measurements exceeding the $DCGL_C$ are investigated further to the extent that it is necessary to determine that the overall average in the survey unit does not exceed the $DCGL_C$. Additionally, consideration should be given to whether any hot spots may exceed the elevated measurement comparison (i.e., hot spot limit) set for such areas.

Scenario B in MARSAME is fundamentally different than that presented in NUREG-1505 and illustrated via an example in Section 7.3. In NUREG-1505, the scenario B null hypothesis is that measurements in the survey unit are indistinguishable from those in the background reference area. Conversely, scenario B in MARSAME is centered on the definition of the "discrimination limit"—the radionuclide concentration that can be reliably distinguished from the action level (or DCGL$_C$) by performing measurements (NRC 2009). Therefore, the null hypothesis in scenario B is that the radionuclide concentration is less than or equal to the action level (DCGL$_C$). The UBGR is equal to the discrimination limit (DL), and the LBGR is equal to the action level. The DL effectively specifies how hard the surveyor needs to look for contamination (e.g., the DL may be set at a regulatory limit; examples discussed in Section 15.3).

MARSAME provides a valuable discussion on the relationship between the UBGR and the required measurement method uncertainty, μ_{MR}, in Section 7.3.1. As all MARSSIM users know, a relative shift (Δ/σ) of 3 is desired to provide a reasonable statistical sample size. The MARSAME demonstrates in Section 7.7.1 that if the measurement variability (σ_M) is negligible relative to the spatial variability (σ_S), which is often the case because the measurement variability can be controlled (minimized) much easier, then σ_M will be $\Delta/10$. Therefore, MARSAME's default recommendation is $\mu_{MR} \leq \Delta/10$ when decisions are being made about the mean of a sampled population.

A simple example illustrating the calculation of the required measurement method uncertainty is provided. Consider the use of gas proportional detectors for performing surface activity measurements on a scrap metal survey unit. Assume that the DCGL$_C$ is 8000 dpm/100 cm² and the LBGR is 2000 dpm/100 cm². The required measurement uncertainty is given by

$$\mu_{MR} = \frac{\Delta}{10} = \frac{8000 - 2000}{10} = 600 \frac{dpm}{100\,cm^2}$$

The actual measurement uncertainty must be less than the required measurement uncertainty. In the present example, the surface activity uncertainty for the gas proportional detector measurements must consider all sources of uncertainty, including counting statistics (related to the gross and background count rates), and the instrument and surface efficiency terms. The propagation of uncertainty for surface activity measurements is provided in Section 17.2.

The final point to address in MARSSIM-type clearance survey designs is how to determine sample locations. The systematic rectangular or triangular grid recommended in the MARSSIM is not particularly well suited for most material survey units. Draft NUREG-1761 suggests that a random-start grid can be used for materials and equipment consisting of many small regularly shaped pieces. These materials can be spread out evenly (perhaps on a concrete floor that has been marked by a two-dimensional grid) and random-start grid can be used to locate samples, ensuring every portion of the batch the same opportunity to be sampled. (Incidentally, MARSAME offers substantially the same guidance.) Another possible method discussed in draft NUREG-1761 for sampling a survey unit of similarly sized small pieces of material is to systematically measure every mth piece. This would require some estimate of the total number of pieces, N, so that N/m would equal or exceed the number, n, required for the statistical tests.

Large, irregularly shaped pieces pose more of a challenge for determining random locations. MARSAME mentions superimposing a grid on top of materials and equipment, using a net or ropes to form a grid to identify measurement locations. In some situations, there may be no choice but to select biased sampling locations; in these cases, rely on process knowledge and the initial assessment to select locations most likely to contain residual radioactivity. This may result in an overestimate for the average radioactivity concentration for the material survey unit. Ensure that the approach used for selecting sampling locations is clearly documented, and that the methodology is specified in advance of the actual sampling (not that we do not trust you).

15.5.4 Scanning-Only Clearance Survey Design

Scan-only surveys rely on scanning to measure materials and equipment. This survey design is appropriate for all types of solid materials defined in Section 15.3.1—for example, hand-held detector scans for large pieces of equipment and scrap metal, while CSMs would be ideal for soil and copper ingots. However, this clearance survey approach is possible only when the survey instrumentation exhibits sufficient scan sensitivity—that is, the scan MDC is less than the $DCGL_C$. Further, scan-only clearance surveys should employ survey instrumentation, such as data loggers, that have the capability of automatically documenting the survey results (preferred over having surveyors manually recording their scanning results).

MARSAME acknowledges the benefit of logging individual measurement data is the ability to statistically evaluate the data. For example, a CSM used to scan copper chop survey unit might yield several thousand independent, 1-s measurements. The mean and standard deviation of this robust data set can be used to calculate the upper confidence limit, and allow comparison to the action level (DCGL$_C$). Additionally, individual results from conventional hand-held (or CSM) scanning can be compared directly to the action level (or elevated measurement comparison). MARSAME Chapter 6 provides guidance on evaluating survey results.

15.5.4.1 Scan-Only Using Conventional Survey Instrumentation

Scanning-only surveys using gas proportional and GM detectors to release materials and equipment have been used for years. A drawback of this survey strategy has been the difficulty in adequately documenting the scan survey results. For instance, a surveyor carefully performs a 100% scan of several pieces of excessed laboratory equipment and does not identify any radiation levels above background levels. What is an acceptable level of documentation for this clearance survey? Perhaps we can turn to ANSI/HPS N13.49 (2001) for help. It states that "surface scans should be reported in terms of gross radiation levels, usually expressed in counts per minute (cpm); background radiation levels of the scanning instrument should be reported in the same units." Sound advice, although addition detail would be welcomed.

Being able to automatically document the scan results for each solid material survey unit is a significant improvement. MARSAME acknowledges this, noting that the "benefit of logging individual measurement results is the ability to statistically evaluate the data." If data logging is not available, a second choice might be to collect a number of surface activity measurements (or media samples) from the material to serve as documentation of the scan results. The number of these measurements in this case should be determined using the DQO process.

MARSAME recommends that the scan coverage should be graded based on the classification—100% of surfaces should be scanned for class 1 materials, 10–100% for class 2 (based on the relative shift in MARSAME Equation 4.1) and also 10–100% for class 3—with a provision that less than 10% scan coverage may be OK for class 3 scans, *and* that they may be solely based on biased measurements. MARSAME Equation 4.1, shown below, illustrates that scan coverage is consistent with the philosophy that survey effort should increase as Δ/σ decreases:

$$\% \text{ Scan} = \frac{(10 - (\Delta/\sigma))}{10} \times 100\%$$

Note that to calculate the % scan coverage, the user must know (or reasonably estimate) the expected radionuclide concentration in the survey unit (to establish the LBGR to calculate the Δ), as well as the measurement standard deviation (σ). So, if the relative shift is 5, then the % scan coverage is 50%. MARSAME requires rounding up to the next 10% coverage, so if MARSAME Equation 4.1 resulted in 33%, the % scan coverage would be 40%.

MARSAME notes that 100% scans for class 1 units means 100%, for example, scanning should be performed on both sides of flat items and all surfaces should be accessible, including "changing the surveyor's grip on the item to ensure all areas are surveyed" (do not let this become your Achilles' heel in achieving 100% coverage). Class 2 scan coverage should be spatially uniform over the entire material survey unit. Scan locations should be determined randomly if the material is expected to have a uniform contamination potential across the survey unit. Otherwise, if there are locations expected to have a higher potential for contamination, then scanning should be biased to include these higher probability locations.

Note that scanning less than 100% of the material survey unit (when actual surfaces scanned are biased to include areas of higher contamination potential) will likely increase spatial variability in the data since the entire population of measurement locations is not being scanned. It is important to consider this effect when analyzing the scan results, the survey unit mean will be a somewhat biased estimate of the population mean.

15.5.4.2 Conveyor Survey Monitors

CSMs can be an effective material release strategy, particularly when significant quantities of bulk material (soil, copper chop, smaller pieces of concrete rubble) are planned for release. CSMs operate by moving materials and equipment past radiation detectors using a conveyor system while automatically storing and analyzing the resulting detector signals via a data acquisition system. MARSAME notes that CSMs may be used to perform *in situ* measurements by stopping the conveyor and taking a longer detector measurement.

CSMs must exhibit sufficient detection sensitivity (based on the MQOs) for the expected radionuclides present in the bulk material; they must be operated in the same configuration used to assess its detection capabilities—thickness of bulk material on conveyor, belt speed, source-to-detector geometry, and so on. The sensitivity of the detection system depends on the proximity of the detector(s) to the radiation source (solid material) and on the speed of the conveyor belt. The fundamental physics and statistical counting parameters associated with measuring a stream of material passing by a CSM detector can be used to estimate the detection sensitivity of a CSM detector system. Section 15.4.2.2 provides guidance for determining the scan MDC for CSMs.

CSMs are commonly outfitted with NaI and plastic scintillators for gamma radiation detection, and gas proportional counters for beta radiation detection. They offer several advantages when compared to conventional surveys by personnel using hand-held detectors, but these advantages come at an increased cost. One example is Ludlum Measurements Model 375P-3500 conveyor radiation monitor used to monitor incoming scrap metal at large scrap yards, foundries, and steel mills. It features a large (3500 in^3) plastic scintillation detector within a lead-shielded housing to identify rogue radiation sources before they can potentially contaminate the facility. Appendix D in MARSAME provides additional information on CSMs, including various options (e.g., segmented gate systems) and applications.

Gross screening of gamma-emitting radionuclides will usually be performed using scintillation detectors such as NaI or plastic scintillators. These detectors may not be ideal based on their modest resolution capabilities; however, their first-rate detection efficiencies and relatively low cost make them a desirable choice for gross gamma measurements. Also, the geometry (volume and shape), encapsulation, and electronic configuration of a scintillation detector determine its overall detection efficiency and background response, thereby defining its signal-to-noise ratio. As pointed out in draft NUREG-1761, it is important to select detectors that balance background response with detection efficiency for the anticipated radionuclide(s). For example, a 3″ × 3″ NaI detector will provide a good signal-to-background ratio for a high-energy gamma emitter such as Co-60, but will have relatively poor performance for the analysis of low-energy gamma emitters such as Am-241, where a thin NaI crystal would work best.

HPGe detectors offer much greater radionuclide assay capability but are fairly expensive to purchase and maintain—especially if one is interested in achieving the same level of detection efficiency offered by larger volume scintillation crystals. HPGe are excellent for gamma-ray spectrometry; their use in a CSM system could be warranted in some instances for nuclide identification following a positive detection during a gross scan.

Beta radiation measurements on bulk materials run through a CSM may be possible depending on the radionuclide, solid material type, and the release limit. Beta detection can be accomplished using thin-window gas-filled detectors, such as proportional and GM detectors and thin-windowed scintillators. Perhaps the best choice for measuring beta emitters will be large-area gas flow-through proportional detectors with thin Mylar entrance windows—think of floor monitor mounted above a conveyor system. Another option may be a large array of small detectors so that each segment monitors a small area (on the belt) while keeping its background to a low level. Smaller detectors can be ganged together in parallel assemblies with common electronics to minimize overall system cost. Draft NUREG-1761 provides a number of useful examples describing CSM system design for various clearance applications.

15.5.5 *In Toto* Clearance Survey Design

In toto is Latin for "totally" or "entirely." Hence, an *in toto* clearance survey approach consists of the solid material being measured in its entirety. Perhaps the most common example is the use of ISGS systems; other *in toto* systems include box counters, tool and bag monitors, and portal monitors.

In toto survey techniques often measure 100% of the materials and equipment in a survey unit, via one or more measurements, depending on the field of view for a measurement. This technique results in averaging the contamination level over the entire material survey unit, so a drawback is the possibility that hot spots will be missed. Thus, this clearance survey is well suited for materials and equipment that do not have a potential for hot spots (i.e., solid materials classified as class 2 or 3). When hot spots are potentially present (e.g., class 1 materials), the impact that they have on the average contamination level should be addressed during the efficiency determination for *in toto* survey techniques. For example, a 10 mCi sealed source inadvertently incorporated into the scrap pile might increase the average radioactivity in the scrap pile waste by some fraction of a pCi/g (depending on this hot spot's source strength and the scrap pile mass). In more severe cases, such as a 1 Ci source (perhaps from an inappropriately disposed teletherapy source), the average might increase by tens of pCi/g. Possible solutions to this problem may include collecting a greater number of collimated measurements to reduce the detector's field of view for each measurement, and/or to perform conventional scanning for hot spots to supplement the *in toto* techniques.

The material's classification will be a likely factor when establishing the material survey unit size—for example, the amount of material comprising class 1 survey units may be smaller than either class 2 or 3 survey units. Another approach may be to maintain consistent survey unit sizes for all material classes, while adjusting the survey coverage based on classification. In each case, the DQO process can be used for making the decision. For example, a tool monitor might be used to assay 100% of the materials in class 1, while less material would be analyzed in class 2 and 3 survey units. The MARSAME requires 100% coverage for class 1 solid materials, while the coverage for class 2 and class 3 materials and equipment is based on the relative shift, as shown in the equation below:

$$\% \text{ Measured or } \% \text{ solid angle coverage} = \frac{(10 - (\Delta/\sigma))}{10} \times 100\%$$

As mentioned in the scan-only discussion, to calculate the percentage of coverage, the user must know (or reasonably estimate) the expected radionuclide concentration and measurement standard deviation in the survey unit. Additionally, for measurements such as *in situ* gamma spectrometry, the % solid angle coverage for each measurement must be known. The basic MARSSIM philosophy applies—that is, the materials and equipment having the greatest potential for contamination should receive the highest degree of survey coverage.

Two common *in toto* clearance survey approaches are covered in more detail below. These include *in situ* gamma spectrometry and volume counters (e.g., tool and bag monitors).

15.5.5.1 In Situ *Gamma Spectrometry*
ISGS measurements for solid materials, particularly in a complex geometry that renders some of the surfaces inaccessible, may be a viable release survey option. This section will discuss some of the considerations and the overall plan for implementing ISGS as a survey tool of solid materials, including discussion of experimental results for applying ISGS to the survey of scrap metal. Chapter 8 provides a description of the ISGS system.

Consider the use of ISGS to release a pallet of scrap metal, appropriately sized to fit the pallet. The clearance survey plan should specify the number and location of measurements, detector geometry, count time, among other considerations. A number of factors affect the required count time, including the efficiency, background level, and desired uncertainty (specified by the MQOs). For this example, the geometry refers to the orientation of the ISGS relative to the pallet of scrap metal. The overall efficiency for this geometry would likely be different for each particular arrangement of scrap metal. Furthermore, the distance the detector is placed from the pallet affects the overall efficiency of the ISGS measurement. The following case study provides an example clearance strategy using various technologies for a large amount of scrap metal.

CASE STUDY   CLEARANCE SURVEY DESIGN FOR SCRAP
METAL DISPOSITION

This case study involves a scrap metal area that has not been active since the late 1970s. An estimated 30,000 tons of scrap metal is stored in the Scrap Metal Park. Some information is known about the radiological (mostly U-238 and U-235) and nonradiological materials associated with the scrap metals, based on process knowledge. Nearly 40% of the scrap was size-reduced (and exists in separate piles) in the early 1980s, but this effort was not completed and the contaminated size reduction equipment was abandoned in the scrap yard (and is now part of the Scrap Metal Park).

Most of the materials consist of piping sections, piping elbows, large duct sections, sheet steel, and other metal parts associated with the plant's processing mission. These pieces were size-reduced to less than a meter long. Other scrap materials include abandoned equipment, storage tanks, 55-gallon steel drums (hoped to be empty), metal lockers, chairs, desks, cabinets, fans, fan blades, and electrical transformers suspected of containing PCB contamination. Overall, there are more than 20 distinct piles or areas of scrap in this area.

The scrap metal is targeted for either clearance (at a scrap metal facility) or disposal at an appropriately licensed facility in the western United States. Once the available process knowledge and historical data associated with the scrap metal have been reviewed, characterization will be performed to fill data gaps. The characterization approach will consist of sample collection from the scrap piles and general radiation monitoring to assess the variations in the radiation fields.

The primary clearance approach combines the use of ISGS (of the scrap metal piles) and a truck monitor (plastic scintillator) as the materials depart the Scrap Metal Park. ISGS measurements using an HPGe detector will be used for demonstrating compliance with the $DCGL_C$ (average contamination level) over the entire scrap metal pile. This detector is sensitive to gamma emissions from the immediate progeny of U-238 (1001 keV from Pa-234 m and 63 keV from Th-234) and U-235 (185 keV)—it is likely that the 1001 keV line will be the key signature for U-238 measurement because of the higher penetrating power of this higher gamma energy. The ISGS system can be calibrated empirically using known radionuclide sources, or through modeling of the detector response to a large scrap metal source (e.g., Canberra's ISOCS). Once the efficiency for the above energies (regions of interest) is determined, the count time for each ISGS measurement is determined to provide an MDC less than the $DCGL_C$ for U-238.

Measurements are collected around the perimeter of each scrap metal pile; measurement locations depend on the size of the pile, ranging from 15 to 40 perimeter locations. The ISGS measurements at each perimeter location were collected at scrap metal pile heights of 1, 2, and 4 m. The use of a scissor-lift or similar lift equipment may be used to position the detector close to the scrap pile at the 2-m and 4-m measurement heights. This clearance survey approach would generate a total of 45–120 *in situ* gamma spectrometry measurements for each scrap metal pile. Depending on the surface area and volume of the pile, the combined field of view provided by all of the ISGS measurements represents some fraction of the entire scrap metal pile. If it is reasonable to assume that the pile was randomly created, then the resulting mean and standard deviation of the ISGS measurements around the outer shell of the scrap metal pile can be used as a good estimate of the entire scrap metal pile.

As a secondary check on the ISGS measurements, the scrap metal can be loaded into transport containers and assessed by a truck monitor as it departs the Scrap Metal Park. One example is the RADOS truck monitor that uses a large area plastic scintillator to detect the gamma radiation levels. It also compensates for the shielding effect of the load. The truck monitor should be calibrated to U-238 distributed over the volume of a transport container. Once the efficiency is obtained, the truck monitor's sensitivity is based on factors such as how fast the truck passes through the monitor and the acceptable false-positive rate (i.e., number of false alarms tolerated). In summary, truck monitors provide a valuable secondary check that smaller volumes of the scrap metal meet the $DCGL_C$, and may catch hot spots that were missed by the ISGS measurements.

15.5.5.2 Volume Counters

Volume counters are particularly well suited to release large quantities of small materials and equipment such as hand tools and other small items. The solid material's shape/size determines whether it can be measured by the volume counter; if it fits, it is counted—not unlike confirming carry-on luggage fits within the "size wise" bins located at the airport gate. MARSAME states that these counters are best suited for measuring class 2 and class 3 materials and equipment; the issue being that potential hot spots in class 1 materials contribute to the average contamination level, but no information would be available on the magnitude of the hot spot contamination level—a system limitation if the clearance criteria include hot spot limits. Another concern may be the inability to measure radioactivity in inaccessible areas (including the potential for self-shielding of the material); in these situations it may be beneficial to disassemble or otherwise open the materials and equipment prior to counting.

Volume counters consist of a 4π-geometry counting chamber, array of detectors, and electronic package for analysis. Shielded configurations are frequently used to reduce the background levels, and these shielded configurations often completely surround the solid material providing a 4π counting geometry. Box counters can be designed to quantify surface activity or total activity, and usually average the activity over the mass of the solid material. A variety of detectors can be used for gamma and beta radiation measurements, the most common counters use plastic and NaI scintillators for gamma radiation assays. Some systems use gas proportional detectors for beta radiation measurements, and it is possible to measure alpha radiation using long-range alpha detection configurations that measure the ionization—that is, the number of ion pairs produced in air—that is proportional to the alpha surface activity level on the materials.

The volume counters are calibrated using appropriate geometries containing sources of known activity. Counting jigs can be used to ensure that consistent source-to-detector geometries are maintained. Some box counters are equipped with a turntable that allows the materials to be counted in different orientations to better measure nonuniformly distributed contamination. The counting time depends on the background level of the counter and the desired MDC needed to achieve MQOs. It is not unusual for count times to be on the order of 30 s or less. MARSAME Appendix D.5 provides additional information on the calibration and operational considerations for using volume counters for the release of materials.

QUESTIONS AND PROBLEMS

1. Some factors in determining an appropriate approach for solid material clearance include stakeholder concerns over free release, cost (and availability) of disposal versus survey costs, and technology advancements (in material processing, disposition modeling, surveying, etc.). Describe how the relative importance of each factor influences the ultimate material disposition decision.

2. Explain the difference between MDC and MQC as the terms apply to MARSAME clearance surveys.

3. Explain three possible clearance survey approaches for scrap metal potentially contaminated with fission and activation products from a nuclear reactor being decommissioned. What role does material preparation (e.g., sizing, cutting, and disassembly) play in each survey approach?

4. Outline the general clearance survey design for piles of soil that were generated from the excavation of leaking pipes from a uranium mill. The soil is potentially contaminated with U-238 and Ra-226, and the preferred soil disposition is to backfill the excavations. The radiological engineer desires to maximize the use of *in situ* gamma spectrometers for the clearance surveys.

SUGGESTED READING

ANSI N13.12 *Surface and Volume Radioactivity Standards for Unconditional Release*
MARSAME
NAS report *The Disposition Dilemma*
NCRP Report No. 141 *Managing Potentially Radioactive Scrap Metal*

CHAPTER 16

Decommissioning Survey Applications at Various Facility Types

O NE PURPOSE FOR DISCUSSING the different facility types is to examine the various ways in which the MARSSIM methodology can be implemented. For example, a facility that produces sealed sources may be expected to have only interior contamination from a number of alpha- and beta-emitting sources. The MARSSIM application might consist of the sign test using the unity rule for alpha and beta surface-activity measurements. Perhaps, another site is contaminated with uranium that might entail the use of the WRS test for land areas. How would MARSSIM be applied to research laboratories contaminated with H-3 and C-14? How about a power reactor being decommissioned?

Some D&D sites became contaminated as a result of various missions conducted over a period of several decades. Detailed HSA, scoping, and characterization surveys are necessary to identify the nature of the contaminants. Consider the Hunter's Point Naval Shipyard in California. This site housed the Naval Radiological Defense Laboratory (NRDL) beginning in 1946 to study the effects of, and to develop counter measures from, nuclear weapons (NRC 2012a). The Defense Lab operated until 1969 and conducted studies related to ship shielding, radioactive waste for deep-sea disposal, animal research, and radiation detection instrumentation development, among other research projects. In particular, the NRDL decontaminated ships involved in nuclear weapons tests in the Marshall Islands. One can imagine the diversity of radioactive contaminants at this site—for example, Cs-137 and Sr-90 from atomic weapon testing, and Ra-226 associated with various luminescent sources (deck markers).

Many of the sites currently being decommissioned are contaminated with uranium, thorium, and radium. These D&D sites present unique challenges in demonstrating compliance with release criteria due to the fact that these radioactive materials are present naturally in the background. Furthermore, they usually have low DCGL values, meaning that the scenario B FSS design (Chapter 7) is a realistic option. In the next two sections, we will consider various types of uranium, thorium, and radium contamination, including uranium ore, processed natural uranium, enriched uranium, depleted uranium (DU), natural thorium, and radium. Each of the above natural radioactive contaminants implies a specific mix of parent and progeny radionuclides.

16.1 URANIUM SITES

It is very much possible that uranium sites are at the top of the list when it comes to the most popular contaminant at D&D sites. The NRC's Site Decommissioning Management Plan (NRC 2012a) lists several sites that are primarily contaminated with uranium. Kerr–McGee's Cimarron Plant site was used to fabricate enriched uranium and mixed oxide fuels for nuclear reactors (enriched uranium), Westinghouse Electric's Hematite Facility manufactured fuel pellets from low-enriched uranium, and ABB Prospects, Inc. Windsor site in Connecticut contains soils, buildings, and equipment that were contaminated with uranium and by-product material from decades of operations. And of course DOE's gaseous diffusion plants in Oak Ridge, Paducah, and Portsmouth are contaminated with various levels of enriched uranium.

In general, uranium contamination can be expected at fuel production facilities, gaseous diffusion plants, uranium mills, and many sites that were used to process uranium for the weapons program and other AEC programs and ended up in the FUSRAP program. Some D&D sites contain uranium contamination as a result of testing DU rounds. For instance, the Jefferson Proving Ground in Indiana has soil contaminated with DU-containing tank penetrator rounds and other unexploded ordinance (UXO).

Depending on the nature of the uranium contamination, the FSS at uranium sites can have varying degrees of complexity. In general, it is anticipated that the sign test will be used for surface-activity assessments, while the WRS test will be the choice for land area FSSs.

16.1.1 Nature of the Contaminant

Uranium ore refers to the entire uranium decay series as it exists in nature (i.e., no processing). The series begins with the alpha-emitting U-238 and is followed by radioactive progeny that decays by emitting alpha, beta, and gamma radiations. Refer to Appendix A for details of the decay characteristics of the uranium decay series and other radionuclides. Noteworthy progeny in the uranium series include Th-234, Pa-234 m, U-234, Th-230, Ra-226, Rn-222, Pb-214, Bi-214, and Po-210. Uranium ore is really not that common at uranium D&D sites—it exists at uranium mills and at facilities that extract rare earths from various ore bodies.

Uranium-235, while it is a member of the actinium series, is an important radionuclide to consider when planning surveys particularly at enriched uranium sites. Uranium-235 represents a relatively small fraction of natural uranium—only 0.7% by weight of natural uranium. As discussed below, U-235 becomes increasingly significant as uranium enrichment is increased (for nuclear weapons and energy programs).

It is necessary to have an understanding of the specific form of uranium used at the site when designing the FSS. This can be determined by reviewing the previous facility history. The uranium fuel cycle provides an effective illustration of the various definitions and challenges presented by uranium. Uranium ore, prior to any processing, will contain the entire U-238 decay series in secular equilibrium ending in stable Pb-206. Once the uranium ore is processed, and the uranium extracted either at a uranium mill or other type of facility, two source terms are created—the uranium (in various chemical forms) and associated tailings. Tailings are comprised predominantly of Th-230 and Ra-226, and the radium progeny (radium is discussed in the next section). It is the tailings that give rise to radon emanations and significant gamma exposure levels from Pb-214 and Bi-214. The resultant processed natural uranium contains naturally occurring uranium isotopic abundances. Within approximately six months of processing, ingrowth of the immediate U-238 progeny—Th-234 and Pa-234 m—which are both beta–gamma emitters, have reached secular equilibrium. This is the reason why many modeling codes assume that Th-234 and Pa-234 m are part and parcel with U-238.

Three distinct FSS approaches are possible given the different contaminant compositions at each step in the uranium fuel cycle. The FSS may be designed for natural uranium ore, uranium that has been processed, or the tailings where Ra-226 and Th-230 are now the principal radionuclides of concern. The next phase of the fuel cycle involves those facilities that chemically process the uranium to various forms ranging from the gaseous uranium hexafluoride (UF_6) to uranium metal. Each of these facilities will continue to be concerned with isotopic abundances and progeny associated with natural processed uranium.

The final step in the cycle is the enrichment phase where the U-235 isotopic abundance is selectively enriched for use as nuclear fuel. As a by-product of the enrichment process, U-234 is also enriched, and due to its lighter weight, becomes the predominant uranium isotope wherever enriched uranium is encountered. The net effect of the enrichment process is a significant alteration of the uranium isotopic abundances. Typically, enriched uranium will contain anywhere from 2% to 3% by weight of U-235 (e.g., power reactor fuel) to greater than 90% U-235 for naval reactors and specialty application fuels, or weapons. The by-product of the enrichment process is DU—containing principally U-238 and its immediate beta progeny—which may then be processed for use in military ordnance, chemical catalysts, or other applications (Abelquist and Vitkus 1997).

The determination of the weight of U-235—which determines the uranium enrichment—is based on its alpha activity present and dividing by the total weight of all the uranium isotopes. Specifically,

$$\% \ EU = \frac{U\text{-}235/2.14E6}{U\text{-}235/2.14E6 + U\text{-}238/3.33E5 + U\text{-}234/6.19E9} \quad (16.1)$$

where U-235, U-238, and U-234 are the activity concentrations in pCi/g, and the numerical values are the respective specific activities for each isotope. Actually, for this equation to yield the % EU, the result needs to be multiplied by 100. (Note that the U-234 weight is usually negligible due to its high specific activity.)

16.1.2 Field Measurements

A FSS challenge encountered after uranium has been processed is determining the relationship of particle emissions (i.e., alpha to beta ratios) for measuring surface activity levels and isotopic abundances for quantifying the total uranium concentrations in soil. For determining surface activity of natural uranium contamination, the alpha to beta activity ratio is approximately 1:1. This is because processed natural uranium contains the following decay series in equilibrium: U-238 (α) → Th-234 (β) → Pa-234 m (β) → U-234 (α), which makes up about 98% of the activity and U-235 (α) → Th-231 (β) that accounts for the other 2% of the activity. As can be seen, the first chain exhibits two alpha emitters and two beta emitters, while the second chain has one alpha emitter and one beta emitter. Therefore, the overall ratio of alpha to beta activity is 1:1.

Once uranium has been processed, this relative relationship may range from approximately 0.6:1 for DU, to 3:1 for low enriched uranium, and even 30:1 for higher enrichments of uranium. On an alpha activity basis, natural uranium is approximately 0.493:0.022:0.485 (percent of the total alpha activity), respectively for U-234, U-235, and U-238. This may be confirmed by performing alpha spectrometry on a processed uranium sample or by substituting these uranium fractions into Equation 16.1 and calculating the natural U-235 weight percent of 0.7%. Isotopic ratios for enriched uranium varies based on percent enrichment—for example, 4% enriched uranium exhibits ratios of approximately 0.81:0.04:0.15 for U-234, U-235, and U-238, respectively. These isotopic ratios are determined by performing alpha spectroscopy on a representative sample, and then taking the ratio of each isotopic concentration to the total uranium concentration. For 4% enriched uranium, the alpha to beta activity ratio would be calculated as follows, recognizing that Th-234 and Pa-234 m are in equilibrium with U-238, and Th-231 is in equilibrium with U-235 (Table 16.1).

Therefore, there are 0.34 betas emitted per alpha emission, or correspondingly, an alpha to beta activity ratio of 2.94:1. These relative radiation emissions are useful for determining weighted instrument and surface efficiencies (refer to Chapters 8 and 10).

TABLE 16.1　Relative Alpha and Beta Ratios from 4% Enriched Uranium

	Relative Radiation Emissions	
Radionuclide	Alpha	Beta
U-238	0.15	—[a]
Th 234	—	0.15
Pa-234 m	—	0.15
U-234	0.81	—
U-235	0.04	—
Th-231	—	0.04
Total	1.0	0.34

[a] Particular radiation not emitted by radionuclide.

The assessment of surface contamination is performed by converting a detector's response to surface activity using instrument and surface efficiency factors determined via calibration. Measurements of surface-activity levels are typically performed using gas proportional, GM, or plastic scintillation detectors with portable ratemeter scalers. The uranium decay series emit both alpha and beta radiation; therefore, it may be feasible for either alpha or beta activity to be measured for determining the residual activity of uranium surface contamination. Because alpha radiation may be attenuated to some variable degree when the surface is porous, dirty, or scabbled, measurement of alpha radiation may not be a reliable indicator of the true surface activity levels. Such surface conditions usually cause significantly less attenuation of beta radiation. A common practice has been to use beta measurements to demonstrate compliance with surface activity guidelines expressed in alpha activity. This approach however becomes untenable at higher uranium enrichments due to high alpha-to-beta ratio. A more recent approach for determining the instrument and efficiencies is to consider the detector's response to all of the radiations present in the decay series (see the case study, "Determining Surface Efficiency and Total Efficiency for 4% Enriched Uranium," in Chapter 10, Section 10.1).

Soil surfaces are usually scanned using a NaI scintillation detector. Typical scan MDCs are provided in Chapter 9 for various uranium enrichments in the soil. Perhaps somewhat interesting, the greater the enrichment becomes, the higher the scan MDC. At first this may seem nonintuitive because the increasing enrichment means more U-235 is present (which is easily detectable at 185 keV). However, as much as the U-235 increases with increasing enrichment, U-234 increases even faster (usually it is 22–30 times greater than U-235). This means that once there is enough U-235 and U-238 present in the soil to make it detectable via, the total uranium concentration is much higher due to the amount of U-234 present, which adds little to the overall scan detectability. Finally, it is very likely that scan MDCs are going to be greater than the required scan MDC, based on the default DCGLs given in Table 4.3. This means that the soil sample size in class 1 survey units will likely be driven higher due to poor scan MDCs.

16.1.3 Laboratory Measurements

Gamma spectroscopy is routinely used to determine the radionuclide concentration of uranium in soil and other environmental media, such as sediment or sludge. Samples of solid materials are processed by drying, mixing, crushing, and homogenizing, and a portion sealed in a calibrated geometry. Net material weights are determined and the samples are typically counted using intrinsic germanium detectors coupled to a pulse height analyzer system. Background and Compton stripping, peak search, peak identification, and concentration calculations are performed using the computer software capabilities inherent in the analyzer system. Total absorption peaks associated with uranium are reviewed for consistency of activity. Energy peaks and radiation yields used for determining the activities are provided in Kocher (1981):

- U-235   144 keV (10.5%) or 185 keV (54%)

- U-238   63 keV (3.8%) from Th-234, assuming secular equilibrium

Chapter 10 discusses some common interferences, such as that from Ra-226 at 186 keV.

The total uranium concentrations in soil samples may be calculated from gamma spectroscopy analyses, based on the secular equilibrium state of the two decay series, or isotopic ratios, such as between U-234 and U-235. For processed uranium at natural isotopic ratios, the total uranium may be determined by multiplying the measured U-238 concentration by 2 (to account for the U-234), and adding to that value, the measured U-235 concentration. For enriched uranium, the isotopic ratio of U-234 to U-235 is established (usually by alpha spectroscopy), and the total uranium is determined by multiplying this ratio by the measured U-235 concentration—to determine the U-234 concentration, and adding to that value the U-235 and U-238 concentrations. For DU, alpha spectroscopy may be used to establish the ratio between U-238 and the total uranium. Therefore, a measure of U-238 may be used to obtain the total uranium concentration using the established ratio; this method assumes that the ratio of U-234 to U-235 is fixed for DU.

Alpha spectroscopy analyses are performed to determine isotopic ratios of uranium in samples—especially to determine the enrichment of uranium, and also to serve as a check on the accuracy of the gamma spectroscopy results. Because of the high cost relative to gamma spectroscopy analyses, alpha spectroscopy is typically limited to a small fraction of the samples analyzed. The results obtained by alpha spectroscopy are generally very accurate. However, because of the small sample size analyzed (1–10 g), it is necessary that the sample be homogenized to ensure results which are indicative of the entire sample.

16.2 THORIUM AND RADIUM SITES

Examples of thorium and radium-contaminated sites include AAR Manufacturing, Inc. that has soil contamination from manufacturing processes and practices, Stepan Chemical Company that used a chemical separation process to manufacture commercial products from the extraction of thorium and rare earth elements from monazite sands, and McClellan (former Air Force base) that has radiological contamination including: (1) radium-226 from aircraft dials and a radium paint shop; (2) buried radiological material from the Radiological Monitoring Lab; and (3) exempt quantities used as laboratory standards (NRC 2012a).

Uranium and thorium contamination are common contaminants at rare earth facilities. A number of the D&D sites being cleaned up by the Department of Defense, along with many that are regulated by State programs, can be described as radium sites. Additionally, the Formerly Utilized Sites Remedial Action Program (FUSRAP) includes sites that have both natural thorium contamination (Th-232 decay series) as well as contamination from uranium mill tailings (Th-230 and Ra-226). Finally, the windblown contamination found at many uranium mills consists of Ra-226 and its progeny. As with the uranium D&D sites, it is anticipated that the sign test will be used for surface-activity assessments, while the WRS test will be the choice for land area FSSs.

16.2.1 Nature of Contaminants

The thorium decay series begins with the alpha-emitting Th-232 and is followed by radioactive progeny that decay by emitting alpha, beta, and gamma radiation. Radionuclides in the thorium series include Ra-228, Ac-228, Th-228, Ra-224, Rn-220, Po-216, Pb-212, Bi-212, Po-212, and Tl-208. Owing to one radionuclide that branches (Bi-212) in the series, the total parent and progeny decays in the thorium series is 10 (including Th-232). Refer to Appendix A for a complete description of the radiological characteristics of this decay series.

Similar to the uranium decay series, natural thorium prior to any processing will contain the entire Th-232 decay series in secular equilibrium ending in stable Pb-208. Extraction processes, such as rare earth separations for zirconium, tantalum, and so on, may result in a thorium-contaminated waste stream that may require remediation. Once the thorium ore has been processed, ingrowth of the entire chain is controlled by the half-life of Ra-228—progeny will reach 90% equilibrium in about 20 years. Therefore, it is frequently the case at thorium-contaminated facilities that the survey design considers the equilibrium status of the thorium series. This can be accomplished by measuring the Th-232 by alpha spectrometry and Ra-228 by Ac-228 (gamma spectrometry). Note that modeling codes (e.g., RESRAD) usually view the Th-232 series in three parts based on the progeny half-lives: (1) Th-232, (2) Ra-228 and Ac-228, and (3) Th-228 and the remainder of the series.

As discussed in the previous section, when uranium has been processed the uranium is separated from the remainder of its progeny, that is, those radionuclides that come after U-234 in the series. Therefore, the tailings include Th-230 and Ra-226 (and Ra-226 progeny) and present entirely different FSS design challenges than the processed uranium.

16.2.2 Field Measurements

An FSS challenge encountered is in determining the state of equilibrium in the thorium decay series. For example, if the series is in secular equilibrium, then the Th-232 series consists of approximately six alphas (Th-232, Th-228, Ra-224, Rn-220, Po-216, 0.36 of Bi-212, and 0.64 of Po-212) and four betas (Ra-228, Ac-228, Pb-212, 0.64 of Bi-212, and 0.36 of Tl-208).

The assessment of surface contamination is performed by converting a detector's response to surface activity using instrument and surface efficiency factors determined via calibration. The thorium and radium decay series emit both alpha and beta radiation; therefore, it may be feasible for either alpha or beta activity to be measured for determining the residual activity of surface contamination. At thorium facilities, the historical regulatory practice (i.e., prior to the NRC's dose-based rulemaking) has been to measure both the alpha and beta surface activity separately, and compare both levels independently to the surface-activity guideline. In this situation, it was desirable to use one detector for alpha measurements and one detector for beta measurements. A zinc sulfide (ZnS) scintillation or gas proportional detector—operated at voltage low enough that only alpha radiation produces sufficient ionization to produce a count—was typically used for alpha surface-activity measurements. Note that this practice has the same inherent alpha attenuation problems that were discussed for uranium measurements—the alpha efficiency must be determined in consideration of field surface conditions. For beta surface-activity measurements, either a GM or gas proportional detector can be used. Because both of these detectors will respond to alpha radiation, alpha activity contributions to beta measurements may be eliminated by using suitable absorber thickness to attenuate alpha radiation. Typically a 3.8 mg/cm^2 Mylar window is used for the gas proportional detectors. If this approach is implemented, it is necessary to select the appropriately weighted beta energy for calibration that is representative of the beta emissions from thorium.

Alternatively, the approach that considers the detector's response to all radiations in the series may be used. That is, the approach to determining the instrument and surface efficiencies for each radionuclide in the thorium series is shown in Table 16.2 for a gas proportional detector. Similarly, Table 8.4 illustrated the determination of the GM detector instrument and surface efficiencies for each radionuclide in the radium series.

The instrument and surface efficiency is used to calculate the static MDC for surface-activity measurements. An early screening $DCGL_W$ published for Th-232 in secular equilibrium with its progeny was 6 dpm/100 cm². (This was the result of using a conservative resuspension factor (RF) value of 1.42×10^{-5}/m for use in the inhalation dose calculation.) How long might we have to count using a gas proportional detector to detect this surface activity level? Let us assume that the background count rate is 310 cpm, and we start with a 5-min measurement; the MDC is calculated by

$$\text{MDC} = \frac{3 + 4.65\sqrt{(310)(5)}}{(5)(1.29)126/100} = 23 \text{ dpm/100 cm}^2$$

While this is certainly a very low static MDC for a gas proportional detector, due in large measure to the high total efficiency, it still falls short

TABLE 16.2   Instrument and Surface Efficiencies for Th-232 Decay Series Using a Gas Proportional Detector

Radionuclide	Average Energy (keV)	Fraction	Instrument Efficiency	Surface Efficiency	Weighted Eff.
Th-232	Alpha	1	0.40	0.25	0.1
Ra-228	7.2 keV beta	1	0	0	0
Ac-228	377 keV beta	1	0.54	0.50	0.27
Th-228	Alpha	1	0.40	0.25	0.1
Ra-224	Alpha	1	0.40	0.25	0.1
Rn-220	Alpha	1	0.40	0.25	0.1
Po-216	Alpha	1	0.40	0.25	0.1
Pb-212	102 keV beta	1	0.40	0.25	0.1
Bi-212	770 keV beta	0.64	0.66	0.50	0.211
Bi-212	Alpha	0.36	0.40	0.25	0.036
Po-212	Alpha	0.64	0.40	0.25	0.064
Tl-208	557 keV beta	0.36	0.58	0.50	0.104
				Total eff. = 1.29	

of the 6 dpm/100 cm². Even when we count for 10 min the static MDC is 16 dpm/100 cm². Re-evaluating the $DCGL_W$ is the best course of action in this situation. NUREG-1720, Re-Evaluation of the Indoor Resuspension Factor for the Screening Analysis of the Building Occupancy Scenario for NRC's License Termination Rule (NRC 2002b) provides guidance on the selection of less conservative resuspension factors. Using a smaller resuspension factor, provided it can be justified, will increase the $DCGL_W$. NUREG-1720 supports a smaller (and more realistic based on the data cited in the report) RF value of 10^{-6}/m for the screening analysis of the inhalation dose calculation.

We certainly could make alpha measurements, taking advantage of the six alpha radiations present in the Th-232 series, as well as the low alpha background. The total instrument efficiency using an alpha-only gas proportional detector would be 6 times 0.40, or 2.4 and the weighted surface efficiency would be 0.25 (the same for each alpha). Let us assume that a 5-min count is made with this detector, and that the background is 2 cpm. The MDC is calculated

$$MDC = \frac{3 + 4.65\sqrt{(2)(5)}}{(5)(2.4)(0.25)126/100} = 4.7 \text{ dpm/100 cm}^2$$

Thankfully, there is a better way to demonstrate compliance with the Th-232 $DCGL_W$ of 6 dpm/100 cm². Soil surfaces at thorium and radium sites are scanned using a NaI scintillation detector. Typical scan MDCs are provided in Chapter 9 for thorium and radium concentrations in the soil. The scan MDCs are low for both the Th-232 and Ra-226 series due to the strong gamma emissions from their progeny. However, the scan MDC for Th-230 is very high due to a low energy, low yield, and gamma radiation emission. While the scan MDCs are low for both Th-232 and Ra-226, unfortunately so are their DCGLs. Therefore, it is possible that scan MDCs are going to be greater than the required scan MDC, based on the default DCGLs given in Table 4.3. While not as likely as it is for uranium, the soil sample size in class 1 survey units will possibly be driven higher due to poor scan MDCs.

16.2.3 Laboratory Measurements

Gamma spectroscopy is routinely used to determine the radionuclide concentration of thorium and radium in the soil and other environmental media, such as sediment or sludge. Samples of solid materials are processed by drying, mixing, crushing, and homogenizing, and a portion sealed in a calibrated geometry. Net material weights are determined and the samples are typically counted using intrinsic germanium detectors coupled to a pulse height analyzer system. Background and Compton stripping, peak search, peak identification, and concentration calculations are performed using the computer software capabilities inherent in the analyzer system. The total absorption peaks associated with uranium are reviewed for consistency of activity. Energy peaks and radiation yields used for determining the activities are found in Kocher (1981):

Th-232	911 keV (27.7%) from Ac-228 (assuming secular equilibrium)
Th-228	239 keV (44.6%) from Pb-212 (assuming secular equilibrium)
Th-230	67 keV (0.37%)
Ra-226	352 keV (37%) from Pb-214 (assuming secular equilibrium)

Chapter 10 discusses some common interferences when interpreting gamma spectroscopy results.

The total thorium concentrations in soil samples may be calculated from gamma spectroscopy analyses, based on the secular equilibrium state of the decay series. The total thorium may be determined by adding the measured Th-232 concentration, based on the Ac-228 progeny, and the Th-228 concentration, based on the Pb-212 progeny—provided that secular equilibrium exists. If equilibrium with a progeny cannot be established, it may be necessary to perform alpha spectroscopy to determine the Th-232 and Th-228 concentrations. Thorium-230, which may be a contaminant of interest with the decommissioning of uranium mill tailings, is difficult to assess by gamma spectroscopy due to its low-energy and radiation yield. This results in a relatively high MDC for the gamma spectrometry analysis of Th-230. Alpha spectroscopy can reduce the MDC for Th-230 in soil by a factor of 10.

Ra-226 can be easily measured in soil if it is reasonable to assume that it is in equilibrium with its progeny Pb-214 (or Bi-214). To ensure that this assumption is valid, many laboratories will keep the soil sample sealed, and sitting on the shelf for a few weeks, so that the Rn-222 and its progeny will achieve secular equilibrium with Ra-226. We prefer not to measure its 186 keV gamma because of interference from the U-235 185 keV gamma.

Alpha spectroscopy analyses are performed to determine isotopic ratios of thorium in samples—especially to determine the equilibrium status of the thorium decay series. Because of the high cost relative to gamma spectroscopy analyses, alpha spectroscopy is typically limited to a small fraction of the samples analyzed. Results obtained by alpha spectroscopy are generally very accurate. However, because of the small sample size analyzed (1—10 g), it is necessary that the sample be homogenized to ensure results which are indicative of the entire sample.

16.3 POWER REACTOR

There are numerous power reactor decommissioning projects currently underway in both the United States and abroad. Domestic reactor decommissioning projects include Dresden Unit 1, Fermi Unit 1, Humboldt Bay, Indian Point Unit 1, La Crosse, Peach Bottom Unit 1, Millstone Unit 1, Rancho Seco, San Onofre Unit 1, Three Mile Island Unit 2, and Zion Units 1 and 2 (NRC 2012a). Decommissioning activities at nuclear power plants require the highest degree of regulatory oversight, certainly more than any other type of D&D project. Not coincidently, these D&D projects have the largest D&D staffs assigned to the FSS, and produce the largest FSS reports on record. In this section, we will consider some of the unique aspects of decommissioning surveys performed at power reactor D&D sites.

Overall, decommissioning surveys at power reactors consist of a large number of surface-activity measurements, especially in comparison to the number of soil samples collected at these sites. This makes intuitive sense given the number of buildings/structures and the magnitude of their surface areas. Buildings might include the containment, auxiliary, fuel, turbine, radwaste, control, and other ancillary buildings. The number of building surface survey units might easily approach several hundred, with total surface areas exceeding 50,000 m². A common challenge in performing these surface-activity measurements is gaining access to all surfaces, for example, high bays, tight space locations, and embedded pipes are just a few examples of the difficult-to-access locations. In general, it is anticipated that the FSS design will be based on the sign test for both building surfaces and land areas (contaminants are not in the background).

16.3.1 Nature of Contaminants

The radionuclide contaminants present at power reactor D&D sites include a number of different fission and activation products, as well as transuranics. The mixture of radionuclides at an operating power reactor will be different than that present during decommissioning due to the decay of short-lived radionuclides. The common fission products include Cs-137 and Sr-90, while the activation products typically include Co-60, Ni-63, Fe-55, Eu-152, Eu-155, C-14, and H-3. If transuranics are present, the likely radionuclides are Pu-238, Pu-239, Pu-240, Pu-241, Am-241, and Np-237. Obviously, the radionuclide mixture varies with the plant location. For example, those materials exposed to a neutron fluence will likely contain activation products. Activated concrete might contain Eu-152, Eu-154, and Eu-155, while the activated rebar consists of Co-60 and Fe-59. Surfaces that have come in contact with the primary cooling system will likely be contaminated with fission products. Furthermore, perhaps certain portions of the plant are more likely than others to exhibit alpha contamination, such as the upender pit. Alternatively, if the HSA indicates that failed fuel incidents were all too common in the plant's history, then the reactor D&D site might have widespread alpha contamination.

16.3.2 Field Measurements

For the most part, the FSS for surface contamination is rather straightforward. The usual strategy is to determine the radionuclide mixture that exists at different locations of the plant, and to calculate the gross activity DCGL based on this mixture (Chapter 6). The one wrinkle that might come up is due to the potential for alpha contamination. One tack that can be taken with alpha contamination is to work it into the gross activity DCGL. This works provided that the alpha relative ratio is known with a reasonable confidence level. The other approach to handling alpha comes into play when it is unreasonable to expect a steady relative ratio, and therefore, separate measurements of alpha and beta surface-activity levels are performed (i.e., unity rule).

As with the decay series discussed previously, the assessment of surface contamination is performed by converting a detector's response to surface activity using instrument and surface efficiency factors determined via calibration. For the most part, the radionuclides at power reactor D&D facilities emit easily measured beta radiation. Those radionuclides that emit low-energy betas (H-3 or Ni-63) or x-rays are usually scaled to radionuclides with more detectable beta radiations.

The surface-activity DCGLs for many of the expected beta-emitting contaminants are relatively high (Tables 4.2 and 4.4). This means that 1-min measurements of surface activity will usually be sufficiently sensitive. In fact, it is likely that the scan MDCs for the floor monitor and gas proportional detector will also be sensitive enough to detect the $DCGL_W$, therefore keeping the sample size to a minimum. Of course, if the power reactor D&D project is fraught with alpha contamination problems, then not only will static surface-activity measurements be longer than 1 min, but samples sizes will likely be driven higher due to poor scan MDCs. The survey instruments used at power reactor D&D projects generally include gas proportional and GM detectors for beta surface-activity measurements, while zinc sulfide (ZnS) scintillation and gas proportional detectors are used for alpha surface-activity measurements. Again, there should be regard for the inherent alpha attenuation problems in field measurements, which relates to the selection of the surface efficiency.

Exterior soil surfaces at power reactor D&D sites are typically scanned using NaI scintillation detectors for Co-60 and Cs-137. Typical scan MDCs for these and other gamma emitters are provided in Chapter 9. A comparison of scan MDCs (Table 9.5) and default soil concentration DCGLs (Table 4.3) indicate that the scan MDCs are very similar to the DCGLs. While not knowing the exact magnitude of the area factors for these contaminants, it is not likely that additional measurements will be necessary due to a poor scan MDC.

16.3.3 Laboratory Measurements

Gamma spectroscopy is routinely used to determine the radionuclide concentration of activation and fission products at power reactor D&D sites present in the soil and the other environmental media, such as sediment or sludge. Samples of solid materials are processed and analyzed as described previously for uranium and thorium decay series. Energy peaks and radiation yields used for determining the activities are (Kocher 1981):

Co-60	1173 keV
Cs-137	662 keV from Ba-137 m (85%)[a]
Cs-134	796 keV (85.4%)
Am-241	59 keV (35.9%)

[a] 85% is calculated based on 94.6% of Cs-137 that beta decays to Ba-137 m, multiplied by 90% radiation yield of 662 keV gamma from Ba-137 m.

Alpha spectroscopy analyses are performed to determine the presence of transuranic radionuclides such as plutonium, curium, and neptunium. Characterization samples representing different parts of the plant are collected and analyzed to assist in determining radionuclide mixtures.

16.4 UNIVERSITY/RESEARCH FACILITIES

Not surprisingly, D&D projects at university and other research facilities greatly outnumber all of the decommissioning projects mentioned in the previous three sections combined. The fact that laboratory space is routinely being turned over from one researcher to another is somewhat responsible for this—and decommissioning surveys support this turnover in laboratory ownership. The Beltsville Agricultural Research Laboratory (Maryland) low-level radioactive burial site is an example of D&D site that includes radioactive contaminants (H-3, C-14, and Ni-63) buried in the pits.

These decommissioning surveys should focus on the likely areas where contaminants may have accumulated, such as on floor surfaces near the benches and hoods where material was used. It may be a challenge to access potentially contaminated building and equipment surfaces due to the presence of hoods, storage cabinets, and equipment; it may be beneficial to remove lab benches and hoods, to access these potentially contaminated surfaces. In addition to the FSS of the floor and wall surfaces, adequate survey consideration should be given to bench tops, hoods, glove boxes, sinks, and sink drain lines. Perhaps some of these surfaces (e.g., bench tops) might actually be incorporated into the random/systematic survey design; while other surfaces should be assessed using judgment surveys—such as beneath tile and other floor coverings, and behind fume hoods.

16.4.1 Nature of Contaminants

Contaminants are identified by reviewing routine health physics surveys, sink disposal logs, and material receipt forms. A review of the broad scope license indicates the radionuclides used in lab research—and commonly include H-3, C-14, I-125, P-32, S-35, and perhaps even gamma-emitting, short-lived microspheres used primarily for tracer studies. These contaminants ultimately result in contaminated bench tops, fume hoods, floors, and sinks. The availability of the sink for waste disposal means that sink drains are likely to be contaminated.

Although dose-based release criteria have been widely used for more than a decade, many labs continue to cleanup to historic criteria—for example, Regulatory Guide 1.86, or the more appropriate NRC 1987 guidance document for nonreactor licenses 5000 dpm/100 cm^2 for beta/gamma emitters and 100 dpm/100 cm^2 for I-125. Conversely, those approved to use the newer screening DCGLs based on a 25 mrem/y dose criterion will have a much easier go of it demonstrates compliance with 1.2E + 08 dpm/100 cm^2 for H-3, 3.7E + 06 dpm/100 cm^2 for C-14, 1.3E + 07 dpm/100 cm^2 for S-35 and 35,000 dpm/100 cm^2 for I-129. And though it has a relatively short half-life, D&D Ver. 2 can be used to determine the DCGL for I-125 (6.9E + 05 dpm/100 cm^2). Once DCGLs are selected, it is necessary to determine how the individual DCGLs will be applied for multiple radionuclides. Specifically, the MARSSIM user must determine the contaminants present in the lab, and their relative ratios to permit calculation of the gross activity DCGL (for surface activity).

The decontamination of laboratories is usually straightforward, including simple techniques such as washing and wiping surfaces down, to more elaborate techniques such as equipment removal, such as hoods or sink drain lines. H-3 is difficult to completely decontaminate because it diffuses into surfaces, and then off-gases from these surfaces. I-125 is easily volatilized, so contamination can be quite widespread in a laboratory, especially on metal surfaces where it plates out. During characterization, media sampling (smears, scrapings, and residue) at judgmental locations help confirm the nature of the radionuclides. Of course, it is possible to rule out the presence of a number of short-lived radionuclides, for example, P-32 (14 day), S-35 (87 day), and I-125 (60 day) due to radioactive decay. A remedial action support survey should be performed immediately following decontamination to check successful remediation activities (Dufault et al. 2000). In general, it is anticipated that the sign test will be used for building surface FSSs.

16.4.2 Field Measurements

Field survey techniques include surface scans, direct measurements of surface activity and smears. ISO-7503-2 is a practical standard that discusses the evaluation of tritium surface contamination, and determination of efficiency (ISO 1988b). Smears for removable surface-activity assessment are particularly important because many of the laboratory contaminants cannot be detected with field survey instrumentation (e.g., H-3). Tritium is usually assessed via wet smears as opposed to the use of a windowless gas proportional detector (due to its mixed results in the field). A GM or gas proportional detector can be used to detect C-14 and even I-125 to a lesser degree—though NaI or CsI detectors perform better for I-125. Perhaps even better would be the use of large-area gas proportional or plastic scintillation detectors.

The static MDC for surface-activity measurements is relatively high for these low-energy beta emitters. Fortunately, the DCGLs for these low-energy beta emitters are even higher (Table 4.2).

In fact, scan MDCs for gas proportional detectors are likely to be less than the DCGLs. Thus, it is anticipated that sample sizes for university and other research facilities will be minimal. It should also be noted that if separate assessments of surface activity are performed at each location—such as GM detector measurements and smears—then the unity rule should be employed.

16.4.3 Laboratory Measurements

While the limited use of gamma spectroscopy and other laboratory analysis techniques might be helpful to assay residue from floor and sink drains, the primary laboratory measurement at university and research facilities is smear counting. The analysis of smears can be performed using either a liquid scintillation counter or a low background gas proportional counter.

The following case study offers a thorough example of a MARSSIM FSS design for a laboratory facility.

CASE STUDY   FSS DESIGN FOR UNIVERSITY LABORATORY

A single laboratory room is being decommissioned and the HSA review of inventory records has commenced. Discussions with the principal investigator indicates H-3 was used twice as much as C-14 for experiments, and I-125 was used occasionally (the last use was several months ago)—so it cannot be ruled out completely due to radioactive decay. The relative ratios used for the FSS survey design are 60% H-3, 30% C-14, and 10% I-125.

The total surface area of lab survey unit is 230 m², and is comprised of class 1 floor, lower walls and bench-top surfaces. The floor surface consists of poured concrete covered with vinyl tile and the walls are painted drywall. The lab bench was decontaminated and remains in the room; hood and storage cabinets were removed.

The FSS strategy is centered on the sign test being used to demonstrate compliance in the survey unit. The average background will be subtracted from each gross measurement of the surface activity. The following DCGLs were approved by the state regulator:

- 1.2E8 dpm/100 cm² for H-3
- 3.7E6 dpm/100 cm² for C-14
- 6.9E5 dpm/100 cm² for I-125

Thus, the surface activity in dpm/100 cm² will be compared to the DCGL to demonstrate compliance. Additionally, moistened smears to measure removable tritium activity levels will be collected at each random and judgmental measurement location. Removable tritium results will be individually compared to 10% of the tritium DCGL$_W$ (the screening DCGL is modeled based on a removable fraction of 10%).

A gas proportional detector will be used for surface-activity measurements. The static MDC must be calculated and shown to be less than 50% DCGL$_W$. The total efficiency is calculated based on the relative ratios—and using an instrument efficiency of 0.21 for C-14, and assuming it to be zero for H-3 and I-125, and using the ISO-7503 surface efficiency of 0.25 for C-14:

$$\varepsilon_T = (0.6)(0) + (0.3)(0.21)(0.25) + (0.1)(0) = 0.016$$

Given the average background counts on vinyl floor (348 cpm) and drywall (286 cpm), the MDC was determined to be 4500 dpm/100 cm² for a 1-min count (and easily satisfied the DQOs since it is about 0.1% of $DCGL_W$).

The expected radiological conditions in the survey unit were obtained from characterization and/or remedial action support surveys. This information was used to determine the LBGR and standard deviation for sign test sample size calculation. A suitable background reference area was identified in a noncontaminated floor in an adjacent lab.

Finally, the integrated FSS consisted of 100% floor, lower wall, and bench scans. A scale map of the lab room (including floor, lower walls, and bench) was prepared to facilitate sample placement. The sign test resulted in 14 sample locations selected randomly; 1-min direct measurement and H-3 smear at each location. Indications of elevated radiation levels will be subject to judgmental measurements, both direct measurement and tritium smears.

QUESTIONS AND PROBLEMS

1. Name five forms of uranium contamination that are possible at D&D sites.

2. When the Th-232 decay series is in secular equilibrium there are six alphas and four betas. Suppose that the Th-232 has been separated within the past dozen years, and that Ra-228 concentration is approximately 60% of the Th-232. Assume that Th-228 is about 80% of the Ra-228. What is the alpha-to-beta ratio for this state of equilibrium?

3. Describe how tritium is handled when determining the total efficiency for a gas proportional detector when the lab radionuclide mixture includes 65% tritium.

CHAPTER 17

FSS Reports and Measurement Uncertainty

A FTER ALL THE REAL decommissioning work has been performed, by planning and implementing the FSSs and demonstrating compliance with survey results, it is not quite the time to coast to the finish line. Rather, it is the time to ensure that all phases of the decommissioning project have been sufficiently documented. It is incredibly important to take credit for all of your hard work—and this can only been done by carefully documenting your work. This chapter discusses the types of information that should be included in the FSS report. In that regard, we will focus on the material that belongs to the FSS report, as well as on how to report survey results.

Do not wait until the FSS is completely finished before starting to write the report. This point will become more apparent as we discuss the content of the report, but is well understood by those who have actually prepared FSS reports previously. Operating history, decommissioning activities, survey procedures, and selection of field and laboratory instrumentation—these and many other topics are known at the onset of the FSS and can be easily written while the FSS is in progress. Also, the results of investigations are much better written and included in the report as close to the time that they occurred as possible. This will minimize the chance for important details to be forgotten or overlooked.

There are a number of excellent sources of information that the D&D contractor should consult when preparing the report. The first one that comes to mind is NUREG/CR-5849. As discussed in Chapter 13, NUREG/CR-5849 actually preceded the MARSSIM, but its guidance on preparing survey reports has not been equaled by the MARSSIM. More recently, the NRC's NUREG-1757 has provided concise information on what should be included in the FSS report. The next section highlights the materials that should be included in the FSS report, drawing on the guidance in Section 9 of NUREG/CR-5849 and Sections 4.4 and 4.5 of NUREG-1757, vol. 2. In fact, NUREG-1757 provides a useful listing of FSS information that should be submitted by the licensee (NRC 2006a), here is a sampling of that list:

- An overview of the results of the FSS.

- A summary of the DCGLs for the facility.

- A discussion on any changes that were made in the FSS from what was proposed in the DP or other prior submittals.

- A description of the method by which the number of samples was determined for each survey unit.

- A summary of the values used to determine the number of samples.

- Survey results for each survey unit including the following:

  - The number of samples taken for the survey unit

  - A description of the survey unit, including a map or drawing of the survey unit

  - Discussion of remedial actions and unique features

    - Measured sample concentrations, in units that are comparable to the DCGLs

    - Statistical evaluation of the measured concentrations

    - Judgmental sample data and discussion of anomalous data

    - Statement that a given survey unit satisfied the $DCGL_W$ and the elevated measurement comparison if any sample points exceeded the $DCGL_W$.

17.1 FSS REPORT CONTENT

The overall size of the report is directly proportional to the scope of the D&D project. The FSS report for a gaseous diffusion plant will undoubtedly have many more volumes than the survey report for a small research laboratory building. Indeed, the Fort St. Vrain FSS report's 32-volumes occupies nearly five shelves on the bookcase, while many other final survey reports fit nicely in a single folder in a filing cabinet. Essentially, the report size is directly proportional to the number of survey units that comprise the D&D site. While the overall size of the report may vary, the survey report content should generally consist of the same types of information.

The first topic to discuss in the final report might include the background for the decommissioning project and a site description of both the interior and exterior areas at the site. Information like why the decommissioning project was being performed is helpful to set the context for the reviewer. The site description is greatly enhanced through the liberal use of maps and diagrams, including maps that illustrate the location of buildings/facilities and other decommissioning points on interest (e.g., creeks, disposal sites, septic fields, etc.) on site. Most importantly, information on the facility operating history, processes performed, waste disposal practices, and decommissioning activities are necessary to justify the classification of site areas and to establish the radionuclides potentially present at the D&D site.

The results of preliminary surveys, such as scoping, characterization, and remedial action support surveys, might well be covered next. The purpose, survey design and results for each of these preliminary surveys should be reported, as well as a brief discussion of the remedial action procedures that were used. This information is essential for justifying the classification of site areas (especially for nonimpacted areas that receive no survey effort), delineation of survey unit boundaries, and establishment of radionuclide relative ratios and standard deviations, to name just a few. The survey design for each of these preliminary surveys should mention the media types evaluated, the field and laboratory instruments used, and include summary tables of the survey results. These results might also include maps of impacted versus nonimpacted areas, and ideally the location and boundaries of each survey unit. The major point of this section is to demonstrate that the MARSSIM user had a solid understanding of the radiological status of the site at the onset of the FSS.

The next area is concerned with the decommissioning release criteria. Release criteria are fundamentally the driving influence behind all of the remediation and decommissioning survey activities. Most readers of this text will be performing and documenting MARSSIM surveys to demonstrate compliance with dose-based release criteria. As discussed in

Chapter 4, dose-based release criteria, such as the NRC's 25 mrem/y dose criterion, are not measurable quantities. That is, to demonstrate compliance with the NRC dose criterion, a pathway modeling must be used to translate the dose to measurable concentrations. Therefore, this FSS report section should describe how these measurable DCGLs were determined. This may include simply referencing the Federal Register Notices that provide default DCGLs, or a more elaborate discussion of the modeling effort used to derive DCGLs. For the latter case, it is likely that the modeling output will fit best in an appendix to the report, while the modeling basis and results (i.e., DCGLs and area factors) belong to the body of the report. A summary table that lists individual DCGLs for each radionuclide, along with appropriate area factors tables for each radionuclide, would make a nice addition to the FSS report.

The FSS procedures and design section is clearly one of the most critical sections of the report, perhaps second only to the FSS results themselves. Again, for MARSSIM users, the survey procedures section should provide an indication that the DQO process served as the basis for the survey design. Indeed, it is hard not to follow the DQO process when the MARSSIM guidance is followed. In general, the information in this section should correspond with the activities described in MARSSIM Chapters 4 and 5. The radionuclides of concern form the basis of the FSS, and therefore, should be clearly restated in this section. For multiple radionuclides, the strategy for dealing with each radionuclide must be explicitly outlined. This might include the use of surrogates and the unity rule (or combination of both) for radionuclides in soil, while gross activity DCGLs might be used for the design strategy for radionuclide contamination on surfaces. This section may have to make reference to the individual DCGLs listed in a previous section, as well as any radionuclide ratios reported earlier in the document.

FSS procedures should include a map showing area classifications and survey units. The selection and description of background reference areas and materials should indicate location and results of all background measurements—the results of the WRS test statistical sample size determination should be reported as it is used to justify the number of background measurements collected. Survey instrument selection should be discussed in detail and include the rationale for selecting instruments and survey techniques, discussion of calibration and operating procedures and determination of MDCs. A complete discussion on the determination of MDCs for static measurements (i.e., for surface-activity measurements and the determination of soil concentrations) and scan MDCs for building surfaces and land areas should be included. Survey procedures for both field and laboratory techniques should be briefly described for those activities performed, including direct measurements of surface activity, scanning, smear and soil sampling and analysis, and exposure rate measurements.

Furthermore, the FSS design section should include the statistical sample size determination for the appropriate statistical test for each survey unit (described in Chapter 13). For class 1 survey units, the comparison of the required scan MDC versus actual scan MDC, and the possible need for additional measurements for the elevated measurement comparison must be reported. The integrated survey strategy should clearly indicate the sample size and type, and scan coverage, for each survey unit. Ideally, the locations of systematic survey locations within each survey unit are indicated on maps. Finally, investigation levels for scanning results, surface-activity levels, and soil concentrations should be clearly stated. Chapter 5 in MARSSIM provides guidance on the selection of appropriate investigation levels.

Common sense dictates that the final section of the report includes an interpretation of the FSS results for each survey unit. This will generally be the largest section of the report, possibly filling numerous volumes, as it provides the survey results for each survey unit. The section should begin, though, by describing the techniques for reducing and evaluating survey data. Any deviations or modification to the FSS procedures or design should be noted. There should also be a general summary of all the survey results, considering the entire site as a whole. The more detailed survey results should be presented on a survey unit-by-survey unit basis.

For each of these survey unit subsections, discuss whether the DQOs were met for that particular survey unit, provide maps indicating survey measurement locations, report the results of preliminary data reviews, including basic statistical quantities (i.e., mean, median, standard deviation), posting plots, and histograms. The results of scans in each survey unit should be documented to record any hot spots identified and their disposition. Any judgment sample data that were collected as a result of scan hits and other biased measurements should be discussed, but not included in the statistical evaluation of the systematic and random samples. The results of statistical tests (WRS or sign) that were applied to the data to determine if the null hypothesis can be rejected should be reported, as well as the results of necessary elevated measurement comparisons. Importantly, survey units that fail the statistical test must be clearly identified, as well as the actions taken to ultimately pass the survey unit. These actions may include remediation and resurvey, resurvey alone, or supplementing the initial sample size with additional samples; documentation of the regulator's concurrence of the particular action taken is recommended.

Finally, a discussion of all survey results—judgmental (e.g., scan results) as well as systematic/random measurements—that triggered investigation levels must be reported. This includes documenting the results of investigations, such as reclassifications, remediation efforts, and resurveys, and any replacement data. Sometimes the initial result that triggered an investigation cannot be confirmed, this outcome must also be clearly documented.

17.2 REPORTING SURVEY RESULTS: MEASUREMENT OF UNCERTAINTIES AND ERROR PROPAGATION

The point of this chapter is to remind the reader that survey documentation should provide complete and unambiguous record of the radiological status of each survey unit, relative to the release criteria (DCGLs). To this end, the individual survey results should be reported as precisely and accurately as possible. According to Upgrading Environmental Radiation Data (EPA 1980), the actual results of the analyses should be reported with an appropriate statement of the uncertainty in the measured value. Do not report data as "less than MDC"; this can jeopardize the power of the statistical tests and makes it difficult to calculate basic statistical quantities for the survey unit, especially when calculating the sum-of-the-ratios for use in the unity rule.

Furthermore, the survey results should be reported in the same units as the DCGL, using the correct number of significant figures. This particular point has a direct application to the sign test. Because the sign test cannot handle values that are exactly equal to the DCGL, these values are essentially thrown out (and the sample size, N, is reduced accordingly). It is as if the data were never collected. Therefore, it is important to keep as much precision in sign test data as warranted by the procedure—do not round your results too early! We will now turn our attention to the matter of measurement uncertainty in individual measurement results.

Measurement uncertainties are classified as either systematic uncertainty or random uncertainty. Systematic uncertainty derives from calibration errors, incorrect detection efficiencies, nonrepresentative survey designs, and "blunders." An example of a systematic uncertainty would be the use of a fixed counting efficiency value even though it is known that the efficiency varies from measurement to measurement but without knowledge of the frequency. If the fixed counting efficiency value is higher than the true but unknown efficiency—as would be the case for calibrating an instrument to a stainless-steel source and then making measurements on wood or concrete—then every measurement result calculated using that efficiency would be biased low.

It is difficult—and sometimes impossible—to evaluate the systematic uncertainty for a measurement process. That is, the systematic error associated with a measurement cannot be estimated or calculated from the data itself, because each and every datum point is affected by the systematic error in the same way—collecting more data and taking the average will not reduce the systematic error. Yet, it is suggested that systematic error bounds be estimated and made small compared to the random uncertainty, if possible. Currie (1984) recommends that if no other information on systematic uncertainty is available, use 16% as an estimate for systematic uncertainties (1% for blanks, 5% for baseline, and 10% for calibration factors).

Random uncertainties refer to fluctuations associated with a known distribution of values, such as the counts obtained on survey meter which are Poisson distributed. Another example of a random uncertainty would be a well-studied and documented chemical recovery or yield, which is known to fluctuate with a regular pattern about a mean. A constant recovery value is used during calculations, but the true value is known to fluctuate from sample to sample with a fixed and known degree of variation. A certain amount of uncertainty is expected in the final value and the degree of uncertainty is relatively well understood.

When performing an analysis with a radiation detector, the result will have an uncertainty associated with it due to the Poisson nature of radioactive decay. To calculate the total uncertainty associated with the counting process, both the background measurement uncertainty and the sample measurement uncertainty must be considered. For example, the net count rate, R_s, may be calculated:

$$R_s = R_{s+b} - R_b = \frac{C_{s+b}}{T_{s+b}} - \frac{C_b}{T_b},$$

where
    R_{s+b} = gross count rate
    R_b = background count rate
    C_{s+b} = number of gross counts (sample)
    T_{s+b} = gross count time
    C_b = number of background counts
    T_b = background count time

The standard deviation of the net count rate, σ_s, can be calculated by propagating the error in both C_{s+b} and C_b (it is customary to assume that the error in count time is negligible).

Assuming that the individual uncertainties are relatively small, symmetric about zero, and independent of one another then the total uncertainty for the final calculated result can be determined by solution of the following partial differential equation (Knoll 2010):

$$\sigma_u = \sqrt{\left(\frac{\partial u}{\partial x}\right)^2 \sigma_x^2 + \left(\frac{\partial u}{\partial y}\right)^2 \sigma_y^2 + \left(\frac{\partial u}{\partial z}\right)^2 \sigma_z^2 + \cdots}$$

where

u = function, or formula, that defines the calculation of a final result as a function of the collected data. All variables in this equation, that is, $x, y, z, \ldots$, are assumed to have a measurement uncertainty associated with them and do not include numerical constants

σ_u = standard deviation, or uncertainty, associated with the final result

σ_x, σ_y, $\ldots$ = standard deviation, or uncertainty, associated with the parameters $x, y, z, \ldots$

This equation is referred to as the error propagation formula, and can be solved to determine the standard deviation of a final result from calculations involving measurement data and their associated uncertainties. When solving the partial derivative for a variable, we treat all the other parameters in the equation as constants. Each partial derivative is known as a sensitivity coefficient, such that the larger it is, the greater the contribution of the parameter's uncertainty to the combined uncertainty.

We can use this equation to compute the standard deviation in the net count rate:

$$\sigma_s = \sqrt{\left(\frac{\partial R_s}{\partial C_{s+b}}\right)^2 \sigma_{C_{s+b}}^2 + \left(\frac{\partial R_s}{\partial C_b}\right)^2 \sigma_{C_b}^2}$$

For an assumed Poisson distribution, the mean equals the variance, so $\sigma_{C_{s+b}}^2$ equals C_{s+b} and $\sigma_{C_b}^2$ equals C_b. Taking the partial derivative, the equation becomes

$$\sigma_s = \sqrt{\left(\frac{1}{T_{s+b}}\right)^2 C_{s+b} + \left(\frac{-1}{T_b}\right)^2 C_b} = \sqrt{\frac{C_{s+b}}{T_{s+b}^2} + \frac{C_b}{T_b^2}}$$

Consider the following example related to the FSS at a reactor facility being decommissioned, where the principal radionuclide is Co-60. Surface-activity measurements are being performed with a GM detector (20 cm² probe area) on a scabbled concrete floor. Earlier floor scanning identified a localized area (<20 cm²) of elevated direct radiation. A 1-min static count with the GM detector will be used to quantify the surface activity at this location. Given the following data, calculate the surface activity associated with this "hot spot," including its random uncertainty at the 95% confidence level.

$C_{s+b} = 528$ counts; $T_{s+b} = 1$ min;

$R_b = 54 \pm 10$ cpm (1σ)—this background count rate was obtained from 10 measurements on a similar surface in a background reference area.

The detector efficiency was determined using Tc-99 (similar to Co-60 beta energy). Repetitive measurements were taken on a nickel-backed calibration source and the total efficiency was computed:

$$\varepsilon_T = 0.15 \pm 0.02 \text{ c/dis } (1\sigma)$$

The first step is to determine the net count rate and its uncertainty:

$$R_s = \frac{C_{s+b}}{T_{s+b}} - R_b = \frac{528 \text{ cpm}}{1 \text{ min}} - 54 \text{ cpm} = 474 \text{ cpm},$$

Since the above net count rate equation is different (i.e., R_b has been determined from repetitive counts) than that shown previously, we must again apply the propagation of errors equation:

$$\sigma_s = \sqrt{\left(\frac{\partial R_s}{\partial C_{s+b}}\right)^2 \sigma_{C_{s+b}}^2 + \left(\frac{\partial R_s}{\partial R_b}\right)^2 \sigma_{R_b}^2}$$

and taking the partial derivatives yields

$$\sigma_s = \sqrt{\left(\frac{1}{T_{s+b}}\right)^2 C_{s+b} + (-1)^2 \sigma_{R_b}^2} = \sqrt{\frac{528}{1^2} + (10)^2} = 25.1 \text{ cpm}$$

Therefore, we can write:

$$R_s = 474 \pm 25 \text{ cpm } (1\sigma)$$

Now, we can calculate the activity of this hot spot, in dpm, total random measurement uncertainty, and the 95% confidence interval for the result. The total activity is computed by dividing the net count rate by the total efficiency (no probe area correction is made because the hot spot is <20 cm²):

$$A = \frac{R_s}{\varepsilon_T} = \frac{474 \text{ cpm}}{0.15 \text{ c/dis}} = 3160 \text{ dpm}$$

Using the equation for error propagation and taking the partial derivatives, the uncertainty in the activity is

$$\sigma_A = \sqrt{\left(\frac{1}{\varepsilon_T}\right)^2 \sigma_{R_s}^2 + \left(\frac{-R_s}{(\varepsilon_T)^2}\right)^2 \sigma_{\varepsilon_T}^2} = 450 \text{ dpm}$$

Finally, we can express the total activity and its uncertainty: $A = 3160 \pm 450$ dpm (1σ), or when stating the result at the 95% confidence level

$$A = 3160 \pm 900 \text{ dpm } (2\sigma)$$

The above result provides the total random uncertainty in the surface activity. Owing to the manner in which the detector was calibrated (on a high backscatter, nickel-backed source) and subsequently used to make measurements on scabbled concrete, it is quite likely that a systematic bias exists in the detector efficiency. That is, the true but unknown detector efficiency for Tc-99 on concrete is probably 10–20% lower than that determined from the calibration source.

The following surface-activity uncertainty assessment was developed by the ORAU/ORISE survey program, and is reproduced here owing to the courtesy of ORAU.

The calculation of surface activity is given by the following equation:

$$A = \frac{R_{S_A+B_A} - R_{B_A}}{\varepsilon_T(PA/100)} = \frac{R_{S_A}}{\varepsilon_T(PA/100)} = \frac{a}{(PA/100)} \quad (17.1)$$

where

$R_{S_A+B_A}$ = gross count rate of suspected contamination
R_{B_A} = background count rate contribution to the gross count rate

R_{S_A} = net count rate of suspected contamination
ε_T = total detector efficiency
$PA/100$ = detector correction factor (PA is physical detector area, in cm²)
a = surface activity uncorrected for detector area

Applying the standard error propagation formula (Knoll 2010), assuming all variables are independent, the error of the surface-activity calculation is given in Equation 17.1 as derived below:

$$\sigma_A = \frac{1}{(PA/100)} \sigma_a \tag{17.2}$$

$$\left(\frac{\sigma_a}{a}\right)^2 = \left(\frac{\sigma_{R_{S_A}}}{R_{S_A}}\right)^2 + \left(\frac{\sigma_{\varepsilon_T}}{\varepsilon_T}\right)^2$$

where

$$\sigma_{R_{S_A}} = \sqrt{\sigma^2_{R_{S_A+B_A}} + \sigma^2_{R_{B_A}}}$$

$$\left(\frac{\sigma_a}{a}\right)^2 = \frac{\sigma^2_{R_{S_A+B_A}} + \sigma^2_{R_{B_A}}}{R^2_{S_A}} + \left(\frac{\sigma_{\varepsilon_T}}{\varepsilon_T}\right)^2$$

$$\sigma_a = a \sqrt{\frac{\sigma^2_{R_{S_A+B_A}} + \sigma^2_{R_{B_A}}}{R^2_{S_A}} + \left(\frac{\sigma_{\varepsilon_T}}{\varepsilon_T}\right)^2} \tag{17.3}$$

Substituting Equation 17.3 into Equation 17.2 and reducing, the error in the surface activity is given below

$$\sigma_A = \frac{1}{(PA/100)} \frac{R_{S_A}}{\varepsilon_T} \sqrt{\frac{\sigma^2_{R_{S_A+B_A}} + \sigma^2_{R_{B_A}}}{R^2_{S_A}} + \left(\frac{\sigma_{\varepsilon_T}}{\varepsilon_T}\right)^2}$$

$$\sigma_A = A \sqrt{\frac{\sigma^2_{R_{S_A+B_A}} + \sigma^2_{R_{B_A}}}{R^2_{S_A}} + \left(\frac{\sigma_{\varepsilon_T}}{\varepsilon_T}\right)^2} \tag{17.4}$$

While Equation 17.4 provides the total propagated error for the surface-activity calculation, the calculation of the total instrument efficiency can be broken down into its core components of instrument and surface efficiency.

17.2.1 Instrument Efficiency

Determining the instrument efficiency involves measuring a calibration source, similar to the contaminant. It is important to note that the instrument efficiency, ε_i, is a 2π emission value. The calculation of instrument efficiency is given by the following equation:

$$\varepsilon_i = \frac{R_{S_i+B_i} - R_{B_i}}{q_{2\pi,sc}} = \frac{R_{S_i}}{q_{2\pi,sc}} \tag{17.5}$$

where

$R_{S_i+B_i}$ = gross count rate of calibration source
R_{B_i} = background count rate contribution to the gross count rate
R_{S_i} = net count rate of calibration source
$q_{2\pi,sc}$ = calibration source 2π emission rate

As discussed in Section 8.4, if the calibration source is larger than the physical area of the detector, the value of ε_i should be multiplied by the factor SA/PA, where SA is the size of the source, in cm², and PA is the physical detector area, in cm². For error propagation, it is assumed that there is no error in the measurements of SA and PA.

Applying the standard error propagation formula, assuming all variables are independent, the error of the instrument efficiency calculation given in Equation 17.5 is derived below:

$$\left(\frac{\sigma_{\varepsilon_i}}{\varepsilon_i}\right)^2 = \left(\frac{\sigma_{R_{S_i}}}{R_{S_i}}\right)^2 + \left(\frac{\sigma_{q_{2\pi,sc}}}{q_{2\pi,sc}}\right)^2$$

where

$$\sigma_{R_{S_i}} = \sqrt{\sigma_{R_{S_i+B_i}}^2 + \sigma_{R_{B_i}}^2}$$

$$\left(\frac{\sigma_{\varepsilon_i}}{\varepsilon_i}\right)^2 = \frac{\sigma_{R_{S_i+B_i}}^2 + \sigma_{R_{B_i}}^2}{R_{S_i}^2} + \left(\frac{\sigma_{q_{2\pi,sc}}}{q_{2\pi,sc}}\right)^2$$

$$\sigma_{\varepsilon_i} = \varepsilon_i \sqrt{\frac{\sigma_{R_{S_i+B_i}}^2 + \sigma_{R_{B_i}}^2}{R_{S_i}^2} + \left(\frac{\sigma_{q_{2\pi,\kappa}}}{q_{2\pi,sc}}\right)^2} \qquad (17.6)$$

17.2.2 Surface Efficiency

The surface efficiency is usually selected as a constant from the rules-of-thumb provided in ISO-7503. While these values are "constants" in one sense, there is no doubt that each surface efficiency factor has uncertainty. It may be appropriate to assume on the order of 25–35% uncertainty associated with the surface efficiency.

Finally, the uncertainties in both the instrument and surface efficiency can be incorporated into the total efficiency uncertainty, and the total propagated error in the surface-activity calculation can be calculated, as shown earlier in Equation 17.4:

$$\sigma_A = A \sqrt{\frac{\sigma_{R_{S_A+B_A}}^2 + \sigma_{R_{B_A}}^2}{R_{S_A}^2} + \left(\frac{\sigma_{\varepsilon_T}}{\varepsilon_T}\right)^2}$$

The EPA document Upgrading Environmental Radiation Data (1980) provides excellent guidance on how to estimate and report the systematic and random components of uncertainty. The general approach is to (1) consider and identify all of the conceivable sources of inaccuracy, (2) from the former set, extract those which have random uncertainties, and (3) for the remaining systematic uncertainties, assign a magnitude to the conceivable limit of uncertainty (e.g., expressed as upper bounds). This document provides examples on how to combine random and systematic uncertainties, and suggests that every reported measurement include (1) the value of the result, (2) the propagated total random uncertainty, and (3) the combined overall uncertainty.

The 95% confidence level is usually calculated based on the approximation 2σ, however, to be more accurate it should be computed based on 1.96σ. Table 17.1 provides $k\sigma$ values that correlate with various confidence intervals about the mean of a normal distribution. These confidence levels represent the probability that the true mean is included in the interval. Simply, standard deviations equate to confidence intervals which are used to express the confidence level in a measurement.

TABLE 17.1   Confidence Intervals about the Mean of a Normal Distribution

Interval ($\mu \pm k\sigma$)	Confidence Level
$\mu \pm 0.674\sigma$	0.500
$\mu \pm 1.00\sigma$	0.683
$\mu \pm 1.65\sigma$	0.900
$\mu \pm 1.96\sigma$	0.950
$\mu \pm 2.58\sigma$	0.990

QUESTIONS AND PROBLEMS

1. State the major sections that should comprise all FSS reports.

2. Calculate the hot-spot activity and its random uncertainty at the 95% confidence level for the following information. Surface-activity measurements are conducted using a gas proportional detector for Th-232 on a concrete surface. A 1-min static count with the gas proportional detector was used to quantify the surface activity at this location. The following data are provided

   $C_{s+b} = 18{,}978$ counts $T_{s+b} = 1$ min

   $R_b = 354 \pm 35$ cpm (1σ) $\varepsilon_T = 1.29 \pm 0.17$ c/dis (1σ)

Practical Applications of Statistics to Support Decommissioning Activities

T HIS CHAPTER DEALS WITH the application of statistics—with an emphasis on statistical tests related to decommissioning activities. Most decommissioning activities either require the use of statistics or would be greatly facilitated by the use of statistics. Examples of decommissioning activities that require some level of statistics include site characterization, final status surveys, and comparisons of data sets with regulators. Of course, the effective use of the MARSSIM requires at least a basic understanding of statistics, and in particular, a working knowledge of statistical hypothesis testing.

Many statistical tests used in the support of decommissioning activities have underlying assumptions that should be evaluated prior to their use. Typical evaluations performed to determine that the data are consistent with statistical test assumptions are described in Chapter 14 of this book and Chapter 8 of MARSSIM (2000a). Some important assumptions include spatial independence of samples, symmetrical data, and normally distributed data. Spatial dependencies may be assessed using posting plots and asymmetry in the data can be evaluated using histograms or quantile plots. Data may be plotted on normal probability paper, and if the result is a straight line, we have evidence of an underlying normal distribution. We will consider tests for normality in the next section.

18.1 TESTS FOR DATA NORMALITY

Tests for normality may be described as a class of goodness-of-fit tests that do not require the distribution to be completely specified (NRC 1988). The purpose of normality tests is to determine whether the assumptions underlying parametric statistical tests (e.g., Student's t test) are valid. For example, the survey design in NUREG/CR-5849 is based on the t test, so an evaluation of the data normality may be necessary to defend the use of that approach. If the normality test suggests that the normality assumption is questionable, then the use of nonparametric techniques is recommended. The assumption of normality may be checked by a number of tests, including the Kolmogorov–Smirnov test, Lillifor's test, Shapiro–Wilk test, and D'Agostino test. We will address the latter two tests in this section. Additionally, the next section will illustrate the use of normality tests in the course of answering D&D problems.

18.1.1 Shapiro–Wilk (*W* Test)

The Shapiro–Wilk (or W test) is an effective means for determining whether a data set is normally distributed. The use of the W test is limited to data sets with 50 or less data points (Gilbert 1987). The W test requires a table of coefficients to compute the test statistic, as well as a table of critical values to compare the test statistic (Shapiro–Wilk only provided these data for up to 50 samples).

The W test is performed by testing the following null hypothesis:

H_0: The population has a normal distribution against the alternative hypothesis

H_a: The population does not have a normal distribution

If the null hypothesis is rejected, we conclude that the data are probably not from a normal distribution.

For sample size n, the W test statistic is calculated by

$$W = \frac{1}{d}\left(\sum_{i=1}^{k} a_i(x_{[n-i+1]} - x_{[i]})\right)^2,$$

where d is the sample variance times $n-1$, or $d=(n-1)s^2$, k is $n/2$ for even n and $(n-1)/2$ for odd n, and the coefficients a_i are functions of sample size (n) and can be found in statistical tables.

Other statistical texts calculate the W test statistic in a similar manner:

$$W = \frac{b^2}{(n-1)s^2},$$

where b^2 is the same quantity as used in the W test statistic equation provided earlier:

$$b^2 = \left(\sum_{i=1}^{k} a_i (x_{[n-i+1]} - x_{[i]}) \right)^2.$$

The W test for normality is based on the ratio of two estimates of the variance of a population; therefore, the test statistic, W, can range from zero to one. It is designed so that small values reflect evidence of nonnormality—therefore, the critical values of the W test are the smaller quantiles of the statistic's distribution (Lurie et al. 2011). That is, the null hypothesis is rejected if the test statistic is less than the critical value (e.g., it is a one-tailed test).

The mechanics of calculating the W statistic are as follows:

1. Rank the data from the smallest to the largest in one column, and next rank data from the largest to the smallest. Then subtract the first column from the second. This facilitates the calculation of $(x_{n-i+1} - x_i)$ in a third column.
2. Obtain the coefficients a_i for $i = 1$ to k (fourth column) and multiply the above differences by these coefficients, and place in the fifth column.
3. Obtain the sum of the fifth column. The W test statistic is then the square of this sum divided by d.

An example is provided to illustrate the Shapiro–Wilk test. Ten measurements of exposure rate were obtained from an on-site area with the following results: 17.7, 18.0, 20.5, 18.3, 25.7, 21.0, 19.0, 17.1, 16.4, and 16.3 μR/h. Are these data normally distributed at the $\alpha = 0.05$ level of significance?

TABLE 18.1   Example of Shapiro–Wilk Normality Test

Rank (Low to High)	Rank (High to Low)	Difference $(x_{n-i+1} - x_i)$	a_i (for $k = 5$)	$a_i \times$ Difference
16.3	25.7	9.4	0.5739	5.395
16.4	21.0	4.6	0.3291	1.514
17.1	20.5	3.4	0.2141	0.728
17.7	19.0	1.3	0.1224	0.159
18.0	18.3	0.3	0.0399	0.012
18.3	18.0	−0.3		Sum = 7.808 (also equal to b)
19.0	17.7	−1.3		
20.5	17.1	−3.4		
21.0	16.4	−4.6		
25.7	16.3	−9.4		

The W test statistic is calculated using the mechanics outlined above. The results of these steps are shown in Table 18.1.

Therefore, the sample standard deviation (s) is 2.83, and the W test statistic is calculated as

$$W = \frac{(7.808)^2}{(10 - 1)(2.83)^2} = 0.846.$$

The null hypothesis of normality is rejected at the α level of significance if W is less than the critical value w_α. The statistical table provides a critical value for $w_{0.05}$ of 0.842. Because 0.846 is not less than 0.842, the null hypothesis is not rejected. This means that at the 5% level of significance, we do not have sufficient proof to reject the H_0, and we conclude that the data are normally distributed. Note that we would reject the null hypothesis in this example if the size of the test was $\alpha = 10\%$.

18.1.2 D'Agostino Test

The D'Agostino test (or D test) is used to test the null hypothesis of normality when the number of data points exceeds 50. Thus, the D'Agostino test complements the Shapiro–Wilk test for normality.

The D test is conducted as follows (Gilbert 1987):

1. The n data points are ordered from the smallest to the largest (recommend using spreadsheet)

2. The D statistic is calculated by

$$D = \frac{\sum_{i=1}^{n}[i - 1/2(n + 1)]x_i}{n^2 s},$$

where s is the sample standard deviation.

3. Transform the D statistic to the Y statistic as follows:

$$Y = \frac{D - 0.28209479}{0.02998598/\sqrt{n}}.$$

Gilbert (1987) warns that one should aim for five-place numerical accuracy in computing D since the numerator in Y is so small.

If n is large and the data are drawn from a normal distribution, then the expected value of Y is zero. For distributions that are not normal, Y will be either less than or greater than zero, depending on the particular distribution. Of course, even for normal distributions, Y is not likely to ever be exactly zero. This situation necessitates a two-tailed test of the hypothesis. The null hypothesis is formally stated as

H_0: The population has a normal distribution ($Y = 0$) versus the alternative hypothesis

H_a: The population does not have a normal distribution ($Y \neq 0$)

The null hypothesis is rejected at the α level of significance if the test statistic, Y, is less than the critical value $Y_{\alpha/2}$ or greater than the critical value $Y_{1-\alpha/2}$.

An example is provided to illustrate the D'Agostino test for normality. Sixty-six soil samples were collected from a background reference area and analyzed for Th-232. The Th-232 concentrations were displayed using a histogram. Are these data normally distributed? The histogram shown in Figure 18.1 is inconclusive, and lends itself to subjectivity in deciding whether the data are normally distributed. The mechanics for conducting the D'Agostino test for normality are shown in the spreadsheet in Figure 18.2.

We will evaluate whether these data are normally distributed at the $\alpha = 0.05$ level of significance. The critical values for the Y statistic were obtained for

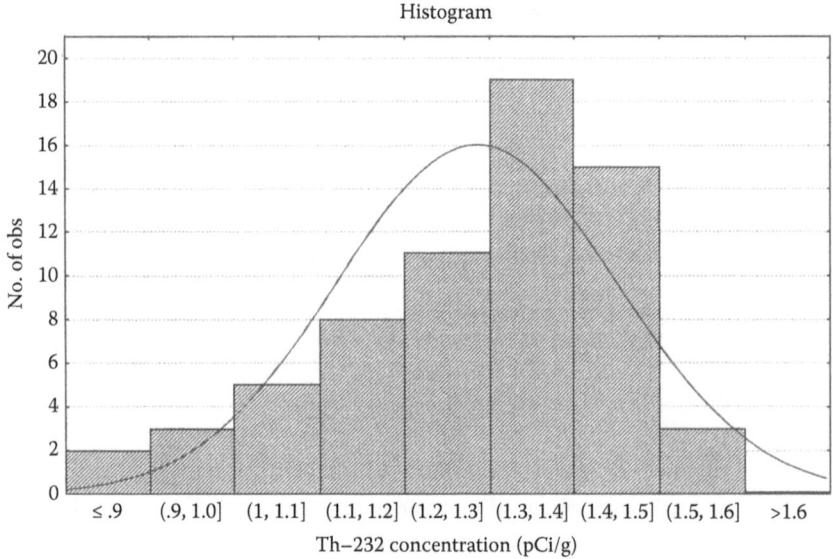

FIGURE 18.1 Histogram of Th-232 concentrations in background.

$n = 70$ (it would be more appropriate to interpolate between $n = 60$ and 70) from Table A8 in Gilbert (1987): $Y_{0.025} = -2.652$ and $Y_{0.975} = 1.176$. Because $Y = 0.954$ is between the two critical values, we cannot reject the null hypothesis, and therefore conclude that the data are normally distributed.

18.2 APPLICATIONS OF STATISTICS IN DECOMMISSIONING: COMPARISON OF DATA SETS

Regulators often find that it is necessary to compare their measurement results to their licensee's results. For example, the NRC describes a mechanism for comparing laboratory analytical results and determining the acceptability of licensee measurements.[*] The first step in this process is to determine the resolution of the procedure, which is defined by the NRC:

$$\text{Resolution} = \frac{\text{NRC value}}{\text{NRC uncertainty}}.$$

The next step is to calculate the ratio of the licensee's value to the NRC's value. If the calculated ratio falls outside of the NRC's criteria for accepting the licensee's measurement, then the NRC concludes that the licensee's

---

[*] Inspection Procedure 84750, "Radioactive waste treatment, and effluent and environmental monitoring," March 1994.

Spreadsheet for D'Agostino Test for Normality					
Background Reference Area					
Th-232	n = 66				
(pCi/g)	s = 0.16	D Statistic (numerator)	(pCi/g)		D Statistic (numerator)
0.886	1	-28.77875	1.319	34	0.6595
0.893	2	-28.1232	1.324	35	1.986
0.949	3	-28.95365	1.333	36	3.3325
0.955	4	-28.16955	1.336	37	4.676
0.990	5	-28.21215	1.338	38	6.021
1.025	6	-28.1875	1.344	39	7.392
1.038	7	-27.507	1.354	40	8.801
1.043	8	-26.5965	1.355	41	10.1625
1.089	9	-26.6805	1.361	42	11.5685
1.093	10	-25.6855	1.372	43	13.034
1.125	11	-25.3125	1.379	44	14.4795
1.137	12	-24.4455	1.383	45	15.9045
1.137	13	-23.3085	1.391	46	17.3875
1.137	14	-22.1715	1.394	47	18.819
1.144	15	-21.164	1.399	48	20.2855
1.154	16	-20.195	1.402	49	21.731
1.159	17	-19.1235	1.405	50	23.1825
1.199	18	-18.5845	1.407	51	24.6225
1.202	19	-17.429	1.417	52	26.2145
1.210	20	-16.335	1.421	53	27.7095
1.223	21	-15.2875	1.439	54	29.4995
1.227	22	-14.1105	1.449	55	31.1535
1.235	23	-12.9675	1.461	56	32.8725
1.248	24	-11.856	1.465	57	34.4275
1.255	25	-10.6675	1.466	58	35.917
1.268	26	-9.51	1.475	59	37.6125
1.274	27	-8.281	1.475	60	39.0875
1.282	28	-7.051	1.476	61	40.59
1.289	29	-5.8005	1.479	62	42.1515
1.302	30	-4.557	1.480	63	43.66
1.312	31	-3.28	1.519	64	46.3295
1.313	32	-1.9695	1.543	65	48.6045
1.316	33	-0.658	1.543	66	50.1475
	sum of D	199.0627			
	(numerator)				
	D =	0.285615674			
	Y =	0.953905628			

FIGURE 18.2   D'Agostino test for normality.

results should be investigated. The acceptance criteria are a function of the resolution, as evidenced in Table 18.2.

If the NRC and licensee results are not in agreement, the inspection procedure suggests that the licensee's calibrations, analytical procedures, and quality control results be reviewed.

TABLE 18.2   Criteria for Accepting
the Licensee's Measurements

Resolution	Ratio
<4	0.40–2.50
4–7	0.5–2.00
8–15	0.6–1.66
16–50	0.75–1.33
51–200	0.80–1.25
>200	0.85–1.18

In this section, we consider the comparison of data sets. In situations where the variances of each distribution are known—usually as a result of being able to use the sample variance as a good estimate of the population variance, σ, we use the standard normal (z) test statistic. When comparing the means of two normal distributions with unknown variances, we use the t test. There are two cases to consider for unknown variances—assuming that the variances are either equal or unequal. When the variances are assumed to be equal, they are combined or pooled to provide a single estimator of the variance. In one test of the comparison of means, we illustrate the case where the variances of each data set are unequal, and therefore the variances are not pooled.

Another case of comparing data sets occurs when the observations from the two distributions are collected in pairs. An example of paired observations would be an NRC inspector and licensee each independently measuring the same soil samples during a decommissioning inspection. Another example would be the assumption of paired measurements of the background and the sample during the formulation of the MDC.

18.2.1 t Test with Unequal Variances: Evaluating Automated Soil Sorter Performance

A recent addition to the arsenal of remediation tools is the soil sorter. The soil sorter is used to minimize waste volume generated during soil remediation. The general operating principle is that soil is passed by a detector on some sort of a conveyor system. Action levels determined during soil sorter calibration are used to separate "clean" soil from soil that has contamination in excess of the preset action level.

Let us assume that soil is being remediated from a site contaminated with low-enriched uranium, where the release criterion is 30 pCi/g of total uranium. As a part of the operation of the soil sorter, the licensee has committed to sampling 10% of the "clean" soil as a quality check on soil sorter performance. This check is performed by collecting five samples for alpha spectroscopy analysis from 10% of the clean soil piles, and determining the total uranium concentration. As long as the total uranium concentration is less than 30 pCi/g, the "clean" soil is used for backfill at other locations on-site.

The project manager would like to speed up the turnaround time on the validation of the "clean" soil piles—as it currently takes 3–4 days to get the results of the alpha spectroscopy analyses. The laboratory manager suggests counting the soil samples for U-235 via gamma spectroscopy, and using the gamma results to validate the "clean" soil piles. According to the lab manager, the turnaround time would be reduced to 1 day, and further stated that studies have already been performed to indicate that a U-235 concentration of 1 pCi/g corresponds to a total enriched uranium concentration of 30 pCi/g at this site. The lab manager proposed that if 10 soil samples are collected from 10% of the piles, and each sample is analyzed by a 30-min count for the 144 keV U-235 gamma line, the results should be just as good as the more precise alpha spectroscopy data. The project manager authorized the immediate evaluation of the lab manager's proposal.

To implement this evaluation, five samples were collected from a fresh "clean" soil pile for alpha spectroscopy analysis, and 10 samples for gamma spectroscopy were also collected from the same soil pile. The resulting data were reported in Table 18.3 as follows:

An appropriate statistic test in this situation is the t test with unequal variances. Because the sample size is relatively small, it is more appropriate to use the t test as opposed to the standard normal z-score (which requires s to be a good estimator of σ). The null hypothesis is stated as follows:

H_0: The means of the alpha spectroscopy and gamma spectroscopy are equal against the alternative hypothesis

H_a: The alpha spectroscopy and gamma spectroscopy results are not equal

The first step is to determine the t' statistic with unequal variances:

$$t' = \frac{\overline{x}_1 - \overline{x}_2}{((s_1^2/n_1) + (s_2^2/n_2))^{1/2}}.$$

TABLE 18.3   Gamma and Alpha Spectrometry Analyses from Soil Sorter

Gamma Spectroscopy (U-235 pCi/g)	Alpha Spectroscopy (U-235 pCi/g)
0.89	0.835
1.05	0.951
0.88	0.872
0.78	0.863
0.82	0.920
0.87	
0.99	$\bar{x}_2 = 0.880$ $s_2 = 0.0466$
0.92	
1.03	
1.09	
$\bar{x}_1 = 0.932$ $s_1 = 0.103$	

The t' statistic does not follow the Student's t distribution when the null hypothesis is true (EPA 2000). However, there is a procedure that can be used for v degrees of freedom that allows the t' distribution to be approximated by the Student's t distribution. The number of degrees of freedom is calculated by

$$v = \frac{((s_1^2/n_1) + (s_2^2/n_2))^{1/2}}{(s_1^2/n_1)^2/(n_1 - 1) + (s_2^2/n_2)^2/(n_2 - 1)}.$$

We can now calculate the t' statistic using our data

$$t' = \frac{0.932 - 0.880}{\left((0.103^2/10) + (0.0466^2/5)\right)^{1/2}} = 1.136,$$

and the number of degrees of freedom is determined as

$$v = \left(\frac{0.103^2}{10} + \frac{0.0466^2}{5}\right)^2 \Big/ \left(\frac{(0.103^2/10)^2}{9} + \frac{(0.0466^2/5)^2}{4}\right) = 12.98.$$

The number of degrees of freedom is rounded off to 12 to provide a level of conservatism.

The critical values of the t statistic at the 5% level of significance for 12 degrees of freedom are $t_{0.025}$ equals -2.179 and $t_{0.975}$ equals 2.179.

Because the *t* statistic (1.136) falls between these critical values, we cannot reject the null hypothesis, and therefore conclude that there are no statistically significant differences ($p > 0.2$) between the alpha spectroscopy and gamma spectroscopy data. Therefore, the validation of the "clean" soil piles may be performed using gamma spectroscopy analyses.

18.2.2 Pairwise *t* Test: Evaluating "Wet" versus Processed Gamma Spectroscopy Results

At some point during the remediation of contaminated soils, the question becomes: "When can I stop digging?" In many cases, it is difficult (if not impossible) to detect the contaminant at guideline levels using field instrumentation. To counter this problem, field laboratories can be set up to measure the soil contamination using gamma spectroscopy on unprocessed or "wet" soils. The gamma spectra are reviewed from the analysis of "wet" soil samples, and a decision is made whether to continue or cease excavation. Normally, a somewhat conservative action level is established to account for uncertainty in these "screening samples"—for example, the decision may be to continue remediation for wet soil samples measuring 25 pCi/g even when the guideline is 35 pCi/g.

The project manager understands that the purpose for these screening samples is to guide the remedial actions, but questions why these data, especially when less than the soil guidelines, are also not used for the final status survey. Thus, the issue is whether the data from the analysis of unprocessed soil samples is of sufficient quality to make decisions concerning compliance with soil guidelines. Hence, it is of interest to evaluate the performance of gamma analyses on "wet" soil versus processed soil. Can we conclude that "wet" results are reasonably similar to the results obtained from the conventional analysis of processed soils in a laboratory?

Let us assume that a site is contaminated with DU—with a guideline of 35 pCi/g—and that gamma spectroscopy is used to quantify the DU concentrations in soil. (Note: A standard protocol for DU determination is to measure the U-238 and multiply an appropriate total U/U-238 ratio determined via alpha spectroscopy.) The standard approach is to process the soil sample via drying, crushing, and homogenizing, and then placing the processed soil into a calibrated geometry (Marinelli beaker) and counting.

The project manager requests that 10 samples recently analyzed "wet" in the field be shipped to the laboratory for routine processing and analysis. Early in the following week, the project manager was presented with

TABLE 18.4 Depleted Uranium Results for Wet and Dry Soil Analyses

"Wet" Soil Analyses (pCi/g DU)	Processed (Dry) Soil Analyses (pCi/g DU)	d_i
4.7	5.3	0.6
21.8	9.6	−12.2
7.1	11.0	3.9
12.5	13.0	0.5
11.7	15.0	3.3
1.3	15.0	13.7
27.0	28.0	1.0
36.8	36.0	−0.8
57.0	75.0	18
90.8	77.0	−13.8

Note: Gamma spectroscopy analyses performed after the soil has been processed can result in the soil concentration either increasing or decreasing from the initial "wet" analyses. Our experience at ORAU indicates that more than two-thirds of the time the radionuclide concentration will increase because the drying reduces the weight of the sample, thus increasing the concentration. The explanation for the "wet" analyses being higher than the processed results is based on the "wet" sample being nonuniform, and perhaps having activity closer to the detector (within the Marinelli) than once it has been homogenized in the laboratory—of course, this effect can also cause the "wet" soil analyses to be much lower than the processed results.

the following data comparing the results of "wet" analyses performed in the field versus processed sample results (Table 18.4).

The project manager decides to apply a pairwise t test for these paired observations to evaluate these results. The null hypothesis is stated as follows:

H_0: The "wet" soil analysis results equal the processed soil results against the alternative hypothesis

H_a: The "wet" soil analysis results do not equal the processed soil results

The first step is to determine the differences (d_i) of each paired observation in the above data. This can be obtained by subtracting the "wet" analysis result from the processed soil result—these results are shown in Table 18.4.

The differences (d_i) are assumed to be normally distributed for the pairwise t test. This assumption can be assessed using the W test at the 5% level of significance (Table 18.5).

TABLE 18.5   Normality Test for the Differences (di) in the Pairwise t Test

Rank (Low to High)	Rank (High to Low)	Difference $(x_{n-i+1} - x_i)$	a_i (for $k = 5$)	$a_i \times$ Difference
−13.8	18	31.8	0.5739	18.25
−12.2	13.7	25.9	0.3291	8.52
−0.8	3.9	4.7	0.2141	1.006
0.5	3.3	2.8	0.1224	0.3427
0.6	1.0	0.4	0.0399	0.0160
1.0	0.6			Sum = 28.13 (also equal to b)
3.3	0.5			
3.9	−0.8			
13.7	−12.2			
18	−13.8			

The mean and sample standard deviation (s) of the differences (d_i) are 1.42 and 9.77, respectively. The W test statistic is calculated as

$$W = \frac{(28.13)^2}{(10 - 1)(9.77)^2} = 0.921.$$

As discussed earlier, the null hypothesis of normality is rejected at the α level of significance if W is less than the critical value w_α—and $w_{0.05}$ equals 0.842. Since 0.921 is not less than 0.842, the null hypothesis is not rejected, and we conclude that the data are normally distributed. We can now proceed to calculate the t statistic for the pairwise t test. (If the differences were not normally distributed, we would use the nonparametric equivalent to the t test—the Mann–Whitney test (also called the Wilcoxon rank sum test).)

The t statistic is calculated as follows:

$$t = \frac{\overline{d_i}}{s_{d_i}/\sqrt{n}} = \frac{1.42}{9.77/\sqrt{10}} = 0.46.$$

The critical values of the t statistic at the 5% level of significance for 9 degrees of freedom are $t_{0.025}$ equals −2.262 and $t_{0.975}$ equals 2.262. Because the t statistic (0.46) falls between these critical values, we cannot reject the null hypothesis, and therefore conclude that there are no statistically significant differences between the "wet" soil analysis data and processed soil sample data. In fact, the p value for this test is $p > 0.6$.

18.2.3 Confirmatory Analyses Using Nonparametric Statistics

A circumstance that arises during many decommissioning inspections and/or confirmatory surveys is the pairwise comparison of licensee and inspector measurements. For instance, the licensee has collected, analyzed, and archived 80 samples from the remediation of a pipe trench. During the decommissioning inspection, the inspector requests 10 of these samples for confirmatory analyses—that is, the inspector will send these selected samples to an independent laboratory for similar analyses. The appropriate statistical test for confirmatory analyses is the pairwise t test of the differences $(d_i\text{'s})$—provided that they are normally distributed. However, if the differences are not normally distributed, we cannot use the pairwise t test; we must consider the use of nonparametric statistics.

Nonparametric statistics require less restrictive assumptions (usually requires an assumption that measured values be independent) concerning the probability distributions from which empirical observations arise. Parametric statistics, such as the t test, require specific assumptions about the probability distribution—the most common assumption being that the observations are normally distributed. (Note: When the parametric assumptions are true, the appropriate parametric methods provide for more powerful hypothesis tests than their nonparametric counterparts. Yet, if parametric tests are used when their basis assumptions are not true, the parametric methods can yield misleading results (NRC 1988).)

An example of the nonparametric Mann–Whitney test is provided. In the scenario offered above, the inspector requested that 10 samples be independently analyzed for the total uranium concentration. Results were obtained and compared to the licensee's reported results in Table 18.6.

The inspector's objective is to make a decision on whether or not his/her independent analyses confirm the results obtained by the licensee. The inspector may generally conclude that there are no huge discrepancies, but realizes that this is not a sufficient conclusion to put in the inspection report. The inspector decides to perform a pairwise t test on these data and remembers to first check the assumption of normality of the differences.

This assumption can be assessed using the Shapiro–Wilk normality test at the 5% level of significance (Table 18.7).

TABLE 18.6   Comparison of Inspector and Licensee Total Uranium Results

Inspector's Data (Total U pCi/g)	Licensee's Data (Total U pCi/g)	d_i
420	293	127
40	51	−11
590	442	148
170	137	33
170	128	42
15	14	1
120	90	30
31	29	2
56	79	−23
50	42	8

TABLE 18.7   Normality Test on Differences between Inspector and Licensee Measurements

Rank (Low to High)	Rank (High to Low)	Difference $(x_{n-i+1} - x_i)$	a_i (for $k = 5$)	$a_i \times$ Difference
−23	148	171	0.5739	98.14
−11	127	138	0.3291	45.42
1	42	41	0.2141	8.778
2	33	31	0.1224	3.794
8	30	22	0.0399	0.8778
30	8			Sum = 157.01 (also equal to b)
33	2			
42	1			
127	−11			
148	−23			

The mean and sample standard deviation (s) of the differences (d_i) are 35.7 and 57.5, respectively. The W test statistic is calculated as

$$W = \frac{(157.01)^2}{(10 - 1)(57.5)^2} = 0.828.$$

As discussed previously, the null hypothesis of normality is rejected at the 5% level of significance if W is less than the critical value ($w_{0.05}$) of 0.842. Since 0.828 is less than 0.842, the null hypothesis is rejected, and we conclude that the data are not normally distributed.

The inspector must now use the nonparametric equivalent to the t test—the Mann–Whitney test. As before, we state the null hypothesis as follows:

H_0: No difference exists between the licensee's and inspector's data—the two populations are equal against the alternative hypothesis

H_a: The licensee's and inspector's data are not equal

The first step is to combine the samples from the two populations and rank the combined sample using ranks 1 to N (low to high), where N is equal to m (licensee measurements) and n (inspector measurements). If there are sample values that are exactly equal, assign to each the average of the ranks (NRC 1988). So, N equals 20 and the combined ranks are as shown in Table 18.8.

The Mann–Whitney test statistic is the sum of the ranks (W') of the inspector's data:

$$W' = \sum_{i=1}^{n} R(x_i) = 108.$$

The critical values of the W' statistic at the 5% level of significance must be determined. For $n = m = 10$, and $\alpha/2 = 0.025$, the lower quantile $w_{0.025}$ equals 79. The upper quantile must be determined by an equation

$$w_{1-\alpha/2} = n(N+1) - w_{\alpha/2} = 10(21) - 79 = 131.$$

TABLE 18.8   Mann–Whitney Test on Inspector versus Licensee Measurement Data

Inspector's Data, x_i (Total U pCi/g)	Combined Rank	Licensee's Data, y_i (Total U pCi/g)	Combined Rank
420	18	293	17
40	5	51	8
590	20	442	19
170	15.5	137	14
170	15.5	128	13
15	2	14	1
120	12	90	11
31	4	29	3
56	9	79	10
50	7	42	6

That is, the upper quantile $w_{0.975}$ equals 131. Because the W' statistic (108) falls between these critical values, we cannot reject the null hypothesis, and therefore conclude that there are no statistically significant differences between the inspector's and licensee's data. Therefore, the inspector can confidently conclude in his/her report that inspection data confirmed the licensee's data.

18.3 CASE STUDY: COMPARING Cs-137 CONCENTRATION IN CLASS 3 AREA WITH BACKGROUND REFERENCE AREA USING BOTH t TEST AND WRS TEST

In this case study, we will consider the comparison of data collected from a class 3 survey unit with an appropriate background reference area. The intent of this exercise is to better illustrate the performance of hypothesis testing—for both the t test and the WRS test. We will also explore what happens when the null hypothesis is stated in different ways. For example, when the null hypothesis is stated in a way that the researcher desires to reject the null hypothesis to prove an effect, the hypothesis test is referred to as "reject-support" testing. Just the opposite occurs when the researcher frames the null hypothesis in a manner where the desire is to accept the null hypothesis—termed "accept-support" testing. The meaning of type I and type II errors are very different depending on how the null hypothesis is phrased. This case study endeavors to examine the workings of hypothesis testing in the context of comparing a class 3 survey unit to a background reference area, where the surveyor desires to demonstrate that the class 3 survey unit is consistent with background levels.

A common situation that occurs during decommissioning is the need to compare the soil concentration from an on-site area to that of the background. It usually takes the form of comparing a site area not expected to be contaminated (i.e., a class 3 area) with a representative background reference area. Of course, the radionuclide of concern in this case is one that exists in the background, perhaps uranium, thorium, radium, or even Cs-137.

Let us assume that a portion of the licensee's land area is going to be released. The only potential contaminant is Cs-137, and it is very unlikely that any contamination exists at all. The licensee proposes to take a number of soil samples from the site area being considered for release, and comparing them to a reference area that is similar to the subject land area. The number of questions that arise even with a seemingly straightforward problem statement is remarkable. How many samples are needed from each area?

What statistical test should be selected? Can the data be assumed normally distributed, or should nonparametric tests be chosen? Exactly what should be tested—Compliance with release criteria? Indistinguishability from the background? How should the null hypothesis be stated?

Suppose that the licensee has previously assessed the data distributions, and has confirmed that Cs-137 concentrations in both the site area and the background reference area are normally distributed (a normality test would be necessary to substantiate this claim). Therefore, the licensee will use the t test, and assume that the variances in both areas are equal. This statistical test is performed by comparing the mean Cs-137 soil concentration in the site area with Cs-137 soil concentration in the background reference area. The number of samples required for this comparison depends on the selected type I and type II decision errors and the estimated standard deviation of Cs-137 in these areas. The two-sample t test will be studied first in determining the sample size for this problem.

18.3.1 Two-Sample t Test

The two-sample t test is used to compare Cs-137 concentrations in a class 3 area with a background reference area. We will start by assuming that the data sets have unknown, but equal variances, and therefore the null hypothesis can be stated as follows:

H_0: The mean Cs-137 soil concentration in the site area ($\bar{x}_s$) ≤ the mean Cs-137 soil concentration in the background reference area ($\bar{x}_b$) against the alternative hypothesis

H_a: The mean Cs-137 soil concentration in the site area ($\bar{x}_s$) > mean Cs-137 soil concentration in the background reference area ($\bar{x}_b$)

This null and alternative hypothesis can be written in a similar fashion by algebraically manipulating the terms

$$H_0 : \bar{x}_s - \bar{x}_b \leq 0;$$

against the alternative hypothesis

$$H_a : \bar{x}_s - \bar{x}_b > 0.$$

With the null hypothesis stated in this manner, a type I error—which results when H_0 is incorrectly rejected—means that one would incorrectly conclude that the concentration in the site area is greater than the concentration in the background reference area. This can be considered the licensee's error of concern. The type II error—which results when H_0 is incorrectly not rejected (remember, we do not "accept" null hypotheses; rather, we "fail to reject" the null hypothesis)—means that one would incorrectly conclude that the concentration in the site area is less than the concentration in the background reference area. This can be considered the regulator's error of concern.

To determine the sample size, the type I and II errors must be specified. Furthermore, not only must the type II error be specified, but this error must also be specifically stated for the net Cs-137 concentration, given by $\bar{x}_s - \bar{x}_b$, for which the alternative hypothesis is accepted (in error). For instance, the licensee might specify a type II error of 0.05 for a $\bar{x}_s - \bar{x}_b$ concentration of 0.15 pCi/g (statisticians sometimes refer to the value of the alternative hypothesis as "effect size"). That is, there is a 5% probability that the null hypothesis will not be rejected, when the Cs-137 concentration in the site area exceeds the background reference area concentration by 0.15 pCi/g. This concentration ($\bar{x}_s - \bar{x}_b$) is given by δ—and is the value at which the specified type II error applies. But it should also be recognized that there is an infinite combination of δs and specified type II errors. It is safe to say that the higher the concentration δ, the smaller the type II error that the regulator will allow.

18.3.1.1 Survey Design

At this point, we can consider the necessary sample size for the t test, assuming equal variances. The sample size is calculated for the number of samples in the site area, n, which also equals the number of samples in the reference area, m, and is given by (EPA 2000)

$$n = \frac{2s^2 (Z_{1-\alpha} + Z_{1-\beta})^2}{\delta^2} + 0.25(Z_{1-\alpha})^2$$

where s is the standard deviation of the radionuclide concentration that is assumed to be equal for both the site area and the reference area (i.e., s^2 is the pooled variance), and the percentiles $Z_{1-\alpha}$ and $Z_{1-\beta}$ are represented by the selected decision error levels, α and β, respectively.

It is instructional to consider a few concepts at this point in the statistical design. First, we must recognize that the null hypothesis is assumed to be true in the absence of strong data to the contrary. So, for this example, the null hypothesis is stated that the concentration in the site area is less than the concentration in the background. Therefore, we can "accept" the null hypothesis because either it is really true or it is false in error, because sufficient data have not been collected to reject the null hypothesis. Hence, for the licensee proposing that the concentration in the site area is less than that in the reference area, it is not only necessary to "accept" (or more correctly, "fail to reject") the null hypothesis but it is also necessary to prove to the regulator that sufficient data were collected to detect the situations where the site area Cs-137 concentration did indeed exceed the background concentration. This proof can take the form of a prospective power curve, which is constructed by plotting the probability of rejecting the null hypothesis versus the δ concentration, given by $\bar{x}_s - \bar{x}_b$, which takes on various values under the null and alternative hypotheses. We will demonstrate the prospective power curve as we continue with our example.

A licensee selects a type I error of 0.05 and a type II error of 0.05 at a δ concentration of 0.15 pCi/g. On the basis of the characterization data from the site area and the background reference area, the standard deviation of Cs-137 in soil is estimated to be 0.18 pCi/g for each area. The sample size for these DQOs is calculated as

$$n = \frac{2(0.18)^2(1.645 + 1.645)^2}{0.15^2} + 0.25(1.645)^2 = 31.8.$$

Therefore, 32 soil samples are required in both the site area (n) and the background reference area (m). Note that Walpole and Myers (1985) offer another way to determine the sample size based on these DQOs (Table A.9, 1985): for a value $\Delta = \delta/s = 0.15/0.18 = 0.83$, and type I = type II = 0.05, the sample size is 31 (for the closest tabulated Δ of 0.85). Equipped with the sample size, decision errors, and standard deviation, we are now prepared to calculate the prospective power curve.

The relevant statistic in this example is δ (equals $\bar{x}_s - \bar{x}_b$), the difference between the concentration in the site area and the background reference area. From the relevant statistic, one can calculate a test statistic that is based on δ and its sampling distribution. By using the t test, we are assuming that δ is normally distributed, with a standard deviation given by s (pooled variance). As emphasized in MARSSIM, it is important not to underestimate the standard deviation because of the loss of statistical power that results. If you are not sure of the standard deviation it is best to collect samples prior to the implementation of the survey to provide a better estimate of standard deviation.

At this point, it is necessary to determine the acceptance and rejection regions for the t test. As we noted earlier, the null hypothesis is stated in such a manner that it lends itself to a one-sided test. Specifically, we note that only large values of δ cause a rejection of the null hypothesis, so the rejection region is located in the upper tail of the t distribution (specifically, in the upper 5% since we have selected a type I error of 0.05). From standard statistical tables, we find that the critical value of the t distribution for alpha equal to 0.05 and degrees of freedom given by $v = n + m - 2 = 62$ to be 1.67. So, the rejection region for the t statistic is given by $t > 1.67$. We now find the corresponding critical value in terms of δ, based on the relationship

$$ t = \frac{\delta}{s((1/n)+(1/m))^{1/2}}. $$

We can solve this equation for δ:

$$ \delta = t \times s\left(\frac{1}{n} + \frac{1}{m}\right)^{1/2} = 1.67 \times 0.18\left(\frac{1}{32} + \frac{1}{32}\right)^{1/2} = 0.075\,\mathrm{pCi/g}. $$

Therefore, in terms of δ, the acceptance region is $\delta \leq 0.075$ pCi/g and the rejection region is $\delta > 0.075$ pCi/g. It is necessary to establish these regions in terms of the relevant statistic so that the prospective power curve may be constructed. We can now focus our attention on the type II error. As mentioned previously, a specific alternative hypothesis (H_a) must be stated before we can evaluate the type II error. To determine the sample size for this experiment, it was determined that it was important to reject H_0 95% of the time (type II error equals 0.05) when the true mean difference δ was 0.15 pCi/g. To construct the power curve, holding the sample size constant, we must calculate the type II error for specified values of δ for which the alternative hypothesis is true. In other words, the power of the test is the probability of rejecting the H_0 given that the alternative hypothesis is true.

To illustrate, let us calculate the type II error for an alternative hypothesis of $\delta = 0.1$ pCi/g. Remember that the null hypothesis is given by a distribution with a mean δ of 0, and its rejection region is $\delta > 0.075$ pCi/g. Note: Type II errors can only occur in the acceptance region because, by definition, they result when H_0 is incorrectly "accepted." So, the type II error can be pictured as the area under the curve of the alternative hypothesis for $\delta_1 = 0.1$ pCi/g, that overlaps the acceptance region of the null hypothesis, and that is bounded on the right by $\delta = 0.075$ pCi/g. The type II error can be written as

$$\text{Type II error} = \Pr(\delta \leq 0.075 \text{ pCi/g when } \delta_1 = 0.1 \text{ pCi/g})$$

To calculate this error, we convert the δ values to the t statistics that correspond to $\delta \leq 0.075$ pCi/g when $\delta_1 = 0.1$ pCi/g:

$$t = \frac{0.075}{0.18(1/32 + 1/32)^{1/2}} = 1.67; \quad t = \frac{0.1}{0.18(1/32 + 1/32)^{1/2}} = 2.22.$$

Therefore, type II error $(\beta) = P(t < 1.67 \text{ when } t = 2.22) = P(t < -0.552) = 0.28$. The power is simply $1 - \beta$, or 0.72. Hence, there is a 72% probability that the null hypothesis (H_0: $\delta \leq 0$) will be rejected when the true δ_1 is 0.1.

Similar calculations using Statistica™ Power Analysis software for a number of different alternative hypotheses are performed to construct Table 18.9.

The prospective power curve is constructed by plotting the power as a function of δ. This was done using the Microsoft Excel spreadsheet application for the data in Table 18.9, and is shown in Figure 18.3.

Obviously, constructing power curves via hand calculations can be a tedious task and is ideally suited for the computer. It should also be noted that power calculations are oftentimes limited by the availability of

TABLE 18.9   Prospective Power Table for a Number of Alternative Hypotheses (δ's)

δ	Power $(1 - \beta)$
0.01	0.0771
0.02	0.1141
0.05	0.2929
0.08	0.5452
0.10	0.7099
0.12	0.8395
0.15	0.9507 (based on design type II error of 0.05)
0.20	0.9970

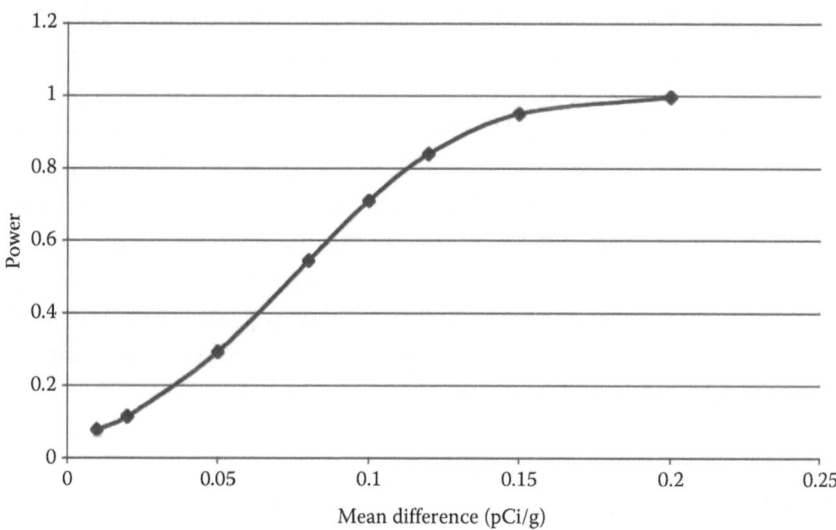

FIGURE 18.3   Prospective power curve for the two-sample *t* test.

detailed statistical distribution tables; that is, many tables only provide *t* values that are in the tails of the distribution (e.g., probabilities less than 20%). Fortunately, there are a number of statistical power analysis software packages available to handle this task. One such software package that is available on the Internet (NCSS, LLC; http://www.ncss.com) for demonstration is titled PASS™. PASS calculates the power for the various input parameters and then provides output options such as plotting power as a function of the alternative hypothesis.

Now that the sampling design has been thoroughly studied, the survey designer has sufficient information to make a decision as to the acceptability of the sampling design. More importantly, the regulator can assess the prospective power curve and quickly see that there is nearly 100% probability of rejecting the null hypothesis when the difference between $\bar{x}_s - \bar{x}_b$ is 0.2 pCi/g, and an 84% probability of rejecting the null hypothesis when the difference is 0.12 pCi/g. As we discussed, for the licensee proposing that the concentration in the site area ($\bar{x}_s$) is less than that in the reference area ($\bar{x}_b$) with this null hypothesis, it is not only necessary to "accept"—or more correctly, "fail to reject"—the null hypothesis but it is also necessary to prove to the regulator that sufficient data were collected to detect the situations where the site area Cs-137 concentration did indeed exceed the background concentration.

Only with the prospective power curve does the regulator know that the sample size is sufficient to detect with 84% probability the situation where the site area exceeds the reference area by 0.12 pCi/g. Let us assume that the regulator deems this risk acceptable and approves the sampling design. However, the survey designer and regulator should recognize that this prospective power curve is accurate only as long as the actual sample size matches that designed, and the prospective standard deviation matches the actual standard deviation. This can only be assessed through the retrospective power analysis—and a good reason why we should restate our null hypothesis. But before we leave this example, let us evaluate the hypothetical survey results using this specific null hypothesis.

18.3.1.2 Survey Implementation and Data Reduction

Let us assume that the survey design was implemented, and that 32 (i.e., $n = m = 32$) soil samples were collected from both the site area and the background reference area. Each sample was processed in the laboratory and analyzed via gamma spectrometry for Cs-137. The basic statistics for each area were as follows:

Class 3 Area (pCi/g, Cs-137)	Background Reference Area (pCi/g, Cs-137)
$\bar{x}_s = 0.52$	$\bar{x}_b = 0.44$
$s_s = 0.25$	$s_b = 0.21$

It is interesting to note at this point that δ is 0.08 pCi/g (i.e., from 0.52 to 0.44 pCi/g), and based on our prospective power curve, there is a 54.5% probability of rejecting the null hypothesis and concluding that the class 3 survey unit is indeed greater than the background. But the power may be even less because the standard deviation assumed in the design was 0.18, whereas the actual standard deviations are 0.21 and 0.25 pCi/g, respectively, for the background and class 3 areas.

Remember, reviewing the power curve at this point is instructional—simply to evaluate what the power is at the δ concentration. The statistical test must be performed to determine whether the null hypothesis is actually rejected or not.

The first calculation is to calculate the pooled standard deviation.

$$s = \sqrt{\frac{(m-1)(s_b)^2 + (n-1)(s_s)^2}{(m-1)+(n-1)}} = \sqrt{\frac{(32-1)(0.21)^2 + (32-1)(0.25)^2}{(32-1)+(32-1)}}$$

$$= 0.23 \text{ pCi/g}.$$

Again, this may be a concern because the assumed standard deviation that we used to calculate the sample size was 0.18 pCi/g—we underestimated the standard deviation, which may have a significant effect on the statistical power, and therefore a concern to the regulator, although this may be a moot point if we reject the null hypothesis anyway.

We calculate the t statistic based on our survey data:

$$t = \frac{0.52 - 0.44}{0.23(1/32 + 1/32)^{1/2}} = 1.39.$$

The critical t value for a type I error of 0.05 and for 62 degrees of freedom is 1.67. Because the t statistic (1.39) falls within the acceptance region, we do not reject the null hypothesis, and conclude that there is insufficient evidence to reject the null hypothesis. However, because we have underestimated the actual pooled standard deviation, it is necessary to construct the retrospective power curve based on the actual standard deviation. This is necessary to ensure that enough data were collected to identify a mean difference (δ) of significance. The first diagnostic is to calculate the sample size using Equation 18.1 with a standard deviation of 0.23 pCi/g—if the new sample size is less than the sample size collected, then the retrospective power is satisfied. Unfortunately, the new sample size based on the increased standard deviation is 52 samples. We can now proceed to calculate the retrospective power curve.

The Statistica Power Analysis software was used for the new standard deviation and for a number of different alternative hypotheses (given by various δ values), to construct Table 18.10.

The retrospective power curve is constructed by plotting the power as a function of δ from Table 18.10. Figure 18.4 illustrates the comparison of the retrospective and prospective power curves.

(Note: The retrospective power curve is less than the prospective power curve due to the fact that the standard deviation was underestimated.)

Both Table 18.10 and Figure 18.4 clearly illustrate the reduced statistical power that results when the standard deviation is underestimated. To illustrate, let us assume that the regulator approved the planned (prospective) 71% probability of rejecting the null hypothesis at a mean difference of 0.1 pCi/g. However, as shown in the table, this probability has been reduced to only 53%. Further, at the designed type II error of 0.05 at 0.15 pCi/g, the type II is actually 0.175—so, instead of a 95% probability

TABLE 18.10 Retrospective Power Table for a
Number of Alternative Hypotheses (δ's)

δ	Prospective Power	Retrospective Power
0.01	0.0771	0.0704
0.02	0.1141	0.0967
0.05	0.2929	0.2163
0.08	0.5452	0.3941
0.10	0.7099	0.5300
0.12	0.8395	0.6625
0.15	0.9507 (design power)	0.8252
0.20	0.9970	0.9637

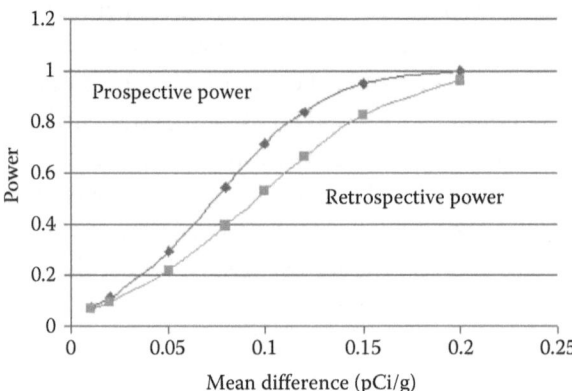

FIGURE 18.4 Retrospective versus prospective power curve for the two-sample *t* test.

at 0.15 pCi/g, the retrospective power analysis shows only a 82.5% probability of rejecting the null hypothesis at 0.15 pCi/g. What is the meaning of all this? Simply, the approved survey design and its implicit statistical power for detecting a specified effect size (concentration of Cs-137 in site area greater than that in the background—mean difference in pCi/g) has not been achieved, and therefore further discussions and possible negotiations with the regulator are necessary. Perhaps the regulator will accept the reduced power of this survey design, but the important point is that without the retrospective analysis it was impossible to determine what mean concentrations might have been missed.

How can this be prevented? First, do not underestimate the standard deviation—conduct a well-planned pilot study to assess the true variability (standard deviation) prior to the survey. Second, do not subject your statistical design to poorly stated null and alternative hypotheses, if possible. In other words, avoid the situation of regulatory approval depending on the prospective power curve—instead, restate the null hypothesis so that the desired result is obtained by rejecting the null hypothesis. Note: Even with the null hypothesis stated in a manner such that rejecting H_0 leads to the desired result, the prospective power curve is still helpful to the survey designer to balance sample size versus the "licensee's decision error." However, with the restated null hypothesis, the prospective power curve should not become part of the package requiring regulatory approval.

18.3.1.3 Survey Design with Different Null Hypothesis

Suppose that the licensee has decided to restate the null hypothesis so that the desired outcome is to reject the null hypothesis (rather than "accepting" the null hypothesis). In fact, the hypothesis testing preference is for the sample data to support the alternative hypothesis—remember that this is termed "reject-support" hypothesis testing convention. So, in formulating the null and alternative hypotheses, a first step is to state the desired outcome of the test, and call this outcome the alternative hypothesis. Again, we will assume that the licensee has previously assessed the data distributions, and has confirmed that Cs-137 concentrations in both the site area and the background reference area are normally distributed. And as before, the licensee will use the t test, and assume that the variances in both areas are equal.

The null hypothesis can be restated as follows:

H_0: The mean Cs-137 soil concentration in the site area ($\overline{x}_s$) is greater than the mean Cs-137 soil concentration in the background reference area ($\overline{x}_b$) against the alternative hypothesis

H_a: The mean Cs-137 soil concentration in the site area ($\overline{x}_s$) is less than or equal to the mean Cs-137 soil concentration in the background reference area ($\overline{x}_b$)

The null and alternative hypothesis can be written similarly as

$$H_0 = \overline{x}_s - \overline{x}_b > 0;$$

against the alternative hypothesis

$$H_a = \bar{x}_s - \bar{x}_b \leq 0.$$

With the restated null and alternative hypotheses, the type I error—which results when H_0 is incorrectly rejected—now means that one would incorrectly conclude that the concentration in the site area is less than the concentration in the background reference area. The type I error is now the regulator's error of concern. The type II error—which results when H_0 is incorrectly not rejected—means that one would incorrectly conclude that the concentration in the site area is greater than the concentration in the background reference area. This can be considered the licensee's error of concern. One key to ensuring that the prospective and retrospective power curves do not necessarily have to become part of the package requiring regulatory approval is to make the type I error the regulator's error of concern. This circumstance arises because the type I error will be the same for the retrospective power curve and the prospective power curve, regardless of actual sample size and standard deviation.

The sample size can be determined once the type I and II errors are specified, and the net Cs-137 concentration at the specified type II error is stated. For instance, the licensee might specify a type II error of 0.05 for a δ concentration of -0.1 pCi/g—this means that there is a 5% chance that the null hypothesis will be "accepted" even when the Cs-137 concentration in the site area is less than that in the background by 0.1 pCi/g. Assuming that the same standard deviation applies as before (and equals 0.18 pCi/g), the sample size is given by

$$n = \frac{2(0.18)^2(1.645 + 1.645)^2}{(-0.1)^2} + 0.25(1.645)^2 = 70.8.$$

Therefore, 71 soil samples are required in both the site area (n) and the background reference area (m). The project manager decides that this sample size is too high, and asks the survey designer to reduce the sample size, even if it means accepting a greater type II error.

To reduce the sample size, the licensee can increase the type II error, specify the type II error at a lower δ concentration (i.e., less than a δ concentration of -0.1 pCi/g), or both. The licensee decides to lower the δ concentration at which the type II error is specified to -0.15 pCi/g. The reduced sample size is calculated as

$$n = \frac{2(0.18)^2(1.645 + 1.645)^2}{(-0.15)^2} + 0.25(1.645)^2 = 31.8.$$

The licensee decides to go with this survey design: 32 soil samples in both the site area and the background reference area. The reader may notice that this is the same sample size that was used for the initial statement of the null and alternative hypotheses, and will be useful for comparison purposes.

At this point, it is necessary to determine the acceptance and rejection regions for the t test. The null hypothesis is still stated in such a way that it lends itself to a one-sided test, but this time the rejection region is in the lower tail of the t distribution because only smaller values of δ cause rejection of the null hypothesis (specifically, in the lower 5% since we have selected a type I error of 0.05). From standard statistical tables, we find that the critical value of the t distribution for alpha equal to 0.05 and degrees of freedom given by $v = n + m - 2 = 62$ to be -1.67. So the rejection region for the t statistic is given by $t < -1.67$. This t value corresponds to a critical value in terms of δ of -0.075 pCi/g. Therefore, in terms of δ, the acceptance region is $\delta > 0.075$ pCi/g, and rejection region is $\delta \leq -0.075$ pCi/g.

To compare the power curves from the two statements of null hypotheses, it is necessary to review what power actually means in each case. (Note: Power always refers to the probability of rejecting the null hypothesis, so it takes on different meaning when the null hypothesis is stated in a different manner.) For the first case, with the null hypothesis stated as H_0: $\bar{x}_s - \bar{x}_b \leq 0$, the power is the probability of concluding that the concentration in the site area is greater than that in the background reference area. For shorthand, let us simply say that it represents the site area not being released. So, as δ is increased, the power increases, as does the probability of not releasing the site area.

For the restated null hypothesis, H_0: $\bar{x}_s - \bar{x}_b > 0$, the power is the probability of concluding that the concentration in the site area is less than that in the background reference area. So, in this case, power has the exact opposite meaning—it is the probability that the site area can be released. Now, as δ is increased, the power decreases, as does the probability of releasing the site area. This leads to some difficulties in comparing the effects of the two statements of hypotheses.

One way to evaluate these data is to compare the power for the first case with the type II error in the second case; so that the net effect is comparing the probability of concluding that the site area is not releasable in each

TABLE 18.11   Probability of Concluding the Class 3 Area Is Not Releasable Based on the Different Null Hypothesis Statements

Cs-137 Concentration ($\delta = \bar{x}_s - \bar{x}_b$)	First Case—Power ($H_0: \bar{x}_s - \bar{x}_b \leq 0$)	Second Case— Type II Error ($H_0: \bar{x}_s - \bar{x}_b \leq 0$)
$\delta = -0.10$	0.00006	0.29
$\delta = -0.05$	0.003	0.71
$\delta = 0$	0.05	0.95
$\delta = 0.02$	0.115	0.98
$\delta = 0.10$	0.72	0.9999

case. Table 18.11 and the corresponding Figure 18.5 help to summarize the comparison between the two hypothesis statements.

Figure 18.5 illustrates the tremendous difference in survey designs depending on how the null hypothesis is stated. The first statement of the null hypothesis is that the Cs-137 in the class 3 area is less than or equal to the background, and results in the prospective power curve (which is the probability of not releasing the class 3 area) shown on the right of Figure 18.5. The second statement of the null hypothesis is that the Cs-137 in the class 3 area is greater than the background, and the probability of concluding that the class 3 area is not releasable is the type II error plot on the left.

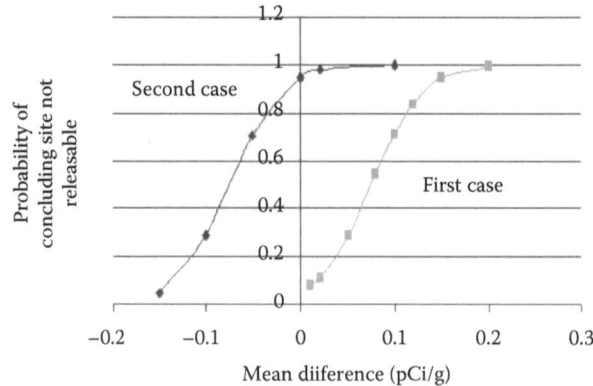

FIGURE 18.5   Prospective power curves for different statements of the null hypothesis.

The results are pretty clear; the restated null hypothesis is much more conservative than the stated form of the hypothesis in the first case. For instance, when the concentrations in the site area and the background reference area are equal (mean difference of zero), there is only a 5% chance of "failing" the site area in the first case, as compared to 95% chance of failing the site area in the second case. This constitutes a significant problem with the second statement of the null hypothesis—that even when the class 3 survey unit concentration equals the background concentration ($\delta = 0$), there is a 95% chance that the class 3 survey unit will fail. This is due to how the null hypothesis is stated and the fact that it takes significant data to the contrary to reject the null hypothesis. This is precisely why regulators would prefer the hypothesis stated as it is in the second case. The benefit of the null hypothesis stated as it is for the second case is that the type I error (regulator's concern) is not affected by an underestimated standard deviation or the collection of less samples than planned. However, this statement of the null hypothesis is simply not workable; the first statement should likely be used and the survey designer will have to provide an assessment of the retrospective power.

Remember that we stated earlier that the null hypothesis should be expressed such that the desired result is to reject H_0 in favor of the alternative hypothesis. This poses a dilemma in the sense that we are striving to reject the H_0 to demonstrate that the Cs-137 concentration in the site area is *not* greater than the concentration in the background reference area. From a purely statistical point of view, the greater the Cs-137 concentration in the background reference area relative to the concentration in the site area, the greater the probability of rejecting the null hypothesis. Of course, this is the desired outcome for the survey designer wishing to demonstrate that the concentration in the site area is no greater than that in the background. However, from a practical standpoint, it is not reasonable to expect the Cs-137 concentration in the background reference area to be greater than that in the site area—if it is, then we would likely conclude that the background reference area that we are comparing the class 3 area to is not representative, and we still have a problem. That is, any discussion of survey design is premature if the background reference area selected is not representative of the site area. Thus the dilemma—how can the null hypothesis be rejected if the condition required for the rejection necessarily voids the initial prerequisite that the background reference area be representative of the site area?

The issue is at what point do we conclude that the class 3 and background reference areas are different? This must be answered to the satisfaction of both the licensee and the regulator. So, it seems that the answer is to state the null hypothesis such that it would be "accepted"—and use the power curve to prove that if there was a difference, that the statistical test would be sufficient to detect the difference. One answer as to what constitutes a reasonable difference is to consider the standard deviation of both the site area and the reference area. A common choice of what constitutes a difference are those results that are outside the 95% confidence level ($\pm 2\sigma$) of the mean. Once this acceptable difference has been agreed upon, it may then be possible to state the null hypothesis such that the desired licensee outcome is to reject the null hypothesis. The example in the next section covers this approach in the context of a nonparametric statistical design.

18.3.2 Mann–Whitney Test: Comparing Cs-137 Concentrations in a Class 3 Area with a Background Reference Area

We now turn to the nonparametric counterpart of the two-sample t test—the Mann–Whitney test or, as it is sometimes referred to, the Wilcoxon rank sum test. As previously mentioned, one of the conditions for using the t test is that the data are normally distributed. In the above two examples, we had made the assumption that the data are normally distributed, which should be verified using a normality test such as the Shapiro–Wilk test. The nonparametric WRS test only requires that the data be continuous and independent.

As in the last example, we will state the null hypothesis such that the desired result is to reject H_0 in favor of the alternative hypothesis. However, in this example, we will assume that the level that constitutes a difference from the background has been determined by the licensee and regulator. This acceptable difference is two times the standard deviation of Cs-137 in the background, or 0.36 pCi/g. The reason for doing this is to better illustrate the problem in stating the null hypothesis in a way that it can hardly be rejected, and therefore, makes it near impossible to release the class 3 area. Thus, the null hypothesis is stated as follows:

H_0: The mean Cs-137 soil concentration in the site area is greater than the mean plus two sigma Cs-137 soil concentration in the background reference area against the alternative hypothesis

H_a: The mean Cs-137 soil concentration in the site area is less than or equal to the mean plus two sigma Cs-137 soil concentration in the background reference area

The null and alternative hypotheses can be written similarly as

$$H_0 = \bar{x}_s > \bar{x}_b + 2\sigma \quad \text{or} \quad \bar{x}_s - (\bar{x}_b + 2\sigma) > 0;$$

against the alternative hypothesis:

$$H_a = \bar{x}_s \leq \bar{x}_b + 2\sigma \quad \text{or} \quad \bar{x}_s - (\bar{x}_b + 2\sigma) \leq 0.$$

Therefore, the survey designer endeavors to reject the null hypothesis in favor of the alternative hypothesis. The type I error—which results when H_0 is incorrectly rejected—means that one would incorrectly conclude that the concentration in the site area is less than the mean plus two sigma Cs-137 concentration in the background reference area. The type I error is the regulator's error of concern. The type II error—which results when H_0 is incorrectly not rejected—means that one would incorrectly conclude that the concentration in the site area is greater than the mean plus two sigma concentration in the background reference area (i.e., the D&D contractor's error of concern).

It is assumed with the WRS test that the difference between the two distributions—if there is a difference at all—is only a difference in the location of the distribution. In other words, it is assumed that there is no difference in the shapes (or variance) of the distributions, only a possible difference in means. The WRS test is conducted by comparing the entire distributions of each sample from the site area and the background reference area—not just a comparison of their means. This comparison is accomplished by combining all the data from both the samples (e.g., class 3 area and background reference area), and ranking the data from 1 to $N = n + m$. The tied values are assigned the mean of the rank positions they would have occupied had there been no ties. The test statistic, W', is calculated by summing the ranks assigned to the background reference area:

$$W' = \sum_{i=1}^{m} R(x_{bi}).$$

For our example, it is necessary to adjust each of the background measurements by adding 0.36 pCi/g to the background value prior to ranking. This is done so that the statistical test is consistent with our statement of the null hypothesis. In other words, we are not interested in whether the class 3 area is greater than the background; rather, we are interested in whether the class 3 area is greater than 0.36 pCi/g (two sigma) plus the background.

The decision on whether to reject or fail to reject the null hypothesis is based on the value of W' and the size of the acceptance region, which is given by the type I error. Large values of W' provide evidence that $\bar{x}_s \leq \bar{x}_b + 2\sigma$, and sufficiently large values of W' will result in the H_0 being rejected. Hence, the decision rule is to reject the null hypothesis if the computed value of W' is greater than the appropriate critical value, w_α, based on n, m, and α.

The sample size for the WRS test can be determined once the type I and type II errors are specified, as well as the $\bar{x}_s - \bar{x}_b$ (or δ) Cs-137 concentration—the background mean plus two sigma concentration of Cs-137 must be greater than the site area concentration for the alternative hypothesis to be true—at the specified type II error. For instance, the licensee might specify a type II error of 0.05 for a δ concentration of 0.10 pCi/g—this means that there is a 5% chance that the null hypothesis will be "accepted" even when the Cs-137 concentration in the site area is only greater than that in the background by 0.10 pCi/g. The WRS test sample size can be calculated using the COMPASS MARSSIM software (and accepting the 20% increase in sample size built into the MARSSIM survey design). In this case, the DCGL$_W$ is 0.36 pCi/g (the agreed-upon increase above background that was acceptable to both parties), and the LBGR is 0.10 pCi/g.

The WRS test sample size in both the survey unit (n) and the background reference area is 19. The prospective power curve in Figure 18.6 shows the probability of passing the class 3 survey unit as a function of the net Cs-137 concentration in the survey unit ($\bar{x}_s - \bar{x}_b$). At a net Cs-137 concentration of 0.36 pCi/g, the probability of passing is given by the type I error of 0.05. The design type II error is shown at the LBGR concentration of 0.10 pCi/g of Cs-137; the type II error is smaller than planned due to the 20% increase in sample size.

18.3.2.1 Survey Implementation and Data Reduction

Assume that the licensee implemented this survey design, collecting 19 samples in both the site area and the background reference area. Each was processed in the laboratory and analyzed for Cs-137 concentration via gamma spectrometry. The results are provided in Table 18.12.

The mean in the class 3 area was 0.46 pCi/g Cs-137, with a standard deviation of 0.13 pCi/g; while the mean in the background reference area

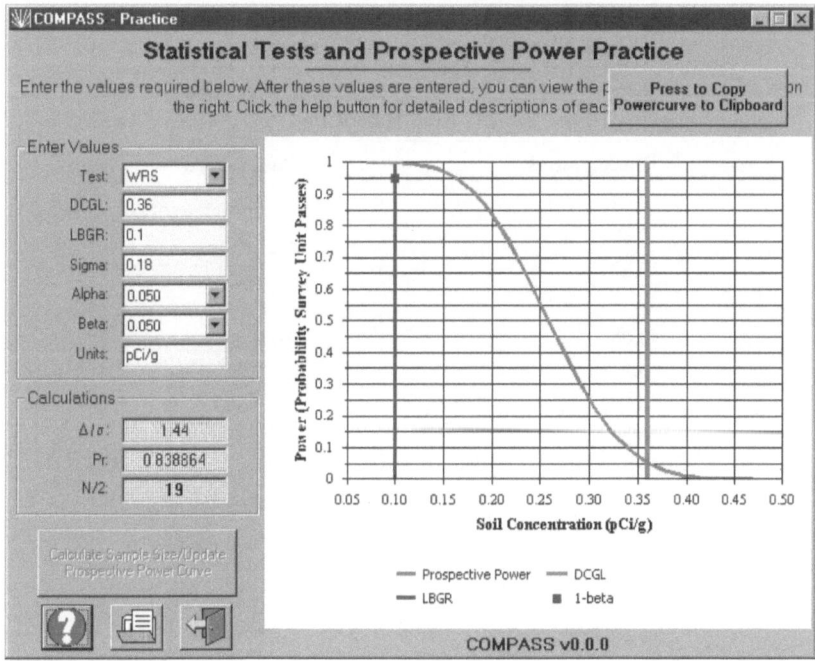

FIGURE 18.6 Prospective power curve for the WRS test of the class 3 area and the background reference area.

was 0.33 pCi/g Cs-137, with a standard deviation of 0.12 pCi/g. Thus, the class 3 area Cs-137 concentration exceeds that in the background by 0.13 pCi/g on average. Since the allowable difference in Cs-137 concentration between the class 3 area and the background reference area is 0.36 pCi/g, it is likely that the class 3 area will pass.

The COMPASS software was used to reduce these data. The first step was to add 0.36 pCi/g to each of the background concentrations. The samples from the two populations were then combined and ranked using ranks 1 to N (low to high), where N is equal to n (class 3 area measurements) plus m (background reference area measurements). If there are sample values that are exactly equal, assign to each the average of the ranks.

COMPASS determined the WRS test statistic—sum of the ranks (W') of the adjusted background reference area measurements—equal to 513. The critical value of the W' statistic (w_α) can be determined for values of $n = m = 19$, and $\alpha = 0.05$ (from MARSSIM Appendix I, Table I.4) and is 427. Because the W' statistic (513) does indeed exceed the critical value (427), the null hypothesis can be rejected. Therefore, the null hypothesis

TABLE 18.12   Cs-137 Concentrations in Soil for
Class 3 Area and Background Reference Areas

**Class 3 Area versus Background Reference Area
Cs-137 Results in pCi/g**

Background	Class 3 Area
0.42	0.26
0.37	0.59
0.52	0.47
0.29	0.33
0.08	0.23
0.32	0.34
0.39	0.54
0.29	0.46
0.34	0.51
0.17	0.63
0.42	0.33
0.39	0.52
0.16	0.55
0.48	0.49
0.26	0.67
0.13	0.34
0.49	0.45
0.35	0.59
0.33	0.43

that the Cs-137 concentration in the site area is greater than the mean plus two sigma Cs-137 concentration in the background reference area is "rejected," and conclude that class 3 satisfies the release criterion.

QUESTIONS AND PROBLEMS

1. Which normality test can be used for a data set that consists of more than 50 values?

2. Given the Cs-137 data in Table 18.12, determine the normality of both the class 3 and background reference area data sets.

3. Briefly describe the reasons why a null hypothesis might be stated such that the desired result is to accept the null hypothesis, versus the standard null hypothesis statements where it is desired to reject the null hypothesis.

CHAPTER 19

International Decommissioning Perspectives

J UST WEEKS FOLLOWING THE March 11, 2011 Fukushima Daiichi nuclear reactor accident, it was clear that there would be serious long-term environmental consequences. While the extent of Cs-137 contamination at Fukushima and the surrounding areas continues to be characterized, initial data indicated very high levels in some areas. As characterization surveys continue to refine the soil contamination profile surrounding Fukushima, it is apparent that large areas of land are significantly contaminated (perhaps as much as 500 square miles), and potentially uninhabitable. Japanese regulatory authorities must establish an adequately protective dose level following the optimization process. A useful document in this regard is ICRP Publication 111 (2009), *Application of the Commission's Recommendations to the Protection of People Living in Long-term Contaminated Areas After a Nuclear Accident or a Radiation Emergency.* The fundamental decision is whether to allow people to live permanently in long-term contaminated areas, and at what release criteria. Initial projections are that the Fukushima cleanup will take 40 years to complete, and the decommissioning costs will be enormous.

Decommissioning continues to rapidly increase throughout the world. Most of the larger industrial countries in the world have devoted significant resources to decommissioning. The Nuclear Decommissioning

Authority in the United Kingdom continues to rival the annual cleanup spending of DOE's Environmental Management program. The NDA has significant decommissioning activities underway at the Sellafield and Dounreay site, the first major closure project in the United Kingdom. And the French are interested in reducing the amount of nuclear power in their energy portfolio.

Most countries recognize the International Atomic Energy Agency as the lead agency in providing decommissioning-related guidance. The IAEA has published extensive guidance on the subject of decommissioning and the performance of final surveys. In fact, IAEA's Compliance Monitoring for Remediated Sites (1999) cites the MARSSIM in Annex I. This document provides case histories of final survey experiences for nearly a dozen countries. The case history for the United States is the final status survey planned for the Cushing Refinery Site in Oklahoma that served as a final test of the MARSSIM methodology.

Decommissioning often occurs when facilities have reached the end of their useful life. Other times the driver for decommissioning is the result of a nuclear accident like Chernobyl or Fukushima. And sometimes political and economic decisions result in D&D commencing long before the end of useful life. A number of countries have decided against nuclear power following the Fukushima Daiichi incident. Germany, for one, recommitted to its decade earlier decision to scrap its nuclear power program following the earthquake, tsunami, and significant release of radioactive materials from the reactor accident in Japan. Others, including utilities in the United States, have decided to prematurely shutdown nuclear reactors due to the extremely low natural gas prices. While this is certainly distressing news to supporters of nuclear power, it nonetheless represents a boon to the international decommissioning industry.

Nuclear cleanup is a complex activity—and it is tremendously expensive and time consuming. Many in the United States have pursued international D&D opportunities throughout the world. Moreover, there seems to be a fair amount of cooperation among countries involved in decommissioning. For example, the IAEA, the Nuclear Energy Agency, and the Commission of the European Communities are among a number of organizations that share experience and technical knowledge about decommissioning. Decommissioning experience has been gained, and shared, in the following technical areas: (1) assessment of the radioactive inventories, (2) decontamination and dismantlement techniques, including remote operations in high radiation fields, (3) radioactive waste management, and (4) the objectives have been to reduce the radiological hazards to decommissioning workers, while simultaneously optimizing the decommissioning operations to reduce the total decommissioning cost.

Decommissioning is certainly a worldwide activity, and is expected to increase due to the number of nuclear power plants expected to shutdown over the next 10 or so years. The IAEA seems to be the unifying force in terms of decommissioning planning. The MARSSIM approach is well on its way to being adopted in other countries. The United Kingdom and Japan have been particularly interested in using the MARSSIM for decommissioning projects in their countries. And while a clearance rulemaking still seems outside of our grasp in the United States, we can hang our hats on the MARSAME. The future of decommissioning looks very robust—will we have the skilled nuclear workforce to do the job?

References

Abelquist, E.W. *Final Status Survey Using MARSSIM Survey Methodologies at the Cushing Refinery Site*, ORISE 97-1138, Oak Ridge, TN: Oak Ridge Institute for Science and Education; 1997.

Abelquist, E.W. *Decommissioning Health Physics: A Handbook for MARSSIM Users*. Institute of Physics Publishing, Bristol, UK. 2001.

Abelquist, E.W. Dose modeling and statistical assessment of hot spots for decommissioning applications. PhD dissertation, University of Tennessee; 2008.

Abelquist, E.W. and Brown, W.S. Estimating minimum detectable concentrations achievable while scanning building surfaces and land areas. *Health Physics.* 76(1):3–10; 1999.

Abelquist, E.W., Condra, R.D., and Laudeman, M.J. Determination of uranium and thorium concentrations in high Z material samples using direct counting method of gamma spectroscopy. *Radiation Protection Management.* 13:42–49; 1996.

Abelquist, E.W. and Vitkus, T.J. Considerations for decommissioning survey technique and implementation at uranium and thorium sites. *Radiation Protection Management.* 14:38–45; July/August 1997.

Albert, J. *Bayesian Computation with R.* New York, NY: Springer; 2007.

American National Standards Institute (ANSI). *Performance Specifications for Health Physics Instrumentation—Portable Instrumentation for Use in Normal Environmental Conditions.* New York, NY: American National Standards Institute; ANSI N42.17A; 1989.

American National Standards Institute. *Radiation Protection Instrumentation Test and Calibration, Portable Survey Instruments.* ANSI N323A-1997; New York, NY: American National Standards Institute; 1997.

American National Standards Institute. *Surface and Volume Radioactivity Standards for Clearance.* ANSI N13.12; New York, NY: American National Standards Institute; 1999.

American National Standards Institute. *Performance and Documentation of Radiological Surveys.* ANSI N13.49; New York, NY: American National Standards Institute; 2001.

American National Standards Institute. *Characterization in Support of Decommissioning Using the Data Quality Objectives Process.* ANSI N13.59; New York, NY: American National Standards Institute; 2008.

American Society for Testing and Materials (ASTM). *Selection and Use of Portable Radiological Survey Instruments for Performing In Situ Radiological Assessments in Support of Decommissioning.* E 1893-97; West Conshohocken, PA; March 1998.

Argonne National Laboratory (ANL). *Data Collection Handbook to Support Modeling the Impacts of Radioactive Material in Soil.* ANL/EAIS-8; Argonne, IL: Argonne National Laboratory; 1993a.

Argonne National Laboratory. *Manual for Implementing Residual Radioactive Material Guidelines Using RESRAD, Version 5.0.* ANL/EAD/LD-2; Argonne, IL: Argonne National Laboratory; 1993b.

Argonne National Laboratory. *User's Manual for RESRAD Version 6.* ANL/EAD-4; Argonne, IL; 2001.

Auster, B.B. Radioactive fork, dear? *U.S. News & World Report*; Vol. 127, Issue 11; September 1999.

Bailey, E.N. *Lessons Learned from Independent Verification Activities.* DCN 0476-TR-02-0, Oak Ridge Institute for Science and Education prepared for the U.S. Department of Energy; July 2008.

Boerner, A.J. The MARSAME methodology: Fundamentals, benefits, and applications. *Waste Management Conference*, Phoenix, AZ; 2012.

Brodsky, A. Exact calculation of probabilities of false positives and false negatives for low background counting. *Health Physics.* 63(2):198–204; August 1992.

Brodsky, A. and Gallaghar, R.G. Statistical considerations in practical contamination monitoring. *Radiation Protection Management.* 8(4):64–78; July/August 1991.

Brosey, B.H. and Holmes, R.D. Site characterization efforts at the Saxton nuclear experimental corporation. *Proceedings from American Nuclear Society Topical Meeting, The Best of D&D...Creative, Comprehensive and Cost-Effective,* Chicago, IL; 1996.

Brown, W.S. and Abelquist, E.A. *Human Performance in Radiological Survey Scanning.* NUREG/CR-6364; Washington, DC: Nuclear Regulatory Commission; 1998.

Cember, H. and Johnson, T. *Introduction to Health Physics.* 4th ed. New York, NY: The McGraw-Hill Companies; 2009.

Chapman, J., Boerner, A., and Abelquist, E. *Spatially-Dependent Measurements of Surface and Near-Surface Radioactive Material Using In Situ Gamma Ray Spectrometry (ISGRS) for Final Status Surveys.* Oak Ridge, TN: Oak Ridge Institute for Science and Education; 2006.

Conover, W.J. *Practical Nonparametric Statistics.* 2nd ed. New York, NY: John Wiley and Sons; 1980.

Currie, L.A. Limits for qualitative detection and quantitative determination. *Analytical Chemistry.* 40(3):586–593; 1968.

Currie, L.A. *Lower Limit of Detection: Definition and Elaboration of a Proposed Position for Radiological Effluent and Environmental Measurements.* NUREG/CR-4007; Washington, DC; 1984.

Cusac, A.M. Nuclear spoons: Hot metal may find its way to your dinner table. *The Progressive*, 62(10):22–27; 1998.

Daniel, W.W. *Introductory Statistics with Applications*. Boston, MA: Houghton Mifflin Company; 1977.

Dehmel J. and Schneider S. Technical considerations for using *in situ* gamma spectroscopy in conducting final status surveys, *Health Physics*. 84(6 Suppl):S136–S140; June 2003.

Dufault, R. et al. Reducing environmental risk associated with laboratory decommissioning and property transfer. *Environmental Health Perspectives*. 108(Suppl 6):1015–1022; December 2000.

Eastern Research Group, Inc (ERG). *Methods for Estimating Fugitive Air Emissions of Radionuclides from Diffuse Sources at DOE Facilities*. Prepared for U.S. Environmental Protection Agency; Morrisville, NC; 2004.

English, R.A., Burdette, R.L., and Kessler, W.E. The new radiological criteria for decommissioning: Importance of transuranic nuclides. *Decommissioning, Decontamination, and Environmental Restoration at Contaminated Nuclear Sites (DDER-'94)*; American Nuclear Society; 1994.

Frame, P.W. and Abelquist, E.W. Use of smears for assessing removable contamination. *Operational Radiation Safety, Supplement to Health Physics*. 76(5); May 1999.

Gogolak, C. *CVG's Informal Environmental Stats Notes #2*, 2000, http://www.gogolak.org/.

Gilbert, R.O. *Statistical Methods for Environmental Pollution Monitoring*. New York, NY: Van Nostrand Reinhold; 1987.

Goles, R.W., Baumann, B.L., and Johnson, M.L. *Contamination Survey Instrument Capabilities*. PNL-SA-1984; Letter to the U.S. Department of Energy; 1991.

Government Accountability Office. *Nuclear Cleanup of Rocky Flats: DOE Can Use Lessons Learned to Improve Oversight of Other Sites' Cleanup Activities*. GAO-06-352; Washington, DC; July 2006.

Green, D.M. and Swets, J.A. *Signal Detection Theory and Psychophysics*. Los Altos, CA: Peninsula Publishing; 1988.

Huizenga, D. *Weapons Complex Monitor*. Washington, DC: ExchangeMonitor Publications, Inc.; October 5, 2012.

International Atomic Energy Agency (IAEA). *Monitoring Programmes for Unrestricted Release Related to Decommissioning of Nuclear Facilities*. Technical report series no. 334; Vienna, Austria; 1992.

International Atomic Energy Agency. *Radiological Characterization of Shut Down Nuclear Reactors for Decommissioning Purposes*. IAEA technical report (9th draft); Vienna, Austria; February 1997.

International Atomic Energy Agency. *Compliance Monitoring for Remediated Sites*. IAEA-TECDOC-1118; Vienna, Austria; October 1999.

International Commission on Radiation Units and Measurements (ICRU). *Gamma-Ray Spectrometry in the Environment*. ICRU Report 53; Bethesda, MD; 1994.

International Commission on Radiological Protection (ICRP). *Report of the Task Group on Reference Man*. ICRP Publication 23. New York, NY: Pergammon Press; 1975.

International Commission on Radiological Protection. *Application of the Commission's Recommendations to the Protection of People Living in Long-Term Contaminated Areas after a Nuclear Accident or a Radiation Emergency.* ICRP Publication 111; Amsterdam: Elsevier; 2009.

International Organization for Standardization (ISO). *Evaluation of Surface Contamination—Part 1: Beta Emitters and Alpha Emitters.* 1st ed. ISO-7503-1; Geneva, Switzerland; 1988a.

International Organization for Standardization. *Evaluation of Surface Contamination—Part 2: Tritium Surface Contamination.* ISO-7503-2; Geneva, Switzerland; 1988b.

International Organization for Standardization. *Reference Sources for the Calibration of Surface Contamination Monitors—Beta Emitters (Maximum Beta Energy Greater than 0.15 MeV) and Alpha Emitters.* ISO-8769; Geneva, Switzerland; 1988c.

Interstate Technology & Regulatory Council. *Real-Time Measurement of Radionuclides in Soil: Technology and Case Studies.* Washington, DC; February 2006.

Johnson, R.R. *Elementary Statistics.* 3rd ed. North Scituate, MA: Duxbury Press; 1980.

Kaplan, S. and Garrick, B.J. On the use of a Bayesian reasoning in safety and reliability decisions—Three examples. *Nuclear Technology;* 44:231–245; 1979.

King, D.A., Altic, N., and Greer, C. Minimum detectable concentration as a function of gamma walkover survey technique. *Operational Radiation Safety,* 102(suppl 1): February 2012.

Knoll, G.F. *Radiation Detection and Measurement.* 4th ed. New York, NY: John Wiley and Sons; 2010.

Kocher, D.C. *Radioactive Decay Data Tables, A Handbook of Decay Data for Application to Radiation Dosimetry and Radiological Assessments.* Washington, DC: U.S. DOE Technical Information Center; DOE/TIC-11026; 1981.

Lurie, D. et al. *Applying Statistics.* U.S. Nuclear Regulatory Commission. NUREG-1475, Rev. 1; Washington, DC; 2011.

Marin, J.M. and Robert, C.P. *Bayesian Core: A Practical Approach to Computational Bayesian Statistics.* New York, NY: Springer; 2007.

Martz, H.F. and Keller-McNulty, S. *Simulation Study to Evaluate the Power of Wilcoxon Rank Sum and Sign Tests for Assessing MARSSIM Surface Activity Compliance.* TSA-1/99-161; Los Alamos, NM; March 1999.

National Academy of Sciences (NAS). *The Disposition Dilemma: Controlling the Release of Solid Materials from Nuclear Regulatory Commission-Licensed Facilities;* Washington, DC; 2002.

National Council on Radiation Protection and Measurements (NCRP). *A Handbook of Radioactivity Measurements Procedures,* 2nd ed. NCRP No. 58. Bethesda, MD; 1985.

National Council on Radiation Protection and Measurements. *Exposure of the Population in the United States and Canada from Natural Background Radiation.* NCRP No. 94. Bethesda, MD; 1987.

National Council on Radiation Protection and Measurements. *Managing Potentially Radioactive Scrap Metal.* NCRP No. 141. Bethesda, MD; 2002.

National Council on Radiation Protection and Measurements. *Approaches to Risk Management in Remediation of Radioactively Contaminated Sites.* NCRP Report No. 146, Bethesda, MD; 2004.

Olsher, R.H. et al. *Alpha RADIAC Evaluation Project.* LA-10729; Los Alamos, NM: Los Alamos National Laboratory; 1986.

Slobodien, M.J. *Decommissioning and Restoration of Nuclear Facilities. Health Physics Society, 1999 Summer School.* Madison, WI: Medical Physics Publishing; 1999.

Sorensen, D. and Gianola, D. *Likelihood, Bayesian, and MCMC Methods in Quantitative Genetics.* New York, NY: Springer; 2002.

Strom, D.J. and Stansbury, P.S. Minimum detectable activity when background is counted longer than the sample. *Health Physics.* 63(3):360–361. September 1992.

Thelin, L. *Radiation Detection Experiment for HP Technicians.* Cintichem Inter-office Correspondence. September 1994.

Turner, J.E. *Atoms, Radiation, and Radiation Protection.* 3rd ed. New York, NY: Wiley-VCH Verlag GmbH & Co. KGaA; 2007.

U.S. Atomic Energy Commission (AEC). *Termination of Operating Licenses for Nuclear Reactors.* Regulatory Guide 1.86, Washington, DC; June 1974.

U.S. Department of Energy (DOE). *Radiation Protection of the Public and the Environment.* DOE Order 5400.5; Washington, DC; 1990.

U.S. Department of Energy. *Decommissioning Handbook.* DOE/EM-0142P; Washington, DC; March 1994.

U.S. Department of Energy. *Application of DOE Order 5400.5 Requirements for Release and Control of Property Containing Residual Radioactive Material.* Air, Water and Radiation Division; EH-412; November 17, 1995.

U.S. Department of Energy. *Implementation Guide for Surveillance and Maintenance during Facility Transition and Disposition.* DOE G 430.1-2; Washington, DC; September 1999.

U.S. Department of Energy. *Implementation Guide for the Control and Release of Property with Residual Radioactive Material.* Draft G441.1-xx; Washington, DC; 2002.

U.S. Department of Energy. *Environment, Safety and Health Bulletin: A Guide to Good Practices for the Control and Release of Property.* DOE/EH-0697. Washington, DC; July 2006.

U.S. Department of Energy. *Radiation Protection of the Public and the Environment.* Office of Health, Safety and Security DOE O 458.1; Washington, DC; February 2011.

U.S. Department of Energy. *Recycle of Scrap Metals Originating from Radiological Areas.* Draft programmatic environmental assessment. DOE/EA-1919; Washington, DC; 2012.

U.S. Environmental Protection Agency (EPA). *Upgrading Environmental Radiation Data.* HPSR-1/EPA 520/1-80-012. Washington, DC: Health Physics Society Committee/Environmental Protection Agency; 1980.

U.S. Environmental Protection Agency. *Limiting Values of Radionuclide Intake and Air Concentration and Dose Conversion Factors for Inhalation,*

Submersion, and Ingestion. Federal Guidance Report No. 11; Washington, DC; 1988.

U.S. Environmental Protection Agency. *External Exposure to Radionuclides in Air, Water, and Soil.* Federal Guidance Report No. 12; EPA 402-R-93-081; Washington, DC; 1993.

U.S. Environmental Protection Agency. *Guidance for Data Quality Assessment, Practical Methods for Data Analysis.* EPA QA/G-9; EPA/600/R-96/084; Washington, DC; 2000.

U.S. Environmental Protection Agency. *Guidance on Choosing a Sampling Design for Environmental Data Collection for Use in Developing a Quality Assurance Project Plan.* EPA QA/G-5S; EPA/240/R-02/005; Washington, DC; 2002.

U.S. Environmental Protection Agency. *Guidance on Systematic Planning Using the Data Quality Objectives Process.* EPA QA/G-4; EPA/240/B-06/001; Washington, DC; 2006.

U.S. Nuclear Regulatory Commission (NRC). *Monitoring for Compliance with Decommissioning Termination Survey Criteria.* NUREG/CR-2082; Washington, DC; 1981a.

U.S. Nuclear Regulatory Commission. *Disposal or Onsite Storage of Thorium or Uranium Wastes from Past Operations.* 46 FR 52061-52063, Washington, DC; 1981b.

U.S. Nuclear Regulatory Commission. *Guidelines for Decontamination of Facilities and Equipment Prior to Release for Unrestricted Use or Termination of License for Byproduct, Source, or Special Nuclear Material.* Policy and Guidance Directive FC 83-23; 1983.

U.S. Nuclear Regulatory Commission. *Guidelines for Decontamination of Facilities and Equipment Prior to Release for Unrestricted Use or Termination of Licenses for Byproduct, Source or Special Nuclear Material.* Washington, DC; 1987.

U.S. Nuclear Regulatory Commission. *Statistical Methods for Nuclear Material Management.* NUREG/CR-4604; Washington, DC; 1988.

U.S. Nuclear Regulatory Commission. *Manual for Conducting Radiological Surveys in Support of License Termination.* NUREG/CR-5849; Washington, DC; 1992a.

U.S. Nuclear Regulatory Commission. *Residual Radioactive Contamination from Decommissioning: Technical Basis for Translating Contamination Levels to Annual Total Effective Dose Equivalent.* NUREG/CR-5512, vol. 1; Washington, DC; 1992b.

U.S. Nuclear Regulatory Commission. *Background as a Residual Radioactivity Criterion for Decommissioning.* NUREG-1501; Washington, DC; 1994a.

U.S. Nuclear Regulatory Commission. *Working Draft Regulatory Guide on Release Criteria for Decommissioning: NRC Staff's Draft for Comment.* NUREG-1500; Washington, DC; 1994b.

U.S. Nuclear Regulatory Commission. *Scenarios for Assessing Potential Doses Associated with Residual Radioactivity.* NMSS Policy and Guidance Directive PG-8-08; May 1994c.

U.S. Nuclear Regulatory Commission. *Branch Technical Position on Site Characterization for Decommissioning.* Draft; Washington, DC; 1994d.

U.S. Nuclear Regulatory Commission. *Measurement Methods for Radiological Surveys in Support of New Decommissioning Criteria.* Draft report for comment. NUREG-1506; Washington, DC; 1995a.

U.S. Nuclear Regulatory Commission. *Radiological Criteria for License Termination.* 10 CFR Part 20, Subpart E. Federal Register 62 FR 39058; July 21, 1997a.

U.S. Nuclear Regulatory Commission. *NMSS Handbook for Decommissioning Fuel Cycle and Materials Licensees.* NUREG/BR-0241; Washington, DC; 1997b.

U.S. Nuclear Regulatory Commission. *A Proposed Nonparametric Statistical Methodology for the Design and Analysis of Final Status Decommissioning Survey.* NUREG-1505; Washington, DC; 1998a.

U.S. Nuclear Regulatory Commission. *Minimum Detectable Concentrations with Typical Radiation Survey Instruments for Various Contaminants and Field Conditions.* NUREG-1507; Washington, DC; 1998b.

U.S. Nuclear Regulatory Commission. *Decision Methods for Dose Assessment to Comply with Radiological Criteria for License Termination.* NUREG-1549; Washington, DC; 1998c.

U.S. Nuclear Regulatory Commission. *Staff Responses to Frequently Asked Questions Concerning Decommissioning of Nuclear Power Reactors.* NUREG-1628; Washington, DC; 1998d.

U.S. Nuclear Regulatory Commission. *Demonstrating Compliance with the Radiological Criteria for License Termination.* Draft Regulatory Guide DG-4006; Washington, DC; 1998e.

U.S. Nuclear Regulatory Commission. *Residual Radioactive Contamination from Decommissioning: Parameter Analysis.* NUREG/CR-5512, vol. 3; Washington, DC; 1999a.

U.S. Nuclear Regulatory Commission. *Comparison of the Models and Assumptions Used in the DandD 1.0, RESRAD 5.61, and RESRAD-Build 1.50 Computer Codes with Respect to the Residential Farmer and Industrial Occupant Scenarios Provided in NUREG/CR-5512.* Draft report for comment. NUREG/CR-5512, vol. 4; Washington, DC; 1999b.

U.S. Nuclear Regulatory Commission. *Radiological Assessments for Clearance of Equipment and Materials from Nuclear Facilities.* NUREG-1640; Washington, DC; 1999c.

U.S. Nuclear Regulatory Commission. *Information Paper on the Viability of Entombment as a Decommissioning Option for Power Reactors.* SECY-99-187; Washington, DC; 1999d.

U.S. Nuclear Regulatory Commission. *Multi Agency Radiation Survey and Site Investigation Manual (MARSSIM).* NUREG-1575; Rev. 1. Washington, DC; 2000a.

U.S. Nuclear Regulatory Commission. *NMSS Decommissioning Standard Review Plan.* NUREG-1727; Washington, DC; 2000b.

U.S. Nuclear Regulatory Commission. *Control of Solid Materials: Results of Public Meetings, Status of Technical Analyses, and Recommendations for Proceeding.* SECY-00-0070; Washington, DC; 2000c.

U.S. Nuclear Regulatory Commission. *Use of Rubblized Concrete Dismantlement to Address 10 CFR Part 20, Subpart E, Radiological Criteria for License Termination.* SECY-00-0041; February 14, 2000d.

U.S. Nuclear Regulatory Commission. *Residual Radioactive Contamination from Decommissioning: User's Manual DandD Version 2.1.* NUREG/CR-5512, vol. 2; Washington, DC; 2001.

U.S. Nuclear Regulatory Commission. *Radiological Surveys for Controlling Release of Solid Materials.* Draft NUREG-1761; Washington, DC; 2002a.

U.S. Nuclear Regulatory Commission. *Re-Evaluation of the Indoor Resuspension Factor for the Screening Analysis of the Building Occupancy Scenario for NRC's License Termination Rule.* NUREG-1720; Washington, DC; 2002b.

U.S. Nuclear Regulatory Commission. *Radiological Assessments for Clearance of Materials from Nuclear Facilities.* NUREG-1640; Washington, DC; 2003a.

U.S. Nuclear Regulatory Commission. *Standard Review Plan for Evaluating Nuclear Power Reactor License Termination Plans.* NUREG-1700, Rev. 1; Washington, DC; 2003b.

U.S. Nuclear Regulatory Commission. *Multi-Agency Radiological Laboratory Analytical Protocols (MARLAP).* NUREG-1576; Washington, DC; July 2004.

U.S. Nuclear Regulatory Commission. *Consolidated Decommissioning Guidance.* NUREG-1757, Vol. 2, Characterization, Survey, and Determination of Radiological Criteria, Rev. 1; Washington, DC; 2006a.

U.S. Nuclear Regulatory Commission. *Consolidated Decommissioning Guidance.* NUREG-1757, Vol. 1, Decommissioning Process for Materials Licensees, Rev. 2; Washington, DC; 2006b.

U.S. Nuclear Regulatory Commission. *Multi-Agency Radiation Survey and Assessment of Materials and Equipment Manual (MARSAME).* NUREG-1575, Supplement 1; Washington, DC; 2009.

U.S. Nuclear Regulatory Commission. *Status of the Decommissioning Program—2012 Annual Report.* SECY-12-0153; Washington, DC; 2012a.

U.S. Nuclear Regulatory Commission. *A Subsurface Decision Model for Supporting Environmental Compliance.* NUREG/CR-7021; Washington, DC; 2012b.

Vitkus, T.J. *Characterization Requirements for Decommissioning of Nuclear Facilities.* International Atomic Energy Agency training course: Decommissioning of Research Reactors and Other Small Nuclear Facilities; March 9–27, 1998.

Vitkus, T.J. *Draft—Technical Bases and Guidance for the Use of Ranked Set Sampling for Demonstrating Compliance with Radiological Release Criteria for Hard-to-Detect Radionuclides.* ORAU Letter to K. Snead, MARSSIM Work Group Chair; December 2011.

Walpole, R.E. and Myers, R.H. *Probability and Statistics for Engineers and Scientists.* New York, NY: Macmillan Publishing Company; 1985.

Solutions to Selected Questions and Problems

CHAPTER 1

1. The underlying survey design framework in MARSSIM is statistical hypothesis testing using nonparametric statistics (WRS and sign tests) to demonstrate compliance with dose- or risk-based release criteria.

2. The major technical advances in decommissioning surveys include (1) the use of dose- or risk-based release criteria and DCGLs; (2) the application of the DQO process; (3) scan sensitivity and the methodology to design final status surveys to deal with potential hot spots; (4) the application of nonparametric statistics in the hypothesis testing framework; and (5) the use of international guidance documents (i.e., ISO-7503) to address surface activity measurements and their technical defensibility.

3. Current issues in decommissioning include the NRC's decommissioning rulemaking that specifies a 25 mrem/year dose criterion for unrestricted release, dose modeling to determine DCGLs, guidance documents for decommissioning surveys (MARSSIM, NUREG-1505), and the proposed clearance rule for release of materials.

4. NUREG-1640 provides clearance survey guidance for both surficially and volumetrically contaminated materials; discusses both equipment reuse and recycle scenarios for steel, copper, aluminum, and concrete; and uses probabilistic methods for determining the parameter values in the modeling to obtain dose factors.

5. The inputs from the DQO process that affect the final status survey sample size include the LBGR (from the relative shift, Δ/σ) and the type I and type II decision errors.

6. The transition in release criteria from Regulatory Guide 1.86 to dose-based criteria (e.g., 25 mrem/year) resulted in

   - Smaller guideline (DCGLs) values for most alpha emitters

   - Larger guideline (DCGLs) values for most beta emitters

   - No more radionuclide groupings like those found in Regulatory Guide 1.86—each radionuclide DCGL is handled separately

   - More consideration in the DCGL value due to dose modeling (if the default DCGL value is deemed too conservative, the user can modify dose modeling parameters)

7. The null hypothesis in the MARSSIM is that the survey unit is "dirty" until proven "clean." In statistical notation, H_0: residual radioactivity in the survey unit exceeds the release criteria.

8. The general steps in the MARSSIM final status survey design include

   - Identify contaminant and DCGLs

   - Classify site areas

   - Identify background reference areas

   - Select instrumentation (both field and laboratory)

   - Select statistical test (WRS or sign test)

   - Select DQOs for determining sample size (relative shift, type I and II decision errors)

   - Assess needs for additional samples due to scan MDC

CHAPTER 2

1. The four phases of a decommissioning project are (1) assessment, (2) development, (3) operations, and (4) closeout.

2. The purpose of each of the five decommissioning surveys is as follows:

Scoping survey. To provide input to the characterization survey design, support the classification of site areas, and identify site contaminants and their variability.

Characterization survey. Primarily to determine the nature and extent of contamination; other characterization objectives may include evaluating decontamination techniques, determining information for site- specific parameters used in pathway modeling and evaluating remediation alternatives.

Remedial action support survey. To provide a real-time assessment of the effectiveness of decontamination efforts.

Final status survey. To demonstrate that residual radioactivity in each survey unit satisfies the release criteria.

Verification survey. To provide an independent, third-party overview of the decommissioning project by providing data to substantiate results of the final status survey.

3. ENTOMB may be an attractive alternative to DECON when no reasonable LLW disposal option exists. This can occur when there is literally no disposal option, or when the costs of disposal are exorbitantly high.

4. The differences between the final status survey and verification are as follows:

   - The final status survey demonstrates that release criteria have been satisfied while the verification survey provides an independent evaluation of the final status survey.

   - The scope of the final status survey is very large, perhaps costing $10 million for power reactor D&D final status surveys while verification is typically 1–3% of the final status survey cost.

   Owing to an obvious conflict of interest, the same organization cannot perform both the final status survey and the verification survey.

5. The remedial action support survey should be used to estimate the contaminant standard deviation in the survey units that have been remediated. The characterization data are obsolete; they no longer represent the radiological condition in those remediated survey units (those media determined to be contaminated by the characterization survey are likely in waste disposal containers).

6. The four conditions necessary for NRC restricted release include

   - The dose to average member of critical group with restrictions in place less than 25 mrem/year

   - The provisions made for legally enforceable institutional controls

   - If controls fail, dose is less than 100 mrem/year and, in rare situations, less than 500 mrem/year

   - The licensee has had public meetings with stakeholders concerning details of restricted release

7. The six principal elements of rubblization are

   - Removing all equipment from buildings

   - Limited decontamination of building surfaces

   - Demolishing the above-grade part of the structure into concrete rubble

   - Leaving the below-grade structure in place

   - Placing the rubble into the below-grade structure

   - Covering, regrading, and landscaping the site surface

   The possible exposure scenarios include

   - Concrete-leaching scenario

   - Resident farmer scenario

   - Excavation scenario

   - Intruder scenario

8. The three D&D program area deficiencies identified during verification include

   - Misapplication of surface activity release criteria, particularly for decay series

   - Improper classification of area

   - Improper instrument selection and determination of scan MDCs

9. The primary driver for the in-process decommissioning inspection was to reduce verification costs associated with the back-end verification surveys. The added benefit was that the in-process decommissioning approach is more consistent with the DQO process.

CHAPTER 3

3. Some of the uses of in situ gamma spectrometry during site characterization include

- Field identification of radionuclide contaminants

- Quick analysis of unprocessed or minimally processed soil samples

- Radionuclide concentrations in soil (if contaminant distributions are known)

- Negative data to confirm absence of contamination

4. Five characterization decisions that can be addressed by characterization surveys include: (1) determining the nature and extent of radiological contamination to decide whether the contamination exceeds release criteria, (2) assessing projected waste volumes generated from remediation activities, (3) evaluating remediation project alternatives (DECON vs. SAFSTOR), (4) providing modeling parameter inputs for RESRAD determination of DCGLs, and (5) evaluating cleanup technologies (e.g., scabbling vs. strippable paint).

CHAPTER 4

3. The gross activity $DCGL_W$ for processed uranium is 13.3 pCi/g (total uranium). The $DCGL_W$ for U-238 is 6.44 pCi/g.

4. The $DCGL_W$ for Th-232+C can take on two different values depending on how the "+C" is defined. Commonly, the "+C" designation indicates the DCGL value for the parent given that the decay progeny are present (as in Table 4.3). Thus, a radionuclide with the "+C" designation will typically have a smaller DCGL than the same radionuclide without this designation because the "+C" indicates that all of the progeny radionuclides are being taken into account. Sometimes, however, the "+C" designation refers to the DCGL for the entire decay series. In the case of Th-232 series, the DCGL for the entire series might be 11 pCi/g, which translates to 1.1 pCi/g for Th-232 alone. The critical point here is to understand the meaning of "+C."

CHAPTER 5

1. The $DCGL_W$ for this particular radionuclide is 735 pCi/g.

2. Four major differences between RESRAD and D&D modeling codes include (1) the manner in which doses are determined, RESRAD reports instantaneous dose rates, while D&D reports the dose averaged over 1 year; (2) RESRAD's capability to get area factors; (3) handling of the groundwater pathway; and (4) general conservatism of the D&D model (it is a screening model).

3. The area factors based on the RESRAD modeling output are as follows:

Area (m²)	Area Factor
2000	1
300	1.07
100	2.30
40	10.3
10	17.1
1	127

CHAPTER 6

4. The gross activity DCGL for the reactor mixture is 2760 dpm/100 cm². Following the slight increase in the Pu-239 fraction, the gross activity DCGL for the reactor mixture is 1160 dpm/100 cm². Even a smaller increase in the Pu-239 fraction would still have a noticeable effect on the gross activity DCGL, and is the reason why separate alpha measurements for transuranics may be appropriate.

5. The modified DCGL for Co-60 is 3.0 pCi/g.

7. The first part of the problem asks for the DCGL for U-238 when it is used as a surrogate for only the uranium isotopes. It is a good idea to mention that the DCGLs for U-238 and U-235 include their short-lived progeny. Equation (I-14) on the MARSSIM page I-32 can be used to calculate the modified DCGL for U-238, once the ratios of each radionuclide to U-238 are known:

$$DCGL_{U\text{-}238,mod} = \cfrac{1}{\left(\cfrac{1}{D_1} + \cfrac{R_2}{D_2} + \cdots + \cfrac{R_n}{D_n}\right)}$$

where D_1 is the $DCGL_W$ for U-238 by itself and D_2 is the $DCGL_W$ for the second radionuclide that is being inferred by U-238. R_2 is the ratio of concentration of the second radionuclide to that of U-238. Therefore, $DCGL_{U\text{-}238,mod}$ can be calculated using the radionuclide mixture defined for this facility and using the individual DCGLs for each uranium isotope. So

$D_1 = 8$ pCi/g U-238 $R_2 = $ U-235/U-238 $= 0.045$

$D_2 = 5$ pCi/g U-235 $R_3 = $ U-234/U-238 $= 1.02$

$D_3 = 7$ pCi/g U-234

and substituting these values into the equation yields

$$DCGL_{U\text{-}238,mod} = \cfrac{1}{\left(\cfrac{1}{8} + \cfrac{0.045}{5} + \cfrac{1.02}{7}\right)} = 3.58 \text{ pCi/g.}$$

Thus, the $DCGL_W$ modified for U-238 is 3.58 pCi/g, down by more than 50% from its individual value of 8 pCi/g. If the unity rule is used for the final status survey design, both U-238 and Tc-99 are measured, and U-238 is compared to a DCGL of 3.58 pCi/g.

Now, the second part of the problem asks for the U-238 $DCGL_W$ modified to account for uranium and Tc-99. There are two ways to calculate this modified U-238 DCGL using the above equation. The first method is to construct the source term from the beginning, while the second is to make use of the intermediate solution calculated above for U-238 modified for the uranium isotopes. We will show both methods, starting with the first.

The source term starting from the beginning must reflect the fractional amount of uranium and Tc-99, so each of the uranium isotopic fractions must be multiplied by 30%. Thus, the source term radionuclide fractions are

Tc-99	0.70
U-238	0.1455
U-234	0.1479
U-235	6.6E–3.

Similar to the first calculation, using the radionuclide mixture defined for this facility and using the individual DCGLs for each uranium isotope and Tc-99, we have

$D_1 = 8$ pCi/g U-238 $R_2 = $ U-235/U-238 $= 0.045$

$D_2 = 5$ pCi/g U-235 $R_3 = $ U-234/U-238 $= 1.02$

$D_3 = 7$ pCi/g U-234 $R_4 = $ Tc-99/U-238 $= 4.81$

$D_3 = 20$ pCi/g Tc-99

and again substituting these values into the equation yields

$$\text{DCGL}_{\text{U-238,mod}} = \frac{1}{\left(\dfrac{1}{8} + \dfrac{0.045}{5} + \dfrac{1.02}{7} + \dfrac{4.81}{20}\right)} = 1.92 \text{ pCi/g}.$$

Thus, the U-238 DCGL$_W$ modified to account for uranium and Tc-99 is 1.92 pCi/g, roughly 25% of its individual DCGL value of 8 pCi/g. The second approach to calculating the U-238 DCGL$_W$ modified to account for uranium and Tc-99 makes use of the intermediate calculation of the U-238 DCGL$_W$ modified for only the uranium isotopes. The same equation is used as before, and it is important to note that the ratio of Tc-99 to U-238 is used, not the ratio of Tc-99 to uranium.

$D_1 = 3.58$ pCi/g U-238 $R_2 = $ Tc-99/U-238 $= 4.81$

$D_2 = 20$ pCi/g Tc-99

and again substituting these values into the equation yields the same value as before:

$$\text{DCGL}_{\text{U-238,mod}} = \frac{1}{\left(\dfrac{1}{3.58} + \dfrac{4.81}{20}\right)} = 1.92 \text{ pCi/g}.$$

The final question to consider in this problem is whether the U-238 $DCGL_W$ should be modified to include all contaminants or just the uranium isotopes. Because the U-238 $DCGL_W$ modified for uranium and Tc-99 (1.92 pCi/g) is still easily detectable, it is recommended that this approach to final status survey design is used. Of course, it is critical to the design that the Tc-99 to uranium ratio is well characterized (consistent), and that periodic samples are analyzed for Tc-99 and uranium to confirm this fact.

CHAPTER 7

1. The basic statistical quantities for the data set include a mean of 249.4 counts, a median of 251.5 counts, and a standard deviation of 11.1 counts. Whenever a mean background is determined, it is likely that the sign test would be performed, since the sign test for surface activity assessment requires that the appropriate mean background is subtracted from each gross measurement. When using the WRS test, the gross measurements are compared to the background measurements; the mean background is not subtracted in this case.

2. The mean Cs-137 concentration in this reference area is 0.62 pCi/g, with a standard deviation 0.26 pCi/g. The variability in these data is probably not sufficient to warrant more than one reference area. An additional piece of information that would assist in this decision would be whether the elevated Cs-137 levels all came from a similar area within the reference area, such as a wooded area. If this was the case, then an additional investigation into differences in Cs-137 levels between wooded and open land areas may be warranted.

   Concerning the decision to add a reference area in context of the DCGL, the greater the DCGL, the greater the tolerance for background variations. However, the background reference areas must be representative of the survey unit being compared.

CHAPTER 8

1. The following detector(s) might be used for the following radiological situations at D&D sites:

- Scans inside a small floor drain opening to a depth of 20 cm for a beta emitter—GM detector—assuming that it fits into the floor drain

- Alpha scans of a large floor area—floor monitor operated in the alpha- only mode

- Beta measurements of surface activity in the presence of an alpha emitter—gas proportional detector operated in the beta-only mode (using an alpha blocker)

- Surface activity measurements in a survey unit contaminated with Co-60, Ni-63, Cs-137, and Am-241—separate measurements of alpha and beta activity using gas proportional detectors in the alpha-only and beta-only modes, respectively

- Surface activity assessment in a survey unit contaminated with H-3, C-14, and Co-60—gas proportional or plastic scintillator, appropriately weighted for the beta mixture (H-3 efficiency is zero)

2. To ensure that the increased cable capacitance does not change the overall response of the detector, it may be necessary to increase the operating voltage or lower the threshold (input sensitivity). Gas proportional and scintillation detectors are more impacted by increases in cable length as compared to GM detectors due to their smaller pulse amplitudes.

5. The instrument efficiency for the GM detector is calculated as follows:

$$\epsilon_i = \frac{\dfrac{1082}{2} - 60}{(20/180)\,15{,}000} = 0.288$$

6. The primary advantage of using the Bicron microrem meter as compared to the pressurized ionization chamber (PIC) is that it is more affordable and much more transportable in the field. The primary advantage of using the PIC is that it is the industry standard for measuring exposure rates, and as such has very flat energy response characteristics.

7. See Table 16.2 for help with determining the instrument efficiency for the thorium series.

CHAPTER 9

1. *Critical level* is the net count rate in a zero-mean count distribution having a probability, given by the selected type I decision error (α), of being exceeded. The critical level is used as a decision tool to decide when activity is present in a sample.

   Detection limit is the number of mean net counts obtained from samples for which the observed net counts are almost always certain to exceed the critical level. The detection limit is positioned far enough above zero so that there is a probability, given by the selected type II decision error (β), that the L_D will result in a signal less than L_C.
   It is generally unacceptable to position the detection limit at the critical level because there would be a 50% probability of committing a type II decision error (β) (i.e., a 50% probability of falsely concluding that activity is not present). In other words, it would be inappropriate to define the detection limit as that activity (net counts) that is detectable only 50% of the time.

2. The MDC is calculated using the following parameters:

   Background: 280 counts in 1 min

   Instrument efficiency (ϵ_i): 0.34

   Surface efficiency (ϵ_s): 0.5

   Physical probe area: 125 cm²

   $$MDC = \frac{2 + 4.65\sqrt{280}}{(0.34)(0.5)(125/100)(1)} = 380 \text{ dpm}/100 \text{ cm}^2$$

3. The problem states that a gas proportional detector (126 cm²) is operated in the alpha-only mode to make measurements of surface activity. The count time is 2 min and the background level is 1 count per minute. The MDC is claimed to be 100 dpm/100 cm², and we assume the MDC equation for paired measurements applies.

$$MDC = \frac{3 + 4.65\sqrt{(2)(1)}}{(\epsilon_{tot})(126/100)(2)} = 100 \text{ dpm}/100 \text{ cm}^2$$

The MDC equation can be solved for total efficiency, which results in $\epsilon_{tot} = 0.038$. While this may seem a bit low, it is certainly a realistic alpha efficiency for porous surfaces and surfaces with overlaying materials (e.g., dust, oil, and paint).

4. The health physicist plans to use a GM detector (the probe area is 20 cm²) for surface activity measurements and wants to determine if a 1-min count time will provide an MDC less than 50% of the DCGL$_W$. The given information includes the background count for a 1-min count time equals 72 counts, and the instrument efficiencies (ϵ_i) for C-14 and S-35 are 0.16 and 0.17, respectively. The DCGLs for each radionuclide are as follows:

    C-14   50,000 dpm/100 cm²

    S-35   78,000 dpm/100 cm²

    First, while the problem states the contaminants of concern as both C-14 and S-35, the survey design need not consider S-35 due to its short half-life (87 days). The MDC for paired measurements can be calculated from the given information:

$$MDC = \frac{3 + 4.65\sqrt{(72)(1)}}{(0.16)(0.25)(1)(20/100)} = 5300 \text{ dpm}/100 \text{ cm}^2$$

    Therefore, the HP can confidently make 1-min measurements and the MDC is less than 50% of the DCGL$_W$. (Note that the surface efficiency (ϵ_s) is based on the default value using the ISO-7503 of 0.25 for low-energy betas (based on the C-14 beta energy).)

5. Using the same parameters as for problem (4), the shortest count time possible that can still achieve a sufficient MDC (i.e., less than 50% of DCGL$_W$) can be determined using the paired measurements MDC equation:

$$MDC = \frac{3 + 4.65\sqrt{(72)(T)}}{(0.16)(0.25)(T)(20/100)} = 25,000 \text{ dpm}/100 \text{ cm}^2$$

The MDC is set equal to 50% of the DCGL$_W$ and solved for T. However, the resulting equation for T, once algebraically simplified, cannot be solved analytically:

$$T - 0.198\sqrt{T} - 0.015 = 0$$

This equation can be solved for T using numerical methods. Perhaps more importantly, the values of T need to consider those which are possible, given the survey instrument used to make surface activity measurements. That is, the shortest "practical" count time must consider not only the value of the MDC obtained but also the available times that counts can be scaled on the instrument. Typical count times include 5 min 2 min, 1 min, 0.5 min, and 0.1 min. Thus, another approach to the solution is to substitute count times less than 1 min into the MDC equation and determine if the resulting MDC is less than 25,000 dpm/100 cm².

For $T = 0.5$ min MDC = 7725 dpm/100 cm²

For $T = 0.1$ min MDC = 19,350 dpm/100 cm²

Therefore, a count time as low as 0.1 min results in a sufficient MDC.

6. A MARSSIM user plans to use a xenon detector for surface activity measurements of I-129 on sealed concrete surfaces using 5-min background measurements and 1-min surface activity measurements in the survey unit. The DCGL$_W$ for I-129 is 1000 dpm/100 cm². The following xenon detector parameters are provided:

Background: 420 counts in 1 min

Total efficiency (ϵ_{tot}): 0.12

Physical probe area: 100 cm²

The MDC using the general equation for different background and sample count times can be calculated:

$$\text{MDC} = \frac{3 + 4.29\sqrt{(420)(1)(1 + 1/5)}}{(0.12)(100/100)(1)} = 640 \text{ dpm/100 cm}^2$$

The MDC using 5-min background counts and 1-min sample counts does not result in an MDC that is the desired sensitivity (i.e., 50% of $DCGL_W$); however, it may be close enough to petition the regulator.

The MDC using paired, 1-min measurements is 820 dpm/100 cm².

7. The impacts for counting background longer than survey unit measurements when using the MARSSIM surface activity assessment are more severe for the WRS test. Simply stated, the WRS test cannot be used because it would require the background distribution—obtained with 5-min counts—to be compared with the 1-min count survey unit distribution. This would severely bias any conclusions obtained using the WRS test.

The sign test for surface activity measurements can still function satisfactorily when the background and sample count times differ. The reason for this is because the background count rate is subtracted from the gross count rate and there is no requirement for the count times to be equal. The sign test is then performed on the calculated surface activities in the survey unit.

11. For the selected performance levels of 95% true-positive rate and 25% false-positive rate, $d' = 2.32$. The MDCR can be calculated as

$$b_i = (1350 \text{ cpm})(2 \text{ s})(1 \text{ min}/60 \text{ s}) = 45 \text{ counts}$$
$$s_i = (2.32)(45)^{1/2} = 15.6 \text{ counts}$$

and

$$MDCR = (15.6 \text{ counts})[(60 \text{ s/min})/(2 \text{ s})] = 467 \text{ cpm}$$

Using a surveyor efficiency of 0.5, and assuming instrument and surface efficiencies of 0.24 (see Table 9.2 for Tc-99) and 0.25 (from ISO-7503 recommendations), respectively, the scan MDC is calculated as

$$\text{Scan MDC} = \frac{467}{\sqrt{0.5(0.24)(0.25)}} = 11,000 \text{ dpm}/100 \text{ cm}^2$$

12. The scan MDC is 6.05 pCi/g for a 1.5″ × 1.25′ NaI scintillation detector for the specified mixture.

CHAPTER 10

1. The total efficiency for the Fe-55 and Co-60 mixture is calculated by multiplying each radionuclide's relative fraction by its instrument and surface efficiency and summing the product for both radionuclides. Both radionuclides have surface efficiencies of 0.25 according to ISO-7503. The total efficiency is calculated as

   Fe-55 $(0.64) \times (0.03) \times (0.25) = 4.8E{-}3$

   Co-60 $(0.36) \times (0.38) \times (0.25) = 3.42E{-}2$

   Total efficiency $= 0.039$

2. The total efficiency for the 40% enriched uranium (4% EU) and 60% Tc-99 must appropriately weight the instrument and surface efficiencies. The instrument efficiency for Tc-99 is 0.39 (Table 8.1), while the case study in chapter 10 shows that the total efficiency for 4% EU is 0.163. Therefore, the total efficiency is calculated as

   Total efficiency $= (0.6) \times (0.39) \times (0.25) + (0.4) \times (0.163) = 0.124$

CHAPTER 11

2. "Hot-spot sensitive" pathways and parameters have area factors that are more conservative (i.e., smaller) than those nonsensitive pathways/parameters whose area factors scale directly with the size of the contaminated area. For instance, the external radiation pathway is hot-spot sensitive—area factors for this pathway are smaller considering the pathway's direct contribution to receptor dose. Conversely, for environmental pathways that are not hot-spot sensitive, the hot-spot dose depends only on the inventory (source term) in the contaminated area, not the size of the contaminated area.

4. While the comparison of the robust t posterior distribution and FSS data indicates comparable distributions at the 95th percentile, the 99th percentile values are significantly different. That is, the robust t posterior distribution has a much larger 99th percentile due to the shape of the distribution (skewed to the right). Based on these posterior distribution results, the hot-spot potential is relatively high in this survey unit; additional sampling for hot spots is warranted.

CHAPTER 12

1. Two components of the standard deviation of the contamination level in the survey unit are the spatial variability and the measurement uncertainty. The measurement uncertainty can be reduced by using more precise measurements or by counting the sample for longer; the spatial variability can be reduced by selecting survey units that are more uniform in their contamination levels. Note that simply collecting more samples does not reduce the spatial variability, rather, it reduces the uncertainty in the mean (i.e., more samples reduce the standard error of the mean, but not the standard deviation itself).

4. A type I error refers to an incorrect rejection of the null hypothesis, while a type II error refers to an incorrect nonrejection of the null hypothesis. It is not a good idea to view the type I error as a "false positive" because the meaning of a type I depends on how the null hypothesis is stated, and it may not make sense in some situations to think of the type I error as a false positive. For example, if H_0 is stated "there is no radioactivity in the sample," then the type I error (incorrect rejection) means that the surveyor falsely concludes that there is radioactivity in the sample—that is, a false positive. However, if H_0 is stated "there is radioactivity in the sample," then the type I error (incorrect rejection) means that the surveyor falsely concludes that there is no radioactivity in the sample—that is, a false negative.

5. For sample size $n = 18$ and a type I error of 0.05, the critical value is 12. Because the test statistic ($S+ = 14$) exceeds the critical value, the null hypothesis can be rejected. If the type I error is reduced to 0.025, the critical value becomes 13. Since the test statistic still exceeds the critical value, the null hypothesis is rejected even at this more restrictive type I error.

CHAPTER 13

1. M is the maximum of the set $(s/\bar{x})_\alpha, (s/\bar{x})_{\beta\gamma}, (s/\bar{x})_\gamma$, and 0.82, and, for the given data, M is based on the alpha surface activity measurements:

$(s/\bar{x})_\alpha = 620/430 = 1.44$, and $n = 45(1.44)^2 = 94$ measurements

This is consistent with the survey block guidance because the maximum number of survey blocks in a 100 m² survey unit is 100 (based on survey blocks 1 m on a side), and the minimum number of survey blocks is 11 (based on survey blocks 3 m on a side).

4. The first step in the solution is to determine the gross activity DCGL for 4% enriched uranium:

$$\text{Gross activity DCGL} = \frac{1}{\left(\dfrac{0.809}{2000} + \dfrac{0.150}{2200} + \dfrac{0.041}{2000}\right)} = 2030 \text{ dpm/100 cm}^2$$

where the fractional amounts of 4% enriched uranium are obtained from Table 8.2.

The gross activity DCGL can be converted to net cpm by multiplying by the two efficiency terms and the probe area (assumed to be 126 cm²) factor, which yields 283 cpm.

The LBGR is set at the expected median concentration in the survey unit: 582–407 or 175 cpm.

The relative shift is given by $(283 - 175)/167$ or 0.647. Rounding this down to 0.6, and using a type I error of 0.05 and type II error of 0.10, the sign test sample size is 52 direct measurements of surface activity.

5. The sample size for the WRS test is based on an LBGR set at the median concentration of Ra-226 of $1.9 - 1.1$ pCi/g, or 0.8 pCi/g. The relative shift is given by $(1.2 - 0.8)/0.6$ or 0.66. Rounding the relative shift down to 0.6, and using type I and type II errors of 0.05, the sample size is 81 soil samples. (Note: for real D&D sites, it is recommended that site-specific modeling parameters are used in an attempt to increase the Ra-226 DCGL. If the DCGL can be increased to 2 pCi/g, the sample size is reduced to 13 samples.)

Based on the survey unit size of 2200 m², the average area bounded by samples (a') is 2200/81 or 27.2 m². The area factor for Ra-226 for this area can be determined from the MARSSIM (Table 5.6). Fitting the area factor data to a power function (area factor $= 54.59 \times (\text{area})^{-0.8437}$), the area factor for 27.2 m² is 3.36. Multiplying this area factor by the DCGL$_W$ (1.2 pCi/g) yields the required scan MDC of 4.0 pCi/g. However, the actual scan MDC (4.5 pCi/g) is not sensitive enough to detect the required scan MDC. It is therefore necessary to calculate the area factor that corresponds to the actual scan MDC: 4.5 pCi/g divided by 1.2 pCi/g = 3.75. Thus, we must determine the area that yields an area factor of 3.75. Using the same equation given above (but solving for the area this time), the area that yields an area factor of 3.75 is 23.9 m². The total sample size in the survey unit is calculated by 2200 m²/23.9 m², or 92 samples. Again, for real D&D sites, the modeling code would be run for site-specific inputs to increase the Ra-226 DCGL$_W$.

CHAPTER 14

1. See Table S.1.

   The critical value for a type I error of 0.10 is 16. Since the test statistic (S+) is greater than the critical value, the null hypothesis is rejected and the survey unit demonstrates compliance with the release criteria.

3. Using the Co-60 data in Table 14.2, the Shapiro–Wilk normality test is performed for a type I error of 0.05 (see Table S.2).
   The sample standard deviation (s) is 1.67, and the W test statistic is calculated as

$$W = \frac{(5.87)^2}{(15 - 1)(1.67)^2} = 0.882$$

   The null hypothesis of normality is rejected at the a level of significance if $W < \omega_\alpha$ the critical value. The statistical table provides a critical value for $\omega_{0.05}$ of 0.881. Since 0.882 ≮ 0.881 (barely), the null hypothesis is not rejected, and it is concluded that the data are normally distributed.

CHAPTER 15

2. MDC vs. MQC is simply the difference between the concentration that is deemed "detectable" (MDC) vs. the concentration that is deemed "quantifiable" (MQC)—recognizing that it takes less activity to be merely *detected* as compared to that needed to be *quantifiable*.

TABLE S.1   Soil Sample Results from Power Reactor Waste Disposal Facility

Sample Number	Co-60 (pCi/g)	Cs-134 (pCi/g)	Cs-137 (pCi/g)	Eu-152 (pCi/g)	Sum-of-Ratios
1	0.7	0.3	1.5	1.7	0.569
2	1.4	0.4	0.4	1.6	0.659
3	0.3	1.3	1	0.8	0.490
4	0.2	0.3	2.7	0.4	0.397
5	0.6	0.4	2.2	1	0.543
6	0.2	0.4	1.9	0.7	0.376
7	2.6	0.1	2	1.5	1.056
8	2	0.1	2	0.7	0.806
9	0.9	0.1	2	2.2	0.689
10	1.1	0.9	1.6	0.8	0.685
11	1.8	0.7	1.4	1.6	0.908
12	0.3	1.1	0.7	1.8	0.542
13	3.8	0.7	0.3	1.2	1.288
14	0.7	0.4	1.7	1.1	0.535
15	0.7	0.2	1.8	0.6	0.452
16	0.1	0.2	3	2.2	0.587
17	0.5	0.6	2.5	2.1	0.705
18	2.8	0.5	0.9	1.3	1.056
19	2.2	1.4	1.3	2.3	1.207
20	0.7	0.6	1.7	0.6	0.513
21	0.2	0.5	1.9	1	0.428
22	0.9	0.3	1.5	1.4	0.587
23	1.1	0.5	1.8	0.6	0.610
24	2	0.8	1.9	1.7	1.035
25	0.1	1	2	0.6	0.453
26	0.4	1	1.4	1.6	0.592
			S+	21	

Specifically, the MDC is the minimum activity concentration that can be detected most of the time ("most" usually taken to be 95%). The MQC on the other hand is the smallest concentration that can be measured with a specified relative standard deviation, often stated as the concentration at which the relative measurement uncertainty is 10% (a fairly accurate measurement). In practice, the MDC is approximately 3–5 times the standard deviation of the blank (σB), whereas the MQC is 10 times the standard deviation of the blank. Therefore, the MQC is usually 2 or 3 times the MDC, that is, we need 2 or 3 times that which is detectable to be considered quantifiable.

TABLE S.2   Shapiro–Wilk Normality Test for Co-60 Soil Samples

Rank (Low to High)	Rank (High to Low)	Difference $(x_{n-I+1} - x_i)$	a_i (for $k = 7$)	$a_i \times$ Difference
0.4	6.6	6.2	0.5150	3.193
0.7	4.8	4.1	0.3306	1.355
0.9	3.9	3.0	0.2495	0.748
1.2	2.9	1.7	0.1878	0.319
1.4	2.6	1.2	0.1353	0.162
1.5	2.3	0.8	0.0880	7.04E−2
1.7	2.2	0.5	0.0433	2.16E−2
1.8	1.8	0		
2.2	1.7	−0.5	Sum = 5.87 (also equal to b)	
2.3	1.5	−0.8		
2.6	1.4	−1.2		
2.9	1.2	−1.7		
3.9	0.9	−3.0		
4.8	0.7	−4.1		
6.6	0.4	−6.2		

3. Three possible clearance survey approaches for scrap metal from a nuclear reactor D&D project include: (1) MARSSIM-type clearance approach where a statistical sample is selected for direct measurements using hand-held survey instruments. Scanning would also be performed with the same instruments to identify hot spots. No sizing or disassembly would be expected. (2) Scan-only clearance approach

TABLE S.3   Alpha-to-Beta Ratio for Th-232 Series

Radionuclide	Equilibrium Fraction	Radiation	Total Result
Th-232	1	1 alpha	1 alpha
Ra-228	0.6	1 beta	0.6 beta
Ac-228	0.6	1 beta	0.6 beta
Th-228	0.8×0.6	1 alpha	0.48 alpha
Ra-224	0.8×0.6	1 alpha	0.48 alpha
Rn-220	0.8×0.6	1 alpha	0.48 alpha
Po-216	0.8×0.6	1 alpha	0.48 alpha
Pb-212	0.8×0.6	1 beta	0.48 beta
Bi-212 36%	0.8×0.6	1 alpha	0.173 alpha
Bi-212 64%	0.8×0.6	1 beta	0.307 beta
Po-212 64%	0.8×0.6	1 alpha	0.307 alpha
Tl-208 36%	0.8×0.6	1 beta	0.173 beta

using a CSM equipped with gas proportional detectors. The scrap metal would likely have to be size reduced so that it fits on the conveyor belt. (3) *In toto,* clearance approach using an *in situ* gamma spectrometer is carried out. In this case, the scrap metal may have to be size reduced to fit a calibrated geometry (e.g., pallet). The DQO process should be used to assess the optimal clearance survey design.

CHAPTER 16

2. The alpha-to-beta ratio for Th-232 in secular equilibrium is six alphas to four betas, for an alpha-to-beta ratio of 1.5. If Ra-228 is 60% of Th-232, and Th-228 is 80% of Ra-228, the alpha-to-beta ratio can be calculated. The assumptions are that Ac-228 is in equilibrium with its parent (Ra-228) and that all of the Th-228 progeny are in equilibrium with Th-228 (see Table S.3). The tally is 3.4 alpha to 2.16 beta, for an alpha-to-beta ratio of 1.57. The alpha-to-beta ratio increased slightly based on this particular state of equilibrium within the Th-232 series.

CHAPTER 17

1. The content of a final status survey report should include the following:

Background for performing the D&D project

Site description

HSA results

   Results of scoping and characterization surveys

   Release criteria—DCGLs and area factors

   Final status survey procedures—field and laboratory measurements (MDCs)

Final status survey design

   Sample sizes from statistical test; need for additional samples in class 1 areas

   Selection of reference areas

Integrated survey strategy—scan coverage, use of investigation levels

Strategies for multiple radionuclides

Grid systems, sample allocation within survey units

Final status survey results and interpretation

Results of statistical tests

Results of investigation levels that were triggered

Conclusions relative to release criteria for each survey unit

2. Using the data given in the problem, the net count rate (R_s) and its uncertainty is calculated as $18\,624 \pm 284$ cpm (2σ). The uncertainty was determined by propagating the error in both the gross counts and the background counting rate. The activity and its uncertainty at the 95% confidence level is determined to be $14\,400 \pm 3820$ (2σ).

CHAPTER 18

1. The D'Agostino test for normality can be used when the number of data points exceeds 50.

Appendix A: Radionuclide and Natural Decay Series Characteristics

TABLE A.1 Thorium Series

A	Element	Z	Half-Life	Radiation Type	Radiation Energy (keV)	End-Point Energy (keV)	Radiation Intensity (%)	Other Radiations
232	Th	90	14.05E9 years	α	3947.2		21.7	α, E, X, γ
				α	4012.3		78.2	
				E CE L	44.573			15.8
228	Ra	88	5.75 years	β⁻ TOT	7.2		100	β⁻, E, X, γ
				E CE M	1.668		37.50	
228	Ac	89	6.15 h	β⁻ TOT	376.7		93	β⁻, E, X, γ
				β⁻	382.3	1158	29.9	
				β⁻	606.9	1731	11.66	
				E CE L	37.294		54	
				E CE M	52.584		14.6	
				γ	338.320		11.27	
				γ	911.204		25.8	
				γ	968.971		15.8	
228	Th	90	1.912 years	α	5340.36		27.2	α, E, X, γ
				α	5423.15		72.2	
				E CE L	65.136		19.3	
224	Ra	88	3.66 days	α	5685.37		94.92	α, E, X, γ
220	Rn	86	55.6 s	α	6288.08		99.886	α, γ
216	Po	84	0.145 s	α	6778.3		99.9981	α, γ
212	Pb	82	10.64 h	β⁻	94.8	335	82.5	E, X, γ
				β⁻ TOT	101.7		100.0	
				β⁻	173.1	574	12.3	
				E AU L	8.150		20.8	
				E CE K	148.1061		32.0	

		Z	Half-life	Radiation	Energy (keV)		%	Emissions
212	Bi	83	25.0 min	XL	10.80		15.1	
				X Kα₂	74.81		10.4	
				X Kα₁	77.11		17.5	
				γ	238.6320		43.3	α
			60.55 min	α	6295.98		26.0	
				α	6335.08		35.0	
				α	6050.78		25.13	α, β, E, X, γ
				α	6089.88		9.75	
				β⁻ TOT	769.6		64.06	
				β⁻	832.5	2248	55.46	
				E AU L	7.780		12.0	
				E CE L	24.510		19.8	
212	Po	84	17.1 ns	α	10180		42.00	
(64.07% from ²¹²Po)			0.299 µs	α	8784.86		100.0	α, γ
			45.1 s	α	11 650		96.83	
208	Tl	81	3.053 min	β⁻	439.6	1286	24.5	β⁻, E, X, γ
(35.93% from ²¹²Po)				β⁻	533.3	1519	21.8	
				β⁻ TOT	557.2		99.4	
				β⁻	647.4	1796	48.7	
				γ	510.77		22.6	
				γ	583.1910		84.5	
				γ	860.564		12.42	
				γ	2614.533		99.16	
208	Pb	82	Stable					

Source: Nuclear data from NuDat, National Nuclear Data Center, Brookhaven National Laboratory, February 23, 1999.

Note: E = atomic electrons; CE = conversion electrons; AU = Auger electrons; TOT = average energy of total beta or positron emitted spectrum; other radiations = primarily low-energy and/or low-yield emissions.

TABLE A.2 Uranium Series

A	Element	Z	Half-Life	Radiation Type	Radiation Energy (keV)	End-Point Energy (keV)	Radiation Intensity (%)	Other Radiations
238	U	92	4.468E9 years	α	4151		21	α, E, X, γ
				α	4198		79	
				E CE L	29.08		15.3	
234	Th	90	24.10 days	β-	27.0	104	19.2	β-, E, X, γ
				β-TOT	44.9		100	
				β-	52.7	196	70.3	
				E CE L	71.275		11.9	
				γ	63.290		4.8	
				γ	92.380		2.8	
				γ	92.800		2.8	
234 m	Pa	91	1.17 min	β-TOT	813		100.20	β-, E, X, γ
				β-	821	2269	98.20	
				γ	1001.03		0.837	
234	Pa	91	6.75 h	β-	137	472	33	β-, E, X, γ
(0.16% from Pa-234 m)				β-	137	472	12.4	
				β-TOT	180		112	
				β-	194	642	19.4	
				E AU L	9.890		72	
				E CE L	21.733		66	
				E CE L	78.103		32	
				E CE K	111.64		10.6	
				X L	13.60		91	

A	Element	Z	Half-life	Radiation	Energy		Intensity	Radiations
234	U	92	245,700 years	X Kα₂	94.6650		11.0	
				X Kα₁	98.4390		17.8	
				γ	131.300		18.0	
				γ	946.00		13.4	
				α	4722.4		28.42	α, E, X, γ
				α	4774.6		71.38	
				E AU L	9.480		10.1	
				E CE L	32.728		20.9	
				X L	13.00		10.9	
230	Th	90	75,380 years	α	4620.5		23.40	α, E, X, γ
				α	4687.0		76.3	
				E CE L	48.4353		17.0	
				X L	12.30		8.6	
				γ	67.6720		0.38	
226	Ra	88	1600 years	α	4784.34		94.45	α, E, X, γ
222	Rn	86	3.8235 days	α	5489.5		99.920	α, γ
218	Po	84	3.10 min	α	6002.35		99.9789	α
214	Pb	82	26.8 min	β⁻	207	671	48.9	β⁻, E, X, γ
(99.98% from ²¹⁸Po)				β⁻-TOT	218	728	101.4	
				β⁻	227		42.2	
				E AU L	8.150		19.7	
				E CE L	36.8400		12	
				X L	10.80		14.2	
				X Kα₁	77.11		10.8	
				γ	295.2240		19.30	

continued

TABLE A.2 (continued) Uranium Series

A	Element	Z	Half-Life	Radiation Type	Radiation Energy (keV)	End-Point Energy (keV)	Radiation Intensity (%)	Other Radiations
218	At	85	1.5 s	γ	351.9320		37.6	
(0.02% from 218Po)				α	6693		89.91	α
214	Bi	83	19.9 min	β-	526	1507	17.02	α, β-, E, X, γ
				β-	540	1542	17.8	
				β--TOT	642		99.8	
				β-	1270	3272	18.2	
				γ	609.312		46.1	
				γ	1120.287		15.10	
				γ	1764.494		15.40	
214	Po	84	164.3 μs	α	7686.82		99.9895	α
(99.979% from 214Bi)								
210	Tl	81	1.30 min	β-	677	1868	24	
(0.021% from 214Bi)				β-	746	2032	10	
				β-	880	2421	10	
				β--TOT	1184		103	
				E AU L	7.970		30	
				E CE L	67		21	
				E CE L	81		20	
				X L	10.60		21	
				γ	296		79	
				γ	799.6		98.96	

A		Z	Half-life	Radiation	Energy (keV)		Intensity (%)	Radiations
210	Pb	82	22.3 years	γ	1316		21	α, β
				β⁻	4.16	16.6	84	
				β⁻ TOT	6.08		100	
				β⁻	16.16	63.1	16	
				E AU L	8.150		35	
				E CE L	30.1515		60.3	
				E CE M	42.5399		14.3	
				X L	10.80		25	
				γ	46.5		4.05	
210	Bi	83	5.013 days	β⁻	389.0	1161.5	100.0	α
210	Po	84	138.376 days	α	5304.33		100.0	α, E, X, γ
(~100% from ^{210}Bi)								
206	Tl	81	3.74 min	E AU L	7.780		29	E, X, γ
(0.000 13% from ^{210}Bi)				E CE K	130.87		10.5	
				γ	216.40		74	
				γ	265.70		86.00	
				γ	453.30		93	
				γ	457.2		22	
				γ	686.50		90	
				γ	1021.50		69	
206	Pb	82	4.199 min	β⁻ TOT	538.0		100.00	β⁻, E, X, γ
206	Pb	82	Stable					

Source: Nuclear data from NuDat, National Nuclear Data Center, Brookhaven National Laboratory, February 23, 1999.

Note: E = atomic electrons; CE = conversion electrons; AU = Auger electrons; TOT = average energy of total beta or positron emitted spectrum; other radiations = primarily low-energy and/or low-yield emissions.

TABLE A.3 Beta Emitters

A	Element	Z	Half-Life	Radiation Type	Radiation Energy (keV)	End-Point Energy (keV)	Radiation Intensity (%)	Other Radiations
3	H	1	12.33 years	β-	5.69	18.5899	100.0	
14	C	6	5730 years	β-	49.47	156.471	100.0	
32	P	15	14.26 days	β-	694.9	1710.2	100.0	
35	S	16	87.51 days	β-	48.63	166.84	100.0	
36	Cl	17	301,000 years	β-	251.20	709.2	98.16	β+, E, X, γ-AN
45	Ca	20	162.61 days	β-	77.2	256.8	100.00	E, X, γ
63	Ni	28	100.1 years	β-	17.425	66.946	100.0	
89	Sr	38	50.53 days	β-	584.6	1495.1	99.99036	E, X, γ
				β- TOT	584.6		100.000 00	
90	Sr	38	28.74 years	β-	195.8	546.0	100.0	
93	Nb	41	16.13 years	E AU L	2.150		79	E, X, γ
				E CE K	11.784		15.0	
106	Ru	44	373.59 days	β-	10.03	39.40	100.0	β-, E, X, γ
147	Pm	61	2.6234 years	β-	61.96	224.6	99.9940	
				β- TOT	61.96		99.9997	
210	Bi	83	5.013 days	β-	389.0	1161.5	100.0	α
228	Ra	88	5.75 years	β-	3.21	12.8	30	β-, E, X, γ
				β	6.48	25.7	20	
				β- TOT	7.2		100	
				β-	9.94	39.2	40.00	
				β-	10.04	39.6	10.00	
				E CE M	1.668		37.50	
241	Pu	94	14.35 years	β-	5.23	20.81	100.00	α, E, X, γ

Source: Nuclear data from NuDat, National Nuclear Data Center, Brookhaven National Laboratory, February 23, 1999.

Note: E = atomic electrons; CE = conversion electrons; AU = Auger electrons; γ-AN = annihilation radiation; TOT = average energy of total beta or positron emitted spectrum; other radiations = primarily low-energy and/or low-yield emissions.

TABLE A.4 Beta-Gamma Emitters

A	Element	Z	Half-Life	Radiation Type	Radiation Energy (keV)	End-Point Energy (keV)	Radiation Intensity (%)	Other Radiations
22	Na	11	2.6088 years	β^+	215.54	545.4	89.84	E, X
				β^+ TOT	215.93		89.90	
				γ	AN	511		≤179.79
				γ	1274.530		99.944	
40	K	19	1.277E9 years	β^-	560.64	1312.1	89.27	β^+, E, X, γ-AN
				γ	1461		10.67	
58	Co	27	9.04 h	E AU L	0.75		124	X, γ
				E CE L	23.963		23.6	
				X $K\alpha_1$	6.93032		16.0	
			70.86 days	β^+	201.1	474.6	14.90	E, X
				E AU K	5.620		49.4	
				X $K\alpha_1$	6.403 84		15.4	
				γ AN	511		≤29.80	
				γ	810.775		99.450	
59	Fe	26	44.503 days	β^-	80.84	273.1	45.3	E, X
				β^- TOT	117.42		100.0	
				β^-	149.10	465.4	53.1	
				γ	1099.251		56.5	
				γ	1291.596		43.2	
60	Co	27	5.2714 years	β^-	95.77	317.86	99.925	β^-, γ
				β^- TOT	96.09		99.99	

continued

TABLE A.4 (continued) Beta-Gamma Emitters

A	Element	Z	Half-Life	Radiation Type	Radiation Energy (keV)	End-Point Energy (keV)	Radiation Intensity (%)	Other Radiations
90	Y	39	3.19 min	γ	1173.237		99.9736	
				γ	1332.501		99.9856	
				E AU L	1.910		11.7	β⁻, E, X, γ
				γ	202.53		97.3	
				γ	479.51		90.74	
			64.10 h	β⁻ TOT	933.6		100.0000	β⁻, E, X, γ
				β⁻	933.7		99.9885	
91	Y	39	49.71 min	β⁻ TOT	603.4		100.0	E, X, γ
			58.51 days	β⁻	604.9	1545.6	99.70	β⁻, γ
94	Nb	41	6.26 min	E AU L	2.150		91	β⁻, E, X, γ
				E AU K	14.00		14.375	
				E CE K	21.95		57.11	
				E CE L	38.24		33.631	
				X Kα₂	16.52100		12.31	
				X Kα₁	16.61510		23.63	
			20,300 years	β⁻	145.9		98.1	
				γ	702.622		97.9	
				γ	871.091		99.90	
95	Nb	41	34.975 days	β⁻	43.36	159.8	99.97	β⁻, E, X, γ
				β⁻ TOT	43.48		100.02	
				γ	765.807		99.81	

A	Z	Element	Half-life	Radiation	Energy (keV)	Emax (keV)	Intensity (%)	Decay
95			86.6 h	γ	235.690		24.9	β^-, E, X, γ
	40	Zr	64.02 days	β^-	109.7	367.8	54.53	β^-, E, X, γ
				β^- TOT	117.4		100.00	
				β^-	120.9	400.3	44.24	
				γ	724.199		44.17	
				γ	756.729		54.46	
99	43	Tc	6.01 h	γ	140.5110		89.06	β^-, E, X, γ
			211,100 years	β^-	84.6	293.5	99.9984	β^-, E, X, γ
				β^- TOT	84.6		100.0000	
103	44	Ru	39.26 days	β^-	30.7		6.61	β^-, E, X, γ
				β^- TOT	63.7		100.0	
				β^-	64.1		92.2	
				E AU L	2.390		80	
				E CE L	36.348		73	
				E CE M	39.133		14.5	
				γ	497.084		91.0	
108	47	Ag	418 years	E AU L	2.500		81	E, X, γ
				E AU K	17.70		13.95	
				X $K\alpha_2$	21.02010		18.07	
				X $K\alpha_1$	21.17710		34.3	
				X $K\beta$	23.80		10.89	
				γ	79.131		6.6	
				γ	433.937		90.5	
				γ	614.276		89.8	

continued

TABLE A.4 (continued) Beta-Gamma Emitters

A	Element	Z	Half-Life	Radiation Type	Radiation Energy (keV)	End-Point Energy (keV)	Radiation Intensity (%)	Other Radiations
			2.37 min	γ	722.907		90.8	
				β-TOT	624		97.2	β^+, β^-, E, X, γ-AN, γ
110	Ag	47	249.79 days	β-	629		95.5	
				β-	21.8	83.5	66.8	β^-, E, X, γ
				β-TOT	68.4		98.2	
				β-	165.5		30.45	
				γ	657.7622		94.0	
				γ	677.6227		10.28	
				γ	706.682		16.33	
				γ	763.944		22.14	
				γ	884.685		72.2	
				γ	937.493		34.13	
				γ	1384.300		24.12	
				γ	1505.040		12.95	
			24.6 s	β-TOT	1185.1		99.4	γ
				β-	1199.3	2892.6	94.91	
129	I	53	1.57E7 years	β-	40.9	154	100.0	E, X
				E AU L	3.430		73	
				E CE K	5.017		79	
				E CE L	34.125		10.7	
				X Kα_2	29.4580		19.9	

Mass	Element	Z	Half-life	Radiation	Energy (keV)	β⁻ E_max (keV)	Intensity (%)	Decay
131	I	53	8.02070 days	X Kα₁	29.7790		37.0	
				X Kβ	33.60		13.2	
				γ	39.578		7.51	
				β⁻ TOT	181.92		100.5	β⁻, E, X, γ
				β⁻	191.58	606.3	89.9	
				γ	364.489		81.7	
134	Cs	55	2.903 h	E AU L	3.550		133	E, X, γ
				E CE L	5.528		77	
				E CE M	10.025		15.8	
				E CE K	91.517		34.9	
				E CE L	121.788		40.6	
				γ	127.502		12.6	
			2.0648 years	β⁻	23.08	88.7	27.29	β⁻, E, X, γ
				β⁻ TOT	157.0		100.07	
				β⁻	210.13	657.9	70.23	
				γ	569.331		15.38	
				γ	604.7210		97.62	
				γ	795.864		85.53	
137	Cs	55	30.04 years	β⁻	174.32	513.97	94.40	β⁻, E, X, γ
				β⁻ TOT	187.87		100.0	
				γ	661.657		85.10	
144	Ce	58	284.9 days	β⁻	50.2	185.1	19.6	β⁻, E, X, γ
				β⁻ TOT	82.1		100.0	
				β⁻	91.1	318.6	76.5	

continued

TABLE A.4 (continued) Beta-Gamma Emitters

A	Element	Z	Half-Life	Radiation Type	Radiation Energy (keV)	End-Point Energy (keV)	Radiation Intensity (%)	Other Radiations
152	Eu	63	9.3116 h	γ	133.5150		11.09	
				β- TOT	687.6		73.1	β-, E, X, γ
				β-	704.0	1864.4	69.6	
				E AU L	4.530		19.3	
				X Kα_1	40.1181		11.0	
				γ	963.390		11.7	
			13.537 years	β-	221.8	695.7	13.7800	β-, E, X, γ
				β- TOT	296.4		27.8	
				E CE K	74.9475		19.6	
				E CE L	114.0449		10.7	
				X Kα_2	39.5224		15.66	
				X Kα^1	40.1181		28.36	
				X Kβ	45.40		11.01	
				γ	121.7817		28.58	
				γ	344.2785		26.5	
				γ	778.9040		12.94	
				γ	964.079		14.60	
				γ	1085.869		10.21	
				γ	1112.069		13.64	
				γ	1408.005		21.01	
154	Eu	63	8.593 years	β-	69.3	248.8	28.6	β-, E, X, γ
				β-	176.2	570.9	36.3	

			Radiation	Energy		Intensity
155	Eu	63	β^- TOT	220.1		100.3
		4.7611 years	β^-	276.6	840.6	16.8
			β^-	695.6	1845.3	10.0
			E AU L	4.840		33.0
			E CE K	72.8319		26.8
			E CE L	114.6954		16.8
			X $K\alpha_1$	42.9962		13.3
			γ	123.0710		40.6
			γ	723.305		20.11
			γ	873.190		12.20
			γ	996.262		10.53
			γ	1004.725		17.91
			γ	1274.435		35.1
		β^-, E, X, γ	β^-	39.2	146.9	47
			β^-	44.5	165.7	25
			β^- TOT	46.8		101
			β^-	70.2	252.2	17.6
			E AU L	4.840		35
			E CE L	10.3884		14
			E CE K	36.306		11.1
			X $K\alpha_1$	42.9962		11.8
			γ	86.545		30.7
			γ	105.305		21.2

Source: Nuclear data from NuDat, National Nuclear Data Center, Brookhaven National Laboratory, February 23, 1999.

Note: E = atomic electrons; CE = conversion electrons; AU = Auger electrons; γ-AN = annihilation radiation; TOT = average energy of total beta or positron emitted spectrum; other radiations = primarily low-energy and/or low-yield emissions.

TABLE A.5 X-Ray Emitters

A	Element	Z	Half-Life	Radiation Type	Radiation Energy (keV)	Radiation Intensity (%)	Other Radiations
41	Ca	20	103 000 years	E AU K	2.970	77.1	
				X Kα_1	3.31380	12.35	
55	Fe	26	2.73 years	E AU K	5.190	60.7	E, X
				X Kα_2	5.88765	8.24	
				X Kα_1	5.89875	16.28	
				X Kβ	6.490	3.29	
59	Ni	28	76 000 years	E AU K	6.070	54.90	E, X
				X Kα_2	6.91530	10.03	
				X Kα_1	6.93032	19.77	
125	I	53	59.402 days	E AU L	3.190	157	E, X, γ
				E CE K	3.6781	80	
				E AU K	22.70	20.0	
				E CE L	30.5527	10.8	
				X L	3.770	15.5	
				X Kα_2	27.20170	39.8	
				X Kα_1	27.47230	74.3	
				X Kβ	31.00	25.8	

Source: Nuclear data from NuDat, National Nuclear Data Center, Brookhaven National Laboratory, February 23, 1999.

Note: E = atomic electrons; CE = conversion electrons; AU = Auger electrons; other radiations = primarily low-energy and/or low-yield emissions.

Appendix B: MARSSIM WRS and Sign Test Sample Sizes (from the MARSSIM Tables 5.3 and 5.5)

TABLE B.1   Values of $N/2$ for a Given Relative Shift (Δ/σ), α and β When the Contaminant is Present in the Background

| | $\alpha=0.01$ | | | | | $\alpha=0.025$ | | | | | $\alpha=0.05$ | | | | | $\alpha=0.10$ | | | | | $\alpha=0.25$ | | | | |
| | β | | | | | β | | | | | β | | | | | β | | | | | β | | | | |
Δ/σ	0.01	0.025	0.05	0.10	0.25	0.01	0.025	0.05	0.10	0.25	0.01	0.025	0.05	0.10	0.25	0.01	0.025	0.05	0.10	0.25	0.01	0.025	0.05	0.10	0.25
0.1	5452	4627	3972	3278	2268	4627	3870	3273	2646	1748	3972	3273	2726	2157	1355	3278	2646	2157	1655	964	2268	1748	1355	964	459
0.2	1370	1163	998	824	570	1163	973	823	665	440	998	823	685	542	341	824	665	542	416	243	570	440	341	243	116
0.3	614	521	448	370	256	521	436	369	298	197	448	369	307	243	153	370	298	243	187	109	256	197	153	109	52
0.4	350	297	255	211	146	297	248	210	170	112	255	210	175	139	87	211	170	139	106	62	146	112	87	62	30
0.5	227	193	166	137	95	193	162	137	111	73	166	137	114	90	57	137	111	90	69	41	95	73	57	41	20
0.6	161	137	117	97	67	137	114	97	78	52	117	97	81	64	40	97	78	64	49	29	67	52	40	29	14
0.7	121	103	88	73	51	103	86	73	59	39	88	73	61	48	30	73	59	48	37	22	51	39	30	22	11
0.8	95	81	69	57	40	81	68	57	46	31	69	57	48	38	24	57	46	38	29	17	40	31	24	17	8
0.9	77	66	56	47	32	66	55	46	38	25	56	46	39	31	20	47	38	31	24	14	32	25	20	14	7
1.0	64	55	47	39	27	55	46	39	32	21	47	39	32	26	16	39	32	26	20	12	27	21	16	12	6
1.1	55	47	40	33	23	47	39	33	27	18	40	33	28	22	14	33	27	22	17	10	23	18	14	10	5
1.2	48	41	35	29	20	41	34	29	24	16	35	29	24	19	12	29	24	19	15	9	20	16	12	9	4
1.3	43	36	31	26	18	36	30	26	21	14	31	26	22	17	11	26	21	17	13	8	18	14	11	8	4
1.4	38	32	28	23	16	32	27	23	19	13	28	23	19	15	10	23	19	15	12	7	16	13	10	7	4
1.5	35	30	25	21	15	30	25	21	17	11	25	21	18	14	9	21	17	14	11	7	15	11	9	7	3
1.6	32	27	23	19	14	27	23	19	16	11	23	19	16	13	8	19	16	13	10	6	14	11	8	6	3
1.7	30	25	22	18	13	25	21	18	15	10	22	18	15	12	8	18	15	12	9	6	13	10	8	6	3
1.8	28	24	20	17	12	24	20	17	14	9	20	17	14	11	7	17	14	11	9	5	12	9	7	5	3
1.9	26	22	19	16	11	22	19	16	13	9	19	16	13	11	7	16	13	11	8	5	11	9	7	5	3
2.0	25	21	18	15	11	21	18	15	12	8	18	15	13	10	7	15	12	10	8	5	11	8	7	5	3
2.25	22	19	16	14	10	19	16	14	11	8	16	14	11	9	6	14	11	9	7	4	10	8	6	4	2
2.5	21	18	15	13	9	18	15	13	10	7	15	13	11	9	6	13	10	9	7	4	9	7	6	4	2
2.75	20	17	15	12	9	17	14	12	10	7	15	12	10	8	5	12	10	8	6	4	9	7	5	4	2
3.0	19	16	14	12	8	16	14	12	10	6	14	12	10	8	5	12	10	8	6	4	8	6	5	4	2
3.5	18	16	13	11	8	16	13	11	9	6	13	11	9	8	5	11	9	8	6	4	8	6	5	4	2
4.0	18	15	13	11	8	15	13	11	9	6	13	11	9	7	5	11	9	7	6	4	8	6	5	4	2

TABLE B.2 Values of N for a Given Relative Shift (Δ/σ), α and β When the Contaminant is Not present in the Background

Δ/σ	$\alpha = 0.01$, β					$\alpha = 0.025$, β					$\alpha = 0.05$, β					$\alpha = 0.10$, β					$\alpha = 0.25$, β				
	0.01	0.025	0.05	0.10	0.25	0.01	0.025	0.05	0.10	0.25	0.01	0.025	0.05	0.10	0.25	0.01	0.025	0.05	0.10	0.25	0.01	0.025	0.05	0.10	0.25
0.1	4095	3476	2984	2463	1704	3476	2907	2459	1989	1313	2984	2459	2048	1620	1018	2463	1989	1620	1244	725	1704	1313	1018	725	345
0.2	1035	879	754	623	431	879	735	622	503	333	754	622	518	410	258	623	503	410	315	184	431	333	258	184	88
0.3	468	398	341	282	195	398	333	281	227	150	341	281	234	185	117	282	227	185	143	83	195	150	117	83	40
0.4	270	230	197	162	113	230	192	162	131	87	197	162	136	107	68	162	131	107	82	48	113	87	68	48	23
0.5	178	152	130	107	75	152	126	107	87	58	130	107	89	71	45	107	87	71	54	33	75	58	45	33	16
0.6	129	110	94	77	54	110	92	77	63	42	94	77	65	52	33	77	63	52	40	23	54	42	33	23	11
0.7	99	83	72	59	41	83	70	59	48	33	72	59	50	40	26	59	48	40	30	18	41	33	26	18	9
0.8	80	68	58	48	34	68	57	48	39	26	58	48	40	32	21	48	39	32	24	15	34	26	21	15	8
0.9	66	57	48	40	28	57	47	40	33	22	48	40	34	27	17	40	33	27	21	12	28	22	17	12	6
1.0	57	48	41	34	24	48	40	34	28	18	41	34	29	23	15	34	28	23	18	11	24	18	15	11	5
1.1	50	42	36	30	21	42	35	30	24	17	36	30	26	21	14	30	24	21	16	10	21	17	14	10	5
1.2	45	38	33	27	20	38	32	27	22	15	33	27	23	18	12	27	22	18	15	10	20	15	12	10	5
1.3	41	35	30	26	17	35	29	24	21	14	30	24	21	17	11	26	21	17	14	9	17	14	11	9	4
1.4	38	33	28	23	16	33	27	23	18	12	28	23	20	16	10	23	18	16	12	8	16	12	10	8	4
1.5	35	30	27	22	15	30	26	22	17	12	27	22	18	15	10	22	17	15	11	8	15	12	10	8	4
1.6	34	29	24	21	15	29	24	21	16	11	24	21	17	14	9	21	16	14	11	8	15	11	9	8	4
1.7	33	28	24	20	14	28	23	20	16	11	24	20	17	14	9	20	16	14	10	8	14	11	9	8	4
1.8	32	27	23	20	14	27	22	20	15	11	23	20	16	12	9	20	15	12	10	8	14	11	9	8	4
1.9	30	26	22	18	12	26	21	18	15	10	22	18	16	12	8	18	15	12	10	6	12	10	8	6	4
2.0	29	26	22	18	12	26	20	18	15	10	22	18	16	12	8	18	15	12	10	6	12	10	8	6	3
2.5	28	23	21	17	12	23	20	17	14	10	21	17	15	11	8	17	14	11	9	5	12	10	8	5	3
3.0	27	23	20	17	12	23	17	17	14	9	20	17	14	11	8	17	14	11	9	5	12	9	8	5	3

Appendix C: Example Decommissioning Inspection Plan for Final Status Survey Program

A generic in-process decommissioning inspection plan to be used during verification activities is provided. This plan is an example of how to illustrate the types of questions that should be covered during a decommissioning inspection. The major elements of this decommissioning inspection plan include the following six areas:

1. General

2. Identification of contaminants and DCGLs

3. Final status survey procedures and instrumentation

4. Analytical procedures for soil samples

5. Miscellaneous inspection activities

6. Instrument comparison activities

The following NRC inspection procedures may be used for guidance, in part, during this inspection:

- Inspection Procedure 83801—Inspection of Final Surveys at Permanently Shutdown Reactors or

- MC 2561—Decommissioning Power Reactor Inspection Program

- MC 2605—Decommissioning Procedures for Fuel Cycle and Materials Licensees

- MC 2602 —Decommissioning Inspection Program for Fuel Cycle Facilities and Materials Licensees

The following documents may be used for guidance during this decommissioning inspection:

- Multi-Agency Radiation Survey and Site Investigation Manual (MARSSIM)

- Multi-Agency Radiation Survey and Assessment of Materials and Equipment Manual (MARSAME)

- License Termination Plan or Decommissioning Plan

- NUREG-1757, vol. 2, chapter 4, Facility Radiation Surveys and Appendix O, Lessons Learned and Questions and Answers to Clarify License Termination Guidance and Plans

- NUREG-1700, rev. 1, Standard review plan for evaluating nuclear power reactor license termination plans

- ANSI N323A, "Radiation Protection Instrumentation Test and Calibration, Portable Survey Instruments"

- ANSI N13.49, "Performance and Documentation of Radiological Surveys"

- ANSI N13.59, "Characterization in support of decommissioning using the data quality objectives process"

1.0 GENERAL

1.1 Visit plant areas to obtain familiarity with the site, surrounding areas, and decommissioning work completed. Review the licensee's plan and schedule for completing further decommissioning activities.

1.2 Review the past operational radiological surveys (HSA) that were used to demonstrate radiological control of the facility. Are there any records of spills or other releases of radioactive material? If so, do the records adequately document the clean-up of these releases of material? What survey equipment was used to demonstrate clean-up was successful?

1.3 Review the results of scoping and characterization surveys for justification of the classification site areas into class 1, class 2, and class 3 areas.

1.4 Review the specific procedures that are being used to remediate contaminated areas. What is the procedure for performing and documenting the remedial action support surveys?

2.0 IDENTIFICATION OF CONTAMINANTS AND DCGLS

2.1 Review the past analytical results to confirm the nature of the contaminants—have all radionuclides been identified? Consider the isotopic ratios of uranium for processed, enriched, or depleted uranium, and the status of equilibrium for the decay series.

2.2 For large radionuclide mixtures, assess the data used to establish the radionuclide ratios. Consider the use of 10 CFR part 61 analyses for waste acceptance criteria used for this purpose. How are hard-to-detect nuclides handled?

2.3 Evaluate how the stated DCGLs are being implemented—for example, use of surrogate measurements and gross activity DCGLs for multiple contaminants, unity rule, averaging conditions for hot spots.

3.0 FINAL STATUS SURVEY PROCEDURES AND INSTRUMENTATION

3.1 Review situations where an area's classification was changed based on accumulated survey data from scoping and characterization surveys. Review documentation to determine whether reclassifications were clearly documented.

3.2 Determine whether the licensee has selected appropriate background reference areas for both material surfaces and land areas. Were sufficient background sample analyses performed to adequately assess the true background level and its variability? Is the number of background samples sufficient to perform the WRS test?

3.3 Review documentation pertaining to detection sensitivity of scanning instrumentation. Does the procedure have sufficient sensitivity to detect the contaminants at the $DCGL_{EMC}$ level? Are the scanning equations appropriate for land areas? For structure surfaces? Review available data that may be used to develop empirical scan MDCs.

3.4 Determine the use of investigation levels for scanning of land areas. Did the licensee perform appropriate follow-up actions based on scan results exceeding the action levels?

3.5 Were the survey instruments selected for surface activity measurements appropriate for the radiations present? Were the calibrations performed using the IS0-7503 guidance (instrument and surface efficiencies)? Are MDCs sufficiently less than the $DCGL_W$?

3.6 When using data from scoping or characterization surveys as final status survey data in an area, what procedures are in place to ensure that the radiological conditions have not changed?

3.7 When using the surrogate approach, was there a sufficient number of samples analyzed for the inferred radionuclide(s)? (MARSSIM recommends 10% of the samples be analyzed for the inferred radionuclide[s].)

3.8 Were advanced survey instruments, such as *in situ* gamma spectrometers, new detector types (CZT, $LaBr_3$, etc.), and positioning/mapping systems (e.g., GIS/GPS) appropriately calibrated and integrated into the final status survey?

3.9 Were clearance survey approaches, instruments and material release procedures consistent with the MARSAME? Were surveys results clearly documented, demonstrating release criteria achieved prior to releasing materials from the site?

4.0 ANALYTICAL PROCEDURES FOR SOIL SAMPLES

4.1 Review the licensee's contract laboratory analytical procedures for radiological analyses, particularly the analysis of soil samples by gamma spectrometry for the gamma-emitting radionuclides. Specifically

- Evaluate the laboratory's sample preparation techniques—for example, geometries used for gamma spectrometry on soil samples, drying and homogenization techniques, holding time for short-lived progeny to reach equilibrium (e.g., for Ra-226).

- Review the protocol the laboratory uses to interpret the gamma spectrometry results, particularly the radionuclide total absorption peaks used to identify various contaminants. Were proper radiation yields and efficiencies applied?

- Review the laboratory QA/QC procedures, including duplicates, blanks, and matrix spikes. Determine the frequency of analysis for each of the QC checks. Determine whether the laboratory participates in some sort of cross-check or performance evaluation program, such as that offered by EML, EPA, and NIST.

5.0 MISCELLANEOUS INSPECTION ACTIVITIES

5.1 Identify any decommissioning program-specific observations concerning the overall performance of the licensee's decommissioning and final status survey program.

5.2 Verify that any commitments made by the licensee have been incorporated into the plan and implemented into the procedures.

5.3 Review the qualifications and training for survey technicians and other project personnel. Qualifications should include, in part, specific training on performing the survey tasks described in the final status survey procedures, data reduction procedures, and training on QA/QC procedures related to the final status survey.

5.4 Collect several archived samples from the licensee and perform confirmatory analyses on these samples for the radionuclides of concern.

6.0 INSTRUMENT COMPARISON ACTIVITIES

6.1 Surface activity measurements using a gas proportional detector and/or GM detector.

- The specific material background radiation levels used to correct the gross counts will be determined for a number of surface types in class 3 or nonimpacted areas. These areas will be considered to be representative of the areas included in the final survey. The surfaces will include various types of concrete, concrete block walls, brick surfaces, dry wall, tiled floors, and steel I-beams. The method used by the licensee to determine background levels will be evaluated, and the results obtained by the licensee and IVC compared.

- Detector calibration will be evaluated, particularly the calibration source radionuclide(s), geometry of the source and source-to-detector spacing, window density thickness, and any other factors considered to have a measurable effect on the resultant efficiency.

- The minimum detectable concentration (MDC) equations will be compared and discussed. Discussion topics will include background and sample count times.

- The interpretation of the raw survey data will be evaluated. Specific discussion items will include subtraction of the appropriate backgrounds, probe area corrections, and any other modifications of the measurement data (e.g., accounting for the detectable fraction when hard-to-detect nuclides are present).

- Following consideration of the above items, side-by-side field measurements will be performed by the licensee and the IVC, within class 1 and class 2 areas, to compare actual surface activity levels. Both gross and net count rates will be compared, to determine the effect of background on the results.

6.2 Exposure rate measurements using a pressurized ionization chamber and NaI-based instrument.

- Background exposure rate levels will be determined for both indoor and outdoor areas within class 3 or nonimpacted areas. These areas will be considered to be representative of the areas included in the final status survey.

- Data will be reported in μR/h. The method used by the licensee to determine background levels will be evaluated, and the results obtained by the licensee and IVC compared.

- Following consideration of the above items, side-by-side field measurements of exposure rate will be performed by the licensee and IVC, within indoor and outdoor affected areas, to determine actual exposure rates.

6.3 Miscellaneous items.

- Smears for removable surface contamination will be discussed— particularly for the assessment of HTDN like H-3 and Fe-55. Topics for discussion will include methodology for sample (smear) collection, for example, moistened smears and laboratory analysis procedures.

- General discussion of instruments and procedures used for scanning—topics will include scanning sensitivity and background levels. Instrumentation and procedures for performing alpha and beta surface activity measurements. Discuss the situation where alpha and beta surface activity may be present and the use of alpha blockers and voltage settings.

6.4 Select 1–5% of the survey units for independent verification activities to include surface scanning, direct measurements of surface activity, and soil sampling.

Index

A

Absorption, self, 182, 209, 219, 224, 236, 283–287, 302
Action level, 32, 44, 297, 485–486, 500, 513, 515
Action levels (ALs), 485–488, 566
Alpha emitters, 108, 215, 219–220, 267, 284, 300, 529
Alpha spectroscopy, 219, 531, 536, 567–569
Am-241, 93–95, 138–140, 142–144, 146, 259–262, 268–271, 433–437
ANSI N13.12, 67, 83–84, 485, 523; see also clearance
ANSI N13.49, 281; see also survey procedures
ANSI N13.59, 41, 54–55, 57; see also characterization survey
Area factors, 114–118, 307–312, 318–324, 329–334, 339–341, 399–401, 407–409
   hot spot, 311–312, 315, 326, 344
   table, 115–118, 273, 399, 401, 429, 436, 438

B

Background
   ambient, 167, 292, 411
   ambient exposure rate, 166, 410–411
   concentrations, 162–163, 165, 177, 179, 388, 511, 578
   count rate, 167, 208–209, 234, 238, 245–246, 249, 411
   detector, 191, 193, 221, 265
   instrument, 166, 411, 502
   levels, 28, 162–164, 166, 252–255, 262, 264–267, 390–391
   measurements, 161–162, 165–166, 178, 290–291, 414, 508, 548
   radiation levels, 166, 410, 502, 515
   reference areas, 161–165, 167 177, 402–406, 412–415, 575–578, 585–587, 589–594
   variability, 88, 164, 169–170, 172, 179, 298, 414
Backscatter, 182, 209, 224, 283–287, 501, 554
Bayesian
   analysis, 349, 352, 376–377
   statistics, 345, 376–378, 382
Beta emitters, 184–186, 266–267, 424, 429–430, 437–438, 529, 541

C

Calibration source, 28, 182, 209–210, 282–284, 286, 293, 556
   area, 210, 251–252
Characterization survey
   activities, 16, 57, 120, 416, 436
   design, 6, 18–19, 42, 48, 51, 54, 492
Class, 63–65, 398–400, 421–425, 450–452, 495–497, 518–519, 588–594
   Class 1, 63–65, 398–400, 421–425, 450–452, 460–461, 463–464, 495–496
   Class 2, 6, 63–65, 392, 419, 422–423, 496–497, 518–519
   Class 3, 6, 19, 64–65, 497, 575–576, 582, 588–594
Classification, 6, 18, 63, 387, 450, 495, 510–511
   material, 485, 494–495

Cleanup criteria, 21–22, 32, 75; *see also* dose-based release criteria

Clearance
of materials, 83, 479, 481, 483, 501–503, 505, 515
survey approaches, 484, 487, 501, 503, 507, 514, 521
survey design, 488, 507, 510
surveys, 479–480, 483–485, 487–489, 491–492, 494–497, 508–509, 518–519

Co-60, 116–117, 129–133, 202, 268–271, 309–312, 393–395, 462–469

COMPASS, 236–237, 290, 425–426, 429–430, 438, 456–457, 592–593

Confidence interval testing, 355–356, 359

Confirmatory analyses, 29–30, 572

Confirmatory survey, 21, 25, 452–453

Contaminant
concentration, 7, 52, 54, 59, 306, 344–345, 351–352
distribution, 8, 20, 61, 63, 306, 344–346, 351–353

Contaminants, significant, 153–154, 156

Contaminated area, 92, 307–310, 317–321, 323–326, 331–335, 340–341, 400–401

Contamination, 6–7, 52–61, 63–66, 76–84, 244–250, 287–288, 484–497
alpha, 24, 203, 207, 245, 247, 501, 538
building surface, 76–77, 82, 423
depth of, 70, 88, 199, 262–264
extent of, 6, 19, 21, 49, 52, 56, 63
groundwater, 61, 74, 90
levels, 14, 48–50, 56–57, 65, 69, 88, 356

Conveyor survey monitors, *see* CSMs

Critical level (L_C), 224–229, 235, 277, 500; *see also* MDC

Cs-137, 101–103, 115–117, 131–132, 138–141, 258–260, 429–431, 433–439
concentrations, 163, 179, 576–577, 584–586, 588–589, 592–594

D
_____

DandD code, 87, 99–102, 108–109; *see also* DandD model

DandD model, 94–95, 100–101, 104–105, 111, 114, 133

Data quality assessment (DQA), 29, 62, 460, 465, 469

Data quality objectives (DQOs), 4, 41, 43, 140–141, 185, 368, 413; *see also* DQO Process

DCFs (dose conversion factors), 94–95, 123–127, 131, 320, 323, 336

DCGL, application of, 137, 139, 141, 143, 145, 147, 149

$DCGL_C$, 146, 485, 488, 496–500, 507–509, 511–515, 520–521

DCGLs (derived concentration guideline levels), 3–4, 78–84, 135–138, 145–146, 401–404, 431–438, 497–499

$DCGL_W$, 238, 416–417, 463, 467, 472, 542–543
contaminant, 394, 427, 431
gross activity, 115–116, 139–140, 152–156, 159, 274, 416, 497–498
gross alpha, 156, 429, 436, 441
gross beta, 427, 429, 431–432, 436–437, 439–441
modified, 140–141, 146, 434

Decision rule, 44–45, 47, 49, 369, 472, 592

Decommissioning, 1–6, 8–39, 41–42, 64–70, 72–76, 525–543, 594–597

Decommissioning projects, 1–2, 15–19, 21, 25–27, 31, 37–39, 67–68

DECON, 11–13, 16–17, 38, 42

Decontamination, 1, 9, 11–12, 14, 17–20, 56, 541

Department of Energy, *see* DOE

Depleted uranium, *see* DU

Derived concentration guideline levels, *see* DCGLs

Detection limit (L_D), 83, 223–225, 228–230, 232–236, 277, 500–503; *see also* MDC

Detection sensitivity, 2, 223–225, 227, 229, 231, 235, 505–506; *see also* MDC

Detector, 185–188, 190–200, 202–204, 209–210, 239, 246–247, 249–252; *see also* instrumentation

Detector efficiency, 189, 202, 208, 236, 245–246, 264, 553–554; *see also* instrument efficiency

Distribution, 52, 226–227, 343–352, 357–360, 363–369, 372, 377
  binomial, 360–362, 364–365
  normal, 347–349, 357, 364–366, 374, 557–558, 560, 563
  Poisson, 230, 361, 363–365, 374–375
  posterior, 345–347, 349–351, 353–354, 377, 382; *see also* Bayesian
  prior, 345–346, 348, 376–377; *see also* Bayesian
  probability, 96, 345–346, 357, 361, 376, 379–380, 572
  t, 347, 359–360, 366–367, 390, 568, 579, 587
  uniform, 314, 346–347, 353

Document reviews, 26–27, 36–37

DOE (Department of Energy), 1, 3, 5, 32, 73–75, 85–86, 479–482

Dose-based release criteria, 2, 4, 67, 69, 71, 73–77, 547–548; *see also* DCGLs

Dose conversion factors, *see* DCFs

DQO Process, 43, 45, 47, 49, 51, 53, 55

E

Efficiency, 182–185, 191–193, 197–199, 209–211, 283–284, 502–503, 519–521; *see also* instrument efficiency, surface efficiency

Elevated area, *see* hot spots

Elevated measurement comparison, *see* EMC

EMC (elevated measurement comparison), 8, 307–308, 423, 428, 435, 467–469, 472

ENTOMB, 11, 14–15, 17, 38

ETF (environmental transport factor), 310, 313, 320, 323

Exposure pathway modeling (dose modeling), 12, 85, 87, 89, 91, 93, 95

Exposure rate measurements, 16, 30, 57–58, 121, 151–152, 181, 215–216

F

Final status survey design, 114, 460, 469; *see also* MARSSIM

FSS (final status surveys), *see* surveys, final status

Fukushima, 67, 595–596

G

Gamma spectroscopy, 136, 196, 276, 299, 530–531, 535–536, 567–570

Gross alpha, 429–432, 436, 438, 440–441, 443, 510

Gross beta, 208, 427, 431, 439–441, 443

Gross measurements, 165, 409–410, 413–414, 416, 418, 429, 436

Guidelines, 4, 68–73, 75–77, 82–83, 161, 388–391, 421–422; *see also* DCGLs

H

Hazard assessment, 31, 85, 119–121, 128–129, 132–133

Historical site assessment, 5, 20, 211, 491, 510

Hot spots, 305–319, 322–330, 332–335, 337–347, 350–354, 398–400, 465–469
  area, 103, 461, 465, 467–468, 473
  doses, 317–318, 326, 329, 331, 333–336, 340–341, 343
  multiple, 306, 344, 351, 468
  size, 248, 251, 258, 264

Human factors, 204, 240–241, 243–244, 247; *see also* surveyor efficiency

Hypothesis testing, 225–226, 355, 359–361, 365–371, 373, 575, 585
  alternative, 170, 225–226, 369, 371–373, 576–581, 583–587, 589–592
  critical values, 170, 172, 367, 466–467, 472, 560–564, 573–575
  decision errors, 47, 50, 53, 228, 233, 265, 444–445
  null hypothesis, 168–170, 172, 368–370, 374–376, 560–564, 567–571, 573–594

I

Independent verification, *see* IV
Independent verification survey, *see*
    confirmatory survey
Instrument, 28, 187–190, 205, 209–211,
    223–225, 289–291, 293–294
    geometry, detector-to-source,
        250–251, 285
    probe area, 184–185, 190–192,
        210, 236–237, 250–253,
        277–278, 430
        physical, 190, 192–193, 210, 251–252,
          277–278, 416–417, 426
Instrument efficiency, 183–184, 208–211,
    213–215, 250–253, 266–268,
    288–289, 556
    scanning, 248, 251–252, 440
    weighted, 211–212, 214, 221, 238, 266,
        288, 438
Instrumentation, 181, 183–185, 187, 203,
    484, 507–509, 514–515
    CSMs (conveyor survey monitors), 204,
        484, 487–488, 495, 501, 505–509,
        514–517
    CZT detectors, 196–198
    gas proportional, 183–188, 192–195,
        205–207, 219–221, 423–426,
        533–535, 541–543
    GM, 185–188, 190–191, 210–212,
        214–216, 221, 251–252,
        504–505
    HPGe detectors, 197–198, 219–220,
        517
    in situ gamma spectrometer, 201,
        450–451, 457
    NaI scintillation, 195–197, 216,
        255–262, 264–266, 294–297,
        401–402, 517
    plastic scintillator, 189, 192, 520–521
    ZnS, 167, 185–187, 189–190, 192,
        210–211, 276, 294
ISO-7503, 181–182, 209, 266–267, 282–289,
    292–294, 430, 438–439
ISO-8769, 182, 210, 293
IV (independent verification), 11, 18, 32,
    452–453; *see also* verification
    surveys

L

Laboratory analyses, 29–30, 140–141,
    164–165, 218–220, 223, 453, 494
LBGR, 168–169, 172–178, 394–395, 404–
    405, 417, 446–449, 511–513
License termination plan, 17, 23, 42,
    87, 114

M

MARSAME, 8–10, 45, 482–486, 488,
    490–491, 499–501, 511–517
MARSSIM
    sample size, 421–422, 443
    survey design, 63, 88, 168, 201, 244,
        424–425, 450–452
    table, 241–242, 253–254, 262, 395,
        405–406, 435, 511–512
MARSSIM, AF, 315–318
MARSSIM, Final Survey Design and
    Strategies, 385, 387, 389, 391, 393,
    395, 397
Mass loading factor, 122, 130, 324–325
Materials
    bulk, 487–488, 490, 516–517
    clearance of, 9–10, 48, 83–84, 190, 362,
        479, 510
    contaminated, 17, 63, 72, 333, 479, 481
    interior surfaces, 69, 490, 492–493
MDC (minimum detectable
    concentration), 140–141,
    223–225, 235–239, 430, 438–439,
    499–503, 534–536
MDC, paired measurements, 233–234,
    237–238, 278, 566
$MDCR_{surveyor}$, 255–256, 258, 260, 263,
    265
Measurement
    locations, 29, 61, 153, 443, 454–455,
        514, 516
    procedures, 223–224, 499, 501
    uncertainty, 356, 500, 545, 547,
        549–553, 555, 557
Measurements, surrogate, *see* surrogate,
    approach
Median concentration, 176, 395, 398, 406,
    432–433, 446–449, 512

Microshield, 257–258, 261, 265, 268–270, 310–312, 315, 335–337
Minimum detectable counting rate (MDCR), 243, 250, 253–259, 265, 267, 269–270, 503–506
Minimum quantifiable concentration, see MQC
Modeling codes, 3, 93–99, 102, 114, 137–138, 257–259, 308–310
MQC (minimum quantifiable concentration), 500–501, 522

N

NaI scintillation detector, see instrumentation, NaI
Normality tests, 560
    D'Agostino test, 560, 562–563, 565
    W test (Shapiro-Wilk test), 560–562, 570–571, 573
NRC (Nuclear Regulatory Commission), 1–6, 69–73, 75–77, 80–82, 90–91, 98–101, 479–483
NUREG-1505, 168–169, 172–174, 177–178, 512
NUREG-1507, 166–167, 224, 254, 256, 282–283, 285–286, 401–402
NUREG-1640, 9–10, 486
NUREG-1720, 100, 109, 535
NUREG-1757, 22–23, 42, 61, 69, 76–77, 163, 546
NUREG-1761, draft, 483–485, 488, 490, 493, 502–508, 513–514, 517–518
NUREG/CR-5849, 5, 17, 34, 384, 386–392, 420–425, 546

O

Observation interval, 239–241, 243, 246–248, 250, 258–260, 262–265, 504–506

P

Parameter values, 23–24, 86, 92, 96–97, 121, 264–265, 269–270
Pathways, 88–89, 93, 95, 100, 106–107, 308–310, 329–335

default parameters, 46, 62, 92, 100, 103, 105, 324–325
direct radiation, 114, 121–122, 151, 216
drinking water, 331–334
exposure, 3–4, 9, 89, 93, 100, 119–121, 154
external radiation, 102, 308–311, 317–319, 329–330, 332–337, 339, 343
inhalation, 94–95, 107–109, 111, 122, 321–322, 329–330, 340–341
water-dependent, 318, 330–332
Power curve, 7, 175, 177, 396–398, 448, 579, 582
    prospective, 396–398, 417–419, 432–433, 447–449, 461–462, 578–586, 588
    retrospective, 169, 178, 461–462, 466, 470–472, 582–584, 586
Power, retrospective, 462, 583–584, 589
Progeny, 82, 105, 112–113, 128, 137–138, 214–215, 532–536
    beta-emitting, 120, 124, 150

R

Ra-226, 81–82, 157–158, 271–273, 295–297, 300–301, 457, 526–528
Radiation dose, 74, 89, 93–94, 103
    direct, 122, 151–152
    ingestion, 95, 124–125, 131–132, 342
    inhalation, 106, 109, 534–535
    inhalation pathway, 322–325
    peak dose, 102, 106–107, 138, 140, 331, 437
    total, 95, 152–154, 156
Radiation doses, external, 95, 125–127, 130, 132, 336–337
Radiological surveys, see surveys
Radionuclide contaminants, 135–136, 537
Radionuclide contamination, see contamination
Radionuclide DCGLs, see DCGLs (derived concentration guideline levels)
Radionuclide ratios, 139, 144, 148–149, 548

Radionuclides
  hard-to-detect, 53, 144, 155, 452
  multiple, 34, 115, 138, 140–141,
      153–154, 156, 497–498
  significant, 152–153, 156
  surrogate, *see* surrogate, approach
Random samples, 369–370, 374, 376, 474, 549
Random uncertainties, *see* error
    propagation
Receptor, 3, 99–100, 110, 310–319, 323–325,
    329–343, 354
  dose, 99, 305–307, 309–313, 319,
      321–323, 330–335, 340–341
  location, 99, 313, 319, 324, 337
Reference area, *see* background, reference
    areas
Regulatory guide, 4, 10, 67–71, 73, 76–77,
    82–83, 485
Relative shift, 7, 394–395, 404–405, 412–413,
    427–428, 431–432, 446–449
Release of materials, *see* clearance of
    materials
Reports (FSS reports), 545, 547, 549, 551,
    553, 555, 557
Residual contamination, *see*
    contamination
RESRAD, 96–98, 110–112, 114, 151–152,
    308–313, 318–325, 329–333
  area factor, 312, 322, 326–328
  code, 90, 98, 100, 113–114, 310–311,
      316, 329
RESRAD-BUILD, 3, 98–100, 110–111,
    335–337, 339–343, 423, 437
RESRAD-BUILD, AF, 337–339
Restricted release, 20, 23–24, 39, 69
Resuspension factor, 3, 20, 24, 94, 100, 106,
    108–110; *see also* NUREG-1720
Retrospective power analysis, *see* power
    curve, retrospective

S

SAFSTOR, 11, 13–14, 17, 42
Samples
  additional, 359, 388, 390, 399, 424,
      468–469, 476–477
  statistical, 376, 446, 510
  systematic soil, 344, 388–389

Sampling
  design, 62, 372, 460, 581–582
  distribution, 357, 371, 578
  double, 474–477
  locations, *see* measurement, locations
  miscellaneous, 51–52, 509–510
  ranked set, 452–453, 455
Scan
  coverage, 8, 510, 515–516, 549
  rate, 247–250, 258–259, 424
  survey, 239, 241, 277, 294
Scan MDC, 246–250, 252–259, 261–277,
    401–402, 439–441, 503–507, 535
  actual, 274, 399, 401, 407–409, 428–429,
      435–436, 441
  alpha scan, 245–250, 267, 439, 443
  beta scan, 248, 439, 443
  required, 273–274, 399, 401–402,
      407–409, 428–429, 435–436,
      442–443
Scanning (scans), 187–189, 239–242,
    244–248, 250–253, 294–295,
    467–468, 514–516
Scenario B, 88, 90–91, 161, 165, 167–170,
    172–179, 511–513
SCM (surface contamination monitor),
    185, 196, 203
Scoping survey, *see* surveys, scoping
Screening values, 46, 77, 80–81, 87–88,
    100, 107
Secular equilibrium, 112–113, 118, 149–150,
    403, 527, 532–534, 536
Self-absorption, *see* absorption, self
Sign test, 392–394, 409–411, 413–415,
    418–421, 431–436, 460–462,
    464–467
  sample size, 395–396, 417, 435,
      441, 443
Signal detection theory, 239–240, 250
Site contaminants, *see* contamination
Soil
  analyses, 20, 66, 181, 219, 298, 570
  concentrations, 31, 81, 140–141, 257,
      277, 323–324, 549
  contamination, 45, 78–79, 90, 102,
      151–152, 156, 162
  density, 96, 258, 265–266
  sample locations, 157, 453, 455

samples, 199–201, 294–295, 298–300,
386–388, 407–408, 435,
453–455
sampling, 26, 152, 200, 281, 398, 549
subsurface, 16, 52, 57, 59–60, 85
Soil concentration, DCGLs, 46, 62, 81,
111–112
Solid materials, *see* clearance surveys
Source
activity, 182, 258, 284
area, 115–118, 133, 249, 407, 425, 437
term, 54, 98–99, 113, 118–121, 126–127,
320–321, 353
SRP (standard review plans), 21–24, 76
Standard review plans, *see* SRP
Statistical
sample size, 7–8, 399, 453, 510
survey designs, 27, 168, 205
Statistics, 355, 357, 375–379, 559, 563,
567–569, 571
central limit theorem, 357–358, 366
chi-square test, 373–374
confidence interval, 345, 355–356,
358–359, 376, 390, 554,
557–558
confidence level, 152, 227, 359–360,
388–391, 421–422, 553–554,
557–558
error propagation, 550–558
Kruskal–Wallis test, 169–172, 174
nonparametric statistics, 367, 382, 392,
465, 572
population, 86–87, 355, 358, 366–369,
376, 385, 560–561
practical applications of, 561, 563, 565,
567, 569, 571, 573
probability, 177, 246–249, 361–363,
371–373, 376–380, 577–584,
587–589
quantile test, 169, 174, 177–178
standard error, 167, 358–359, 371–372
Structure surfaces, 244–245, 250, 266,
387–388, 411, 436, 443
Surface activity
alpha measurements, 33, 149–150, 167,
185–186, 189, 212, 246
assessments, 283–284, 286, 290–292,
383, 410–411, 413, 415

beta measurements, 33, 149–151, 186,
211–212, 430–431, 440–441, 533
guidelines, 69–74, 83, 162, 212, 485, 530
levels, 4, 30, 70, 94–95, 104, 120,
181–182
measurements, 28–30, 119–120,
181–182, 188–190, 209–210,
275–278, 509–511
Surface activity measurements, gross, 292,
410–411, 414
Surface contamination monitor, *see* SCM
Surface efficiency, 209–211, 266–268,
283–286, 288–290, 293–294,
438–440, 556–557
weighted, 212, 214–215, 238, 288–289,
439, 535
Surface emission rate, 182, 209–210, 221,
284, 293
Surface material backgrounds, 161,
164–166, 292–293, 410–411, 425;
see also background
Surface materials, 58, 167, 287–288,
409–410, 412–413, 415, 418–420
Surface soil, 79–80, 452
Surrogate
approach, 136, 140–146, 151–152,
156–157, 159, 434, 497–498
inferred radionuclides, 54, 141–143,
145, 149, 159, 498
measured radionuclide, 54, 141,
145–146, 498–499
ratio, 54, 142–143, 146, 149–150, 159,
436, 498
Survey
instruments, 166–167, 181–182, 184–
185, 187–190, 196–197, 224–225,
508–509
measurements, 28–30, 162, 187,
205–206, 225
Survey instrumentation, *see*
instrumentation
Survey meter, *see* instrumentation
Survey procedures, 2, 21, 26, 35, 37,
281, 421
Survey unit, 7, 168, 386–387, 412, 420–421,
446, 510–511
area, 322, 325, 329, 341–342, 345,
399–400, 407

Survey unit (*Continued*)
  building surface, *see* structure surfaces
  material, 34, 485, 487–488, 490, 496,
    510–512, 514
  size, 103, 110, 309, 334, 342, 460, 490
Surveyor efficiency, 243, 253, 255, 258,
    260, 262–265, 504–505
Surveys, 4–6, 16–19, 57–59, 63–66,
    483–485, 491–497, 507–509
  characterization, 16, 18–21, 41–43,
    45–47, 49–51, 53–59, 63–66
  decommissioning, 4–5, 10, 19, 181–182,
    187, 537, 540
  final status, 4–5, 17–23, 37–39, 63–65,
    459–461, 473–475, 545–549
  MARSSIM, 5, 201, 402, 475, 547
  remedial action support, 18, 394, 404,
    460, 469, 543, 547
  scoping, 19, 54, 57–58, 64, 216, 496
Systematic uncertainty, 550–551, 557; *see
    also* error propagation

T

T test, 386–389, 391–392, 560, 566–567,
    569–572, 574–579, 584–585
Test statistic, 170, 178, 368–369, 374–375,
    465–467, 472, 560–563
Th-232, 71–73, 111–113, 157–158, 174–176,
    388–391, 402–404, 532–536
Thorium, *see* Th-232
Total efficiency, 137, 155, 215, 282–285,
    288–290, 416–417, 423–424
  weighted, 155, 211–212, 303, 417, 423
Transuranics, 51, 71, 73, 76, 142–143,
    155–156, 537
Two-sample, 402–407, 409–415, 419–421,
    427–428, 471–472, 511–512,
    590–593; *see also* WRS test
Type I error, 172, 176–177, 368–373,
    397–398, 444–445, 475–477,
    586–587
Type II decision errors, 7, 233–235, 239,
    392–394, 403–405, 412–413,
    443–444

Type II error, 175–177, 233–236, 370–373,
    397–398, 444–448, 577–580,
    586–587

U

U-234, 78–80, 111–114, 122–127, 147–148,
    213–214, 416–417, 527–531
U-235, 71–72, 78–80, 122–127, 147–148,
    300–301, 368–369, 527–531
U-238, 78–80, 111–114, 122–128, 147–148,
    403–404, 407–409, 527–531
Unity rule, 139–140, 152–153, 155–158,
    403–406, 431, 434–436, 440–441
Unrestricted release, 2, 4, 15, 23–24, 86,
    100, 384
Uranium, 70–73, 79–80, 111–114, 117–126,
    147–151, 213–214, 525–533; *see
    also* U-238, U-235, U-234
  DU (depleted uranium), 60, 150, 186,
    526, 528–529, 531, 569–570
  enriched uranium, 79–80, 147–148, 213,
    288–290, 303, 526, 528–531
  low enriched, 213–214, 288–289, 529
  processed natural, 79, 84, 111–114,
    117–118, 271–272, 415–416,
    526–528
  total, 79–80, 143, 147, 531, 566
Uranium contamination, 33, 98, 149, 213,
    403, 526, 543
Uranium enrichment, 527–528, 531
Uranium mills, 220, 295, 523, 526–527, 532

V

Verification process, 22, 25–27, 31–35,
    38–39
Verification surveys, 22, 25, 30–31, 34
Volume counters, 519, 521–522

W

WRS test, 157–158, 402–407, 409–415,
    419–422, 427–428, 511–512,
    590–593